Simulation-Based Optimization
of Antenna Arrays

Other World Scientific Titles by the Author

Multi-Objective Design of Antennas Using Surrogate Models
by Slawomir Koziel and Adrian Bekasiewicz
ISBN: 978-1-78634-147-1

Simulation-Driven Aerodynamic Design Using Variable-Fidelity Models
by Leifur Leifsson and Slawomir Koziel
ISBN: 978-1-78326-628-9

Simulation-Driven Design Optimization and Modeling for Microwave Engineering
edited by Slawomir Koziel, Xin-She Yang and Qi-Jun Zhang
ISBN: 978-1-84816-916-6

Simulation-Based Optimization of Antenna Arrays

Slawomir Koziel
Stanislav Ogurtsov

Reykjavik University, Iceland

World Scientific

NEW JERSEY · LONDON · SINGAPORE · BEIJING · SHANGHAI · HONG KONG · TAIPEI · CHENNAI · TOKYO

Published by

World Scientific Publishing Europe Ltd.
57 Shelton Street, Covent Garden, London WC2H 9HE
Head office: 5 Toh Tuck Link, Singapore 596224
USA office: 27 Warren Street, Suite 401-402, Hackensack, NJ 07601

Library of Congress Cataloging-in-Publication Data
Names: Koziel, Slawomir, author. | Ogurtsov, Stanislav, author.
Title: Simulation-based optimization of antenna arrays / by Slawomir Koziel, Reykjavik University, Iceland, and Stanislav Ogurtsov, Reykjavik University, Iceland.
Description: New Jersey : World Scientific, [2018] | Includes bibliographical references and index.
Identifiers: LCCN 2018040145 | ISBN 9781786345981 (hc : alk. paper)
Subjects: LCSH: Antenna arrays--Computer-aided design. | Antenna arrays--Computer simulation.
Classification: LCC TK7871.67.A77 K69 2018 | DDC 621.382/4--dc23
LC record available at https://lccn.loc.gov/2018040145

British Library Cataloguing-in-Publication Data
A catalogue record for this book is available from the British Library.

Copyright © 2019 by World Scientific Publishing Europe Ltd.

All rights reserved. This book, or parts thereof, may not be reproduced in any form or by any means, electronic or mechanical, including photocopying, recording or any information storage and retrieval system now known or to be invented, without written permission from the Publisher.

For photocopying of material in this volume, please pay a copying fee through the Copyright Clearance Center, Inc., 222 Rosewood Drive, Danvers, MA 01923, USA. In this case permission to photocopy is not required from the publisher.

For any available supplementary material, please visit
https://www.worldscientific.com/worldscibooks/10.1142/Q0179#t=suppl

Desk Editors: V. Vishnu Mohan/Jennifer Brough/Koe Shi Ying

Typeset by Stallion Press
Email: enquiries@stallionpress.com

To our families:
Dorota, Anna Halina, Irina, Stefaniya, and Daria.

Preface

Design of contemporary antennas and antenna arrays is heavily based on full-wave electromagnetic (EM) simulation models. Although necessary to ensure evaluation reliability, EM analysis incurs considerable computational expenses. Due to structural complexity and large electrical size of arrays, individual simulations may be time consuming, whereas repetitive simulations performed as parts of design procedures are often unmanageable. In this context, simulation-driven design of antenna arrays is a challenging endeavor. Due to lack of proper algorithmic tools it is typically addressed in a simplistic manner, e.g., rudimentary decomposition or parameter sweeping, which hinders accounting for performance-degradation effects, thus leading to suboptimal results. Accelerated and automated design can be realized using surrogate-assisted techniques. This book is focused on demonstrating how surrogate models, both data-driven and physics-based, can be exploited to expedite antenna arrays design procedures such as parametric optimization, statistical analysis, or fault detection. The presented algorithms are illustrated using a large variety of examples or real-world printed-circuit antenna and antenna array structures. In many cases, numerical results are supported by experimental validation of fabricated prototypes. Furthermore, the book contains introductory material concerning numerical optimization, both conventional (gradient-based and derivative-free, including metaheuristics) and surrogate-based, as well as a considerable selection of customized procedures developed specifically to handle antenna array problems. Recommendations concerning practical aspects of surrogate-assisted multi-objective antenna optimization are also given.

Slawomir Koziel and Stanislav Ogurtsov
Reykjavik, Iceland

April 2018

About the Authors

Slawomir Koziel received the M.Sc. and Ph.D. degrees in electronic engineering from Gdansk University of Technology, Poland, in 1995 and 2000, respectively. He also received the M.Sc. degrees in theoretical physics and in mathematics, in 2000 and 2002, respectively, as well as the PhD in mathematics in 2003, from the University of Gdansk, Poland. He is currently a Professor with the School of Science and Engineering, Reykjavik University, Iceland. His research interests include CAD and modeling of microwave and antenna structures, simulation-driven design, surrogate-based optimization, space mapping, circuit theory, analog signal processing, evolutionary computation and numerical analysis.

Stanislav Ogurtsov received his degree of Physicist from Novosibirsk State University, Novosibirsk, Russia, in 1993, and his Ph.D. degree in Electrical Engineering from Arizona State University, Tempe, USA, in 2007. He is currently with the School of Science and Engineering, Reykjavik University, Iceland. His research interests include computational electromagnetics, EM-based CAD of antennas and RF/microwave/light-wave circuits, as well as microwave/antenna measurements. He co-authored over 150 peer-reviewed publications.

Acknowledgments

We would like to acknowledge the efforts of all those students, researchers, and colleagues who have helped us during the research work presented in this book. We would also like to thank Dassault Systèmes, France, for making CST Microwave Studio available for our research purposes.

Contents

Preface   vii

About the Authors   ix

Acknowledgments   xi

1. Introduction   1

2. Antenna Array Fundamentals   11

     2.1. Introduction 11
     2.2. Antenna Array Radiation Figures 14
     2.3. Array Factor Analysis 17
     2.4. Mutual Coupling and Antenna Array Reflection Response 22
     2.5. Mutual Coupling and Antenna Array Design Validation 26
     2.6. Simulation-Based Antenna Array Design 29
     2.7. A Concept of Antenna Array Simulation-Based Optimization 30
     2.8. Challenges of Antenna Array Simulation-Based Optimization 31

3. Fundamentals of Numerical Optimization 33
- 3.1. Formulation of the Optimization Problem 34
- 3.2. Gradient-Based Optimization 35
  - 3.2.1. Optimization using descent methods ... 37
  - 3.2.2. Newton and quasi-Newton methods 42
  - 3.2.3. Qualitative comparison of descent methods 46
  - 3.2.4. Remarks on constrained optimization 46
- 3.3. Derivative-Free Optimization 51
  - 3.3.1. Pattern search methods 53
  - 3.3.2. Hooke–Jeeves direct search 54
  - 3.3.3. Nelder–Mead algorithm 55
- 3.4. Summary 57

4. Global Optimization: Population-Based Metaheuristics 59
- 4.1. Fundamentals of Population-Based Metaheuristics 61
- 4.2. Evolution Strategies 65
- 4.3. Genetic Algorithms 68
  - 4.3.1. Algorithm flow and representation 68
  - 4.3.2. Crossover 69
  - 4.3.3. Mutation 70
  - 4.3.4. Selection 71
  - 4.3.5. Elitism 72
  - 4.3.6. Selected topics 72
- 4.4. Evolutionary Algorithms 74
- 4.5. Particle Swarm Optimization 76
- 4.6. Differential Evolution 77
- 4.7. Firefly Algorithm 79
- 4.8. Other Methods 81
- 4.9. Summary 81

5. Fundamentals of Surrogate-Based Modeling and Optimization — 85

- 5.1. Surrogate-Based Optimization — 86
- 5.2. Surrogate Modeling: Data-Driven Surrogates — 91
  - 5.2.1. Modeling flow for data-driven surrogates — 92
  - 5.2.2. Design of experiments — 92
  - 5.2.3. Data-driven modeling techniques — 94
    - 5.2.3.1. Polynomial regression — 95
    - 5.2.3.2. Radial basis functions — 96
    - 5.2.3.3. Kriging — 96
    - 5.2.3.4. Artificial neural networks — 98
    - 5.2.3.5. Support vector regression — 99
    - 5.2.3.6. Other approximation methods — 100
  - 5.2.4. Model validation — 101
- 5.3. Surrogate Modeling: Physics-Based Surrogates — 102
- 5.4. Optimization with Data-Driven Surrogates — 109
  - 5.4.1. Optimization by means of response surfaces — 109
  - 5.4.2. Sequential approximate optimization — 110
  - 5.4.3. SBO with kriging surrogates: Exploration versus exploitation — 113
  - 5.4.4. Summary — 115
- 5.5. SBO Using Physics-Based Surrogates — 115
  - 5.5.1. Space mapping — 116
  - 5.5.2. Approximation model management optimization — 119
  - 5.5.3. Manifold mapping — 119
  - 5.5.4. Shape preserving response prediction — 120
  - 5.5.5. Adaptively adjusted design specifications — 121
  - 5.5.6. Feature-based optimization — 124
  - 5.5.7. Summary — 130

6. Antenna Models for Simulation-Based Design — 133

- 6.1. Low-Fidelity Antenna Models in Simulation-Based Optimization — 133
- 6.2. Coarse-Discretization as a Basis of Low-Fidelity Models — 136
- 6.3. Other Simplifications of Low-Fidelity Models — 139
- 6.4. Automated Selection of Model Fidelity — 143

7. Element Design: Case Studies — 145

- 7.1. EM-Driven Design of a Planar UWB Dipole Antenna with Integrated Balun — 145
  - 7.1.1. Antenna geometry — 146
  - 7.1.2. Optimization procedure — 146
  - 7.1.3. Numerical results — 148
  - 7.1.4. Experimental validation — 148
- 7.2. Design of Compact UWB Slot Antenna — 148
  - 7.2.1. Antenna structure — 149
  - 7.2.2. Optimization algorithm — 151
  - 7.2.3. Numerical results and experimental validation — 152
- 7.3. Optimization of Slot-Ring Coupled Patch Antenna — 155
  - 7.3.1. Antenna structure — 156
  - 7.3.2. Low-fidelity model selection — 156
  - 7.3.3. Results, benchmarking, and experimental validation — 159
- 7.4. Low-Cost Modeling and Optimization of Ring Slot Antenna — 160
  - 7.4.1. Modeling methodology — 161
  - 7.4.2. Ring slot antenna structure — 164

		7.4.3.	Application examples and experimental validation	166
	7.5.	Multi-objective Design of Planar Yagi Antenna		168
		7.5.1.	Antenna structure and problem statement	169
		7.5.2.	Design procedure and numerical results	170
		7.5.3.	Experimental validation	173
	7.6.	Design of Microstrip Patch Antennas		174
		7.6.1.	Recessed microstrip line fed MPA: Geometry and model setup	176
		7.6.2.	Recessed microstrip line fed MPA: Optimization procedure	177
		7.6.3.	Recessed microstrip line fed MPA: Numerical results and experimental validation	177
		7.6.4.	Slot-energized MPA: Geometry and model setup	177
		7.6.5.	Slot-energized MPA: Design with optimization, results, and validation	179

8. Microstrip Antenna Subarray Design Using Simulation-Based Optimization   183

	8.1.	Design Method: Optimization Algorithm	184
	8.2.	Design Formulation and EM Models of Microstrip Antenna Subarrays	188
		8.2.1. MPAS I (one-side configuration)	190
		8.2.2. MPAS II (two-side configuration)	190
	8.3.	Optimization Results	193
		8.3.1. MPAS I (one-side configuration)	194
		8.3.2. MPAS II (two-side configuration)	197
	8.4.	Validation by Measurements	199
	8.5.	Summary	201

9. Antenna Array Models for Simulation-Based Design and Optimization   203

   9.1. Array Factor-Based Models of Antenna Array Apertures 205
   9.2. Computational EM Models of Antenna Arrays 208
   9.3. Simulation-Based Superposition Models 212

10. Design of Linear Antenna Array Apertures Using Surrogate-Assisted Optimization   213

   10.1. Optimum Design of Array Factor Models Using Smart Random Search and Gradient-Based Optimization 214
     10.1.1. Problem formulation 214
     10.1.2. Optimization methodology 215
     10.1.3. Case study 1: Linear end-fire array optimization 216
       10.1.3.1. Sidelobe reduction with two variables 216
       10.1.3.2. Peak directivity maximization with two variables 218
       10.1.3.3. Sidelobe reduction with different phase shifts 219
       10.1.3.4. Sidelobe reduction with different phase shifts and spacing 222
   10.2. Null Controlled Pattern Design 223
   10.3. 20-Element Broadside Array Design for Pattern Nulls and Sector Beam 229
   10.4. Phase–Spacing Optimization of Linear Arrays Using Simulation-Based Surrogate Superposition Models 231
     10.4.1. Array aperture geometry and design problem outline 231
     10.4.2. Array factor model for the radiation response estimation 233

	10.4.3.	Design using optimization of simulation-based surrogates	234
		10.4.3.1. Problem formulation: Objective function	234
		10.4.3.2. Optimization of the array factor model	235
		10.4.3.3. Correction of the simulation-based low-fidelity model	235
		10.4.3.4. Optimization algorithm	237
	10.4.4.	Optimization results	238
		10.4.4.1. Optimization with non-uniform spacing and phases	239
		10.4.4.2. Optimization with uniform spacing and non-uniform phases	244
	10.4.5.	Optimized designs as phased array apertures	246
10.5.	Summary		251

11. Design of Planar Microstrip Antenna Arrays Using Variable-Fidelity EM Models — 253

11.1.	Planar Antenna Array Design Problem	254
11.2.	Design Optimization Methodology	255
	11.2.1. Surrogate-based optimization	255
	11.2.2. Surrogate-based optimization for array design	257
11.3.	Implementation and Numerical Results	260
11.4.	Rapid Optimization of Radiation Response	264
	11.4.1. Design case: 49-element microstrip array	264
	11.4.2. Utilized models	266
	11.4.3. Optimization with non-uniform amplitude excitation	268
	11.4.4. Optimization with non-uniform phase excitation	269
11.5.	Summary	270

12. Design of Planar Microstrip Array Antennas Using Simulation-Based Superposition Models — 273

- 12.1. Design Problem and Array Models 274
  - 12.1.1. Design problem and antenna array geometry 274
  - 12.1.2. Superposition models and discrete EM models 274
- 12.2. Design Optimization Using Surrogates 276
  - 12.2.1. Design problem formulation: Objective function 276
  - 12.2.2. Low-fidelity model: Model correction ... 277
  - 12.2.3. Optimization algorithm 278
- 12.3. Results 279
  - 12.3.1. 16-Element Cartesian lattice antenna array 280
  - 12.3.2. 100-Element Cartesian lattice antenna array 282
  - 12.3.3. 100-Element hexagonal antenna array 286
- 12.4. Summary 287

13. Design of Planar Arrays Using Radiation Response Surrogates — 291

- 13.1. Design Optimization Methodology 291
  - 13.1.1. Problem formulation 292
  - 13.1.2. Response correction of array factor model 292
  - 13.1.3. Optimization flow 294
- 13.2. Case Study I: 100-Element Microstrip Patch Antenna Array 295
- 13.3. Case Study II: 28-Element Microstrip Patch Antenna Array 298
- 13.4. Summary 300

14. Simulation-Based Design of Corporate Feeds for Low-Sidelobe Microstrip Linear Arrays — 301

14.1. Sidelobe Reduction in Arrays Driven with Corporate Feeds Comprising Equal Power Split Junctions 303
    14.1.1. Approach justification 303
    14.1.2. Design task, feed elements, and feed architectures 305
    14.1.3. Fast models of corporate feeds 307
    14.1.4. Realization and simulation-based optimization of aperture-feed structures...................... 313
    14.1.5. Validation 322

14.2. Design of Low-Sidelobe Arrays Implementing Requited Excitation Tapers: The Case of Corporate Feeds Comprising Unequal-Split Junctions 328
    14.2.1. Approach justification 329
    14.2.2. Design process 330
        14.2.2.1. Modeling and optimization of unequal-split junctions 334
        14.2.2.2. Feed redesign for sidelobe minimization 336
    14.2.3. Realization of the design process: Numerical results and measurements ... 338
        14.2.3.1. Example of a taper-oriented design 339
        14.2.3.2. Example of an SLL-oriented design 343

14.3. Summary.......................... 355

15. Design of Linear Phased Array Apertures Using Response Correction and Surrogate-Assisted Optimization — 357

15.1. Optimization Methodology 358
    15.1.1. Element optimization 358

15.1.2. Correction of the array factor model ... 359
15.1.3. Design for scanning 360
15.2. Case Study: 16-Element Linear Phased Array .. 362
15.2.1. Element optimization 362
15.2.2. Optimal excitation taper for the major lobe pointing broadside 364
15.2.3. Optimization for major lobe scanning 365
15.2.4. Experimental validation of the optimal design 365
15.3. Optimal Design as the Phased Array 367
15.4. Summary 370

16. Fault Detection in Linear Arrays Using Response Correction Techniques — 373

16.1. Fault Detection in Array Antennas 374
16.2. Fault Detection Methodology 376
16.2.1. Surrogate model construction 376
16.2.2. Fault detection using fast enumeration 381
16.3. Numerical Results 385
16.3.1. On–off faults 385
16.3.2. Partial faults 389
16.4. Summary 390

17. Surrogate-Assisted Tolerance Analysis of Microstrip Linear Arrays with Corporate Feeds — 393

17.1. Manufacturing Tolerances in Linear Antenna Arrays with Corporate Feeds 394
17.2. Local Surrogate Modeling of Antenna Array Apertures 395
17.2.1. Radiation pattern surrogate of array elements 395
17.2.2. Local modeling of array aperture: Array factor model 396

		17.2.3. Local modeling of array aperture: Model correction using response features	396
	17.3.	Local Surrogate Modeling of Corporate Feeds	399
	17.4.	Case Study: 12-Element Microstrip Array	400
		17.4.1. Array structure	400
		17.4.2. Results and discussion	403

18. Discussion and Recommendations: Prospective Look   **409**

References   417

Index   441

Chapter 1

Introduction

Computer-aided design (CAD) of antenna arrays is becoming a common practice these days. In a significant part, this is due to development of computational electromagnetics which delivers reliable tools for setting up and simulation of models of antenna array elements, feeds, and array apertures. The development of computers also tremendously contributes to the state of the art of antenna array CAD so that antenna array models comprising millions and tens of millions of unknowns (mesh cells, knots, etc.) can be evaluated on regular desktops in a reasonable time. The use of a graphics processing unit (GPU) for certain discrete types of electromagnetic solvers can speed up simulations in a few folds. Therefore, configured designs of array components and even the entire antenna array structures, to certain extend, including their installation environment, such as connectors, radomes, covers, finite grounds of printed apertures, can be simulated by full-wave discrete solvers prior manufacturing at a high-fidelity level of description.

Antenna theory provides overall understanding of electromagnetic interactions within antenna structures, behavior of basic antenna array quantities, including their dependency on geometry parameters, properties of materials, and frequency. Importantly, antenna theory also provides design guidelines and working formulas for setting up prototypes of array elements, feeds, and for dimensioning radiating apertures. On the other hand, computationally inexpensive analytical models can only provide — in most cases — an approximate estimation of the optimum design. This is particularly the case when certain interactions within the array elements, feed networks, and with the array environment (e.g., housing, installation fixture, connectors) have to be taken into account. Reliable evaluation of design characteristics and sensitivity to tolerances before manufacturing are particularly important for integrated antenna array structures. For these reasons, full-wave electromagnetic (EM) simulation plays an essential role in antenna array verifications.

An important part of the antenna array design process is the adjustment of geometry and material parameters to ensure that the array responses satisfy design specifications in terms of certain quantities such as sidelobe level, beamwidth, locations of pattern nulls, peak realized gain, reflection loss, efficiency, etc. (Balanis, 2005; Mailloux, 2005; Hansen, 2009). For these purposes, reliable knowledge of antenna array quantities at different points of the design space is necessary, which requires high-fidelity modeling and simulation of antenna array models.

Full-wave EM simulations of realistic and finely discretized antenna array models are computationally expensive: evaluation for a single combination of design parameters may take up to several hours. While such computational costs are acceptable from the design validation standpoint, it is usually prohibitive for tuning or optimization which normally requires a large number of EM simulations of the structure of interest. Such a large number of EM simulation is, in part, a consequence of the fact that realistic models of antenna array apertures and array feeds are typically described by many design variables. As far as accurate simulated data about antenna array figures are expected, e.g., simulated excitation amplitudes at the inputs of low-sidelobe apertures, radiation patterns with sidelobe levels of 30 dB and lower, the use of circuit decomposition approaches to relief a computational load in this situation is quite restricted, unfortunately, due to underlying EM interactions within the structures.

Significant numbers of designable parameters in the antenna array models, especially, at the stage of implementation, and essentially multi-objective nature of array design problems motivate the use of numerical optimization for solving antenna array design tasks. At the stage of prototyping, the objective function is usually computed using explicit formulas, e.g., an array factor, while at the stage of implementation, the objective function is evaluated by means of computational electromagnetics.

Automation of the antenna array design process can be realized by formulating an adjustment of array components or/and an adjustment of the entire aperture-feed circuit as an optimization problem with the objective function supplied by an EM solver (Special Issue, 2007; Bandler et al., 2006). Unfortunately, most of the conventional optimization techniques, including gradient-based methods (Nocedal and Wright, 2006), e.g., conjugate-gradient, quasi-Newton, sequential quadratic programming, etc., and derivative-free (Kolda et al., 2003), e.g., Nelder–Mead, pattern search techniques, require a large number of objective function evaluations to converge to an improved design. For many realistic EM models, where evaluation time per design reaches a few hours with contemporary computing facilities, the cost of such an optimization process may be unacceptably high.

Another practical problem of conventional optimization techniques, dealing with discrete EM models, is numerical noise of the simulated data. This noise is partially a result of adaptive meshing techniques used by most contemporary discrete solvers: even a small change of design variables may result in a change of the mesh topology, and, consequently, discontinuity of the EM-simulated responses as a function of designable parameters. Such numerical noise is particularly an issue for gradient-based methods that normally require smoothness of the objective function.

The aforementioned challenges result in a situation where a common approach to simulation-driven antenna array design is based on repetitive parameters sweeps (usually, one parameter at a time) about an initial design which is set up using other means such as formulas, analytical and circuit models, etc. This approach is usually more reliable in case of simulated responses of antenna arrays than a brute-force optimization using built-in optimization capabilities of commercial simulation tools; however, it is also very laborious, time consuming, and demanding significant supervision of the designer. Moreover, such a parameter-sweep-based optimization process does not guarantee optimal results because only a limited number of parameters can be handled that way. It is also difficult to utilize correlations between the parameters properly. Finally, truly optimal values of the designable variables can be quite counter-intuitive.

In recent years, population-based search methods (also referred to as metaheuristics) (Yang, 2005, 2008) have gained considerable popularity. This group of methods includes, among others, genetic algorithms (GA) (Back et al., 2000), particle swarm optimizers (PSO) (Kennedy, 1997), differential evolution (DE) (Storn and Price, 1997), ant colony optimization (Dorigo and Stutzle, 2004). Most of metaheuristics are biologically inspired algorithms designed to alleviate certain difficulties of the conventional optimization methods, in particular, handling problems with multiple local optima (Yang, 2010a).

Probably the most successful application of the metaheuristic algorithms for antenna arrays resided so far in array factor optimization, e.g., Ares-Pena et al. (1999), Haupt (1995), Jin and Rahmat-Samii (2005, 2006, 2007), Petko and Werner (2007), Bevelacqua and Balanis (2007), Selleri et al. (2008), Li et al. (2010), Rajo-Iglesias and Quevedo-Teruel (2007) and Roy et al. (2011). In these problems, the cost of evaluating the single element response is not of the primary concern or the response of a single element is already available, e.g., with pre-assigned (e.g., isotropic) or pre-simulated array elements. However, application of metaheuristics to EM-simulation-driven design of array apertures and feeds is not practical because corresponding computational costs would be tremendous: typical GA, PSO or DE algorithms require hundreds, thousands or even tens of thousands objective function evaluations to yield a solution (Ares-Pena et al., 1999; Haupt, 1995; Jin and Rahmat-Samii, 2007; Petko and Werner, 2007; Bevelacqua and Balanis, 2007; Pantoja et al., 2007;

Selleri *et al.*, 2008; Li *et al.*, 2008; Rajo-Iglesias and Quevedo-Teruel, 2007; Roy *et al.*, 2011).

The problem of high-computational cost of conventional EM-based optimization can be alleviated to some extent by the use of adjoint sensitivity (Director and Rohrer, 1969), which is a computationally cheap way to obtain derivatives of the system response with respect to its design parameters. Adjoint sensitivities can substantially speed up microwave design optimization while using gradient-based algorithms (Bandler and Seviora 1972; Chung *et al.*, 2001). This technology was also demonstrated for antenna optimization (Jacobsson and Rylander, 2010; Toivanen *et al.*, 2009; Zhang *et al.*, 2012). It should be noted, however, that adjoint sensitivities are not yet widespread in commercial EM solvers. To our knowledge this feature is available only in CST Microwave Studio (CST, 2016) and ANSYS (HFSS, 2016). In addition, a reliable utilization of adjoint sensitivities is limited by numerical noise of EM-simulated responses (Koziel *et al.*, 2013b).

Surrogate-based optimization (SBO) (Forrester and Keane, 2009; Queipo *et al.*, 2005) is one of the most recent and promising ways to realize computationally efficient simulation-based design of antenna arrays (Koziel and Ogurtsov, 2014a,b, 2015a–d). The main idea of SBO is to shift the computational burden of the optimization process to a surrogate model. This surrogate model is a cheap representation of the optimized structure (Bandler *et al.*, 1994; Queipo *et al.*, 2005; Koziel *et al.*, 2006; Koziel and Ogurtsov, 2011a). In a typical scenario, the surrogate model is used as a prediction tool to find approximate location of the original (high-fidelity or fine) model. After evaluating the high-fidelity model at this predicted optimum, the surrogate is updated in order to improve its local accuracy. The key prerequisite of the SBO paradigm is that the surrogate is much faster than the high-fidelity model. Also, in many SBO algorithms, the high-fidelity model is only evaluated once per iteration. Therefore, the computational cost of the design process with a well working SBO algorithm may be significantly lower than those with most of conventional optimization methods.

There are two major types of surrogate models. The first one comprises function-approximation models constructed from sampled high-fidelity simulation data (Simpson et al., 2001). A number of approximation and interpolation techniques are available, including artificial neural networks (Haykin, 1998), radial basis functions (Wild et al., 2008), kriging (Forrester et al., 2009), support vector machines (Smola and Schölkopf, 2004), Gaussian process regression (Angiulli et al., 2007; Jacobs, 2012), or multi-dimensional rational approximation (Shaker et al., 2009). If the design space is sampled with sufficient density, the resulting model becomes reliable so that the optimal design can be found just by optimizing the surrogate. In fact, approximation methods are usually used to create multiple use library models of specific components. The computational overhead related to such models may be very high. Depending on the number of designable parameters, the number of training samples necessary to ensure decent accuracy might be hundreds, thousands or even tens of thousands. Moreover, the number of samples quickly grows with the dimensionality of the problem (so-called curse of dimensionality). As a consequence, globally accurate approximation modeling is not suitable for ad-hoc (one-time) antenna array optimization. Iteratively improved approximation surrogates are becoming popular for global optimization (Couckuyt, 2013). Various ways of incorporating new training points into the model (so-called infill criteria) exist, including exploitative models (i.e., models oriented towards improving the design in the vicinity of the current one), explorative models (i.e., models aiming at improving global accuracy), as well as model with balanced exploration and exploitation (Forrester and Keane, 2009).

Another type of surrogates, so-called physics-based surrogates, are constructed from underlying low-fidelity (or coarse) models or the respective structures. Because the low-fidelity models inherit some knowledge of the system under consideration, usually a small number of high-fidelity simulations are sufficient to configure a reliable surrogate. The most popular SBO approaches using physics-based surrogates that proved to be successful in microwave engineering are space mapping (SM) (Bandler et al., 1994, 1995, 2003, 2004a,b), tuning SM (Cheng et al., 2010; Koziel et al., 2011b), as well as various response correction methods (Echeverria and Hemker, 2005; Koziel and Leifsson, 2012b, 2016). To ensure computational efficiency, it is important to have the low-fidelity model considerably faster than the high-fidelity model. For that reason, circuit equivalents or models based on analytical formulas are preferred (Bandler et al., 2004a). The aforementioned methods (particularly SM) were mostly used for design of microwave filters or transmission-line-based components. Unfortunately, in case of antennas and antenna array apertures, reliable circuit equivalents are rarely available. For radiating structures, a universal way of obtaining their low-fidelity models is with low-fidelity simulations. Such low-fidelity models are relatively expensive even for a single antenna. This poses additional challenges in terms of optimization.

The central topic of this book is the simulation-based antenna array CAD methods where numerical optimization serves as a mean to conduct the automated design process, implement designs, and/or tune already configured designs of antenna array elements, apertures, feeds, and entire aperture-feed circuits. In the view of the outlined challenges faced by simulation-based antenna array design methods, a substantial part of this book is devoted to numerically efficient realization of optimization techniques, in particular, the ones utilizing surrogate models. We believe that the methods described and demonstrated in this book will contribute to the development of practical and reliable antenna array CAD tools. The book provides a useable description of optimization methods which are relevant to EM-simulation-based antenna and antenna array design. We also expect that a variety of application examples help in explaining how the described methods work for different components and at different levels of antenna array circuits. Finally, we hope that this book will help an interested reader in forming his vision about the benefits, complexity, and challenges of the described methods in the context of antenna array CAD.

We begin, in Chapter 2, by giving a motivation for formulating the simulation-based antenna array design task as an optimization problem. We also outline basic antenna array figures of interest as well as relations which, to our experience, are typically used for simulation-based design of antenna arrays.

In Chapter 3, we provide background information about conventional numerical optimization techniques, including both gradient-based and derivative-free methods.

Chapter 4 is focused on introducing numerical methods for global optimization.

Chapter 5 introduces fundamentals of SBO, techniques for surrogate-based modeling, as well as surrogate-based algorithms for solving computationally expensive microwave and antenna engineering problems.

Chapter 6 overviews basic requirements imposed on low-fidelity models of antennas. Basic options for setting such models are described. The accuracy–speed tradeoff which is inherent to simulated low-fidelity models of antennas is illustrated on examples.

Chapters 7 and 8 describe and demonstrate application of the surrogate-based methods, which are described in Chapter 5, to antenna and antenna subarray design problems. Experimental validation is provided for a number of obtained designs.

Chapter 9 discusses antenna array models in the context of simulation-based design and optimization.

Chapter 10 explains and demonstrates design of linear antenna array apertures conducted by means of numerical optimization. We start with design problems of the initial stage of the design process where dimensioning of the array aperture and search for the excitation taper are realized as optimization of the array factor. Further, we demonstrate application of the surrogate-based methods, discussed in Chapter 5, and simulated physics based-surrogate models, discussed in Chapter 9, to design of linear antenna array apertures. Application of models discussed in Chapter 9 to evaluations of the optimal designs for phased array applications is demonstrated.

Chapter 11 considers optimization of planar array apertures using surrogate-based algorithms and simulated variable-fidelity models.

Chapter 12 considers optimization of planar array apertures using surrogate-based algorithms and simulation-based superposition models for design of excitation tapers with respect to EM interactions within the array aperture.

Chapter 13 discusses and illustrates a response correction technique which is specifically developed to conduct fast design of fixed-beam planar array apertures yet at the high-fidelity level of description.

Chapter 14 introduces and demonstrates two alternative processes for design of low-sidelobe microstrip linear antenna arrays with integrated apertures and corporate feeds. The discussed design procedures include identification of the optimal feed architectures, implementation and optimization of the corporate feeds, as well as optimization of the entire integrated aperture–feed structure for the required sidelobe levels. Experimental validation of selected designs is provided.

Application of response correction and surrogate-assisted optimization to design of excitation tapers of linear phased array apertures is described in Chapter 15. Experimental validation of the optimal design is provided.

Surrogate-assisted fault detection in linear arrays is addressed in Chapter 16 using response correction techniques.

Chapter 17 addresses the problem of computationally efficient tolerance analysis of microstrip linear arrays with microstrip corporate feeds using surrogate-assisted techniques.

Finally, Chapter 18 provides a conclusive discussion on surrogate-assisted CAD methods for antenna array problems, summarizes the key features of the surrogate-assisted methods, emphasizes the challenges and open problems, as well as gives recommendations for practical utilization of these methods for solving real-world antenna array design tasks.

Chapter 2

Antenna Array Fundamentals

2.1. Introduction

An antenna consisting of a single radiating or receiving element is often incapable of meeting technical requirements imposed on radiation pattern, null locations, beamwidth, sidelobe levels, high gain, beam steering, etc. On the other hand, arrays of antennas can be designed to meet various and demanding specifications (Elliott, 2003; Balanis, 2005; Volakis, 2007; Mailloux, 2005; Hansen, 2009). Antenna array analysis and synthesis methods, including the theoretical background of these methods, have become highly elaborated subjects substantially contributing to the state of the art of radar and communication technologies (Elliott, 2003; Balanis, 2005; Mailloux, 2005; Schantz, 2005; Volakis, 2007; Hansen, 2007, 2009). Different types of reliable, low cost, low profile antenna array elements and components of antenna circuits are available these days because of the development of manufacturing technologies, material science and engineering (Petoza, 2007; Volakis et al., 2010; Haupt, 2010; Gao et al., 2014). Highly sophisticated antenna systems have become a part of everyday life (Mailloux, 2005; Hansen, 2009; Haupt, 2010).

Evolution of computational electromagnetics (Taflove and Hagness, 2005; Pan, 2003; Jin, 2014; Rautio and Harrington, 1987; Harrington, 1993) has resulted in a number of software packages, custom as well as commercial, for modeling, full-wave analysis, computer-aided design (CAD), and numerical validation of components of antenna circuits (Burke and Poggio, 1981; Makarov, 2002; HFSS, 2016; CST Microwave Studio, 2016; FEKO, 2015; Sonnet EM, 2016; XFDTD, 2016). Discrete electromagnetic solvers of these packages are based on various computational methods (Taflove and Hagness, 2005; Jin, 2014; Rautio and Harrington, 1987; Harrington, 1993). In some cases, different solvers are available as simulation options for a particular project within a single software package, e.g., CST (2016) and FEKO (2015). These packages, when run on contemporary computers, can simulate discrete full-wave electromagnetic models of antenna components and circuits comprising millions of unknowns (elements, knots, or mesh-cells) in reasonable time intervals even on general-purpose desktops. Today, built-in numerical optimization routines are available in most of commercially available packages for electromagnetic analysis. Integration of full-wave discrete electromagnetic modelers/solvers with circuit evaluators and co-simulations of other types (thermal, mechanical) is also available in some cases, e.g., CST (2016) and HFSS (2016). At the same time, systematic and automated processes of antenna array simulation-based design, from design specifications up to layouts, have not become practices as microwave CAD (Gupta *et al.*, 1981; Okoshi, 1985; Steer *et al.*, 2002) which shares many physics backgrounds, formal methods, and simulation tools with the former.

According to our understanding, the reasons for this situation are the following. First, real-life antenna array design problems are essentially multi-objective (Mailloux, 2005; Hansen, 2009; Gao *et al.*, 2014) so that several objectives (or performance figures of interest) should be handled to satisfy certain requirements, e.g., sidelobe level of the radiation pattern should be suppressed up to a certain level, whereas the major lobe should be left intact, and/or the reflection coefficient at the common input of the array feed should be minimized over the operating frequencies. Additional requirements can be considered such as those for half-power beamwidth, null-to-null beamwidth, locations of radiation pattern nulls, shape of the beam, back radiation level, peak realized gain, signal-to-noise ratio, etc.

It is worth noticing that elaborated methods of analytical and semi-analytical radiation pattern synthesis (Kraus, 1988; Elliott, 2003; Balanis, 2005; Hansen, 2009; Haupt, 2010) are definitely indispensable at the initial stage of prototyping, i.e., for determining the required excitation amplitudes (array element weights). However, such methods have only one or two parameters as variables. Furthermore, as a matter of fact, works on simulation-based design of the array feeds realizing required sets of excitation tapers, including questions of optimality and sensitivity of the feed, have only started to appear (Koziel and Ogurtsov, 2017; Ogurtsov and Koziel, 2017a,b). Moreover, certain antenna array characteristics can only be evaluated for a given design with full-wave methods at the implementation stage (Koziel and Ogurtsov, 2017; Ogurtsov and Koziel, 2017a,b) and using discrete full-wave electromagnetic (EM) simulations which can be quite computationally demanding even for an antenna-feed structure containing one radiator and a simple feed (see, e.g., Sec. 6.2 in Chapter 6).

Furthermore, antenna array models are normally described using multiple parameters such as element spacings, dimensions of the antennas including detail of energizing elements, dimensions of the feeds, and material parameters. In addition, quantitative influence of the environment such as extensions of the ground planes of printed-circuit arrays, connectors, and radome covers as well as sensitivity of the radiation and reflection responses on the dimensional parameters is not available prior full-wave simulation. Therefore, adjustment of antenna array models, even by parts, becomes a challenging design task, which is hardly realizable through parameter sweeping where only one or two parameters are varied at a time.

Finally, development of the whole design process, resulting in a systematic CAD technique or tool, is a multi-disciplinary activity utilizing various skills and knowledge from microwave engineering, electromagnetics, and antenna analysis to mathematical modeling, numerical optimization, and fast computing.

In this chapter, first, we overview antenna array basics which, according to our understanding and experience, are useful for setting up antenna array models and conducting, primarily simulation-driven, CAD. Next, we present our vision of contemporary challenges for such antenna array CAD processes.

For a complete antenna array theory and contemporary practice of antenna array analysis and design, an interested reader is suggested to consider the available texts on these subjects, e.g., Kraus (1988), Elliott (2003), Balanis (2005), Mailloux (2005), Volakis (2007) and Haupt (2010).

2.2. Antenna Array Radiation Figures

An electric field produced by an aperture comprising N sources, e.g., radiation elements (antennas), at a certain observation point, can be written as

$$\mathbf{E}(\mathbf{r}) = \sum_{n=1,\ldots,N} a_n \mathbf{E}_n(\mathbf{r}), \tag{2.1}$$

where $\mathbf{E}_n(\mathbf{r})$ is a vector of electric field due to the nth element at the observation point $\mathbf{r}$, and a_n is the complex excitation coefficient (weight) of element n (see Fig. 2.1).

A general meaning of a_n, consistent with the superposition principle, is the complex amplitude (phasor) of the incident wave exciting element n. Electric field $\mathbf{E}_n(\mathbf{r})$ listed in (2.1) is already normalized, that is, it is evaluated with a unit excitation, e.g., using a full-wave discrete solver. Otherwise, it should be normalized to its

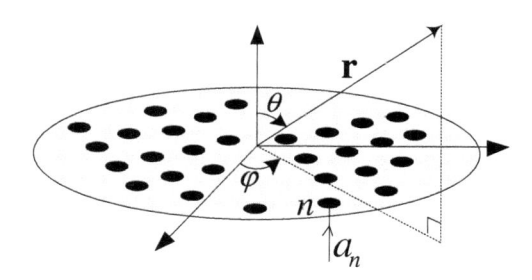

Fig. 2.1. Antenna array aperture.

excitation phasor $a_{n,e}$ as

$$\mathbf{E}(\mathbf{r}) = \sum_{n=1,\dots,N} a_n \mathbf{E}_n(\mathbf{r})/a_{n,e}. \quad (2.2)$$

Acquisition of fields $\mathbf{E}_n(\mathbf{r})$ for design of linear and planar antenna array apertures is discussed in Chapters 10, 12, and 15. It is worth to mention that (2.1) is a generic relation, assuming linear responses of elements to their excitations, and it is useable also for evaluation of the total near-field (Kowalski and Jin, 2000) as well as for total far-field (Koziel and Ogurtsov, 2015a,d). In the former case, $\mathbf{E}_n(\mathbf{r})$ stands for a near-field due to element n. In the latter, case $\mathbf{E}_n(\mathbf{r})$ stands for a far-field due to element n. For the far-field case, it is convenient to work with radiation directions θ and φ so that $\mathbf{r} = [r, \theta, \varphi]^T$ and distance r to an observation point in the far-zone is considered as a parameter which is common to all radiation directions. Hence, (2.1) for a far-field becomes

$$\mathbf{E}(\theta, \varphi) = \sum_{n=1,\dots,N} a_n \mathbf{E}_n(\theta, \varphi). \quad (2.3)$$

The use of (2.3) for setting up reusable simulation-based models of antenna arrays is discussed Chapters 10, 12, and 15. Equation (2.3) is quite convenient for numerical evaluation of radiation patterns, realized gains versus scan angle, and beam broadening of phased apertures. In particular, if $\mathbf{E}_n(\theta, \varphi)$ are pre-simulated far-fields of the array embedded elements, then the effects of elements coupling on the radiation figures are accounted with (2.3).

The electric field $\mathbf{E}(\theta, \varphi)$ depends on the distance with r^{-1}. Therefore, in the far-zone, it is convenient to work with the radiation intensity (W/unit solid angle), which does not depend on the distance,

$$U(\theta, \varphi) = \frac{r^2}{2\eta}|\mathbf{E}(\theta, \varphi)|^2 \approx \frac{r^2}{2\eta}(|E_\theta(\theta, \varphi)|^2 + |E_\varphi(\theta, \varphi)|^2), \quad (2.4)$$

where E_θ and E_φ are, respectively, the θ- and φ-components of the electric far-field, and the radial component E_r is assumed to vanish in the far-zone. Here η stands for the intrinsic impedance of the medium, $\approx 377\,\Omega$ for air and vacuum. The total radiated power, in terms of the radiation intensity, is given by

$$P_r = \int_0^{2\pi} \int_0^{\pi} U(\theta, \varphi) \sin\theta \, d\theta \, d\varphi. \qquad (2.5)$$

The figure of directivity is "the ratio of the radiation intensity in a given direction from the antenna to radiation intensity averaged over all directions" according to IEEE Standard Definitions of Terms for Antennas (IEEE Standard, 2013), i.e.,

$$D = \frac{U(\theta, \varphi)}{U_{\text{ave}}} = 4\pi \frac{U(\theta, \varphi)}{P_r}. \qquad (2.6)$$

If the direction is not specified, the direction of maximum radiation intensity is implied (IEEE Standard, 2013), i.e.,

$$D_0 = D_{\max} = \frac{U_{\max}}{U_{\text{ave}}} = 4\pi \frac{U_{\max}}{P_r}, \qquad (2.7)$$

where D_0 stands for maximum directivity in this case. Directivity can be also directly written in terms of the far-field as

$$D = \frac{S(\theta, \varphi)}{S_{\text{ave}}} = \frac{\frac{1}{2\eta}|\mathbf{E}(\theta, \varphi)|^2}{\frac{1}{4\pi r^2} \int_0^{2\pi} \int_0^{\pi} \frac{1}{2\eta}|\mathbf{E}(\theta, \varphi)|^2 r^2 \sin\theta \, d\theta \, d\varphi}, \qquad (2.8)$$

where $S(\theta, \varphi)$ and S_{ave} are, respectively, the power density (W/m^2) toward the direction of interest (at distance r) and the power density averaged over the surface of the far-zone sphere of radius r. Thus,

$$D = 4\pi \frac{|\mathbf{E}(\theta, \varphi)|^2}{\int_0^{2\pi} \int_0^{\pi} |\mathbf{E}(\theta, \varphi)|^2 \sin\theta \, d\theta \, d\varphi}. \qquad (2.9)$$

Definitions (2.4)–(2.9) apply to antennas comprising single radiators as well. Gain of antenna is defined as

$$G = 4\pi \frac{U(\theta, \varphi)}{P_{\text{in}}} \qquad (2.10)$$

with P_{in} standing for the power entering the antenna. Realized gain (IEEE Standard, 2013), also referred to as absolute gain (Balanis,

2005), is defined as

$$G_r = 4\pi \frac{U(\theta, \varphi)}{P_{\text{inc}}} \tag{2.11}$$

with P_{inc} standing for the power incident at the antenna input, or the total power incident at the inputs of the antenna array aperture. If the direction is not specified for gain and realized gain, then the direction of maximum radiation intensity is implied. Directivity, gain, and realized gain are related through

$$G_r = e_m G = e_m e_r D. \tag{2.12}$$

In (2.12), e_m stands for the reflection (mismatch) efficiency, which, for a single input antenna, is related to the reflection coefficient at the antenna input Γ by

$$e_m = 1 - |\Gamma|^2, \tag{2.13}$$

and e_r stands for the antenna radiation efficiency, that is the radiated fraction of the power entering the antenna. Powers of (2.6), (2.10), and (2.11) are related through efficiencies as follows:

$$P_r = e_r P_{\text{in}} = e_m e_r P_{\text{inc}} = e_{\text{tot}} P_{\text{inc}}, \tag{2.14}$$

where e_{tot} is termed as the total efficiency.

It is quite common practice to measure (list) directivity, gain, realized (absolute) gain, and efficiencies in dB. Often, directivity defined with (2.6) is listed in dBi and referred as directivity relative to the isotropic radiator (Balanis, 2005). Definitions and descriptions of many other antenna figures the reader can find in Krauss (1988), Elliott (2003), Balanis (2005), and Mailloux (2005).

2.3. Array Factor Analysis

The application of Eqs. (2.1)–(2.3) for antenna arrays is rather straightforward though it can be associated with considerable computational effort spent on evaluating the contributions of the array embedded elements. Therefore, it is beneficial to have also simpler means for evaluation of antenna array radiation figures. Such means are needed to prototype array apertures and feeds (Koziel and Ogurtsov, 2017; Ogurtsov and Koziel, 2017a,b), approximately evaluate their radiation figures, and obtain fast predictors for simulation-based optimization.

Certain simplifying assumptions about radiation and interactions of the array elements have to be made. The elements of the array are assumed to be identical and identically oriented. The current distributions in the elements are also assumed to be the same, up to the amplitude. Also, the distance r to the observation point in the far-zone is much larger than the maximal extension of the array aperture. With the above assumptions, (2.3) can be rewritten as

$$\mathbf{E}(\theta, \varphi) = \mathbf{E}_e(\theta, \varphi) \sum_{n=1,\ldots,N} a_n e^{j\psi_n}, \qquad (2.15)$$

where ψ_n is the phase shift due to location of the element n, and $\mathbf{E}_e(\theta, \varphi)$ is the far-field of the element driven by a unit excitation (e.g., $a_{n,e} = 1$ in (2.2)) if it would be originating from the origin of the used coordinate system. Summation of (2.15) is referred as the array factor (Krauss, 1988; Elliott, 2003; Balanis, 2005; Mailloux, 2005),

$$\mathrm{AF}(\theta, \varphi) = \sum_{n=1,\ldots,N} a_n e^{j\psi_n}. \qquad (2.16)$$

With the reference to Fig. 2.2, the phase shift can be written as

$$\psi_n = \beta\, r_n \cos\gamma_n = \frac{2\pi}{\lambda} r_n \cos\gamma_n \qquad (2.17)$$

where $\beta(\mathrm{m}^{-1})$ and $\lambda(\mathrm{m})$ are, respectively, the propagation constant and the wavelength in the medium, typically air or vacuum, at the frequency of interest. In general

$$\cos\gamma_n = \sin\theta \sin\theta_n \cos(\phi - \phi_n) + \cos\theta \cos\theta_n, \qquad (2.18)$$

where $(r_n, \theta_n, \varphi_n)$ is the location of the element n. For planar apertures, e.g., as for that outlined in Fig. 2.3, all $\theta_n = 90°$, therefore,

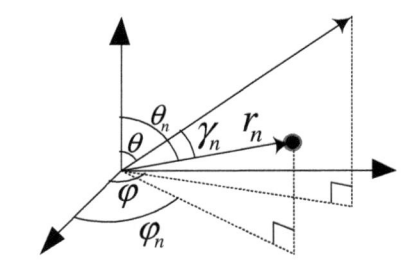

Fig. 2.2. Coordinate system nomenclature for far-field calculation.

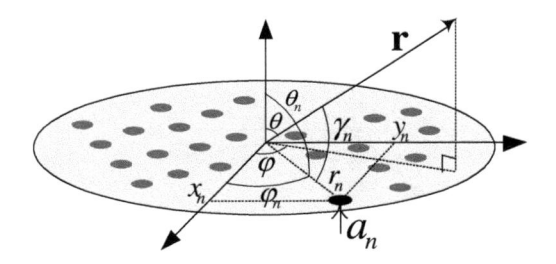

Fig. 2.3. Planar array aperture (all $\theta_n = 90°$).

(2.18) simplifies to
$$\cos \gamma_n = \sin \theta \cos(\phi - \phi_n). \tag{2.19}$$

Thus for planar apertures, e.g., as for that of Fig. 2.3, the phase shift (2.17) can be written also as

$$\psi_n = \beta \sin \theta (\cos \varphi \, x_n + \sin \varphi \, y_n) \tag{2.20}$$

with x_n and y_n standing for the Cartesian coordinates of the element n. In particular, for a single ring circular array outlined in Fig. 2.4, i.e., when all $r_n = a$, the phase shift can be listed as

$$\psi_n = \beta a \sin \theta \cos(\phi - \phi_n). \tag{2.21}$$

For a linear array depicted in Fig. 2.5, the phase shift does not depend on the azimuth angle φ, so that

$$\psi_n = \beta \cos \theta \, z_n \tag{2.22}$$

with z_n standing for the coordinate of the array element n.

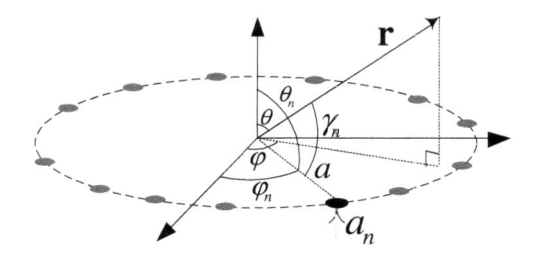

Fig. 2.4. Ring array aperture (all $r_n = a$ and $\theta_n = 90°$).

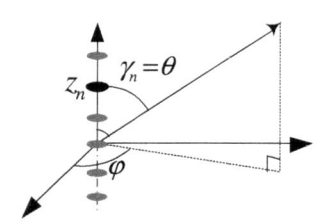

Fig. 2.5. Linear array aperture.

Equation (2.15) allows conducting design and/or analysis of antenna arrays by designing the contributions of the element and the array, including geometry and excitations of the latter, separately by

$$\mathbf{E}(\theta, \varphi) = \mathbf{E}_e(\theta, \varphi) \mathrm{AF}(\theta, \varphi). \tag{2.23}$$

Thus, for an array of identical antenna elements, the radiation intensity, $U(\theta, \varphi)$, is the product of the element radiation intensity, $U_e(\theta, \varphi)$, and the array factor magnitude squared,

$$U(\theta, \varphi) = U_e(\theta, \varphi) |\mathrm{AF}(\theta, \varphi)|^2. \tag{2.24}$$

Directivity of an array is defined with (2.9) assuming isotropic radiation element,

$$D_A(\theta, \varphi) = 4\pi \frac{|\mathrm{AF}(\theta, \varphi)|^2}{\int_0^{2\pi} \int_0^{\pi} |\mathrm{AF}(\theta, \varphi)|^2 \sin\theta \, d\theta \, d\varphi}; \tag{2.25}$$

and the maximum directivity of an array is defined as

$$D_{A,0} = 4\pi \frac{|\mathrm{AF}|_{\max}^2}{\int_0^{2\pi} \int_0^{\pi} |\mathrm{AF}(\theta, \varphi)|^2 \sin\theta \, d\theta \, d\varphi}. \tag{2.26}$$

For linear arrays, definitions (2.25) and (2.26) turn into

$$D_A(\theta) = \frac{2|\text{AF}(\theta)|^2}{\int_0^\pi |\text{AF}(\theta)|^2 \sin\theta \, d\theta}, \qquad (2.27)$$

$$D_{A,0} = \frac{2|\text{AF}|_{\max}^2}{\int_0^\pi |\text{AF}(\theta)|^2 \sin\theta \, d\theta}. \qquad (2.28)$$

Directivity (2.27) can be cast in a more convenient form by evaluating the integral analytically, taking into account (2.22), and noticing that the excitation coefficients a_n are complex in general, i.e., that

$$a_n = |a_n|e^{j\alpha_n} = b_n e^{j\alpha_n}, \qquad (2.29)$$

where the real number b_n is the amplitude of the excitation coefficient. The result is

$$D_A(\theta) = \frac{|\sum_{n=1}^N b_n e^{j(2\pi z_n \cos\theta + \alpha_n)}|^2}{\sum_{n=1}^N b_n^2 + 2\sum_{n=1}^N b_n (\sum_{k=n+1}^N b_k \cos(\alpha_n - \alpha_k) \text{Sinc}(2\pi(z_n - z_k)))}, \qquad (2.30)$$

where z_n and z_k are coordinates of elements n and k respectively, normalized to the wavelength. If the distances between all elements are integer numbers of the half-wavelength, $z_n - z_k = 0.5\lambda m_{nk}$, where m_{nk} are integers, and they are not necessarily all the same, when (2.30) simplifies, due to the properties of the Sinc function, to

$$D_A(\theta) = \left|\sum_{n=1}^N b_n e^{j(2\pi z_n \cos\theta + \alpha_n)}\right|^2 \bigg/ \sum_{n=1}^N b_n^2. \qquad (2.31)$$

It is worth to notice that the numerator of (2.31) is $|\text{AF}(\theta)|^2$. Furthermore, a remarkable property of linear arrays, in which the elements are uniformly spaced integer number M of half-wavelengths, is that their maximum directivity with respect to angle θ as well as maximal magnitude of the array factor does not depend on the scan angle. Directivity of such arrays with elements excitations of the form

$$a_n = b_n e^{j\alpha_n} = b_n e^{j(\alpha_{0n} + n\delta)} \tag{2.32}$$

can be written as

$$D_A(\theta) = \left| \sum_n b_n e^{j(\pi M n \cos\theta + \alpha_{0n} + n\delta)} \right|^2 \bigg/ \sum_n b_n^2, \tag{2.33}$$

where δ is the progressive phase shift, summations in (2.33) are performed over all array elements, and integers n in (2.33) can take negative values as well. The maximum directivity with respect to angle θ

$$D_{A,0} = \max\{\theta : D_A(\theta)\}, \tag{2.34}$$

where $D_A(\theta)$ is given by (2.33), does not depend on the progressive phase shift δ.

2.4. Mutual Coupling and Antenna Array Reflection Response

Interaction of an antenna with its environment is referred to as coupling. In particular an interaction of neighboring antennas, depicted in Fig. 2.6, and antennas within the same antenna array through their near-fields is referred to as mutual coupling. Such interactions are typically undesirable. First, quantitative evaluation of mutual coupling requires full-wave simulations, and complicates the antenna models and, consequently, the design processes. Second, for phased array apertures, mutual coupling causes changes in the reflection coefficients with the scan direction and it can result in severe degradation of the performance parameters of the scanned array beam, e.g., drop of the peak realized gain at certain directions.

Coupling through the feed of an antenna array, depicted in Fig. 2.6(c), can be characterized using microwave circuit analysis to certain extend; however, such analysis needs reliable description of the array aperture scattering parameters (usually simulated with an EM solver) as input data.

Undesirable effect of the through-feed-coupling, even for fixed-beam arrays, is that the incident waves driving the array elements, entries a_k in Fig. 2.6(c), can substantially deviate from the nominal values (incident waves a_k at the inputs of the feed connected to the matched loads, Figs. 2.6(b) and 2.6(d)). The nominal values of the incident waves a_k are determined at the prototyping stage of the design process. It can be done using analytical, therefore, computationally inexpensive models of the array aperture involving the array factor. In this case, the mutual coupling, depicted in Figs. 2.6(a) and 2.6(b), is either not accounted for at all or accounted for using analytical models only for certain types of radiators and array lattices (Allen and Diamond, 1966; Balanis, 2005; Hansen, 2009).

Mutual coupling can be also evaluated with full-wave methods at relatively low cost by imposing certain simplifications, e.g., unit

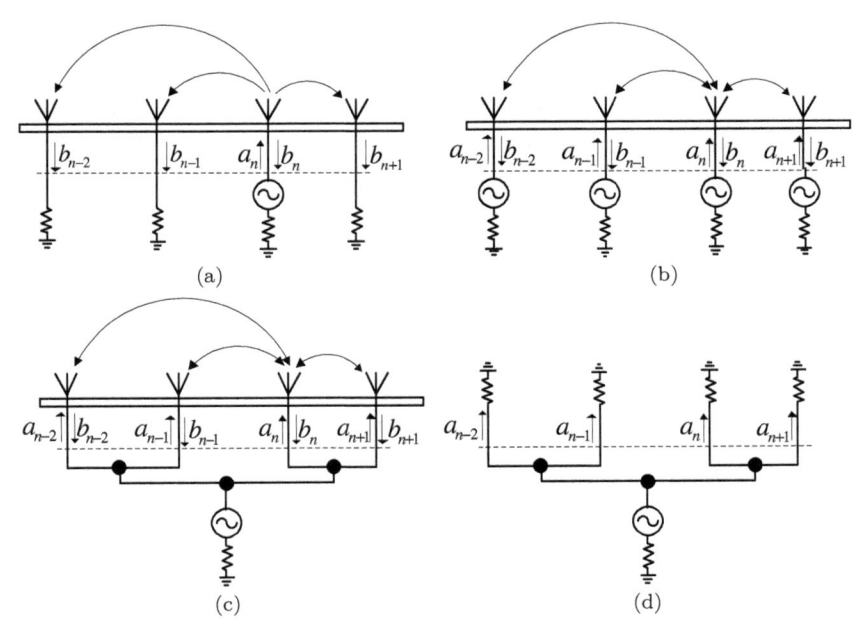

Fig. 2.6. Mutual coupling of the array elements: coupling through near-fields (a) with only one element n active and (b) with simultaneous excitation of all elements where all elements are terminated on matched loads; (c) coupling through near-fields and through a connected corporate feed; (d) a feed implementing excitation $\{a_n\}$ which is required for the aperture (b).

cells, periodic boundary conditions (Hansen, 2009; Vouvakis and Schaubert, 2011). Subsequently, discrete simulations of the array aperture model with the determined excitation set $\{a_k\}$ are used for validation and/or turning.

Most accurate description of mutual coupling due to the near-field interactions for the fixed- and scanning-beam array apertures can be done using full-wave superposition models (Kowalski and Jin, 2000; Koziel and Ogurtsov, 2015a,d). As a result of through-feed-coupling reliable designs of the feeds and array apertures with low-sidelobe levels (e.g., 20 dB and lower) may require simulation-based tuning associated with repetitive runs of the computationally expensive high-fidelity models at different points of the design space.

Accurate quantitative evaluation of the reflection response of antenna arrays in the presence of mutual coupling effects is only possible based on full-wave EM simulations of the array aperture. A quite transparent and generic description of coupling and its impact on the reflection response of the array aperture can be provided using scattering parameters. Namely, for scenario of Fig. 2.6(a), i.e., when only one input n is being energized, one has

$$\begin{aligned} b_1 &= S_{1n}a_n \\ &\vdots \\ b_n &= S_{nn}a_n \\ &\vdots \\ b_N &= S_{Nn}a_n, \end{aligned} \quad (2.35)$$

and for scenario of Fig. 2.6(b), i.e., with simultaneous excitation of all ports, one has

$$\begin{aligned} b_1 &= S_{11}a_1 + \cdots + S_{1n}a_n + S_{1N}a_N \\ &\vdots \\ b_n &= S_{n1}a_1 + \cdots + S_{nn}a_n + S_{nN}a_N \\ &\vdots \\ b_N &= S_{N1}a_1 + \cdots + S_{Nn}a_n - S_{NN}a_N, \end{aligned} \quad (2.36)$$

where $\mathbf{b} = [b_1 \ b_2 \ \ldots \ b_n \ \ldots \ b_N]^T$, $\mathbf{a} = [a_1 \ a_2 \ \ldots \ a_n \ \ldots \ a_N]^T$, S_{ij} are, respectively, a vector of scattering waves, a vector of incident waves, and simulated scattering parameters. In matrix form (2.36) becomes $\mathbf{b} = \mathbf{Sa}$.

With simultaneous excitation of the inputs, the reflected signal at input n, also referred as active reflection coefficient or scan reflection coefficient (Hansen, 2009) in the case of phased array antennas, is

$$\Gamma_n = \frac{b_n}{a_n} = (S_{n1}a_1 + \cdots + S_{nn}a_n + S_{nN}a_N)\frac{1}{a_n}$$
$$= S_{nn} + \frac{1}{a_n}\left(\sum_{m \neq n} S_{nm}a_m\right). \tag{2.37}$$

Thus the apparent input impedance at input n is

$$Z_{in,n} = Z_{0n}\frac{1+\Gamma_n}{1-\Gamma_n} = Z_{0n}\frac{1 + S_{nn} + \frac{1}{a_n}(\sum_{m \neq n} S_{nm}a_m)}{1 - S_{nn} - \frac{1}{a_n}(\sum_{m \neq n} S_{nm}a_m)} \tag{2.38}$$

with Z_{0n} standing for the characteristic impedance of the transmission line at input n. Ones sees that the reflection coefficient of (2.37) and apparent impedance of (2.38) both depend on coupling terms S_{nm} (property of the aperture itself) as well as on the applied excitation taper $\{a_n\}$ (an applied stimulus). The impedance matrix, relating the total voltages and currents at the inputs, is readily available (Gupta et al., 1981) with

$$\mathbf{Z} = \mathbf{Z}_0^{1/2}(\mathbf{I}+\mathbf{S})(\mathbf{I}-\mathbf{S})^{-1}\mathbf{Z}_0^{1/2}, \tag{2.39}$$

where $\mathbf{I}$ is the identity matrix, and

$$\mathbf{Z}_0^{1/2} = \mathrm{diag}[\sqrt{Z_{01}}, \ldots, \sqrt{Z_{0n}}, \ldots, \sqrt{Z_{0N}}]. \tag{2.40}$$

In case of identical input transmission lines with line impedance Z_0, one has

$$\mathbf{Z} = Z_0(\mathbf{I}+\mathbf{S})(\mathbf{I}-\mathbf{S})^{-1}. \tag{2.41}$$

2.5. Mutual Coupling and Antenna Array Design Validation

Simulation-based validation of antenna array design steps in the presence of mutual coupling can be outlined as follows. Consider the design process with major steps outlined in the diagram of Fig. 2.7. First, the array aperture is prototyped, i.e., the number of array elements N, spacings, and the set of required excitation coefficients $\{a_n\}$ are determined. This step is performed using the array factor as the fast model of the array aperture so that certain radiation figures, e.g., sidelobe level, half-power beamwidth, etc., have been satisfied for specific design requirements. Implementation technology is typically implied at this step or accounted for in rather simplistic manner,

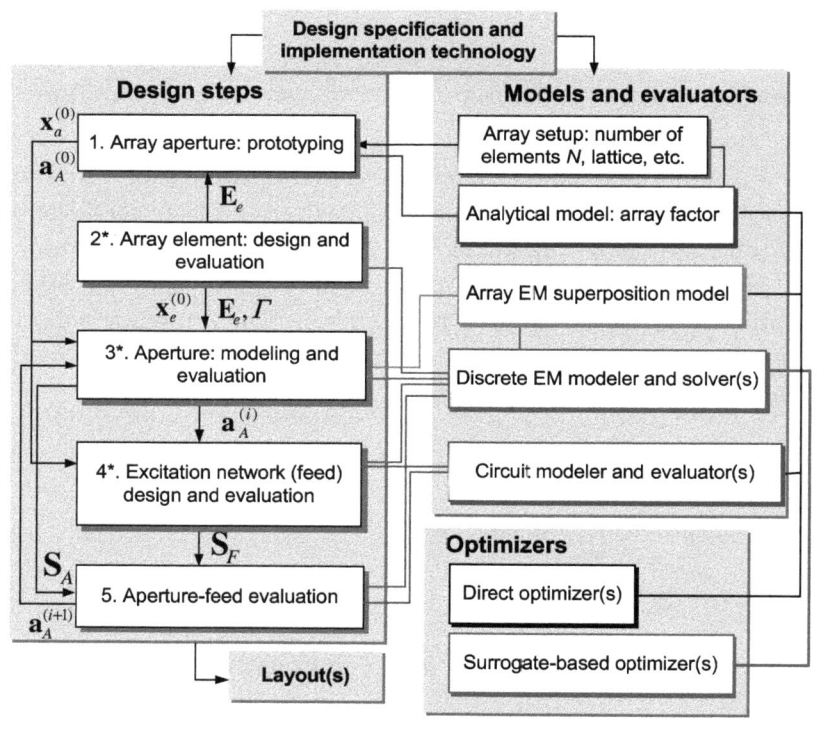

Fig. 2.7. Antenna array design process: an outline. * steps can involve validation by measurements.

e.g., by taking into account the element power pattern and element reflection coefficient which can be evaluated using either an EM model of an isolated element or an element simulated with periodic boundary conditions or analytical formulas available for certain types of elements (e.g., wire dipoles).

At the next step, the aperture including the elements (antennas) is realized, i.e., modeled, in certain manufacturing technology (e.g., as microstrip circuit) so that the aperture scattering matrix $\mathbf{S}_A$ (size of $N \times N$) and the far fields $\mathbf{E}_n(\theta, \varphi)$ are obtained with full-wave simulations; therefore, mutual coupling of the elements due to near-field interactions has been accounted for. Necessary tuning of the aperture can be required at this step to have the active reflections coefficients (2.37) and, primarily, the radiation figures, initially considered with a simplified model, in accordance with expectations. Due to many dimensional parameters involved in the array aperture EM model and because this model is expected to be accurate necessary adjustments can be typically performed as simulation-based optimization. In case of phased apertures, computational and optimization loads at this step can be quite substantial.

Another important step of the process is design of the feed implementing the aperture excitation $\{a_n\}$, which can be original (determined at the prototyping step) or modified at the previous step. This step is usually performed using microwave circuit techniques and needs to include EM modeling and discrete simulations of the feed. The scattering matrix of the feed $\mathbf{S}_F$ (size of $(N+1) \times (N+1)$) is available upon completion of the design. Simulation-based tuning of the feed is typically needed in case of low-sidelobe patterns, i.e., when transmission coefficients (output waves) of the feed are substantially different and need to be accurately realized, and/or in a case of feed matching and improving the feed impedance bandwidth. It is worth to notice that feed designs are typically configured for the feed ports terminated on matching loads as depicted in Fig. 2.6(d).

Given the array aperture scattering matrix $\mathbf{S}_A$ (size of $N \times N$) and the feed scattering matrix $\mathbf{S}_F$ (size of $(N+1) \times (N+1)$), the excitation of the array aperture elements, which are driven though the feed as depicted in Fig. 2.8, can be found as the solution of the

28 Simulation-Based Optimization of Antenna Arrays

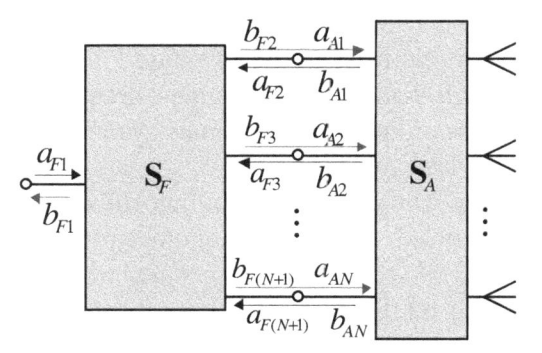

Fig. 2.8. Antenna array excitation through the feed. Here $\mathbf{S}_F$ can stand for the total S-matrix of the excitation network which may include active elements and adjustable components (e.g., phasers, time-delay units, etc.).

following coupled equations for $\mathbf{b}_F$:

$$\mathbf{b}_F = \mathbf{S}_F \mathbf{a}_F, \tag{2.42}$$
$$\mathbf{b}_A = \mathbf{S}_A \mathbf{a}_A. \tag{2.43}$$

Excitation of the feed input and connection between the waves are, respectively

$$\mathbf{a}_{F1} = 1, \tag{2.44}$$
$$\mathbf{a}_{F(2:N+1)} = \mathbf{b}_{A(1:N)}, \tag{2.45}$$
$$\mathbf{a}_{A(1:N)} = \mathbf{b}_{F(2:N+1)}. \tag{2.46}$$

Equations (2.42)–(2.46) can be cast in an $(N+1) \times (N+1)$ linear system

$$(\mathbf{I} - \mathbf{S}'_F \mathbf{S}'_A) \mathbf{b}_F = \mathbf{s}''_F, \tag{2.47}$$

where the entries of (2.47) mean

$$\mathbf{S}'_F = \begin{bmatrix} 0 & S_{F12} & S_{F13} & \cdots & S_{F1(N+1)} \\ 0 & S_{F22} & S_{F23} & \cdots & S_{F2(N+1)} \\ \vdots & \vdots & \vdots & \vdots & \vdots \\ 0 & S_{F(N+1)2} & S_{F(N+1)3} & \cdots & S_{F(N+1)(N+1)} \end{bmatrix}, \tag{2.48}$$

$$\mathbf{S}'_A = \begin{bmatrix} 0 & 0 & 0 & \cdots & 0 \\ 0 & S_{A11} & S_{A12} & \cdots & S_{A1N} \\ \vdots & \vdots & \vdots & \vdots & \vdots \\ 0 & S_{AN1} & S_{AN2} & \cdots & S_{ANN} \end{bmatrix}, \qquad (2.49)$$

$$\mathbf{s}''_F = \begin{bmatrix} S_{F11} \\ S_{A21} \\ \vdots \\ S_{F(N+1)1} \end{bmatrix}. \qquad (2.50)$$

In accordance with the superposition principle, the waves $\mathbf{a}_A$ of (2.46) energize the array elements in the presence of coupling, so that they can be used to evaluate the far-field at the high-fidelity level of description, e.g., using (2.1)–(2.3), or with approximate models of the far-field, e.g., using (2.15), (2.30).

2.6. Simulation-Based Antenna Array Design

EM simulations, typically performed with discrete full-wave solvers, are essential for development of antenna array circuits. EM simulations can be intensively involved through the entire process of an antenna array design, as outlined in Fig. 2.7, where EM simulations are utilized in four out of five major designs steps for adjusting the responses of antenna array circuits. Namely, EM simulation can be executed for element design, array aperture implementation and tuning, and necessarily for design, and validation of excitation (power distribution) network and its components. Realistic models of these circuits are typically dependent on many adjustable parameters. In such situations numerical optimization is often the only feasible mean to adjust models of antenna array circuits. In addition, realistic models of antenna array circuits, e.g., subarray and array apertures, power distribution networks, can be quite graded and have EM sizes of a few characteristic wavelengths; therefore, such models are computationally expensive. Surrogate-based optimization allows adjustments of computationally expensive models with acceptable computational cost yet at the high-fidelity level of description

(Bandler et al., 1994; Booker et al., 1999; Koziel et al., 2008a; Forrester et al., 2008).

The diagram of Fig. 2.7 also outlines that design of antenna arrays is a complex and iterative process. It requires intelligent interaction of involved models, evaluation and adjustment tools. To our understanding, a successful development of antenna technology strongly relies on availability and reliability of dedicated CAD tools in which simulation-based optimizers play an essential role. In subsequent chapters, we discuss the development and application of such optimizers at different steps of antenna array design.

2.7. A Concept of Antenna Array Simulation-Based Optimization

A concept of simulation-based optimization for engineering problems, antenna array design in particular, is outlined in the diagram of Fig. 2.9 where the optimization setup consists of two major parts: the optimizer (optimization algorithm) and the model. For simulation-based optimization, the model is usually defined and evaluated using discrete full-wave modeler/software solver. High-fidelity description of the model and its accurate responses, e.g., scattering parameters, radiation patterns, etc., will be denoted as $\mathbf{f}$. A candidate design is represented by a vector of adjustable parameters, design variables, $\mathbf{x}$. Given operation conditions and constraints, the optimizer searches for the best possible design $\mathbf{x}^*$. The best design $\mathbf{x}^*$ should maximize/minimize a merit function U encoding design requirements and constraints. The optimizer repeatedly suggests the candidate design $\mathbf{x}$ which is verified by the model returning its response at this design $\mathbf{f}(\mathbf{x})$. Based on the value

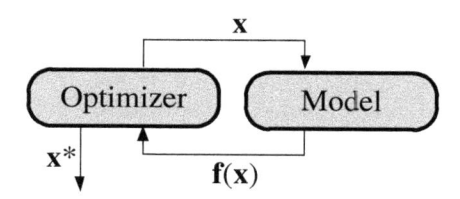

Fig. 2.9. Simulation-based optimization loop: a concept.

of the merit function $U(\mathbf{f}(\mathbf{x}))$, the optimizer either accepts or rejects $\mathbf{x}$ and subsequently determines a new candidate design. This process continues, producing a sequence of design improvements $\{\mathbf{x}^{(i)}\}$ with i standing for iteration of the optimization loop, until satisfying given termination criteria or getting stuck in a local maximum/minimum.

It worth to notice that for verification of candidate designs and, especially, for verification of the best design, the high-fidelity description $\mathbf{f}$ is essential for having reliable designs. At the same time, the use of $\mathbf{f}$ for predicting of a new candidate design can be very computationally demanding. A simplified version of the model $\mathbf{f}$ is referred to as a coarse or low-fidelity model $\mathbf{c}$. Optimization involving model $\mathbf{c}$ runs faster, however, with sacrificing accuracy and, consequently, reliability of the generated design. As a mean to work around this issue, a properly corrected (relative to the model f) coarse model referred to as the surrogate model $\mathbf{s}$ can be constructed and utilized. The construction and the use of surrogate models to conduct numerically efficient yet reliable simulation-based designs of antenna arrays and their components will be discussed and illustrated in subsequent chapters of this book.

2.8. Challenges of Antenna Array Simulation-Based Optimization

Simulation-based optimization of antenna arrays and their components faces the following challenges:

- Antenna array circuits, e.g., array apertures, feeds, have the size of several characteristic wavelengths and, in addition, are highly graded. One consequence of this is that high-fidelity EM simulations of such circuits are computationally expensive: a single simulation run can take a few hours to complete. Another consequence is that fast simulations of coarsely discretized EM models of such circuits loose accuracy and often are unreliable.
- Realistic antenna array circuits, even at the component level, can be described with a large (for simulation-based design) number of variables, typically 10–20.
- During a search for an improved design, discrete models of the circuits need to be repeatedly simulated at different points of

the design space. Even with an acceptable run time of a single simulation, e.g., tens of minutes or hours, the total accumulated time can be quite long.
- Antenna and antenna array problems are associated with simultaneous handling of design requirements imposed on several response figures, e.g., sidelobe level of the array, back-to-front ratio, efficiency, signal-to-noise level, reflection coefficient at the feed input, impedance bandwidth at a certain level, location of maximum of radiated power in frequency, peak realized gain, array aperture footprint, feed footprint, etc.
- Certain level of numerical noise is inherent to the simulation results. In addition, the response figures can be quite sensitive to the discretization fidelity of the model. These issues lead to serious obstacles for the gradient-based optimization algorithms.
- Quick and computationally inexpensive models of antenna arrays, e.g., formula-based one, cannot in general provide all required information about the array performance. For instance, array factor models can deliver estimates of the radiation pattern and sidelobe level, however, they provide no reliable information about mutual coupling and active reflection coefficients.

The methods presented in this book are intended to alleviate the difficulties and address the design challenges mentioned above. In particular, we discuss the ways of incorporating variable fidelity computational models, correction techniques, and iterative design procedures into computationally efficient and reliable algorithmic frameworks that facilitate optimization of antenna elements, array apertures, and the array feeds. At the same time, it should be emphasized that while the methods presented and discussed in this book address the key challenges to certain extent, some important fundamental problems of simulation-based antenna array circuit optimization are yet to be elaborated on and researched further so that they can become applicable in practice. One of such challenges is automatic adjustment of fidelity of the discrete EM model so that it results in the optimum combination of the optimization algorithm speed and reliability.

Chapter 3

Fundamentals of Numerical Optimization

The main topic of this book is automated, simulation-driven design of antenna arrays. The primary tools utilized in the design process are therefore numerical optimization algorithms. In the following three chapters including this one, we discuss the fundamentals of numerical optimization. The material covered here is not supposed to be a systematic and exhaustive presentation of the subject (interested readers may refer to numerous textbooks such as Nocedal and Wright, 2006; Yang, 2010a; Conn et al., 2009) but it should give the reader sufficient background information necessary to understand the remaining parts of the book. In this chapter, we cover conventional methods, including gradient-based algorithms for both unconstrained and constrained optimization, as well as derivative-free techniques such as pattern search. Chapter 4 is a brief exposition of global optimization using population-based metaheuristics. Methods of this class are probably the most popular algorithms for design optimization of antenna arrays these days, although, as pointed out later, their application is very limited for handling realistic design problems involving expensive simulated electromagnetic models. As already mentioned in Chapter 2, design of contemporary antenna arrays heavily relies on full-wave electromagnetic (EM) analysis which is typically carried out using methods of computational electromagnetics. Because of considerable CPU cost of accurate EM simulations, direct optimization of electromagnetic models might be prohibitive. Surrogate-assisted methods alleviate this difficulty by shifting the majority of operations to a fast replacement model (also referred to as the surrogate). This class of techniques is discussed in Chapter 5.

3.1. Formulation of the Optimization Problem

The optimization problem to be solved is formulated as

$$\mathbf{x}^* = \arg\min_{\mathbf{x} \in X} f(\mathbf{x}) \tag{3.1}$$

where $f(\mathbf{x})$ is a scalar objective function, whereas $\mathbf{x} \in X \subseteq R^n$; X is the objective function domain (also referred to as a design/search space); $\mathbf{x}^*$ is the design optimum to be found. The problem (3.1) can also be formulated with explicit constraints as

$$\mathbf{x}^* = \arg\min\{x : f(x)\} \tag{3.2}$$

subject to

$$g_k(\mathbf{x}) \leq 0, \quad k = 1, \ldots, N, \tag{3.3}$$

$$h_k(\mathbf{x}) = 0, \quad k = 1, \ldots, M. \tag{3.4}$$

The most typical type of constraints are lower and upper bounds for design variables (box constraints), $\mathbf{l} \leq \mathbf{x} \leq \mathbf{u}$ (the inequalities are understood here as component-wise). Geometry constraints are normally introduced to ensure physical consistency of the design as well as to control the antenna size. Some of the performance figures may also be controlled through constraints (e.g., minimization of the antenna array sidelobe level while ensuring sufficient matching at the array inputs).

As previously mentioned, the algorithms described in this chapter are formulated for scalar objective functions. On the other hand, it should be emphasized that for practical antenna and antenna array design problems, the objective function is always of the form $U(\mathbf{f}(\mathbf{x}))$, i.e., it is a composition of the merit function U and a vector-valued system response $\mathbf{f}(\mathbf{x})$ (see also Chapter 2). Typical responses include

Fundamentals of Numerical Optimization 35

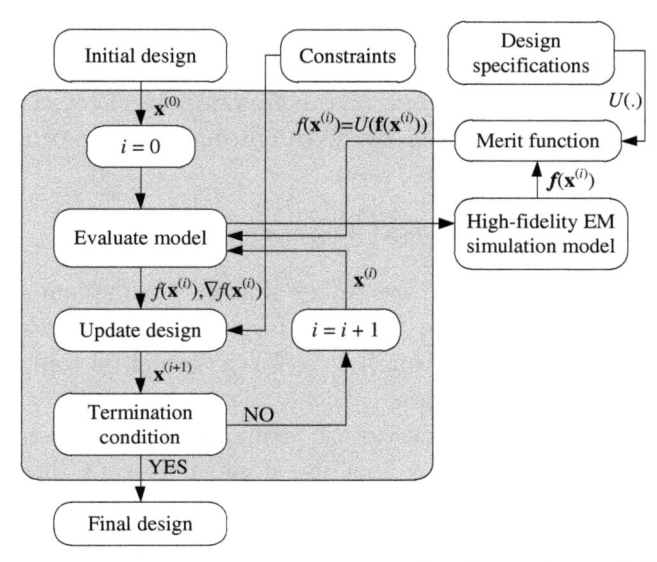

Fig. 3.1. Direct simulation-driven optimization flowchart. The candidate designs generated by the algorithm are evaluated through high-fidelity EM simulations. The objective function is calculated based on given performance requirements. The evaluation is for verification purposes but also to provide the optimizer with information to search for better designs (subjected to constraints, if any). The search process may be guided by the model response only, or the response and its derivatives (if available).

reflection coefficients versus frequency, radiation pattern cuts (e.g., radiation intensity versus elevation angle), peak realized gain, etc. The merit function translates the design specifications into a single number representing the performance of the system. Figure 3.1 shows the generic flowchart of simulation-driven optimization of antennas and antenna arrays.

3.2. Gradient-Based Optimization

Gradient-based techniques belong to the most popular optimization methods (Nocedal and Wright, 2006). The search process is based on the gradient $\nabla f(\mathbf{x})$ of the objective function $f(\mathbf{x})$

$$\nabla f(\mathbf{x}) = \left[\frac{\partial f}{\partial x_1}(\mathbf{x}) \quad \frac{\partial f}{\partial x_2}(\mathbf{x}) \quad \cdots \quad \frac{\partial f}{\partial x_n}(\mathbf{x}) \right]^T. \qquad (3.5)$$

Assuming that the function $f(\mathbf{x})$ is sufficiently smooth (i.e., at least first-order continuously differentiable), the gradient $\nabla f(\mathbf{x})$ determines the local behavior of the function in the vicinity of $\mathbf{x}$. This can be better observed by investigating the first-order Taylor expansion of f around $\mathbf{x}$. We have

$$f(\mathbf{x} + \mathbf{h}) \cong f(\mathbf{x}) + \nabla f(\mathbf{x})^T \cdot \mathbf{h} < f(\mathbf{x}) \qquad (3.6)$$

for sufficiently small $\mathbf{h}$ and assuming that $\mathbf{h}$ is a descent direction, i.e., $\nabla f(\mathbf{x})^T \cdot \mathbf{h} < 0$. Note that the largest local reduction of the function value is obtained for $\mathbf{h} = -\nabla f(\mathbf{x})$ which is referred to as the steepest descent direction.

There are two basic ways of utilizing the gradient in the optimization process: (i) by moving along a descent direction, or (ii) by exploiting a local approximation model of the objective function constructed with ∇f. As explained later, moving along the steepest descent direction may not be the best strategy.

Although it is possible to conduct the optimization process utilizing only the gradient information (and many practical techniques do so), more comprehensive information about the local behavior of the function can be obtained from the second-order derivatives, specifically, the Hessian $\mathbf{H}(\mathbf{x})$

$$\mathbf{H}(\mathbf{x}) = \begin{bmatrix} \frac{\partial^2 f}{\partial x_1 \partial x_1}(\mathbf{x}) & \cdots & \frac{\partial^2 f}{\partial x_1 \partial x_n}(\mathbf{x}) \\ \vdots & \ddots & \vdots \\ \frac{\partial^2 f}{\partial x_n \partial x_1}(\mathbf{x}) & \cdots & \frac{\partial^2 f}{\partial x_n \partial x_n}(\mathbf{x}) \end{bmatrix}. \qquad (3.7)$$

In particular, the gradient and Hessian together allow us to identify the local minimizer of a function. The points for which the gradient ∇f vanishes are referred to as the *stationary points* of f. Vanishing of the gradient is a necessary condition for the minimum: if $\mathbf{x}^*$ is a local minimizer of f then $\nabla f(\mathbf{x}^*) = 0$. Unfortunately, this is not a sufficient condition: the stationary point may or may not be a local minimizer as shown in Fig. 3.2. This is determined by the properties of the Hessian. In particular, if $\mathbf{x}^*$ is a stationary point

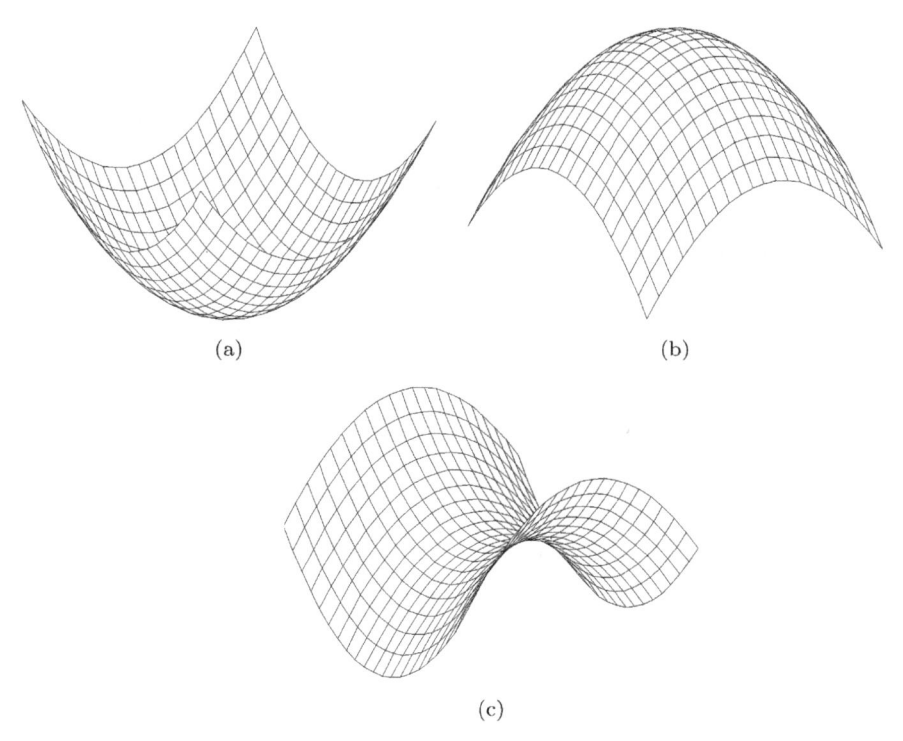

Fig. 3.2. Graphical illustration of stationary points in a two-dimensional search space: (a) local minimizer, (b) local maximizer, and (c) a saddle point. The Hessian is positive/negative definite in the local minimum/maximum, and it is neither positive nor negative definite in the saddle point.

and $\mathbf{H}(\mathbf{x}^*)$ is positive definite (Nocedal and Wright, 2006), then $\mathbf{x}^*$ is a local minimizer of f. On the other hand, if $\mathbf{x}^*$ is a local minimizer of f then $\mathbf{H}(\mathbf{x}^*)$ is positive semidefinite.

3.2.1. *Optimization using descent methods*

Here, we consider iterative methods generating a sequence of vectors $\mathbf{x}^{(i)}$, $i = 0, 1, 2, \ldots$, that are supposed to approximate the solution of (3.1). The fundamental requirement is that $f(\mathbf{x}^{(i+1)}) < f(\mathbf{x}^{(i)})$, i.e., an improvement with respect to the cost function value is enforced between iterations. The new vector $\mathbf{x}^{(i+1)}$ is found starting from $\mathbf{x}^{(i)}$ by moving along a descent direction, along which f decreases (at least locally).

An important performance measure of an iterative method is the convergence rate which describes how quickly we approach the minimizer $\mathbf{x}^*$. Linear convergence is defined by a condition $\|\mathbf{x}^{(i+1)} - \mathbf{x}^*\| \leq c_1 \|\mathbf{x}^{(i)} - \mathbf{x}^*\|$ with $0 < c_1 < 1$. Quadratic convergence is defined by $\|\mathbf{x}^{(i+1)} - \mathbf{x}^*\| \leq c_2 \|\mathbf{x}^{(i)} - \mathbf{x}^*\|^2$ with $0 < c_2 < 1$. Finally, superlinear convergence, defined as $\|\mathbf{x}^{(i+1)} - \mathbf{x}^*\|/\|\mathbf{x}^{(i)} - \mathbf{x}^*\| \to 0$ for $i \to \infty$, is faster than a linear one but not as good as a quadratic, Fig. 3.3. Note that the aforementioned conditions are supposed to be satisfied when we are close to the optimum.

Let $\mathbf{x}^{(0)}$ be the initial solution. A generic descent algorithm works as follows (Nocedal and Wright, 2006):

1. Set $i = 0$;
2. Find a search direction $\mathbf{h}_d$;
3. Find a step length α;
4. Set $\mathbf{x}^{(i)} = \mathbf{x}^{(i)} + \alpha \mathbf{h}_d$;
5. Set $i = i + 1$;
6. If the termination condition is not satisfied, go to 2;
7. END.

The two critical stages of the iteration include finding a descent direction $\mathbf{h}_d$ (Step 2), and finding an appropriate length of the step taken along this direction (Step 3). As mentioned before, a descent direction is a vector $\mathbf{h}_d$ for which $\nabla f(\mathbf{x})^T \cdot \mathbf{h}_d < 0$ (cf. (3.6)). The step length $\alpha \mathbf{h}_d$ has to be determined to ensure that $f(\mathbf{x}^{(i+1)}) < f(\mathbf{x}^{(i)})$. This is realized using a separate algorithm as explained later. There are various termination conditions that can be utilized in practical implementations such as convergence in argument, i.e., $\|\mathbf{x}^{(i+1)} - \mathbf{x}^{(i)}\| \leq \varepsilon_1$, vanishing of the gradient, i.e., $\|\nabla f(\mathbf{x}^{(i)})\| \leq \varepsilon_2$, convergence in the function value, i.e., $f(\mathbf{x}^{(i)}) - f(\mathbf{x}^{(i+1)}) \leq \varepsilon_3$, or a combination of the above.

A procedure of finding the appropriate length of the step along the descent direction is referred to as a line search (Antoniou and Lu, 2007). More specifically, one considers a one-dimensional function of

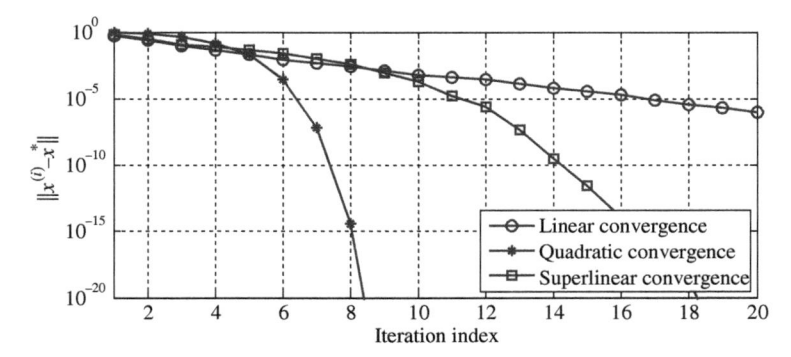

Fig. 3.3. Conceptual illustration of various convergence rates of optimization algorithms: linear (○), quadratic (∗), and superlinear (□).

a length parameter α

$$\varphi(\alpha) = f(\mathbf{x} + \alpha \mathbf{h}_d). \tag{3.8}$$

There are two types of line search that are utilized in practice. In a traditional (and more intuitive) *exact line search*, the step length α_e is found as $\alpha_e = \arg\min\{\alpha > 0: f(\mathbf{x} + \alpha \mathbf{h}_d)\}$. On the other hand, *soft line search* only requires satisfaction of certain (and usually rather loose) conditions such as the Wolfe conditions (Fletcher, 1987):

$$\varphi(\alpha_s) \leq \lambda(\alpha_s) \quad \text{where } \lambda(\alpha) = \varphi(0) + \rho \varphi'(0) \cdot \alpha \text{ with } 0 < \rho < 0.5 \tag{3.9}$$

and

$$\varphi'(\alpha_s) \geq \beta \cdot \varphi'(0) \quad \text{with } \rho < \beta < 1. \tag{3.10}$$

The first condition ensures sufficient decrease of the cost function f. The second condition allows for avoiding steps that are too short. Both conditions are illustrated in Fig. 3.4.

It can be shown that the descent methods with line search satisfying (3.9), (3.10) are convergent to at least a first-order stationary point of f (i.e., where the function gradient vanishes), assuming that the function f is sufficiently smooth (here, uniformly continuous). It is also worth mentioning that soft line search is, in practice, more efficient than exact line search: although the latter

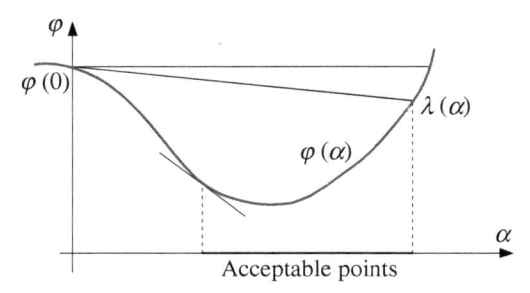

40 Simulation-Based Optimization of Antenna Arrays

Fig. 3.4. Illustration of the Wolfe conditions. The lower bound of the acceptable point interval is determined by (3.10), whereas the upper bound is obtained from (3.9).

normally requires a smaller number of iterations to converge, the cost per iteration is considerably higher than for soft line search, and so is the overall computational cost of the optimization process.

At first glance, selecting the steepest descent direction $\mathbf{h} = -\nabla f(\mathbf{x})$ as the search direction seems to be a good idea. Unfortunately, a steepest-descent method (where, $\mathbf{h} = -\nabla f(\mathbf{x})$ at each iteration) performs poorly when close to the optimum. This can be illustrated using a famous examples of the Rosenbrock function $f(\mathbf{x}) = f([x_1 x_2]^T) = (1 - x_1)^2 + 100(x_2 - x_1^2)^2$ (Rosenbrock, 1960). Figure 3.5(a) shows how the steepest descent method fails to find the optimum of this function: due to a narrow and curved valley, the steps taken in subsequent iterations of the steepest-descent method (with exact line search) become shorter and shorter as they are perpendicular to each other. On the other hand, the steepest-descent method may be useful in the initial stages of the optimization process (away from the optimum), which leads to various hybrid methods.

Better efficiency of the optimization process can be obtained by choosing — as the search direction — a linear combination of the previous direction $\mathbf{h}_{\text{prev}}$ and the current gradient, i.e.,

$$\mathbf{h} = -\nabla f(\mathbf{x}^{(i)}) + \gamma \mathbf{h}_{\text{prev}}. \tag{3.11}$$

This idea is exploited by the so-called conjugate-gradient methods (Yang, 2010a). The difference between various conjugate-gradient methods is in the strategy of updating the search direction. An important example is the Fletcher–Reeves method (Fletcher and Reeves, 1964) with

$$\gamma = \frac{\nabla f(\mathbf{x})^T \nabla f(\mathbf{x})}{\nabla f(\mathbf{x}_{\text{prev}})^T \nabla f(\mathbf{x}_{\text{prev}})}. \tag{3.12}$$

Another example is a Polak–Ribiére method (Polak and Ribiére, 1969) with the updating formula

$$\gamma = \frac{(\nabla f(\mathbf{x}) - \nabla f(\mathbf{x}_{\text{prev}}))^T \nabla f(\mathbf{x})}{\nabla f(\mathbf{x}_{\text{prev}})^T \nabla f(\mathbf{x}_{\text{prev}})}. \tag{3.13}$$

It should be noted that the direction $\mathbf{h}$ obtained from (3.11) is not guaranteed to be downhill (e.g., when the soft line search is used to find the iteration step). A safeguard is to default to $\mathbf{h} = -\nabla f(\mathbf{x})$ in case $\nabla f(\mathbf{x})^T \mathbf{h} \geq 0$. Figure 3.5(b) shows the Fletcher–Reeves conjugate-gradient method (with soft line search) optimizing the Rosenbrock function. Conjugate-gradient methods exhibit the linear convergence rate (Crowder and Wolfe, 1972).

An alternative implementation of descent methods is by utilizing trust-region framework (Conn et al., 2000) rather than line search. Here, the iteration step is determined by means of the local expansion

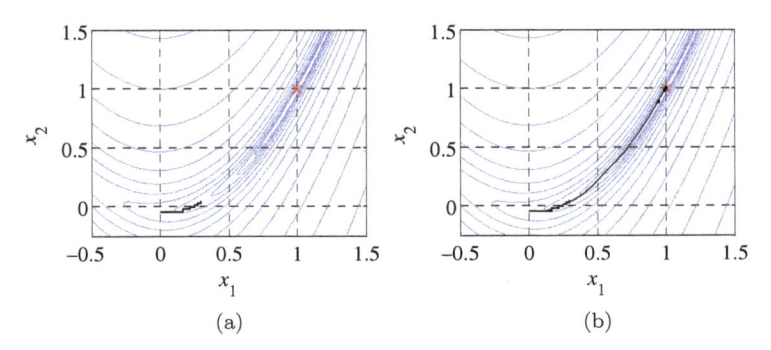

Fig. 3.5. Optimization of the Rosenbrock function (Rosenbrock, 1960): (a) steepest-descent methods where exact line search fails to find the optimum at $\mathbf{x}^* = [1\ 1]^T$ (marked ×): the narrow and curved valley containing the minimum slows down the algorithm convergence preventing the method from reaching $\mathbf{x}^*$; (b) conjugate-gradient algorithm with the Fletcher–Reeves update method and soft line search: the optimum is found in less than 30 iterations.

model of f created using first- (or, optionally, higher-)order derivatives. The first-order Taylor expansion of f is defined as

$$f(\mathbf{x} + \mathbf{h}) \approx q(\mathbf{h}) \quad \text{with } q(\mathbf{h}) = f(\mathbf{x}) + \nabla f(\mathbf{x})\mathbf{h}^T. \tag{3.14}$$

The second-order Taylor expansion (Bischof et al., 1993) is given by

$$f(\mathbf{x} + \mathbf{h}) \approx q(\mathbf{h}) \quad \text{with } q(\mathbf{h}) = f(\mathbf{x}) + \nabla f(\mathbf{x}) \cdot \mathbf{h}^T + \tfrac{1}{2}\mathbf{h}^T \mathbf{H}(\mathbf{x})\mathbf{h}. \tag{3.15}$$

The model $q(\mathbf{h})$ is a good approximation of the function f in the vicinity of $\mathbf{x}$, i.e., for sufficiently small vectors $\mathbf{h}$ (assuming that the function f is sufficiently smooth). In the trust-region framework, the candidate iteration step size $\mathbf{h}_{\text{tr}}$ is determined as

$$\mathbf{h}_{\text{tr}} = \arg\min_{\mathbf{h}, \|\mathbf{h}\| \leq \delta} q(\mathbf{h}), \tag{3.16}$$

where $\delta > 0$ is the trust-region radius.

The step $\mathbf{h}_{\text{tr}}$ obtained using (3.16) is only accepted if it leads to the improvement of the cost function, i.e., $f(\mathbf{x} + \mathbf{h}_{\text{tr}}) < f(\mathbf{x})$. Otherwise it is rejected and the search continues from $\mathbf{x}$ using a reduced trust-region radius δ. The value of δ is adjusted using a so-called *gain ratio* defined as

$$r = \frac{f(\mathbf{x}) - f(\mathbf{x} + \mathbf{h})}{q(\mathbf{0}) - q(\mathbf{h})}. \tag{3.17}$$

A typical adjustment scheme would be $\delta \leftarrow 2\delta$ if $r > 0.75$ and $\delta \leftarrow \delta/3$ if $r < 0.25$. Because both models (3.14) and (3.15) are first-order consistent with f at $\mathbf{x}$ (i.e., $q(0) = f(\mathbf{x})$ and $\nabla q(0) = \nabla f(\mathbf{x})$), the condition $f(\mathbf{x} + \mathbf{h}_{\text{tr}}) < f(\mathbf{x})$ would be satisfied for sufficiently small δ.

3.2.2. Newton and quasi-Newton methods

Performance of gradient-based search can be improved by exploiting higher-order derivatives. In particular, utilization of the Hessian gives rise to a family of Newton and quasi-Newton methods outlined in this section.

Assuming that the cost function f is at least twice continuously differentiable, it can be locally represented using its second-order

Taylor approximation (3.15). If the Hessian $\mathbf{H}(\mathbf{x})$ of f at $\mathbf{x}$ is positive definite, then the model $q(\mathbf{h})$ has a unique minimizer for $\mathbf{h}$ such that $\nabla q(\mathbf{h}) = 0$, i.e.,

$$\nabla f(\mathbf{x}) + \mathbf{H}(\mathbf{x})\mathbf{h} = 0. \tag{3.18}$$

This observation is a basis of the Newton's method (Antoniou and Lu, 2007) where the next iteration point is obtained by solving (3.18) as follows:

$$\mathbf{x}^{(i+1)} = \mathbf{x}^{(i)} - [\mathbf{H}(\mathbf{x}^{(i)})]^{-1}\nabla f(\mathbf{x}^{(i)}). \tag{3.19}$$

In order for the algorithm (3.19) to be well defined, the Hessian $\mathbf{H}(\mathbf{x})$ has to be non-singular. Moreover, if the Hessian is positive definite for all iterations and the starting point is sufficiently close to the optimum, the method usually converges very quickly (quadratically) to a minimum.

It should be emphasized that the basic Newton algorithm as formulated in (3.19) does not work well in practice for the following reasons: it is not globally convergent for many problems, it may converge to a maximum or a saddle point, and — most importantly — the problem (3.19) may be ill-conditioned. Furthermore, the Newton method requires analytical second-order derivatives.

The fundamental difficulty, i.e., the Hessian not being positive definite, can be alleviated by using the so-called damped Newton method, where (3.19) is replaced by

$$\mathbf{x}^{(i+1)} = \mathbf{x}^{(i)} - [\mathbf{H}(\mathbf{x}^{(i)}) + \mu\mathbf{I}]^{-1}\nabla f(\mathbf{x}^{(i)}), \tag{3.20}$$

where $\mathbf{I}$ is the identity matrix. By choosing sufficiently large $\mu > 0$ the matrix $\mathbf{H}(\mathbf{x}) + \mu\mathbf{I}$ can be made positive definite because of being dominated by the term $\mu\mathbf{I}$. The step $\mathbf{h}_\mu$ found as a solution to the problem $[\mathbf{H}(\mathbf{x}) + \mu\mathbf{I}]\mathbf{h}_\mu = -\nabla f(\mathbf{x})$ is a minimizer of the following model:

$$\begin{aligned} q_\mu(\mathbf{h}) &= q(\mathbf{h}) + \tfrac{1}{2}\mu\mathbf{h}^T\mathbf{h} \\ &= f(\mathbf{x}) + \nabla f(\mathbf{x}) \cdot \mathbf{h}^T + \tfrac{1}{2}\mathbf{h}^T[\mathbf{H}(\mathbf{x}) + \mu\mathbf{I}]\mathbf{h}. \end{aligned} \tag{3.21}$$

Additionally, $\mathbf{h}_\mu$ is a descent direction for f at $\mathbf{x}$. Moreover, large steps are penalized in (3.20) because of the term $\mu\mathbf{h}^T\mathbf{h}/2$. It is also interesting to note that for very large μ, the algorithm defaults to the steepest descent method because $\mathbf{h}_\mu \approx -\nabla f(\mathbf{x})/\mu$.

In practical implementations of the damped Newton's method, large values of μ are utilized at the early stages of the iteration process (3.20), whereas μ is reduced as the algorithm converges. When close to the optimum (and with small value of μ), the algorithm gets back to the original Newton's method which allows for finding the minimizer of f. One of the most popular implementations of this concept is the Levenberg–Marquardt (LM) algorithm (Bates and Watts, 1988), where the value of μ is increased at the beginning of each iteration until $\mathbf{H}(\mathbf{x}) + \mu \mathbf{I}$ becomes positive definite, and then μ is updated based on the value of the gain ratio r (3.17). The typical updating formula is: $\mu \leftarrow \mu \cdot \max\{1/3, 1 - (2r - 1)^3\}$ if $r > 0$, and $\mu = 2\mu$ otherwise (Nielsen, 1999). The flow of the LM algorithm is shown in Fig. 3.6.

Damped Newton methods resolve all of the issues of the basic Newton's algorithm but one: a potentially high-cost of obtaining second-order derivatives of the objective function f. This issue can be addressed by the so-called quasi-Newton methods, where the exact Hessian (or, even better, its inverse) is approximated using appropriate updating formulas. As of now, the best and the most popular updating formula is the Broyden–Fletcher–Goldfarb–Shanno (BFGS) one (Nocedal and Wright, 2006). BFGS preserves positive symmetry and definiteness of the approximation. It is defined as follows for a Hessian approximation $\mathbf{B}$:

$$\mathbf{B}_{\text{new}} = \mathbf{B} + \frac{1}{\mathbf{h}^T \mathbf{y}} \mathbf{y}\mathbf{y}^T - \frac{1}{\mathbf{h}^T \mathbf{u}} \mathbf{u}\mathbf{u}^T, \tag{3.22}$$

where

$$\mathbf{h} = \mathbf{x}_{\text{new}} - \mathbf{x}, \mathbf{y} = \nabla f(\mathbf{x}_{\text{new}}) - \nabla f(\mathbf{x}), \quad \mathbf{u} = \mathbf{B}\mathbf{h}.$$

Approximation of the inverse of Hessian $\mathbf{D}$ is defined as

$$\mathbf{D}_{\text{new}} = \mathbf{D} + \kappa_1 \mathbf{h}\mathbf{h}^T - \kappa_2(\mathbf{h}\mathbf{v}^T + \mathbf{v}\mathbf{h}^T), \tag{3.23}$$

```
x = x⁽⁰⁾; μ = μ⁽⁰⁾; found = false; k = 0;
while ~found & k ≤ k_max
    while f''(x)+μI not positive definite
        μ = 2*μ;
    end
    Solve (f''(x)+μI)h_dn = -f'(x); Compute gain factor r;
    if r > δ
        x = x + h_dn; μ = μ*max{1/3,1−2(2r−1)³};
    else
        μ = μ*2;
    end
    k = k+1; Update found;
end
```

Fig. 3.6. Pseudocode of the Levenberg–Marquardt algorithm.

```
x = x⁽⁰⁾; D = D⁽⁰⁾; k = 0; n_f = 0;
while ||f'(x)|| > ε & k < k_max & n_f < n_f.max
    h_qn = -Df'(x);
    [t,k_f] = soft_line_search(x,h_qn);
    n_f = n_f + k_f;
    x_new = x + t·h_qn; k = k + 1;
    if h_qn^T f'(x_new) > h_qn^T f'(x)
        Update D;
    end
    x = x_new;
end
```

Fig. 3.7. Pseudocode of the quasi-Newton algorithm with BFGS update. Initialization of the inverse Hessian by the identify matrix is recommended. Updating of the matrix **D** is realized using (3.23).

where

$$\mathbf{h} = \mathbf{x}_{\text{new}} - \mathbf{x}, \quad \mathbf{y} = \nabla f(\mathbf{x}_{\text{new}}) - \nabla f(\mathbf{x}), \quad \mathbf{v} = \mathbf{B}\mathbf{y},$$
$$\kappa_2 = \frac{1}{\mathbf{h}^T \mathbf{y}}, \quad \kappa_1 = \kappa_2(1 + \kappa_2(\mathbf{y}^T \mathbf{v})).$$

The quasi-Newton algorithm with the BFGS formula uses soft line search with loose conditions (e.g., $\rho = 0.0001$, and $\beta = 0.9$, cf. (3.9), (3.10)). The algorithm flow is shown in Fig. 3.7.

Quasi-Newton algorithms belong to the most efficient methods for unconstrained optimization. On the other hand, conjugate-gradient methods may be better when the number of design variables is large: Newton-like methods rely on matrix operations while conjugate-gradient ones only use vector operations, which is cheaper; also Newton-type methods require more storage ($n \times n$ matrix versus a few vectors for conjugate gradients).

3.2.3. Qualitative comparison of descent methods

Table 3.1 shows a qualitative comparison of the descent methods discussed in this chapter. It contains the most important characteristics of algorithms such as the computational complexity and convergence rates, but also some practical remarks and observations.

3.2.4. Remarks on constrained optimization

The algorithms discussed so far in this chapter allow unconstrained optimization. On the other hand, vast majority of real-world

Table 3.1. Qualitative comparison of gradient-based descent methods.

Method	Complexity	Convergence	Remarks
Steepest descent	$O(n)$	Linear	Not used in practice; fails for many problems; slow convergence
Conjugate gradient	$O(n)$	Linear	Reliable and computationally cheap but relatively slow convergence
Newton's	$O(n^3)$	Quadratic	Best convergence rate but requires analytical Hessian; not convergent for many problems; may converge to a maximum or saddle point; may be ill-conditioned
Damped Newton's	$O(n^3)$	Superlinear	Alleviates many of Newton's method drawbacks but still requires Hessian data
Quasi-Newton (BFGS updating)	$O(n^2)$	Superlinear	One of the best methods for unconstrained optimization to date

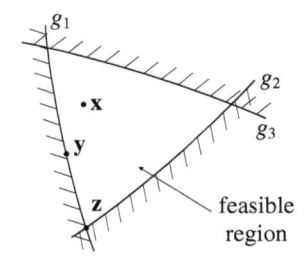

Fig. 3.8. Illustration of the feasible region for an example problem with three inequality constraints $g_k, k = 1, 2, 3$. None of the constraints is active at point **x**, constraint g_1 is active at points **y**, and two constraints g_1 and g_3 are active at point **z**.

problems are constrained. Typical constraints include lower/upper bounds for design parameters but also linear and nonlinear equality and inequality constraints. The purpose of this section is to provide the reader with some remarks concerning constrained optimization methods. More detailed exposition of the subject can be found in numerical optimization textbooks, e.g., Bertsekas (1982) and Nocedal and Wright (2006).

The formulation of a constrained optimization problem has been presented in Sec. 2.1, see (3.2)–(3.4). The set of points satisfying the constraints (3.3) and (3.4) is referred to as a feasible region (cf. Fig. 3.8). Constrained optimization seeks for the minimum of the function f within its feasible domain.

The concepts of the optimum and optimality conditions are more complex than for the unconstrained case. The first-order necessary conditions, the so-called Karush–Kuhn–Tucker or KKT conditions (Kuhn and Tucker, 1951), for an optimum of a constrained problem (3.2)–(3.4), state that — at the local optimum $\mathbf{x}^*$ of f — there exist constants $\mu_1, \ldots, \mu_N$, and $\lambda_1, \lambda_2, \ldots, \lambda_M$, such that

$$\nabla f(\mathbf{x}^*) + \sum_{k=1}^{N} \mu_k \nabla g_k(\mathbf{x}^*) + \sum_{k=1}^{M} \lambda_k \nabla h_k(\mathbf{x}^*) = 0 \qquad (3.24)$$

and

$$g_k(\mathbf{x}^*) \leq 0, \quad \mu_k g_k(\mathbf{x}^*) = 0, \quad k = 1, \ldots, N, \qquad (3.25)$$

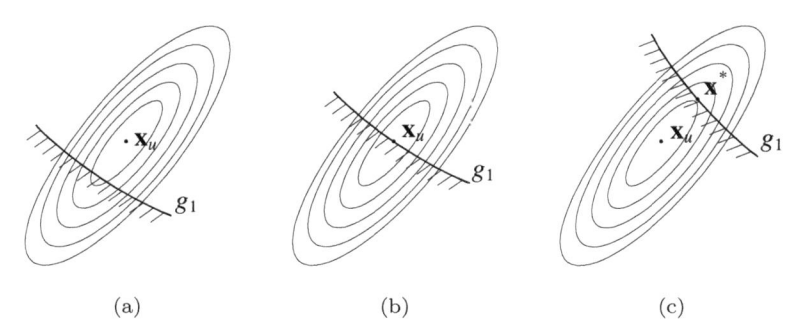

48   Simulation-Based Optimization of Antenna Arrays

(a)   (b)   (c)

Fig. 3.9. Strongly and weakly active constraints ($\mathbf{x}_u$ is the unconstrained minimum, $\mathbf{x}^*$ is constrained minimum): (a) constraint g_1 is inactive at $\mathbf{x}_u$ because $g_1(\mathbf{x}_u) > 0$ (therefore, $\mathbf{x}_u$ is both unconstrained and constrained minimizer of f); (b) g_1 is weakly active at $\mathbf{x}_u$ as $g_1(\mathbf{x}_u) = 0$ and $f'(\mathbf{x}_u) = \lambda g_1(\mathbf{x}_u)$ with $\lambda = 0$ ($\mathbf{x}_u$ is both unconstrained and constrained minimizer of f); (c) g_1 is strongly active: $g_1(\mathbf{x}_u) < 0$ ($\mathbf{x}_u$ is infeasible); $f'(\mathbf{x}^*) = \lambda g_1(\mathbf{x}^*)$ with $\lambda > 0$ (i.e., $\mathbf{x}^*$ is a constrained minimizer of f).

where

$$\mu_k \geq 0, \quad k = 1, \ldots, N \tag{3.26}$$

and provided that all the functions (both objective and constraints) are continuously differentiable. The function at the left-hand side of (3.26) is referred to as the Lagrangian function, whereas coefficients μ and λ are called Lagrange multipliers.

Figure 3.9 shows examples of unconstrained and constrained minima of the function f for a case with a single inequality constraint. Depending on the location of the unconstrained minimum and the constraint, it may be inactive, weakly active, or strongly active at the minimum. In the last case, the constrained minimum of the function is different from the unconstrained one.

One can distinguish two general approaches to handling constrained optimization. Assuming that the unconstrained minimum of f is not in the interior of the feasible region, the constrained minimum must be on its boundary. So-called active set methods search for the optimum along the boundary. Inequality constraints are handled by keeping track of the set of active constraints while moving downhill along the boundary.

Fundamentals of Numerical Optimization   49

There is another group of methods that is based on approaching the optimal solution in an iterative way, either from within the feasible region (so-called interior point methods) or by possible use of infeasible points (however, not by moving along the feasible region boundary). In the latter approaches, the objective function is modified and corresponding unconstrained optimization problems are solved at each iteration.

For the rest of this section we will discuss a few specific techniques for constraint handling. One of the simplest methods is the penalty function approach, where the original problem (3.2) with constraints (3.3), (3.4) is replaced by

$$\arg\min_{\mathbf{x}} \phi(\mathbf{x}, \boldsymbol{\alpha}, \boldsymbol{\beta}) = \arg\min_{\mathbf{x}} \left\{ f(\mathbf{x}) + \sum_{k=1}^{N} \alpha_k \bar{g}_k^2(\mathbf{x}) + \sum_{k=1}^{M} \beta_k h_k^2(\mathbf{x}) \right\}, \quad (3.27)$$

where $\boldsymbol{\alpha} = [\alpha_1, \ldots, \alpha_N]$, $\boldsymbol{\beta} = [\beta_1, \ldots, \beta_M]$, $\alpha_k, \beta_k \gg 1$, are the penalty factors, and $\bar{g}_k(\mathbf{x}) = \max\{0, g_k(\mathbf{x})\}$ (inequalities only contribute to φ if they are active). In practice, one solves a sequence of problems (3.27) for increasing values of penalty factors using the solution of the previous iteration as the starting points for the next one. An example illustrating the penalty function approach is shown in Fig. 3.10. We attempt to minimize the function $f(x, y) = x^2 + 4y^2$ with an equality constraint $h_1(x, y) = (x + 2)^2 + (y + 2)^2 - 2$. The unconstrained minimum is $[0\ 0]^T$, whereas the constrained minimum is $[1.389\ -0.725]^T$. The sequence of Figs. 3.10(a)–(f) shows the contour plots of the penalized cost function (3.27) for the increasing value of the penalty factor β_1 (by one order of magnitude in each iteration).

Another technique for constraint handling is the barrier method. It replaces the original constrained problem (3.2) by the following formulation (for simplicity, only inequality-type constraints are assumed here):

$$\arg\min_{\mathbf{x}} \phi(\mathbf{x}, \mu) = \arg\min_{\mathbf{x}} \left\{ f(\mathbf{x}) - \mu \sum_{k=1}^{N} \ln(-g_k(\mathbf{x})) \right\}, \quad (3.28)$$

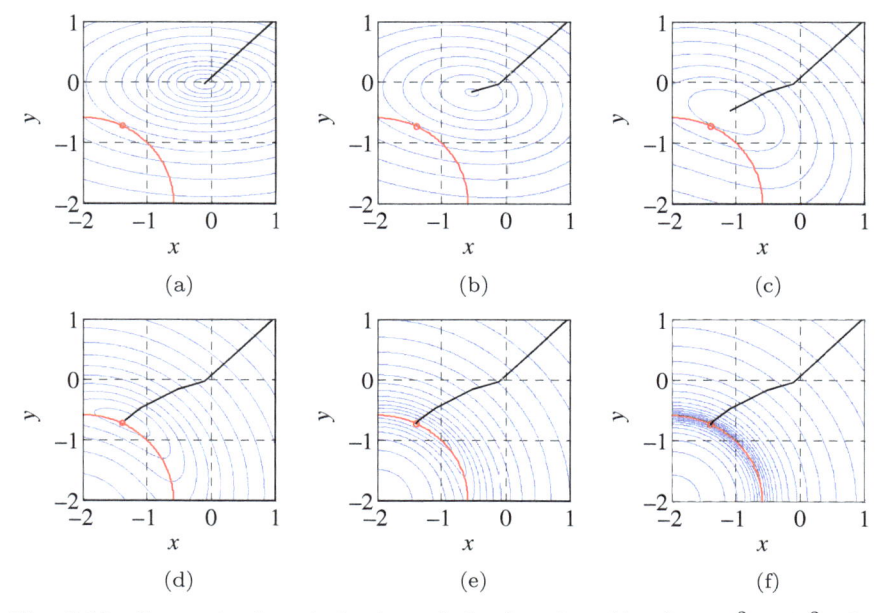

50 Simulation-Based Optimization of Antenna Arrays

Fig. 3.10. Constrained optimization of the function $f(x,y) = x^2 + 4y^2$; the equality constraint is $h_1(x,y) = (x+2)^2 - (y+2)^2 - 2$. The unconstrained minimum is $[0\ 0]^T$, the constrained minimum is $[1.389\ -0.725]^T$. Plots (a)–(f) show the contour plots of the penalized cost function (3.25) for the increasing values of the penalty factor β_1. Constrained minimum marked using a circle.

where $\mu > 0$ is a barrier parameter. The search starts from a feasible point. A sequence of unconstrained problems is solved with μ decreasing to zero so that the minimum of $\varphi(\mathbf{x}, \mu)$ converges to a solution of f.

Other approaches for constrained optimization are augmented Lagrangian methods (Tapia, 1978). These are a class of algorithms that are similar to penalty methods in the sense that the original constrained problem is solved as a sequence of suitably formulated unconstrained tasks. The unconstrained objective is the Lagrangian of the constrained problem with an additional penalty term (so-called augmentation). Assuming — for the sake of simplicity — equality-only constraints $h_k(\mathbf{x}), k = 1, \ldots, M$, the augmented Lagrangian algorithm generates a new approximation of the constrained solution to (3.2) as

$$\mathbf{x}^{(i+1)} = \arg\min_{\mathbf{x}} \phi^{(i)}(\mathbf{x})$$

$$= \arg\min_{\mathbf{x}} \left\{ f(\mathbf{x}) + \frac{\mu^{(i)}}{2} \sum_{k=1}^{M} h_k^2(\mathbf{x}) - \sum_{k=1}^{M} \lambda_k h_k(\mathbf{x}) \right\}. \quad (3.29)$$

The starting point is the previous approximation $\mathbf{x}^{(i)}$. In each iteration, the coefficients μ_k are updated as follows: $\lambda_k \leftarrow \lambda_k - \mu^{(i)} h_k(\mathbf{x}^{(i+1)})$. The coefficients λ_k are estimates of the Lagrange multipliers (cf. (3.24)), and their accuracies increase as the optimization process progresses. The value of μ is increased in each iteration, however, unlike in the penalty method, it is not necessary to ensure $\mu \to \infty$.

One of the most popular techniques for solving constrained optimization problems is nowadays sequential quadratic programming (SQP; Han, 1977; Bertsekas, 1982). At each iteration of SQP, the following quadratic programming subproblem is utilized to compute a search direction $\mathbf{h}^{(i)}$:

$$\mathbf{h}^{(i)} = \arg\min_{\mathbf{h}} \left\{ f(\mathbf{x}^{(i)}) - \nabla f(\mathbf{x}^{(i)})^T \mathbf{h} + \frac{1}{2} \mathbf{h}^T \mathbf{H}(\mathbf{x}^{(i)}) \mathbf{h} \right\} \quad (3.30)$$

so that

$$g_k(\mathbf{x}^{(i)}) + \nabla g_k(\mathbf{x}^{(i)})^T \mathbf{h} \leq 0, \quad k = 1, \ldots, N \quad (3.31)$$

and

$$h_k(\mathbf{x}^{(i)}) + \nabla h_k(\mathbf{x}^{(i)})^T \mathbf{h} = 0, \quad k = 1, \ldots, M. \quad (3.32)$$

Here, $\mathbf{x}^{(i)}$ is a current iteration point, whereas $\mathbf{H}$ is a symmetric, positive definite matrix (preferably an approximation of the Hessian of the Lagrangian of f, cf. (3.24)). The new approximation $\mathbf{x}^{(i+1)}$ is subsequently obtained using line search.

3.3. Derivative-Free Optimization

Derivatives of the objective function provide a lot of information that can be utilized in the optimization process as elaborated earlier in this chapter. Unfortunately, despite the strong theoretical foundations of the gradient-based techniques and their good performance on smooth objective functions, they cannot be applied to a growing number of practical optimization problems. An obvious reason for that is derivative information may not be available or it might be too expensive to compute (e.g., through finite differentiation of an expensive objective function evaluated with simulations). Nowadays, another and even more important issue arises which is that gradient-based search does not perform well when the objective function is noisy. This is normally the case for objective functions evaluated using computer simulation such as finite-element analysis. One of the sources of the numerical noise is utilization of adaptive meshing techniques in many commercial solvers. Adaptive meshing may lead to considerable changes of the mesh topology even for very slight changes of the geometry parameters of the structure under analysis. Another issue is termination of the simulation process upon achieving certain (required) resolution of the solution in terms of convergence, e.g., based on the residual energy for the finite-volume transient solvers.

It should also be emphasized that majority of gradient-based optimization algorithms (in particular those discussed in Sec. 3.2) are local methods. At the same time, increased complexity of contemporary engineering systems, including antenna structures (and, to even higher extent, antenna arrays), often results in non-convex, multi-modal, and rugged functional landscapes. For such cases, global optimization may be necessary to find a satisfactory design.

The aforementioned difficulties have significantly increased popularity of derivative-free optimization techniques where the search process does not rely on derivative data (Conn et al., 2009). These techniques include local methods, i.e., mostly pattern search algorithms (Kolda et al., 2003), trust-region methods based on interpolation and approximation models (Conn et al., 2000), global optimization methods such as population-based metaheuristics (Yang, 2010a), as well as surrogate-based optimization (SBO) algorithms (Queipo et al., 2005; Koziel et al., 2011a).

This book contains a separate chapter on metaheuristics (Chapter 4). Surrogate-assisted techniques are outlined in Chapter 5. In the remaining parts of this chapter, we briefly present the concept of the pattern search (Sec. 3.3.1) as well as two specific algorithms: the Hooke–Jeeves direct search (Sec. 3.3.2), and the Nelder–Mead algorithm (Sec. 3.3.3).

3.3.1. *Pattern search methods*

Pattern search is a popular class of optimization algorithms in which the search is restricted to a predefined grid. The objective function is evaluated on a stencil determined by a set of directions that are suitable from either geometric or algebraic viewpoints. The initial grid is subsequently modified (in particular, refined) during the optimization run. In the simplest case, the grid may be rectangular, i.e., spanned by the vectors aligned with the axes of the coordinate system. Computationally more efficient versions of the pattern search methods may rely on the grids generated using so-called positive spanning directions (Davis, 1954) ensuring local improvement of the objective function (assuming its continuity) for a sufficiently refined grid while using only $n+1$ neighboring points (versus $2n+1$ points for n dimensional rectangular grids).

A generic illustration of the pattern search is conceptually illustrated in Fig. 3.2 (Kolda et al., 2003). For the sake of simplicity, the optimization process in Fig. 3.11 is restricted to the rectangular grid and thus explores a grid-restricted vicinity of the current design. If the operations performed on the current grid fail to improve the objective function, the grid is refined to allow smaller steps. A typical termination criterion involves reaching a required resolution (e.g., a user-defined minimum stencil size). There are numerous variations of the pattern search algorithms. For a review see, e.g., Kolda et al. (2003).

Pattern search methods are known to be relatively robust, however, they usually exhibit slow convergence compared to the gradient-based algorithms. On the other hand, they do not need derivative information; more importantly, they are relatively immune to numerical noise. Interest in pattern search methods has recently

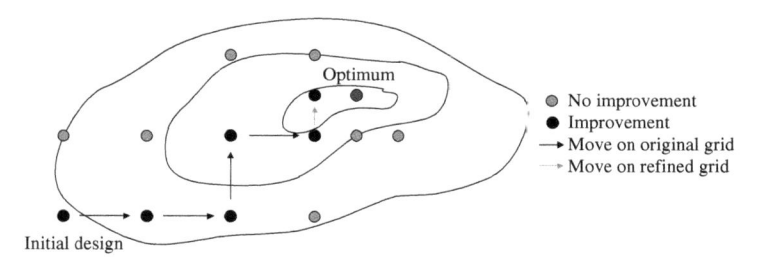

54   Simulation-Based Optimization of Antenna Arrays

Fig. 3.11. Conceptual illustration of the pattern search process, here using a rectangular grid. The exploratory movements are restricted to the grid around the initial design. In the case of failure of making a successful move, the grid is refined to allow smaller steps (Koziel and Ogurtsov, 2014a). Practical implementations of pattern search routines typically use more sophisticated strategies (e.g., grid-restricted line search).

been revived not only due to the development of rigorous convergence theory (e.g., Conn *et al.*, 2009) but also because these methods easily benefit from parallel computing.

3.3.2.   Hooke–Jeeves direct search

The Hooke–Jeeves direct search (Hooke and Jeeves, 1961) is a pattern search method. Actually, it is the first method to use the term of "direct search" and to exploit the idea of a pattern. The method is based on two types of steps: exploratory and pattern. The algorithm flow is as follows:

1. Set $j = 0$; select a base point $\mathbf{x}^j$ and the search directions $\mathbf{d}_k$, $k = 1, \ldots, n$.
2. (Exploratory move) Perform local search by sequential exploration of all search directions in an opportunistic manner, i.e., $\mathbf{x}^j + \mathbf{d}_1$, then $\mathbf{x}^j + \mathbf{d}_1 + \mathbf{d}_2$ (if $\mathbf{x}^j + \mathbf{d}_1$ resulted in a cost function improvement) or $\mathbf{x}^j - \mathbf{d}_1$ otherwise, etc. The search is continued until all search directions have been used.
3. In case there is no cost function improvement in Step 2, the step size is reduced and the search restarts from the previous best point.
4. In case there was an improvement in Step 2, a pattern move is executed as follows. Let $\mathbf{x}_{\text{tmp}}^{j+1} = \mathbf{x}^{j+1} + \alpha(\mathbf{x}^{j+1} - \mathbf{x}^j)$ be a temporary

head, where $\mathbf{x}_{j+1}$ is the best point found in Step 2, and $\alpha \geq 1$ is an acceleration factor (typically, $\alpha = 2$).
5. If $f(\mathbf{x}_{\text{tmp}}^{j+1}) < f(\mathbf{x}^{j+1})$, set $\mathbf{x}^{j+1} = \mathbf{x}_{\text{tmp}}^{j+1}$. Continue expansion as in Step 4 as long as the cost function value is decreased.
6. If $f(\mathbf{x}_{\text{tmp}}^{j+1}) \geq f(\mathbf{x}^{j+1})$ and the termination condition is not satisfied, set $j = j+1$ and go to 2.

The termination condition may be based on reducing the search vectors below a user-defined threshold. It should be noted that the Hooke–Jeeves procedure is intrinsically serial. It might be beneficial in situations where an optimum is far from the initial guess. In case the initial search vectors are large, the procedure may be able to avoid some local minima. To some extent it is also robust against noisy cost functions.

3.3.3. Nelder–Mead algorithm

The Nelder–Mead algorithm is a very popular derivative-free downhill simplex procedure for unconstrained optimization (Nelder and Mead, 1965). The search process is based on manipulating a finite set of candidate solutions forming a simplex. A simplex or n-simplex is a convex hull of a set of $n+1$ affinely independent points $\mathbf{x}^{(i)}$ in a vector space (normally, we consider n-simplex in an n-dimensional space). The set $X = \{\mathbf{x}^{(i)}\}_{i=1,\ldots,k}$ is affinely independent if the vectors $\mathbf{v}^{(i)} = \mathbf{x}^{(i+1)} - \mathbf{x}^{(1)}, i = 1, \ldots, k-1$, are linearly independent. A convex hull $H(X)$ of X is defined as a set of all convex combinations of the points from X, i.e., $H(X) = \{\sum_{i=1,\ldots,k} a_i \mathbf{x}^{(i)} : \mathbf{x}^{(i)} \in X, a_i \geq 0, \sum_{i=1,\ldots,k} a_i = 1\}$. Examples of 1-, 2-, and 3-simplex are, respectively, a line segment, a triangle, and a tetrahedron (see Fig. 3.12).

The Nelder–Mead algorithm follows the procedure that consists of processing and updating the simplex vertices based on the corresponding objective function values (see also Fig. 3.13):

- Order the vertices so that $f(\mathbf{x}^{(1)}) \leq f(\mathbf{x}^{(2)}) \leq \cdots \leq f(\mathbf{x}^{(n+1)})$. An auxiliary point $\mathbf{x}^{(0)}$ is then defined to be center of gravity of all vertices but $\mathbf{x}^{(n+1)}$;

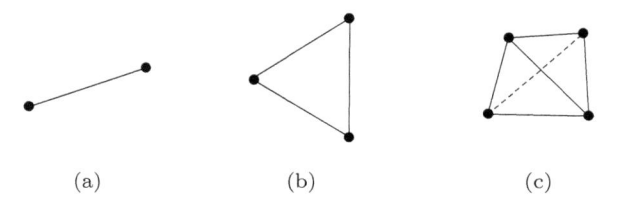

56   Simulation-Based Optimization of Antenna Arrays

Fig. 3.12. The concept of simplex: (a) 1-simplex, (b) 2-simplex, and (c) 3-simplex.

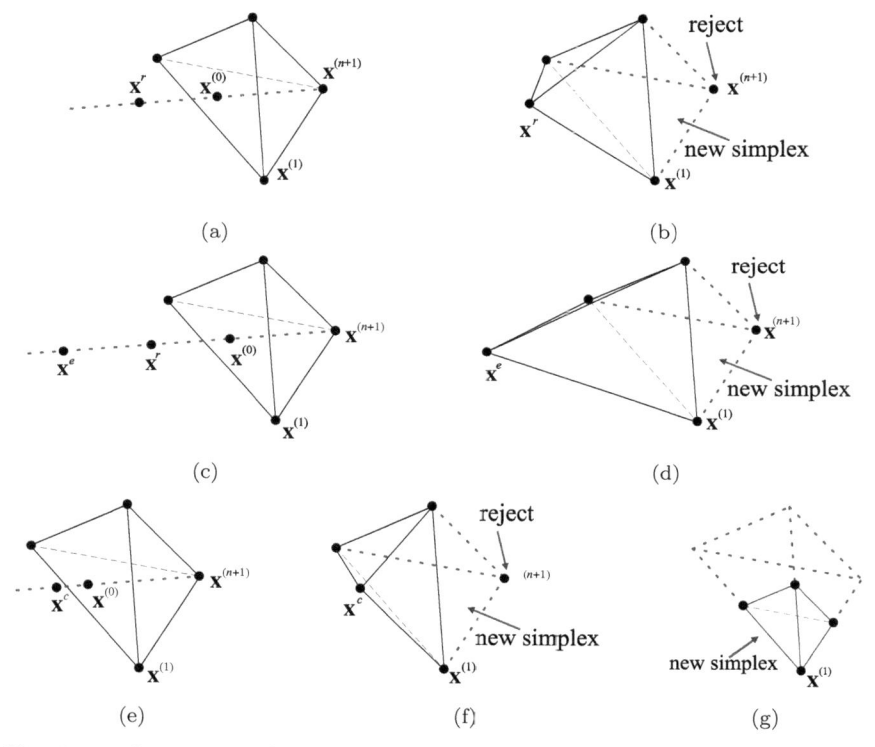

Fig. 3.13. Operations of the Nelder–Mead algorithm: (a) reflection: $\mathbf{x}^r = (1+\alpha)\mathbf{x}^{(0)} - \alpha\mathbf{x}^{(n+1)}$, (b) acceptance of the reflection point $\mathbf{x}^r$ if $f(\mathbf{x}^{(1)}) \leq f(\mathbf{x}^r) < f(\mathbf{x}^{(n)})$, (c) expansion: $\mathbf{x}^e = \rho\mathbf{x}^r + (1-\rho)\mathbf{x}^{(0)}$, (d) acceptance of the expansion point $\mathbf{x}^e$ if $f(\mathbf{x}^e) < f(\mathbf{x}^r)$, (e) contraction: $\mathbf{x}^c = (1+\gamma)\mathbf{x}^{(0)} - \gamma\mathbf{x}^{(n+1)}$, (f) acceptance of the contraction point $\mathbf{x}^c$ if $f(\mathbf{x}^c) \leq f(\mathbf{x}^r)$, and (g) shrinking of the simplex if neither of the above operations led to the objective function improvement.

- Compute a *reflection* $\mathbf{x}^r = (1+\alpha)\mathbf{x}^{(0)} - \alpha\mathbf{x}^{(n+1)}$ (a typical value of α is 1); If $f(\mathbf{x}^{(1)}) \leq f(\mathbf{x}^r) < f(\mathbf{x}^{(n)})$ reject $\mathbf{x}^{(n+1)}$ and update the simplex using $\mathbf{x}^r$;
- If $f(\mathbf{x}^r) < f(\mathbf{x}^{(1)})$ compute an *expansion* $\mathbf{x}^e = \rho\mathbf{x}^r + (1-\rho)\mathbf{x}^{(0)}$ (a typical value of ρ is 2); If $f(\mathbf{x}^e) < f(\mathbf{x}^r)$ reject $\mathbf{x}^{(n+1)}$ and update the simplex using $\mathbf{x}^e$;
- If $f(\mathbf{x}^{(n)}) \leq f(\mathbf{x}^r) < f(\mathbf{x}^{(n+1)})$ compute a *contraction* $\mathbf{x}^c = (1+\gamma)\mathbf{x}^{(0)} - \gamma\mathbf{x}^{(n+1)}$ (a typical value of γ is 0.5); If $f(\mathbf{x}^c) \leq f(\mathbf{x}^r)$ reject $\mathbf{x}^{(n+1)}$ and update the simplex using $\mathbf{x}^c$;
- If $f(\mathbf{x}^c) > f(\mathbf{x}^r)$ or $f(\mathbf{x}^r) \geq f(\mathbf{x}^{(n+1)})$ shrink the simplex: $\mathbf{x}^{(i)} = \mathbf{x}^{(1)} + \sigma(\mathbf{x}^{(i)} - \mathbf{x}^{(1)}), i = 1, \ldots, n+1$ (a typical value of σ is 0.5).

The outlined procedure describes one iteration of the algorithm which is continued until the termination condition is met. Typically, the termination condition is set as a reduction of the simplex size below a user-defined threshold. The Nelder–Mead algorithm is characterized by a relatively low computational cost of only a few objective function evaluations per iteration; however, its convergence is rather slow.

3.4. Summary

In this chapter, several types of conventional optimization techniques have been outlined including gradient-based and derivative-free methods. The discussed techniques are primarily local ones so that the quality of the final solution very much depends on the starting point. From the antenna design viewpoint, these methods may be useful for solving simple tasks such as tuning of individual elements of antenna arrays. Clearly, direct handling of expensive EM simulation models of antennas and arrays can be prohibitive when using conventional methods. However, these methods become indispensable within surrogate-based optimization frameworks (cf. Chapter 5), specifically, for optimizing fast surrogate models of antennas. On the other hand, gradient-based optimization may be a practical solution when cheap sensitivity information is available, e.g., with adjoint sensitivity (Georgieva et al., 2002; Koziel and Bekasiewicz, 2015a).

Derivative-free methods are suitable for handling analytically intractable (noisy, discontinuous) functions. As mentioned before, the methods considered here are local so that they are only suitable for finding the local minimum, typically, the closest to the initial design. Global optimization can be realized using, among others, metaheuristic algorithms, described in Chapter 4.

In this book, we refer to direct optimization methods as those that directly handle the original (here, EM simulation-based) objective function. Despite their usefulness, direct methods have an important disadvantage of the high cost. Conventional optimization algorithms (e.g., gradient-based schemes with numerical derivatives) require tens, hundreds or even thousands of objective function calls per run, depending on the number of design variables. At the same time, high-fidelity EM models of antennas and antenna arrays tend to be very expensive, up to a few hours per simulation, depending on the structure complexity and used computers. Therefore, direct solving of the problem (3.1) by coupling the optimization algorithm with the EM solver as the objective function evaluator is usually impractical. It should be emphasized that despite unprecedented development and availability of computational resources (both hardware- and software-wise), the issue of high EM simulation cost is just as problematic today as it was a decade or two ago. Designers handle more and more complex systems, e.g., by including environmental components such as connectors or housing into the computational model (Kempel, 2007), but they also utilize more accurate models, e.g., with finer discretization of the structure. From this perspective, the most promising way of handling computationally expensive models in antenna and antenna array design is surrogate-assisted optimization. Introduction to surrogate-based methods is provided in Chapter 5.

Chapter 4

Global Optimization: Population-Based Metaheuristics

From practical perspective (with our main focus being design of antenna arrays), one of the most important features of the optimization methods presented in Chapter 3 is their locality. Gradient-based algorithms as well as pattern search derivative-free techniques normally allow for finding an optimum that is located in the vicinity of the initial solution, or, more generally, in the region of attraction of the nearest local optimum. Unfortunately, the objective functions that one needs to deal with in a large number of real-world engineering problems are characterized by multiple local optima. Moreover, the functional landscape of the problem at hand is often unknown in terms of the nonlinearity of the objective function, importance of particular variables, but also the number and the location of the optima. In many cases, finding a reasonably good starting point is a non-trivial task by itself. In such situations, straightforward application of local algorithms often leads to unsatisfactory results; therefore, global search may be necessary. Design optimization of antenna arrays, in particular, radiation pattern synthesis (e.g., for minimum sidelobe level) or design of sparse arrays, normally requires global optimization. Still, local methods are useful when good initial point has been identified.

Global optimization is a broad subject with numerous methods and algorithms developed over the last few decades (Goldberg, 1989; Horst et al., 2000; Horst and Tuy, 1996; Hansen and Walster, 2003). The algorithms for global optimization can be roughly categorized into deterministic (Horst and Tuy, 1996; Zhigljavsky and Žilinskas, 2008) and stochastic ones (Goldberg, 1989; Talbi, 2009). The most successful deterministic methods include cutting plane methods, branch and bound methods, and interval methods (Luenberger, 2003; Jaulin et al., 2001; Hansen and Walster, 2003). Stochastic methods include simulated annealing (van Laarhoven and Aarts, 1989), Monte-Carlo methods (Kroese et al., 2011), as well as population-based approaches (Goldberg, 1989; Talbi, 2009).

This chapter does not attempt to give a comprehensive introduction to global optimization. Instead, we focus on population-based metaheuristics as the most popular solution approaches to global optimization problems nowadays (Talbi, 2009; Yang, 2008), including optimization of antennas (e.g., Deb et al., 2014; Kerkhoff and Ling, 2007; Kibria et al., 2014) and antenna arrays (Karamalis et al., 2009; Jang et al., 2016; Elragal et al., 2011; Datta and Misra, 2009). There are several reasons for such popularity. The most important reasons include simple implementation, versatility, ability to handle "difficult" functions (discontinuous, noisy), as well as suitability for parallelization. Although this book is focused on (local) surrogate-assisted design methods, global optimization is also of interest primarily as a tool for finding promising regions of the design space that is executed at the level of simplified representations such as an array factor model.

The chapter is organized as follows. In Sec. 4.1, we briefly introduce population-based metaheuristics. The outline of the most popular classes of algorithms, including evolution strategies, genetic and evolutionary algorithms, particle swarm optimizers, differential evolution, and firefly algorithm is provided in Secs. 4.2–4.7. Some more recent techniques are briefly mentioned in Sec. 4.8. The chapter is concluded in Sec. 4.9 with remarks concerning general characteristics of population-based algorithms and their applicability for design optimization of antennas and antenna arrays.

4.1. Fundamentals of Population-Based Metaheuristics

Metaheuristics are derivative-free global optimization methods. They have been developed out of practical necessity of handling optimization problems that were challenging (or even unmanageable) for conventional algorithms, including gradient-based routines. The most important challenges included non-differentiable, discontinuous or noisy objective functions, as well as the presence of multiple local optima. The last issue is particularly troublesome because it makes the optimization outcome very much dependent on the initial solution when local methods are used.

Majority of metaheuristic algorithms are based on mimicking of natural processes (e.g., biological or social systems). It should be emphasized right away that there is usually little (or none) rigorous explanation why a specific scheme should work and why it should be better than other schemes. Available performance comparisons are exclusively based on numerical experiments (e.g., Datta and Misra, 2009; Deb et al., 2014) and may be contradictory with each other depending on the inclination of a particular author towards a particular technique.

Most metaheuristics process sets (or populations) of potential solutions to the optimization problem at hand often referred to as individuals or agents. In the course of the optimization process, the method-specific interactions between individuals as well as selection methods are applied in order to identify and explore promising regions of the design space. Some older metaheuristic algorithms (e.g., simulated annealing; Kirkpatrick et al., 1983; van Laarhoven and Aarts, 1987) process a single solution rather than a set of individuals.

The first population-based optimization methods were developed in late 1960s (evolution strategies (ES), Beyer, 2001). A sudden increase of their popularity begun in late 1980s and in 1990s (genetic algorithms (GAs), Goldberg, 1989; evolutionary algorithms (EAs), Back et al., 2000). Today, metaheuristics are standard techniques widely used and, sometimes, overused in almost every engineering discipline. Apart from the already mentioned algorithms, other popular methods include particle swarm optimizers (PSO) (Kennedy et al., 2001), differential evolution (DE) (Price et al., 2005), and, more recently, firefly algorithm, cuckoo search, bat algorithm, and others (Yang, 2010a,b, 2014). There are many new (at least by the name) techniques proposed every year (Duman et al., 2012; Cheng and Prayogo, 2014; Tamura and Yasuda, 2011; Doğan and Ölmez 2015, etc.), however, it seems that these are mostly minor variations of the existing methods.

Below, we outline and discuss a generic flow of a typical population-based metaheuristic algorithm (see Fig. 4.1 for a flow diagram). We will use the symbol P to denote a set of individuals (population) processed by the algorithm:

1. Initialize population P;
2. Evaluate population P;
3. Select parent individuals S from P;
4. Create a new population P from the parent individuals S by applying recombination operators;
5. Introduce local perturbations in individuals of P by applying mutation operators;
6. If termination condition is not satisfied go to 2;
7. END.

In most algorithms, the population is initialized randomly. In Step 2, the individuals in the population are evaluated, and the corresponding values of the objective function determine their "fitness". In many algorithms, such as GAs and EAs, an important step is selection of the subset of individuals (so-called parent individuals) to form a new population. Depending on the algorithm, the selection process can be deterministic (i.e., by picking up the best individuals, ES) or partially random where probability of being selected depends on the fitness value but there is a chance to survive even for poor individuals, EAs. One of the best and the most popular selection procedures is a tournament selection where the selection is executed in a deterministic fashion (based on the higher "fitness" value) from

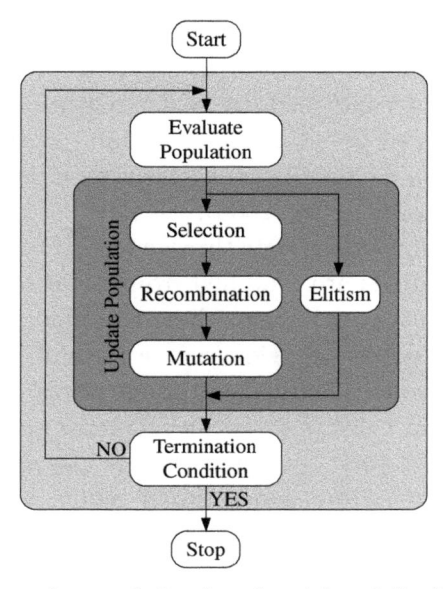

Fig. 4.1. Generic flow of a population-based metaheuristic algorithm. Evaluation of the individuals is followed by a selection process where parent individuals are chosen to create an updated population. The selected individuals undergo the recombination and/or mutation processes. In some implementations, elitism is also utilized to prevent the best individual (or individuals) from extinction by identifying and inserting that individual directly into the new population with bypassing the recombination procedures. Termination condition may be based on the maximum number of iterations or certain convergence criteria (e.g., reduction of the population diversity below a user-defined threshold).

a randomly chosen candidate set. In particular, the candidate set may consist of just two individuals (binary tournament selection). This kind of procedure favors better individuals yet gives a chance for worse ones to survive.

The amount of preference given to the best individuals to survive is referred to as selection pressure. It is one of the important factors controlling the performance and convergence properties of the population-based search algorithm. If it is too low, the algorithm may converge slowly and the search may turn into a random search; if it is too high, the best individuals quickly take over the population and one may face a premature convergence which usually negatively affects the algorithm performance (the search concludes before a global optimum can be identified). The opposite mechanism is production of a new information (or "genetic material" in GAs), which is controlled by, e.g., mutation operators. The selection pressure can be adjusted by certain control parameters of the algorithm; e.g., in tournament selection it is adjusted by the candidate set size (Tan et al., 2005).

In certain metaheuristic algorithms, e.g., PSO or DE, no selection is used, i.e., individuals are modified from iteration to iteration but never die out. This is the case when the algorithm is based on mimicking certain social behavior (e.g., a flock of birds in PSO, or a swarm of fireflies) so that individuals in the population are relocated according to certain rules (e.g., being attracted to the best solution found so far) rather than replacing the old population by the new one according to a certain generational model and recombination operators.

In general, there are two fundamental types of operators that can be used to modify individuals in the population: exploratory ones (e.g., crossover in EAs or ES) and exploitative ones (e.g., mutation in GAs). The purpose of exploratory operators is to combine information contained in the parent individuals to create an offspring. An example of such an operator used by evolutionary algorithms with natural (floating-point) representation is an arithmetic crossover, where a new individual $\mathbf{c}$ can be created as a convex combination of the parents $\mathbf{p}_1$ and $\mathbf{p}_2$, i.e., $\mathbf{c} = \alpha \mathbf{p}_1 + (1-\alpha)\mathbf{p}_2$, where $0 < \alpha < 1$ is a random number (Eiben and Smith, 2003). The purpose of this type of operators is to allow making large "steps" in the design space and, consequently, explore the new and promising regions (Michalewicz, 1996).

Exploitative operators introduce small perturbations (e.g., $\mathbf{p} \leftarrow \mathbf{p} + \mathbf{r}$, where $\mathbf{r}$ is a random vector selected according to a normal probability distribution with zero mean and a certain problem-dependent variance). These operators allow exploitation of a given region of the design space improving local search properties of the algorithm (Deb, 2001; Coello Coello et al., 2007). In some algorithms, e.g., PSO, the difference between both types of operators is not as clear: modification of the individual may be based on the best solution found so far by that given individual as well as the best solution found by the entire population (Clerc, 2006; Kennedy et al., 2001). On the other hand, the operators used by algorithms such as PSO should be interpreted as exchange of information between individuals that affects their behavior rather than the individuals themselves.

In Secs. 4.2–4.7, we outline several types of metaheuristic algorithms, specifically, evolution strategies, genetic algorithms, evolutionary algorithms, PSO, differential evolution, and firefly algorithm. In Sec. 4.8, we briefly mention a few more recent algorithms, whereas Sec. 4.9 summarizes the chapter and provides a short discussion of population-based methods in the context of design optimization of antennas and antenna arrays.

4.2. Evolution Strategies

Evolution strategies (ES) are probably the first population-based search methods ever conceived. They were developed in mid-1960s in Germany by I. Rechenberg and H.P. Schwefel and applied for solving continuous optimization problems, specifically, aerodynamic shape optimization (Rechenberg, 1965; Schwefel, 1968).

There are several variations of ES available (Hansen et al., 2015). The initial versions were simple schemes with only mutation operators. Later on, recombination operators were introduced to allow exchange of information between individuals (Schwefel and Rudolph, 1995). More recent versions of ES utilize covariance matrix adaptation (CMA-ES) to shape the mutation distribution so as to adjust it to a particular landscape of the optimized cost function (Hansen and Kern, 2004). A review of various ES algorithms can be found in Hansen et al. (2015).

Evolution strategies use a natural representation (i.e., floating-point numbers) and process a population of μ individuals so that, in each iteration, λ offsprings are created from the set of parent individuals using problem-specific recombination operators. The new set of parent individuals is chosen either from offsprings (so-called (μ, λ)-ES) or from offsprings and parents (so-called (μ, λ)-ES). Each

66 Simulation-Based Optimization of Antenna Arrays

```
initialize population P_μ = {(x⁽¹⁾,s⁽¹⁾),...,(x⁽μ⁾,s⁽μ⁾)};
while ~termination_condition
    select a parent population P_λ = {(x⁽¹⁾,s⁽¹⁾),...,(x⁽λ⁾,s⁽λ⁾)} from P_μ;
    for j = 1:λ
        mutate s⁽ʲ⁾;
        mutate x⁽ʲ⁾ using updated value of s⁽ʲ⁾;
    end
    select a new population P_μ from P_λ ((μ,λ)-selection) or from
        P_λ ∪ P_λ ((μ+λ)-selection);
end
```

Fig. 4.2. Pseudocode of the basic version of the ES algorithm.

individual contains both the solution data (design parameters) and the set of strategy parameters that determine its further mutation process.

The basic flow of the ES algorithm is shown in Fig. 4.2. Evolutionary strategies use random selection. The new population is chosen using the deterministic truncation selection based on objective function values of individuals. The mutation operators for unconstrained real-valued search space are defined as

$$(\mathbf{x}, \mathbf{s}) \leftarrow \begin{cases} \mathbf{s} \leftarrow \mathbf{s} \cdot e^{\tau N(0,1)}, \\ \mathbf{x} \leftarrow \mathbf{x} + \mathbf{s} \cdot N(0, \mathbf{I}), \end{cases} \quad (4.1)$$

where $N(0,1)$ and $N(0, \mathbf{I})$ are $(0,1)$ normally distributed random scalars and vectors ($\mathbf{I}$ stands for the identity matrix of the size equal to the number of design variables n), and τ is the rate of self-adaptation (learning parameters). Parameter $\mathbf{s}$ controls the strength of the design parameter mutation. It is suggested (Schwefel, 1988) that τ is proportional to $n^{-1/2}$. In more advanced implementations of ES, the identity matrix $\mathbf{I}$ in $N(0, \mathbf{I})$ may be replaced by a general covariance matrix with evolving entries (Hansen et al., 2015) that allows for taking into account correlations between variables, thus leading to more efficient search for the optimum.

Later versions of ES introduced recombination to allow exchange of information between individuals (Beyer, 2001), cf. Fig. 4.3. The two most typical recombination operators used by ES are the

```
initialize population P_μ = {(x⁽¹⁾,s⁽¹⁾),...,(x⁽μ⁾,s⁽μ⁾)};
while ~termination_condition
    for j = 1:λ
        select a parent set P_ρ = {(y⁽¹⁾,t⁽¹⁾),...,(y⁽ρ⁾,r⁽ρ⁾)} from P_μ;
        create (x⁽ʲ⁾,s⁽ʲ⁾) by recombining (y⁽¹⁾,t⁽¹⁾) to (y⁽ρ⁾,r⁽ρ⁾);
        mutate s⁽ʲ⁾;
        mutate x⁽ʲ⁾ using updated value of s⁽ʲ⁾;
    end
    select a new population P_μ from P_λ ((μ,λ)-selection) or from
        P_λ ∪ P_λ ((μ+λ)-selection);
end
```

Fig. 4.3. Pseudocode of the ES algorithm with recombination.

arithmetical averaging

$$\mathbf{x} = \frac{1}{\rho} \sum_{j=1}^{\rho} \mathbf{y}^{(j)}, \quad (4.2)$$

where $\mathbf{y}^{(j)}$ are parent individuals, and coordinate-wise random transfer

$$\mathbf{x} = [x_1 \ldots x_n]^T = [y_1^{(r_1)} \ldots y_n^{(r_n)}]^T. \quad (4.3)$$

Where r_1 to r_n are randomly selected from the set $\{1, 2, \ldots, \rho\}$.

In the literature, one can find some recommendations and rules of thumb concerning practical usage of ES (Beyer and Schwefel, 2002). These include:

- Typical μ/λ ratios in (μ, λ)-selection in continuous search spaces are from $1/7$ to $1/2$.
- Using $(\mu + \lambda)$-selection in conjunction with variation operators (allowing reaching any point of the search space) guarantees stochastic convergence to a global solutions.
- Evolution is usually performed on a phenotypic level (i.e., directly on floating-point representation in case of continuous optimization).
- Mutation should respect the reachability condition (each point of the search space should be reachable in a finite number of steps).

Recombination is applied whenever possible and useful. The main goal of recombination is conservation of the common components of the parents.

4.3. Genetic Algorithms

Genetic algorithms (GAs) are probably the most famous class of population-based metaheuristics around. They have been developed in late 1970s, mostly by J. Holland, K. De Jong and D. Goldberg (Holland, 1975; De Jong, 1975; Goldberg, 1989). Because the initial idea was based on mimicking biological evolution, GA processed populations of individuals on a genotypic level of binary strings while evaluating solutions on a phenotypic level pertinent to a problem at hand (e.g., floating-point representation for continuous optimization). Nowadays, integer or floating-point representation is preferred. GAs traditionally emphasize the role of crossover operator. They are considered to be good heuristics for combinatorial problems, although they may be used for continuous optimization. On the other hand, some modern metaheuristics (especially PSO or DE) are generally better for the latter task.

4.3.1. Algorithm flow and representation

The overall flow of a genetic algorithm is very similar to a generic structure of population-based metaheuristics as shown in Fig. 4.1. Typically GAs process individuals that are represented in certain internal format (referred to as *genotype*). Recombination and mutation operations are performed on a genotype; however, assessment of individual's fitness is performed on decoded individuals (referred to as *fenotype*).

Traditional GAs utilize bit-string representation (Goldberg, 1989). Gray coding is often utilized instead of standard binary coding because small changes of the genotype imply small changes of the fenotype with the Gray coding (for standard coding, the fenotype changes depend on the position of the bit in the string).

Nowadays, a natural representation is preferred, e.g., floating-point representation is used for continucus optimization problems, integer representation is used for combinatorial optimization (e.g., traveling salesman problem), etc. It is recommended that a representation is tailored to the problem in order to facilitate genetic operations or ensure feasibility of individuals (Michalewicz, 1996; Caorsi *et al.*, 2000; Lai and Jeng, 2006; Ding *et al.*, 2008).

4.3.2. Crossover

Crossover is a fundamental operator in GAs. It designed to exchange information between individuals. Figure 4.4 shows several variations of two-parent crossover for bit-string representation. Crossover is explorative, that is, it allows for making large steps and thus helps discovering the promising area in the search space. Typical crossover rate, p_c, is high (e.g., 0.8), so that most of the parents exchange their data. Crossover may be generalized to involve more parents; however, no significant advantages of this approach have been reported (Coello Coello *et al.*, 2007). It should be noticed that crossover does not produce new genetic information; therefore, it cannot be used as the only operator in a genetic algorithm.

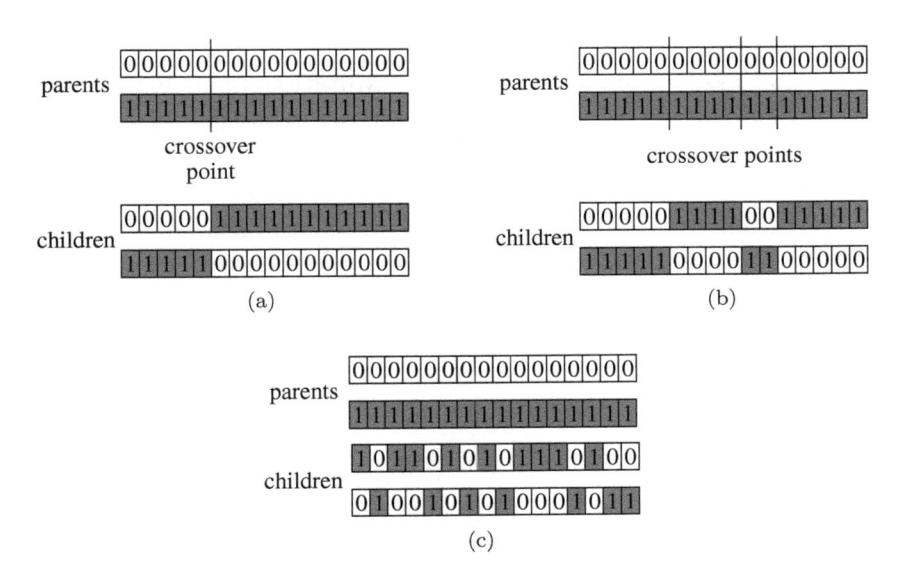

Fig. 4.4. Crossover operators for bit-string GAs: (a) one-point crossover, (b) multi-point crossover, and (c) uniform crossover (bits are changed randomly).

Crossover operators for floating-point representation are arranged differently. For the sake of presentation, we will denote the parent vectors as $\mathbf{x} = [x_1 \ldots x_n]^T$ and $\mathbf{y} = [y_1 \ldots y_n]^T$, whereas the child (offspring) will be denoted as $\mathbf{z} = [z_1 \ldots z_n]^T$. The following three basic types of crossover operators can be distinguished (Eiben and Smith, 2003):

- Discrete: $z_i = x_i$ or y_i (random selection);
- Intermediate: $z_i = ax_i + (1-a)y_i$ with $0 \le a \le 1$ (a selected randomly);
- Arithmetic: $\mathbf{z} = a\mathbf{x} + (1-a)\mathbf{y}$ with $0 \le a \le 1$ (a selected randomly).

Variations and mixtures of the above operators are also used in practical implementations of GAs.

4.3.3. Mutation

The purpose of mutation is to introduce small, random changes into the genotype. The mutation rate p_m is normally small, typically between $1/N$ and $1/L$ (where N and L are the population size and the chromosome length, respectively). Bit-flip mutation operator for traditional GAs is shown in Fig. 4.5. Mutation is an exploitative (local) operator, in other words, it facilitates optimization within a promising area using available information encoded in the chromosome. Mutation produces new genetic information and therefore, in principle, may be used as the only operator in a genetic algorithm. However, it is recommended to use both mutation and crossover in general (Goldberg, 1989).

The mutation operator for floating-point representation (we assume that the individual is denoted as $\mathbf{x} = [x_1 \ldots x_n]^T$) can be

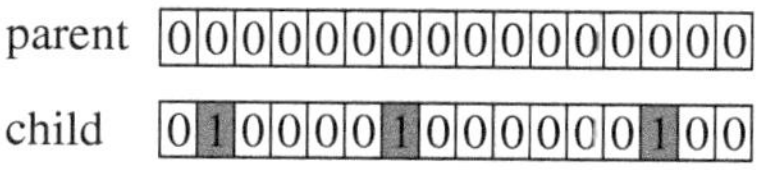

Fig. 4.5. Mutation operator for bit-string GAs. Each gene is flipped independently with a probability p_m.

written as $x_i \to x'_i = x_i + \Delta x_i$, where Δx_i is a random deviation that may be drawn with uniform or non-uniform (e.g., normal) distribution; in both cases, x'_i is eventually curtail to the range of x_i (Eiben and Smith, 2003). Non-uniform distributions that promote smaller changes are typically preferred, e.g., (Michalewicz, 1996),

$$\Delta x_i = \begin{cases} (x_{i.\,\max} - x_i) \cdot (2(r - 0.5))^\beta & \text{if } r > 0.5, \\ (x_{i.\,\min} - x_i) \cdot (2(0.5 - r))^\beta & \text{otherwise,} \end{cases} \quad (4.4)$$

where $r \in [0, 1]$ is a random number; β is normally larger than 1 (e.g., 3) and may increase in later stages of the algorithm run. In a particular implementation of (4.4), mutation is arranged to ensure satisfaction of the box constraints, i.e., $x_{i.\min} \leq x_i + \Delta x_i \leq x_{i.\max}$ for $i = 1, \ldots, n$.

4.3.4. Selection

Selection is another important GA component. It is a process of choosing individuals that will become parents of a new population. Selection is based on the quality of individuals, traditionally referred to as fitness. Normally, better individuals are given higher value of fitness; for minimization of the objective function f, the fitness of an individual **x** could be $-f(\mathbf{x})$. Genetic algorithms utilize partially stochastic selection, i.e., better individuals are favored but there is an element of randomness that allows weak individuals to be selected as well (although with lower probability). On one hand, this sort of strategy comes from the overall background of GAs (mimicking of natural evolution); on the other hand, it helps avoiding premature convergence of the algorithm.

Chronologically, one of the first selection schemes used in GAs was the roulette wheel selection (Coello Coello et al., 2007), where the probability of an individual to be selected was proportional to its fitness value. An obvious disadvantage of this scheme is a risk of premature convergence: a single highly fit individual may quickly take over the population. Another disadvantage is considerable reduction of the selection pressure when all fitness values are on a similar level (this can be controlled by fitness scaling; Goldberg, 1989).

One of the most popular selection schemes today, free from the aforementioned drawbacks, is a tournament selection which works as follows (Bäck, 1996):

1. Randomly pick up a subset of k individuals from the current population (k is a tournament size).
2. Select an individual with the highest fitness value within the subset.
3. Repeat Steps 1 and 2 N times in order to create the parent set.

The selection pressure can be adjusted by changing the tournament size, from pure random selection for $k = 1$, to fully deterministic selection for $k = N$ (the latter would lead to instant convergence of the algorithm).

4.3.5. *Elitism*

An obvious issue related to stochastic selection is that the best individual (or individuals) is not guaranteed to survive. Even if it does, it may be altered by recombination and/or mutation operators. Elitism is a technique frequently used in GAs and other types metaheuristics (Zitzler and Thiele, 1999; Lim and Isa, 2014; Yang, 2014) to avoid losing the best individual(s) found so far: the set of best individuals (typically one) is stored and deterministically inserted into the new population (cf. Fig. 4.6). Various versions of this technique have been considered; for example, there are algorithms where individuals may survive within the population up to a certain number of iterations (Tan *et al.*, 2005).

4.3.6. *Selected topics*

In this section, we focus on two practical issues pertinent to GAs, namely, various population models and the controlling the selection pressure.

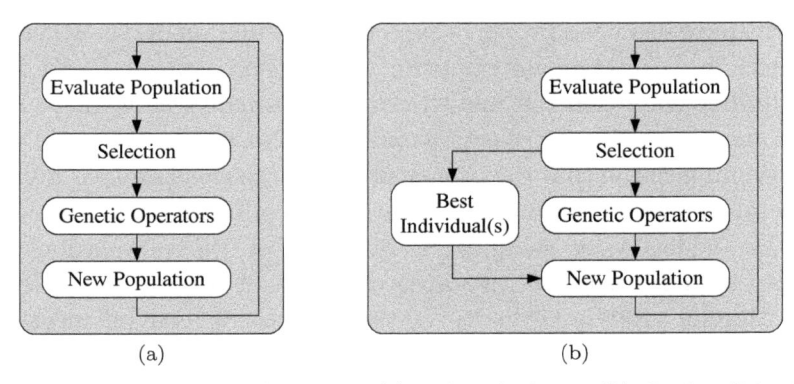

Fig. 4.6. Flowchart of a GA without (a) and with elitism (b). In the elitist GA, the best individual (or individuals) is found during the selection process and directly inserted into the new population bypassing the genetic operations.

There are three basic approaches to replacing the old population by a new one (Coello Coello et al., 2007):

1. Generational models: each individual survives for exactly one generation, and the entire set of parents is replaced by the offspring.
2. Steady-state models: one offspring is generated per iteration, and one member of the population is replaced.
3. Generation gap: the fraction of the population is replaced varying from 1.0 (as in generational models) to $1/N$ (as in steady-state models).

The most popular model is a generational one; however, other models may exhibit certain advantages. One of these is lower computational costs of the algorithm iteration for the generation gap models and, obviously, the steady-state one.

Selection pressure has been already mentioned on a few occasions in this chapter. It is basically the amount of preference given to the better individuals to survive. It is mainly controlled by the selection scheme utilized in the algorithm. Selection pressure is critical for controlling the operation and performance of GAs. In particular, if it is too high, it may lead to premature convergence, i.e., the situation when all individuals become (almost) identical before the search space has been properly explored. Production of a new genetic material is a mechanism opposite to selection pressure, and it is mainly dependent on the mutation rate. Normally, we want to keep a balance between the selection pressure and production of a new genetic material so that the premature convergence is avoided but the evolution is not a random search either. In practice, such a balance can be achieved by monitoring diversity of the population and adjusting the mutation rate accordingly (self-adaptation) (Hesser and Männer, 1991). Diversity of the population can be measured using standard deviation of individuals (details depend on a particular representation used in the algorithm). The mutation rate can be reduced at the final stages of the algorithm run in order to improve exploitation at the expense of exploration and, therefore, obtain more precise localization of the optimal solution.

4.4. Evolutionary Algorithms

The term of evolutionary algorithms has been conceived in early 1990s (see, e.g., Michalewicz, 1996). It refers to the entire class of population-based search methods that are distinguished from other types of optimization algorithms by the following features:

- the use of population;
- the use of stochastic search operators;
- the use of stochastic selection.

Genetic algorithms are often considered to be special case of evolutionary algorithms. As a matter of fact, genetic algorithms using floating-point representation are considered by some to be evolutionary algorithms but not genetic algorithms.

There is a little of theory explaining why evolutionary algorithms work. Early, semi-heuristic results for bit-string-based genetic algorithms (so-called schemata theorem by J. Holland; see Holland, 1975) suggesting that genetic algorithms work by discovering, exploiting, and combining "building blocks" (groups of closely interacting genes) have been later criticized partially because of success of algorithms using natural representations (Eiben and Smith, 2003;

Liu et al., 2014a; Ding and Wang, 2013; Alander et al., 2004; Yeung et al., 2008). The results obtained using Markov Chain analysis, where GAs are treated as stochastic processes (Horn, 1993; David, 1994; Suzuki, 1995), consider algorithm convergence to a global solution in a probabilistic sense; however, practical significance of such results is minor.

Evolutionary algorithms (EAs) can be considered as robust problem solvers in the following sense (cf. Fig. 4.7):

- EAs do not perform as good as problem-specific methods for problems for which the latter methods were designed.
- EAs perform better than a random search "on average".
- EAs perform better than problem-specific methods "on average".

These features are generally common for other types of population-based metaheuristics as well (cf. Secs. 4.5–4.8).

On a practical note, it should be emphasized that "standard" off-the-shelf implementation of an evolutionary algorithm may not perform well when applied to a particular engineering problem. This is in line with so-called no free lunch (NFL) theorems (Wolpert and Macready, 1997, 2005) that basically say — when referred to evolutionary methods — that performance of all algorithms is the same when averaged over all problems. Consequently, a general recommendation is that EAs should be individually designed for solving any given task in order to maximize performance and

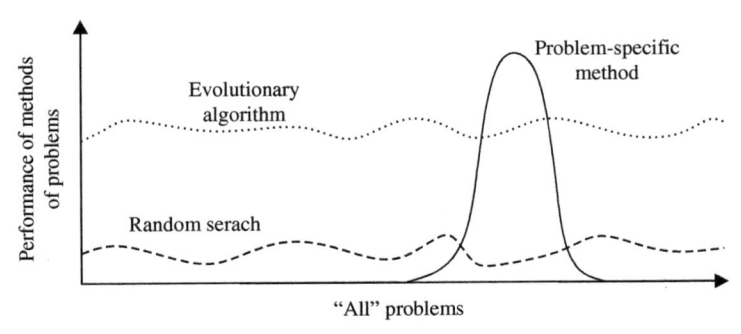

Fig. 4.7. Robustness of evolutionary algorithms versus pure random search and problem-specific methods.

reliability. There are various (and problem-specific) ways of doing so, including development of appropriate representation, tailored recombination, and mutation operators.

It is normally better to preserve feasibility of solutions by suitably defined genetic operations rather than to enforce extinction of infeasible individuals by assigning them low fitness values. A famous example is a partially-mapped crossover operator for the TSP problem as described by Goldberg (1989).

4.5. Particle Swarm Optimization

As of today, particle swarm optimization (PSO) belongs to the most popular metaheuristic algorithms. PSO is based on mimicking the swarm behavior such as fish or bird schooling in nature (Kennedy, 1997; Kennedy *et al.*, 2001). The first version of PSO was developed in mid-1990s. Since then, many variations of the algorithm have been applied to virtually every area of optimization, computational intelligence, and engineering design including antenna engineering (Lizzi *et al.*, 2008; Poli *et al.*, 2014; Pavlidis *et al.*, 2005; Rocca *et al.*, 2012; Manica *et al.*, 2009).

A PSO algorithm processes a set of individuals (traditionally referred to as swarm) that are iteratively relocated within the design space and interact with other individuals by means of social influence and social learning. Individuals (or particles) are vectors in a multi-dimensional space and are represented by their positions and velocities. Particles are relocated based on their velocities but also information about the best positions found so far (both individual and overall for the swarm). The movement of a swarming particle consists of two major components: a stochastic component and a deterministic component. Each particle is attracted toward the position of the current global best $\mathbf{g}$ and its own best location $\mathbf{x}_i^*$ found during the optimization run. Additionally, there is a stochastic component controlled by suitable algorithm parameters.

Let $\mathbf{x}_i$ and $\mathbf{v}_i$ denote the position the velocity vectors of the ith particle. The velocity and position are updated as follows:

$$\mathbf{v}_i \leftarrow \chi[\mathbf{v}_i + c_1\mathbf{r}_1 \bullet (\mathbf{x}_i^* - \mathbf{x}_i) + c_2\mathbf{r}_2 \bullet (\mathbf{g} - \mathbf{x}_i)], \qquad (4.5)$$

$$\mathbf{x}_i \leftarrow \mathbf{x}_i + \mathbf{v}_i. \tag{4.6}$$

Here, $\mathbf{r}_1$ and $\mathbf{r}_2$ are vectors with components being uniformly distributed random numbers between 0 and 1; • denotes componentwise multiplication. The control parameter χ is typically equal to 0.7298 (Kennedy et al., 2001); c_1 and c_2 are acceleration constants determining how much the particle is directed towards good positions (they represent a "cognitive" and a "social" components, respectively); usually, $c_1, c_2 \approx 2$. The sum of $c_1 + c_2$ should not be smaller than 4.0; as the values of these parameters increase, χ gets smaller as does the damping effect.

PSO normally works with random initialization (using uniform sampling), which increases the chance of identifying promising regions of the search space. This is especially important for multi-modal problems. The initial velocity of a particle can be taken as zero.

There are many different variants and extensions of the standard PSO algorithm (Clerc and Kennedy, 2002; Kennedy et al., 2001; Fong et al., 2016). For example, various modifications concerning the parameter χ (also called the inertia weight) have been considered to better control the convergence rate of PSO (Bansal et al., 2011; Nickabadi et al., 2011). There were also multiple attempts to combine PSO with other metaheuristics such as GAs (Kao and Zahara, 2008), simulated annealing (Da and Xiurun, 2005), or ant colony optimization (Shelokar et al., 2007) but also with local search methods (Noel, 2012).

4.6. Differential Evolution

Differential evolution (DE) is currently one of the most popular population-based metaheuristic algorithms for continuous global optimization. DE was developed in late 1990s by Storn and Price (1997). It is a vector-based evolutionary algorithm, and can be considered as a further development to genetic algorithms. It is a derivative-free stochastic search algorithm with self-organizing tendency. The design parameters are represented as n-dimensional vectors (also called agents). The new agents are allocated in the search space by combining the positions of other existing agents. More specifically, an intermediate agent is generated from two agents randomly chosen from the current population. This temporary agent is then mixed with a predetermined target agent. The new agent is accepted for the next generation if and only if it yields reduction in objective function. In other words, DE is using a sort of deterministic selection.

The basic DE algorithm uses a random initialization. A new agent $\mathbf{y} = [y_1 \ldots y_n]$ is created from the exisiting one $\mathbf{x} = [x_1 \ldots x_n]$ as follows:

1. Three agents $\mathbf{a} = [a_1 \ldots a_n]$, $\mathbf{b} = [b_1 \ldots b_n]$ and $\mathbf{c} = [c_1 \ldots c_n]$ are randomly extracted from the population ($\mathbf{a} \neq \mathbf{b} \neq \mathbf{c} \neq \mathbf{x}$).
2. A position index $p \in \{1, \ldots, N\}$ is determined randomly (N is the population size).
3. The new agent $\mathbf{y}$ is determined as follows (separately for each of its components $i = 1, \ldots, n$):

    a. select a random number $r_i \in (0, 1)$ with uniform probability distribution;
    b. if $i = p$ or $r_i < CR$ let $y_i = a_i + F(b_i - c_i)$, otherwise let $y_i = x_i$; here, $0 \leq F \leq 2$ is the differential weight and $0 \leq CR \leq 1$ is the crossover probability, both defined by the user;
    c. if $f(\mathbf{y}) < f(\mathbf{x})$ then replace $\mathbf{x}$ by $\mathbf{y}$; otherwise reject $\mathbf{y}$ and keep $\mathbf{x}$.

The above DE scheme is often referred to as *DE/rand/1/bin*, meaning that the donor vector $\mathbf{a}$ is selected randomly ("rand"), there is one pair of vectors selected to calculated the mutation differential ("1"), and that the recombination is binomial ("bin"). Some of DE variations, using the terminology explained above are as follows (Mezura-Montes *et al.*, 2006a,b, 2010).

- *DE/rand/p/bin* — explained above; p denotes the number of pairs of vectors used to calculate the mutation differential (in particular, p can be equal to 1);

- *DE/rand/p/exp* — as above but with exponential crossover (Noman and Iba, 2008);
- *DE/best/p/bin, DE/best/p/exp* — a donor vector is the best solution found so far in the population;
- *DE/current-to-rand/p/bin, DE/current-to-best/p/exp* — variants with arithmetic recombination;
- *DE/current-to-rand/p/bin* — a variant with combined arithmetic–discrete recombination.

It should be emphasized that although DE resembles some other stochastic optimization techniques, unlike traditional EAs, DE perturbs the solutions in the current generation vectors with scaled differences of two randomly selected agents. As a consequence, no separate probability distribution has to be used, and thus the scheme presents some degree of self-organization. Additionally, DE is simple to implement, uses very few control parameters, and has been observed to perform satisfactorily in multi-modal optimization problems (Chakraborty, 2008; Rocca et al., 2011a).

Many research studies have been focused on the choice of the control parameters F and CR as well as the modification of the updating scheme of the step 3b above. For details, interested reader is referred to Price *et al.* (2005).

4.7. Firefly Algorithm

Firefly algorithm (FA) is one of the most recent yet already popular metaheuristics. It was developed in 2007 by Yang (2008) by modeling behavior of firefly using a few idealized rules, namely: (i) fireflies are attracted to each other and (ii) attractiveness is proportional to their brightness (i.e., the less brighter one moves towards the brighter one); (iii) the brightness of a firefly depends on the objective function landscape (e.g., for maximization problem the brightness may be proportional to the objective function value). Figure 4.8 shows the algorithmic flow of the firefly algorithm.

Formally, the updating formula for the ith firefly is as follows:

$$\mathbf{x}_i \leftarrow \mathbf{x}_i + \beta_0 e^{-\gamma r_{ij}^2}(\mathbf{x}_j - \mathbf{x}_i) + \alpha \varepsilon_i, \tag{4.7}$$

```
objective function f(x), x = [x₁ ... x_d]ᵀ;
initialize population x⁽ⁱ⁾, i = 1, ..., N;
define light absorption coefficient γ;
while ~termination_condition
    for i = 1:N
        for j = 1:N
            if I(x⁽ⁱ⁾) < I(x⁽ʲ⁾)
                move firefly i towards j
            end
            vary attractiveness with distance r proportionally to exp(-γr);
            evaluate new solutions and update light intensity;
        end
    end
    rank the fireflies and find the global best g*;
end
```

Fig. 4.8. Flow of the firefly algorithm. Light intensity $I(\mathbf{x}^{(i)})$ is determined by $f(\mathbf{x}^{(i)})$ (e.g., $I(\mathbf{x}^{(i)}) = -f(\mathbf{x}^{(i)})$ for a minimization problem). Termination condition is typically based on the maximum number of iterations.

where $r_{ij} = ||\mathbf{x}_i - \mathbf{x}_j||$ is the Euclidean distance between the ith and jth firefly (this can be replaced by other type of norm), β_0 is the attractiveness parameter at a zero distance, $\alpha \in (0, 1]$ is a randomization parameter, whereas ε_i is a random vector drawn with Gaussian or uniform probability distribution. Note that (4.7) is essentially a random walk towards brighter fireflies (Yang, 2008). The convergence rate of the algorithm critically depends on the parameter γ (taken in the range 0.1 to 10 in most applications).

It should be noted that FA is, as a matter of fact, quite closely related to PSO. In particular, by setting $\gamma \to 0$, FA becomes a special case of accelerated PSO (Yang, 2008). Also, by removing the inner loop in Fig. 4.8, and replacing the light intensity by the global best g^*, one essentially gets the standard PSO (Weyland, 2015). On the other hand, for $\gamma \to \infty$, FA turns into random search because the influence of other fireflies on relocation of any given firefly becomes negligible. In practice, FA is normally operating between these two extreme cases thus arguably better than both random search and PSO (Yang, 2008).

4.8. Other Methods

Metaheuristic optimization has been constantly growing over the last three decades. With the major methods and algorithms developed before 1980s (e.g., ES, GAs, EAs) and 1990s (PSO, DE, Ant Colony optimization), the number of new algorithms as well as improvement of the existing ones proposed in the literature has been accelerating. Examples include bee algorithm (Yang, 2005), bacteria foraging algorithm (Gazi and Passino, 2004), harmony search (Geem et al., 2001), bat algorithm (Yang, 2010a), gravitational search algorithm (Rashedi et al., 2009). Most of these techniques are based on observations of specific natural or social phenomena; however, the general concepts are usually similar (cf. Sec. 4.1). A good exposition of some of the recent methods can be found in Yang (2010a). At the same time, it can be argued that the differences between many of the recently proposed techniques are quite negligible. In particular, the majority of them are actually small variations of the PSO algorithm (Weyland, 2015). This, in conjunction with the very limited (or non-existing) theoretical justification, makes development of the new algorithmic variations rather questionable.

4.9. Summary

The purpose of this chapter was to outline the fundamentals of global optimization by means of population-based metaheuristics. We presented the overall concept as well as described several major techniques, including evolutionary strategies, genetic and evolutionary algorithms, PSO, differential evolution, and a firefly algorithm. We also discussed some practical aspects of population-based optimization, such as balancing selection pressure to control the algorithm convergence.

Although a large variety of population-based metaheuristic algorithms are available in the literature, all of them share similar basic principles as mentioned earlier in the chapter. In the initial period of development of these methods (late 1960s through late 1990s), several rather distinct classes of techniques were proposed. In the recent years, however, the differences between various new algorithms seem to be minor (despite giving them catchy names) and so is the performance enhancement.

A majority of the methods described here are for unconstrained optimization (apart from box constraints that can be easily handled by appropriate definitions of the recombination operators). Constrained metaheuristic optimization is a research area in its own (Mezura-Montes *et al.*, 2006a,b; Koziel and Michalewicz, 1998) and exceeds the scope of this chapter.

There are several important properties of population-based methods that make them attractive tools for solving a variety of real-world problems. These include global search properties, capability of handling "difficult" functions (non-differentiable, discontinuous, noisy, multi-modal, etc.), as well as easy implementation and parallelization. Also, most major metaheuristics can (and have been) extended to allow multi-objective optimization (Abbas *et al.*, 2001; Aljibouri *et al.*, 2000; Babu and Jehan, 2003; Binh and Korn, 1997; Deb *et al.*, 2002; Goudos *et al.*, 2009; Jin and Rahmat-Samii, 2007; Kukkonen and Lampinen, 2004; Madavan, 2002; Mezura-Montes *et al.*, 2006b; Sivasubramani and Swarup, 2011).

We conclude this chapter with a few remarks concerning application of population-based metaheuristics for design optimization of antennas and antenna arrays. The fundamental difficulty is the significant computational cost of metaheuristics: with typical population sizes of 20 to over 100 and the number of iterations from a few dozen to many hundreds, the cost varies from about a thousand to tens of thousands of objective function evaluations. Clearly, there are cases where reasonably good analytical models are available (e.g., antenna array design with analytical array factor models, Aljibouri *et al.*, 2008; Jin and Rahmat-Samii, 2005; Rocca *et al.*, 2011b) and utilization of metaheuristics is well justified. Unfortunately, majority of real-world antenna design cases require utilization of full-wave electromagnetic simulations. Electromagnetic simulations, when repeatedly performed at a high-fidelity level of description, make the computational cost of metaheuristic optimization prohibitive. This difficulty can be alleviated to some extent by surrogate-assisted methods where optimization is executed at the level of a fast representation of the structure under design (Bandler et al., 2004a; Forrester and Keane, 2009; Queipo et al., 2005). The surrogates may be data-driven, i.e., constructed by approximating sampled EM-simulation results (Koziel and Ogurtsov, 2013a; Koziel et al., 2013a) or physics-based (usually derived from coarse-discretization electromagnetic simulations, Jacobs and Koziel, 2014; Koziel, 2009). Exposition of these methods will be given in Chapter 5.

Chapter 5

Fundamentals of Surrogate-Based Modeling and Optimization

Design optimization of antennas and antenna arrays is challenging for several reasons. One of these is that the process involves high-fidelity EM simulation models necessary for reliable performance evaluation of the structure under design. Another reason is a large number of designable parameters: these might be dimensions of radiating elements and feeds, parameters of substrates as well as excitation amplitudes and/or phases. Other challenges might include typically graded composition of the array structures, presence of connectors and radomes. Furthermore, global optimization is often necessary because functional landscapes pertinent to antennas and antenna arrays tend to be multi-modal with multiple local optima. All these challenges make conventional numerical optimization techniques (see Chapters 3 and 4 for details) unsuitable for handling real-world antenna design problems. In particular, the computational cost of such algorithms is typically measured in hundreds (for local methods) or thousands and tens of thousands (for population-based metaheuristics) of objective function evaluations. A notable exception is gradient-based search with adjoint sensitivities (Ghassemi et al., 2013; Koziel and Ogurtsov, 2012c; Koziel et al., 2013b, 2014b; Koziel and Bekasiewicz, 2015a, 2016b) where the optimization process can be conducted in reasonable time even for relatively large number of designable parameters. Unfortunately, adjoints are not yet widespread in computational electromagnetics, especially in terms of availability of this technology through commercial simulation software packages (CST, 2016; HFSS, 2016).

Probably the most promising approaches in terms of expedited EM-driven design are those exploiting surrogate models. Surrogate-assisted algorithms allow considerable reduction of the optimization cost by executing majority of operations at the level of a fast replacement model (surrogate) and only occasional reference to the high-fidelity one. In this chapter, we provide an introduction to surrogate-based optimization (SBO). We discuss the SBO concept and various surrogate modeling techniques. Several approximation- and physics-based surrogate-assisted optimization methods are overviewed. In this chapter, we focus on generic algorithms, whereas specific techniques tailored to handle antenna and antenna array design problems will be presented in the following chapters along with illustrative examples.

5.1. Surrogate-Based Optimization

Development of optimization methods has always been driven by practical necessity oriented towards handling specific classes of objective functions. For example, gradient-based algorithms require that the function is sufficiently smooth, e.g., at least continuously differentiable up to a certain order (Nocedal and Wright, 2006). Population-based metaheuristics are designed to handle multi-modal and even discontinuous functions; however, the function evaluation cost is supposed to be low due to a large number of objective functions required for convergence of metaheuristics algorithms.

Contemporary engineering design is heavily based on computer simulation models. The continuing shift towards simulation-driven design, observed over the last two decades or so, has been closely related to and stimulated by rapid development of computational hardware and software tools. At the same time, analytical models are becoming incapable to adequately represent complex systems; this issue—again—makes simulation-based techniques indispensable.

Clear advantages of simulation models include potentials for design automation and cost reduction of prototyping. The downside of simulation models — from the point of view of simulation-driven design — is the high cost. In case of antennas, typical simulation times are from a few minutes for models of simple structures to many hours for finely discretized models of complex antenna arrays (Afshinmanesh et al., 2008; Bekasiewicz and Koziel, 2015; Chamaani et al., 2010; Hanninen, 2012; Koziel et al., 2015). Obviously, hundreds and thousands of simulations required by conventional optimization routines add up to impractically high computational costs.

Surrogate-based optimization (SBO) techniques (Queipo et al., 2005; Koziel et al., 2011a; Koziel and Leifsson, 2013a; Forrester and Keane, 2009; Koziel and Ogurtsov, 2014a; Leifsson and Koziel, 2015a; Yelten et al., 2012) seem to be the most promising approaches for solving design optimization problems involving expensive simulation models. The key concept behind surrogate-assisted methods is to replace direct optimization of the high-fidelity simulation model by an iterative process yielding a sequence $\mathbf{x}^{(i)}$, $i = 0, 1, \ldots$, of approximate solutions to the original optimization problem (3.1). The designs $\mathbf{x}^{(i)}$ are obtained using a fast (yet reasonably accurate) representation of the high-fidelity model, referred to as a surrogate. Normally, the surrogate model is updated at each iteration using the available high-fidelity model data. In case of local methods this could be a high-fidelity model response evaluated at the most recent design. In case of a global search there might be several designs selected using an appropriate infill strategy (Forrester et al., 2008; Koziel et al., 2011a; Koziel and Ogurtsov, 2014a).

A popular way to generate the new design is optimization of the current surrogate model. In this case, the surrogate-based optimization process can be written as (Koziel et al., 2011a, Fig. 5.1)

$$\mathbf{x}^{(i+1)} = \arg\min_{\mathbf{x}} U(s^{(i)}(\mathbf{x})), \qquad (5.1)$$

where $s^{(i)}$ is the surrogate model at the ith iteration. The initial design $\mathbf{x}^{(0)}$ is often obtained using engineering insight or by optimizing an available lower-fidelity model.

The most important advantage of SBO over conventional methods is computational savings that can potentially be obtained. Assuming that the surrogate model is very fast, the cost of solving (5.1) can be negligible so that the overall optimization cost is merely determined by the number of the high-fidelity model evaluations. On the other hand, the number of iterations required by a typical SBO algorithm is substantially smaller than those of majority of conventional methods when solving (3.1) (Koziel et al., 2006). Thus SBO procedures' costs tend to be low.

There are various ways to build a surrogate model; most of them are problem-dependent. In certain (and rather simple) cases, the surrogate models can be obtained from analytical models. Another option is utilization of simplified physics models such as equivalent circuit representations in microwave engineering (Bandler et al., 2004a). Finally, the surrogates can be constructed as data-driven models also referred to as response surface approximation (RSA) models (Liu et al., 2014b). The RSA models can be constructed beforehand using either extensive sampling of the design space (Koziel and Ogurtsov, 2011b) or sequential sampling techniques (Forrester and Keane, 2009).

In case of antennas and antenna arrays, the range of possible surrogate models is rather limited. Because reliable equivalent network models are hardly available, the surrogate model is usually constructed using coarse-discretization simulations (Koziel and Ogurtsov, 2012a; Koziel et al., 2016). For antenna arrays, analytical array factor (AF) models are often utilized to evaluate the radiation pattern (Mailloux, 2005). Unfortunately, AF models are of limited accuracy particularly when coupling between radiators is strong. In addition, the AF models do not give any reliable information about reflection coefficients. Another modeling option is simulation-based superposition models (Koziel and Ogurtsov, 2015) that are constructed using individually simulated reflection and radiation responses of the array-embedded radiating elements. Still, the most generic way of representing antenna/antenna array at the lower-fidelity level is coarse-mesh EM analysis. An important practical challenge here is finding a proper balance between the model accuracy and speed (Koziel and Ogurtsov, 2012a; Leifsson and Koziel, 2015b). In practical antenna design cases, the achievable time evaluation ratio between the high- and low-fidelity models is between 5 to 50 or so. One of the consequences is that the computational cost of the surrogate model evaluation when solving (5.1) cannot be neglected. As a matter of fact, it may become a significant contributor to the overall optimization cost, which leads to additional challenges concerning the development and implementation of the SBO algorithms (Koziel et al., 2011a; Koziel and Ogurtsov, 2014a; Forrester and Keane, 2009). More detailed discussion concerning low-fidelity models for antenna and antenna arrays is provided in Chapter 6.

Reliability of the SBO process depends, among others, on the surrogate model accuracy. From the viewpoint of local optimization, it is important that the surrogate is consistent with the high-fidelity model at the current design $\mathbf{x}^{(i)}$, i.e., $s^{(i)}(\mathbf{x}^{(i)}) = f(\mathbf{x}^{(i)})$ (zeroth-order consistency, Alexandrov and Lewis, 2001), and $J[s^{(i)}(\mathbf{x}^{(i)})] = J[f(\mathbf{x}^{(i)})]$ (first-order consistency), where $J[\cdot]$ stands for the model Jacobian. Assuming first-order consistency and embedding the surrogate-based algorithm in the trust-region framework (Conn et al., 2000)

$$\mathbf{x}^{(i+1)} = \arg \min_{\mathbf{x}, \|\mathbf{x}-\mathbf{x}^{(i)}\| \leq \delta^{(i)}} U(s^{(i)}(\mathbf{x})), \qquad (5.2)$$

ensures convergence of the algorithm (5.1) to a local optimum of f (Alexandrov et al., 1998) if the models involved are sufficiently smooth.

The trust-region radius $\delta^{(i)}$ in (5.2) is updated in each iteration based on a gain ratio $\rho = [U(f(\mathbf{x}^{(i+1)})) - U(f(\mathbf{x}^{(i)}))]/[U(s(\mathbf{x}^{(i+1)})) - U(s(\mathbf{x}^{(i)}))]$, i.e., the actual versus surrogate-predicted objective function improvement (Conn et al., 2000); $\delta^{(i)}$ is increased if the improvement is consistent with the prediction and decreased otherwise. Given the first-order consistency, the high-fidelity model objective is guaranteed to improve for sufficiently small values of δ.

Clearly, first-order consistency condition can only be satisfied if high-fidelity model gradients are available. Furthermore, additional

90 *Simulation-Based Optimization of Antenna Arrays*

assumptions concerning the smoothness of the functions involved are necessary for convergence (Echeverría and Hemker, 2008). Convergence of various SBO algorithms can be ensured under other scenarios of, e.g., space mapping (Koziel et al., 2006, 2008b), manifold mapping (Echeverría and Hemker, 2008), and surrogate management framework (Booker et al., 1999).

It is worth to mention at this point that SBO can be implemented either as a local or as a global search algorithm. In case of local optimization, trust-region-like convergence safeguards are often used (cf. (5.2)), and at least zeroth-order consistency conditions are ensured by appropriate construction of the surrogate model (Koziel et al., 2010a). For global search, the surrogate is constructed in the larger portion of the design space. The new candidate solutions are typically found using global methods such as evolutionary

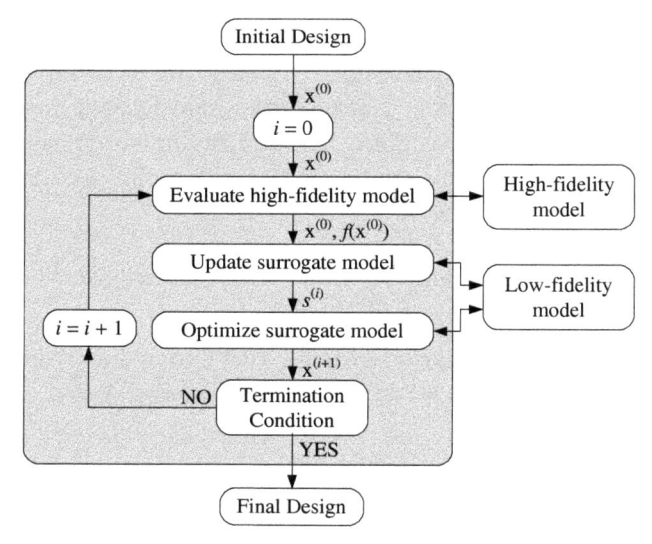

Fig. 5.1. The algorithmic flow of the surrogate-based optimization (SBO) process. The surrogate model s is iteratively updated and optimized to yield a sequence $\mathbf{x}^{(i)}$, $i = 0, 1, \ldots$ of approximate solutions to the original problem $\mathbf{x}^* = \arg\min\{\mathbf{x} : U(f(\mathbf{x}))\}$. The high-fidelity model f is evaluated for verification and to provide data needed to update the surrogate. The computational benefits of SBO are due to: (i) typically small number of iterations (compared to conventional techniques), and (ii) low cost of each iteration (as the search is executed at the level of cheap surrogate).

algorithms. The surrogate model itself is updated using appropriate statistical infill criteria based either on the expected improvement of the objective function or on minimization of the (global) modeling error (Forrester and Keane, 2009; Liu et al., 2014b).

5.2. Surrogate Modeling: Data-Driven Surrogates

Approximation models (also referred to as data-driven models, response surface approximation models, etc.) are undoubtedly the most popular class of surrogates. A large variety of modeling methods have been already developed (Simpson et al., 2001; Søndergaard, 2003; Forrester and Keane, 2009; Couckuyt, 2013; Lophaven et al., 2002; Gorissen et al., 2010a); yet there are certain features common to all types of approximation models. The common features can be identified as follows:

- They are data-driven, i.e., the model is constructed merely from sampled training data so that no problem-specific knowledge is required.
- They are generic, therefore, applicable to a wide range of problems and easily transferrable between various areas.
- They are typically based on explicit analytical formulas.
- They are cheap to evaluate.

Despite these merits, data-driven surrogates have an important disadvantage. That is, a large number of training data points is necessary to ensure good predictive power of the model. This number rapidly grows with the number of the parameters representing the so-called curse of dimensionality (Forrester and Keane, 2009; Gorissen et al., 2010b). In addition, the number of necessary data points quickly grows with the parameter ranges. In case of antennas and antenna arrays, data-driven modeling is essentially limited to a few ($<$5–6) dimensions unless the ranges of geometry parameters are considerably restricted. On the other hand, local approximation models can be quite useful as auxiliary optimization tools.

In this section, we outline the surrogate modeling flow as well as discuss popular approximation techniques and model validation methods.

5.2.1. *Modeling flow for data-driven surrogates*

A surrogate is constructed by approximating sampled data of the high-fidelity model. There are four steps involved in the modeling process:

- Design of experiments (DOE), which is allocation of the training points in the design space using a strategy of choice. Due to computational budget constraints as well as potential numerical problems of handling large amounts of training data, the number of samples is normally limited in practice. Nowadays, space-filling DOEs are preferred (Simpson *et al.*, 2001; Forrester and Keane, 2009), cf. Sec. 5.2.2.
- Training data acquisition at the locations assigned at the first stage.
- Identification of the surrogate model. This step aims at finding the model parameters which are either determined by solving a suitably defined minimization problem, e.g., using kriging (Kleijnen, 2009), neural networks (Rayas-Sanchez, 2004), or by solving an appropriate regression problem analytically, e.g., with polynomial approximation (Queipo *et al.*, 2005).
- Validation of the model. This step estimates the generalization error of the surrogate, i.e., this step estimates the predictive power at the designs not seen during the identification stage.

In practice, surrogate model construction is carried out iteratively, by repeating the above steps until the model accuracy reaches the required level (Koziel *et al.*, 2014a). At each iteration, a new set of so-called infill samples are added using a given strategy referred to as infill criteria (Forrester and Keane, 2009) and the model is re-identified (Bekasiewicz *et al.*, 2014b). Figure 5.2 shows the block diagram of the surrogate modeling process.

5.2.2. *Design of experiments*

Design of experiments (DOE) is a strategy for allocating the training data points in the design space (Giunta *et al.*, 2003; Santner *et al.*, 2003; Koehler and Owen, 1996). The high-fidelity model

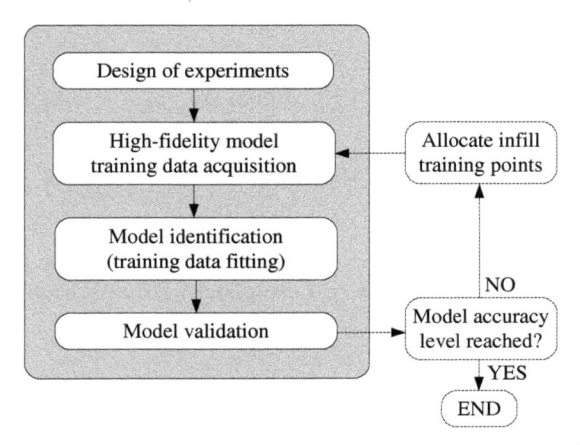

Fundamentals of Surrogate-Based Modeling and Optimization 93

Fig. 5.2. A flowchart of data-driven surrogate model construction. Optional iterative model enhancement involving infill training points is indicated using dashed-lines.

data are subsequently acquired at these locations to construct the surrogate. As mentioned before, the training points may be selected beforehand or using the feedback about the model performance, e.g., by allocating more points in the regions where the system exhibits more linear behavior (Couckuyt, 2013; Devabhaktuni et al., 2001).

The most traditional DOE strategies are factorial designs with the samples allocated in the corners, edges or faces of the design space. An example of factorial DOE is shown in Fig. 5.3(a). Factorial designs allow for estimating the main effect and interactions between design variables without excessive number of samples. The most "economical" factorial DOE is a so-called star distribution (Fig. 5.3(a)) which is often used in conjunction with space mapping (Cheng et al., 2006). In factorial DOEs, the samples are normally spread as far from each other as possible. This is quite helpful to reduce the effects of uncertainties in evaluating the system response (e.g., through physical measurements).

Nowadays most of the data comes from deterministic computer simulations. Thus a majority of modern DOE approaches attempt to allocate the training points uniformly within the design space (Queipo et al., 2005). Such approaches are useful for constructing surrogates when the knowledge about the system is limited. Popular

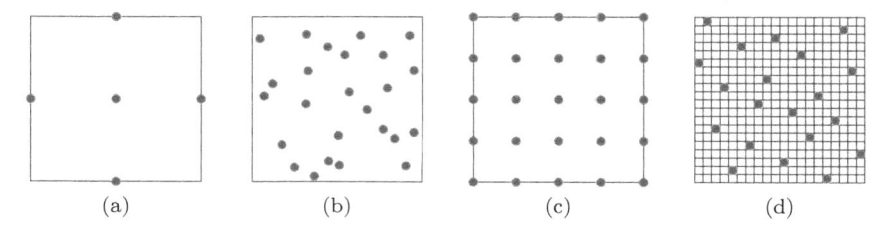

Fig. 5.3. Popular DOE techniques: (a) factorial designs (here, star distribution); (b) random sampling; (c) uniform grid sampling; (d) Latin hypercube sampling (LHS).

space-filing DOEs include pseudo-random sampling (Guinta et al., 2003; Fig. 5.3(b)), uniform grid sampling (Fig. 5.3(c)), and Latin hypercube sampling (LHS, McKay et al., 1979; Fig. 5.3(d)). Although simple, pseudo-random sampling does not ensure good uniformity. Grid sampling is only applicable for low-dimensional spaces. LHS and its improvements (e.g., Beachkofski and Grandhi, 2002; Leary et al., 2003; Ye, 1998; Palmer and Tsui, 2001) offer good uniformity and flexibility. In order to allocate p samples with LHS, the range for each parameter is divided into p bins which yields a total number of p^n bins in the design space for n design variables. The samples are randomly selected in the design space so that (i) each sample is randomly placed inside a bin, and (ii) there is exactly one sample in each bin for all one-dimensional projections of the p samples and bins.

An example realization of 20-sample LHS in two-dimensional space has been shown in Fig. 5.3(d). Other uniform sampling techniques that are commonly used include orthogonal array sampling (Queipo et al., 2005), quasi-Monte Carlo sampling (Giunta et al., 2003), or Hammersley sampling (Hammersley, 1960). Uniform sample allocation can also be obtained through optimization (Leary et al., 2003).

5.2.3. Data-driven modeling techniques

There are many data-driven (or function approximation based) modeling techniques available including methods such as polynomial regression (Queipo et al., 2005), radial-basis functions (Wild et al., 2008), kriging (Forrester and Keane, 2009), neural networks

(Haykin, 1998), support vector regression (Gunn, 1998), Gaussian process regression (Rasmussen and Williams, 2006), and multi-dimensional rational approximation (Shaker et al., 2009). Selected methods are briefly outlined below. More detailed information can be found in the references listed above as well as in review works, e.g., in Simpson et al. (2001).

Throughout this section we denote the training data set as $\{\mathbf{x}^{(i)}, f(\mathbf{x}^{(i)})\}$, $i = 1, \ldots, p$, where $\mathbf{x}^{(i)}$ are data points, $f(\mathbf{x}^{(i)})$ are corresponding high-fidelity number evaluations, and p is the number of the training points.

5.2.3.1. *Polynomial regression*

Probably the most popular yet simple approximation technique is polynomial regression (Queipo et al., 2005). Low-order polynomials are quite useful particularly for local modeling or to construct rough initial surrogates indicating the trends of the function of interest.

The surrogate model is of the form

$$s(\mathbf{x}) = \sum_{j=1}^{K} \beta_j v_j(\mathbf{x}), \tag{5.3}$$

where v_j are the (polynomial) basis functions and β_j are the model parameters that can be found as a least-square solution to the linear system $\mathbf{F} = \mathbf{X}\boldsymbol{\beta}$. $\mathbf{F} = [f(\mathbf{x}^{(1)})\ f(\mathbf{x}^{(2)})\ \ldots\ f(\mathbf{x}^{(p)})]^T$, $\mathbf{X}$ is a $p \times K$ matrix containing the basis function values at the sample points, and $\boldsymbol{\beta} = [\beta_1\ \beta_2\ \ldots\ \beta_K]^T$. If $p \geq K$ and $\text{rank}(\mathbf{X}) = K$, the model parameter can be determined in the least-squares sense through the pseudoinverse $\mathbf{X}^+$ of $\mathbf{X}$ as $\boldsymbol{\beta} = \mathbf{X}^+\mathbf{f} = (\mathbf{X}^T\mathbf{X})^{-1}\mathbf{X}^T\mathbf{f}$ (Golub and Van Loan, 1996).

One of the simplest examples of a regression model is a second-order polynomial one defined as

$$s(\mathbf{x}) = s([x_1\ x_2\ \ldots\ x_n]^T) = \beta_0 + \sum_{j=1}^{n} \beta_j x_j + \sum_{i=1}^{n}\sum_{j \leq i}^{n} \beta_{ij} x_i x_j, \tag{5.4}$$

with the basis functions being monomials: 1, x_j, and $x_i x_j$.

In general, the basis functions do not need to be polynomials; in a particular application they could be exponential or sinusoidal.

5.2.3.2. Radial basis functions

Another popular technique for constructing data-driven surrogate models is radial basis function (RBF) interpolation (Forrester and Keane, 2009; Wild et al., 2008). The RBF model is defined as a linear combination of radially symmetric functions ϕ

$$s(\mathbf{x}) = \sum_{j=1}^{K} \lambda_j \phi(\|\mathbf{x} - \mathbf{c}^{(j)}\|), \tag{5.5}$$

where $\boldsymbol{\lambda} = [\lambda_1 \ \lambda_2 \ \ldots \ \lambda_K]^T$ is the vector of model parameters, and $\mathbf{c}^{(j)}$, $j = 1, \ldots, K$, are the known basis function centers. The parameters can be calculated as $\boldsymbol{\lambda} = \boldsymbol{\Phi}^+ \mathbf{F} = (\boldsymbol{\Phi}^T \boldsymbol{\Phi})^{-1} \boldsymbol{\Phi}^T \mathbf{F}$, where $\mathbf{F} = [f(\mathbf{x}^{(1)}) \ f(\mathbf{x}^{(2)}) \ \ldots \ f(\mathbf{x}^{(p)})]^T$, and the $p \times K$ matrix $\boldsymbol{\Phi} = [\Phi_{kl}]_{k=1,\ldots,p;\, l=1,\ldots,K}$, with the entries defined as

$$\Phi_{kl} = \phi(\|\mathbf{x}^{(k)} - \mathbf{c}^{(l)}\|). \tag{5.6}$$

If $p = K$ and the centers of the basis functions coincide with the data points and are all different, then $\boldsymbol{\Phi}$ is a regular square matrix, and the model coefficients can be found as $\boldsymbol{\lambda} = \boldsymbol{\Phi}^{-1} \mathbf{F}$.

A popular basis function is a Gaussian one, $\phi(r) = \exp(-r^2/2\sigma^2)$, where σ is the scaling parameter. Other basis functions include multiquadric $\phi(r) = (1 + \sigma^2 r^2)^{1/2}$, inverse quadratic $\phi(r) = (1 + \sigma^2 r^2)^{-1}$, inverse multiquadric $\phi(r) = (1 + \sigma^2 r^2)^{-1/2}$, and thin plate spline $\phi(r) = r^2 \ln(r)$.

5.2.3.3. Kriging

Kriging belongs to the most popular surrogate modeling techniques today (Matheron, 1963; Journel and Huijbregts, 1981; Simpson et al., 2001; Kleijnen, 2009; O'Hagan, 1978). A basic formulation of kriging assumes that the function of interest is of the form

$$f(\mathbf{x}) = \mathbf{g}(\mathbf{x})^T \boldsymbol{\beta} + Z(\mathbf{x}), \tag{5.7}$$

where $\mathbf{g}(\mathbf{x}) = [g_1(\mathbf{x}) \; g_2(\mathbf{x}) \; \ldots \; g_K(\mathbf{x})]^T$ are known (e.g., constant) functions, $\boldsymbol{\beta} = [\beta_1 \; \beta_2 \; \ldots \; \beta_K]^T$ are the unknown model parameters (hyperparameters); $Z(\mathbf{x})$ is a realization of a normally distributed Gaussian random process with zero mean and variance σ^2. The regression part $\mathbf{g}(\mathbf{x})^T \boldsymbol{\beta}$ is a trend function for f, and $Z(\mathbf{x})$ takes into account localized variations. The covariance matrix of $Z(\mathbf{x})$ is given as

$$\text{Cov}[Z(\mathbf{x}^{(i)}) Z(\mathbf{x}^{(j)})] = \sigma^2 \mathbf{R}([R(\mathbf{x}^{(i)}, \mathbf{x}^{(j)})]), \quad (5.8)$$

where $\mathbf{R}$ is a $p \times p$ correlation matrix with $R_{ij} = R(\mathbf{x}^{(i)}, \mathbf{x}^{(j)})$. $R(\mathbf{x}^{(i)}, \mathbf{x}^{(j)})$ is the correlation function between the data points $\mathbf{x}^{(i)}$ and $\mathbf{x}^{(j)}$.

The most popular correlation function is a Gaussian one

$$R(\mathbf{x}, \mathbf{y}) = \exp\left[-\sum_{k=1}^{n} \theta_k |x_k - y_k|^2\right], \quad (5.9)$$

where θ_k are the correlation parameters to be determined; x_k and y_k are the kth components of the vectors $\mathbf{x}$ and $\mathbf{y}$, respectively. The kriging predictor (Simpson et al., 2001; Journel and Huijbregts, 1981) is defined as

$$s(\mathbf{x}) = \mathbf{g}(\mathbf{x})^T \boldsymbol{\beta} + \mathbf{r}^T(\mathbf{x}) \mathbf{R}^{-1} (\mathbf{F} - \mathbf{G}\boldsymbol{\beta}), \quad (5.10)$$

where $\mathbf{r}(\mathbf{x}) = [R(\mathbf{x}, \mathbf{x}^{(1)}) \; \ldots \; R(\mathbf{x}, \mathbf{x}^{(p)})]^T$, $\mathbf{F} = [f(\mathbf{x}^{(1)}) \; f(\mathbf{x}^{(2)}) \; \ldots \; f(\mathbf{x}^{(p)})]^T$, and $\mathbf{G}$ is a $p \times K$ matrix with $G_{ij} = g_j(\mathbf{x}^{(i)})$. We also have $\boldsymbol{\beta} = (\mathbf{G}^T \mathbf{R}^{-1} \mathbf{G})^{-1} \mathbf{G}^T \mathbf{R}^{-1} \mathbf{F}$. Model fitting is accomplished by maximum likelihood for θ_k (Journel and Huijbregts, 1981).

One of the most attractive properties of kriging is that the random process $Z(\mathbf{x})$ gives information about the approximation error of the model. That information can be used for improving the surrogate by allocating additional training samples at the locations where the estimated model error is the highest (Forrester and Keane, 2009; Journel and Huijbregts, 1981). This feature is also utilized in various global optimization methods (Couckuyt, 2013, and references therein).

5.2.3.4. Artificial neural networks

Artificial neural networks (ANNs) (Haykin, 1998) are a large subject and an extensive research area. One of important applications of ANNs is classification. However, from the point of view of this chapter, ANNs can be considered as rather complex nonlinear regression models.

The key component of ANN is the neuron (or single-unit perceptron) (Haykin, 1998; Minsky and Papert, 1969). A neuron realizes a nonlinear operation (cf. Fig. 5.4(a)), where w_1 through w_n are regression coefficients, β is the bias value of the neuron, and T is a user-defined slope parameter. The most common neural network architecture is the multi-layer feed-forward network shown in Fig. 5.4(b).

There are two fundamental stages of ANN construction: selection of the network architecture and the network training. For example, for a multi-layer ANN (cf. Fig. 5.4(b)) one needs to select the number of hidden layers as well as the number of neurons in these layers. Both affect the network flexibility, i.e., its ability to approximate the training data. Training of the network, i.e., assignment of the values to the neuron weights is realized as nonlinear least-squares regression problem. A popular technique for solving this regression problem is the error back-propagation algorithm (Simpson *et al.*, 2001;

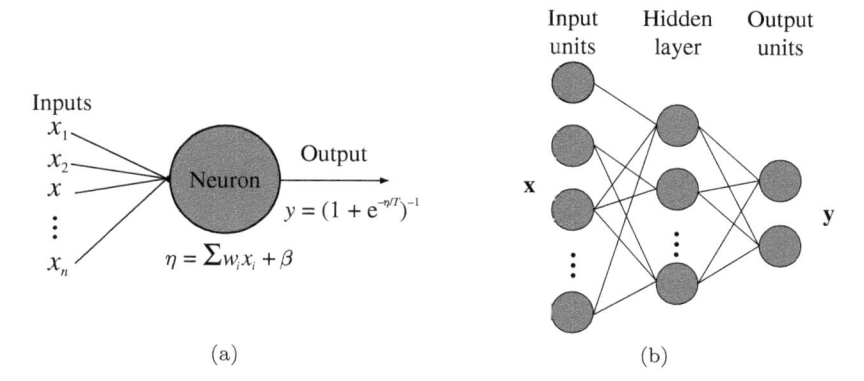

Fig. 5.4. Basic concepts of artificial neural networks: (a) structure of a neuron; (b) two-layer feed-forward neural network architecture.

Haykin, 1998). For large networks with complex architectures, the use of global optimization methods might be necessary (Alba and Marti, 2006). It should be mentioned that finding the optimal ANN architecture is a non-trivial problem yet of primary importance: the network that is too simple may not be able to approximate the training data well; on the other hand, if it has too many degrees of freedom, it may exhibit poor generalization capability.

5.2.3.5. *Support vector regression*

Support vector regression (SVR) (Gunn, 1998) features good generalization capability (Angiulli et al., 2007) and easy training by means of quadratic programming (Smola and Schölkopf, 2004). SVR has gained popularity in various areas including electrical engineering and aerodynamic design (Pattipati et al., 2011; Rojo-Alvarez et al., 2005; Yang et al., 2005; Meng and Xia, 2007; Xia et al., 2007; Andrés et al., 2012; Zhang and Han, 2013).

In this section, we formulate SVR for a vector-valued function **f**. We denote the training samples by $r^k = [r_1^k \ r_2^k \ \ldots \ r_m^k]^T = \mathbf{f}(\mathbf{x}^k)$, $k = 1, 2, \ldots, N$. SVR is supposed to approximate r^k at all points $\mathbf{x}^k$. For linear regression, we aim at approximating a training data set, here, the pairs $D_j = \{(\mathbf{x}^1, r_j^1), \ldots, (\mathbf{x}^N, r_j^N)\}$, $j = 1, 2, \ldots, m$, by a linear function $f_j(\mathbf{x}) = \mathbf{w}_j^T \mathbf{x} + b_j$. The optimal regression function is given by the minimum of the functional (Smola and Schölkopf, 2004)

$$\Phi_j(\mathbf{w}, \xi) = \frac{1}{2}\|\mathbf{w}_j\|^2 + C_j \sum_{i=1}^{N}(\xi_{j.i}^+ + \xi_{j.i}^-). \qquad (5.11)$$

Here, C_j is a user-defined value, and $\xi_{j.i}^+$ and $\xi_{j.i}^-$ are the slack variables representing upper and lower constraints on the system output. The typical cost function used in SVR is an ε-insensitive loss function defined as $L_\varepsilon(y) = 0$ for $|f_j(\mathbf{x}) - y| < \varepsilon$ and $L_\varepsilon(y) = |f_j(\mathbf{x}) - y|$ otherwise. The value of C_j determines the trade-off between the flatness of f_j and the amount up to which deviations larger than ε are tolerated (Gunn, 1998).

Here, we describe nonlinear regression employing the kernel approach in which the linear function $\mathbf{w}_j^T \mathbf{x} + b_j$ is replaced by the

$\Sigma_i \gamma_{j.i} K(\mathbf{x}^k, \mathbf{x}) + b_j$, where K is a kernel function. The SVR model is defined as

$$s(\mathbf{x}) = \left[\sum_{i=1}^{N} \gamma_{1.i} K(\mathbf{x}^i, \mathbf{x}) + b_1 \ldots \sum_{i=1}^{N} \gamma_{m.i} K(\mathbf{x}^i, \mathbf{x}) + b_m \right]^T \quad (5.12)$$

with the parameters $\gamma_{j.i}$ and b_j, $j = 1, \ldots, m$, $i = 1, \ldots, N$ obtained according to a general SVR methodology. In particular, Gaussian kernels of the form $K(\mathbf{x}, \mathbf{y}) = \exp(-0.5 \cdot \|\mathbf{x} - \mathbf{y}\|^2 / c^2)$ with $c > 0$ can be used, where c is the scaling parameter. Both c as well as parameters C_j and ε can be adjusted to minimize the generalization error calculated using cross-validation (Queipo et al., 2005).

5.2.3.6. *Other approximation methods*

The approximation techniques outlined in Secs. 5.2.3.1–5.2.3.5 are the most popular ones; however, there are a number of other interesting methods as well. One of them is moving least squares (MLS) (Levin, 1998). In MLS, the error contribution from each training point $\mathbf{x}^{(i)}$ is multiplied by a weight ω_i that depends on the distance between $\mathbf{x}$ and $\mathbf{x}^{(i)}$. This improves the model flexibility but also increases computational complexity as the model has to be identified for each $\mathbf{x}$ separately. A typical choice for the weights is

$$\omega_i(\|\mathbf{x} - \mathbf{x}^{(i)}\|) = \exp(-\|\mathbf{x} - \mathbf{x}^{(i)}\|^2). \quad (5.13)$$

Gaussian process regression (GPR) (Rasmussen and Williams, 2006) is a surrogate modeling technique that addresses the approximation problem from a stochastic point of view. From this perspective, and since Gaussian processes are mathematically tractable, it is relatively easy to compute error estimations for GPR-based surrogates in the form of uncertainty distributions. Under certain conditions, GPR models can be shown to be equivalent to large neural networks while requiring much less regression parameters (Rasmussen and Williams, 2006). In recent years, GPR has been successfully applied for modeling of antenna structures (Jacobs, 2012, 2016; Jacobs and Koziel, 2014).

Another technique to mention is co-kriging (Forrester *et al.*, 2007; Toal and Keane, 2011). Co-kriging is an extension of kriging interpolation (cf. Sec. 5.2.3.3) that allows for combining information from computational models of various fidelity levels. The fundamental benefit of such combination is that — by exploiting knowledge embedded in the lower-fidelity model — the surrogate can be created at much lower computational cost than for the models exclusively based on high-fidelity data. Co-kriging is a rather recent method with relatively few applications in engineering (Toal and Keane, 2011; Huang and Gao, 2012; Laurenceau and Sagaut, 2008; Koziel *et al.*, 2013c; Koziel and Bekasiewicz, 2016c).

5.2.4. *Model validation*

Upon identification, the surrogate model has to be validated in order to assess its predictive power also referred to as generalization capability that is an ability to predict the response of the high-fidelity model at the locations others than those utilized for training. The most common validation approach is a split sample method (Queipo *et al.*, 2005), where a part of the available data set (the training set) is used to construct the surrogate, whereas another part (the test set) serves purely for model validation. The method is straightforward; however, the error estimation depends strongly on how the set of data samples is partitioned.

More reliable estimation of the generalization error of the surrogate can be obtained by means of cross-validation (Queipo *et al.*, 2005; Geisser, 1993). In this approach, the available data set is split into K subsets, and each of these subsets is sequentially used as testing set for a surrogate constructed on the other $K - 1$ subsets. The prediction error can be estimated with all the K error measures obtained in this process. The primary advantage of cross-validation is that the error estimation is less biased than with the split-sample method.

As mentioned before, the entire cycle of constructing the surrogate and its validation may be iterated. The process of adding new (infill) samples based on infill criteria (Forrester and Keane, 2009) is often referred to as sequential sampling. It is normally continued until the prescribed surrogate accuracy level is reached. In the modeling context (as opposed to optimization), the objective is improvement of the global accuracy of the surrogate, and the new samples are inserted at the locations where the estimated modeling error is the highest. It should also be mentioned that finding infill point locations normally requires solving global optimization problems (Couckuyt, 2013); such problems are often handled with population-based metaheuristics.

5.3. Surrogate Modeling: Physics-Based Surrogates

Data-driven surrogates are perfect solutions for many practical modeling problems. Their major advantages are versatility and low evaluation cost. Unfortunately, application of data-driven surrogates in antenna array design is rather limited. Their fundamental issue is high cost of training data acquisition. Other issues include high dimensionality of design spaces for real-world antenna structures as well as wide parameter ranges that are necessary for the models configured over reasonable ranges of operating conditions (e.g., frequencies, major lobe direction). In this section, we discuss physics-based surrogates that seem to be more useable for the antenna and antenna array area.

Physics-based surrogates are constructed by suitable correction or enhancement of an underlying low-fidelity model. In case of antennas and antenna arrays, the low-fidelity model is normally obtained from EM analysis performed with coarse-discretization simulations. In a very simple case, it could also be an equivalent network, or even an analytical model. The correction process aims at improving alignment between the low-fidelity model and the high-fidelity model, either locally or across the entire design space.

As the low-fidelity model embeds certain knowledge about the system of interest (here, an antenna), the models are normally well correlated. Consequently, only a limited amount of high-fidelity data is normally sufficient to ensure reasonable accuracy of the surrogate model. For the same reason, feature-based surrogates typically feature good generalization. The latter translates into improved

Fundamentals of Surrogate-Based Modeling and Optimization 103

computational efficiency of the SBO process (Koziel et al., 2011a; Koziel and Ogurtsov, 2014a).

In this section, we outline the main concepts of physics-based surrogate modeling. More detailed treatment of the subject can be found in the literature (Koziel et al., 2011a; Koziel and Ogurtsov, 2014a; Bekasiewicz et al., 2014b). The majority of the optimization techniques we discuss in this book in the context of antenna array optimization exploit physics-based surrogates.

The low-fidelity (coarse) model of the system of interest will be denoted as $c(\mathbf{x})$. In order to explain the physics-based surrogate modeling concept, we discuss a simple case of multiplicative response correction in the context of the SBO algorithm (5.1). For ensuring convergence of a sequence $\{\mathbf{x}^{(i)}\}$ of approximate solutions to the original problem (3.1), produced by (5.1) it is imperative to have the surrogate and the high-fidelity model well (at least locally) aligned. The surrogate $s^{(i)}(\mathbf{x})$ at iteration i can be constructed with response correction

$$s^{(i)}(\mathbf{x}) = \beta_k(\mathbf{x})c(\mathbf{x}), \qquad (5.14)$$

where $\beta_k(\mathbf{x}) = \beta_k(\mathbf{x}^{(i)}) + \nabla \beta(\mathbf{x}^{(i)})^T(\mathbf{x} - \mathbf{x}^{(i)})$, and $\beta(\mathbf{x}) = f(\mathbf{x})/c(\mathbf{x})$. This construction ensures both the zeroth-order consistency, $s^{(i)}(\mathbf{x}^{(i)}) = f(\mathbf{x}^{(i)})$, and the first-order consistency, $\nabla s^{(i)}(\mathbf{x}^{(i)}) = \nabla f(\mathbf{x}^{(i)})$ (Alexandrov and Lewis, 2001). Quality of a surrogate constructed with (5.14) is illustrated in Fig. 5.5.

Response correction can also be realized in a multi-point fashion where high-fidelity data at several designs are utilized for model alignment. For example (Koziel and Leifsson, 2013b), the surrogate can be defined for vector-valued models as

$$\mathbf{s}^{(i)}(\mathbf{x}) = \mathbf{\Lambda}^{(i)} \circ \mathbf{c}(\mathbf{x}) + \mathbf{\Delta}_r^{(i)} + \boldsymbol{\delta}^{(i)} \qquad (5.15)$$

with column vectors $\mathbf{\Lambda}^{(i)}$, $\mathbf{\Delta}_r^{(i)}$, and $\boldsymbol{\delta}^{(i)}$ and $\circ$ denoting component-wise multiplication. The global response correction parameters $\mathbf{\Lambda}^{(i)}$ and $\mathbf{\Delta}_r^{(i)}$ are obtained as

$$[\mathbf{\Lambda}^{(i)}, \mathbf{\Delta}_r^{(i)}] = \arg \min_{[\mathbf{\Lambda}, \mathbf{\Delta}_r]} \sum_{k=0}^{i} \|\mathbf{f}(\mathbf{x}^{(k)}) - \mathbf{\Lambda} \circ \mathbf{c}(\mathbf{x}^{(k)}) + \mathbf{\Delta}_r\|^2. \qquad (5.16)$$

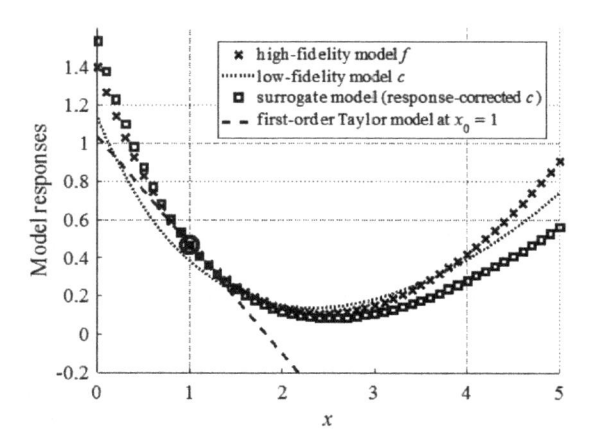

104 *Simulation-Based Optimization of Antenna Arrays*

Fig. 5.5. Visualization of the response correction (5.14) for the example analytical functions c (low-fidelity model) and f (high-fidelity model). The correction is established at $x_0 = 1$. It should be emphasized that the surrogate exhibits good alignment with the high-fidelity model in a relatively wide vicinity of x_0, especially compared to the first-order Taylor model set up using the same data from f (the value and the gradient at x_0).

The response scaling (5.16) is supposed to improve alignments of the models for all previous iterations. The (local) additive term $\boldsymbol{\delta}^{(i)}$ is then defined as

$$\boldsymbol{\delta}^{(i)} = \mathbf{f}(\mathbf{x}^{(i)}) - [\boldsymbol{\Lambda}^{(i)} \circ \mathbf{c}(\mathbf{x}^{(i)}) + \boldsymbol{\Delta}_r^{(i)}]. \tag{5.17}$$

Equation (5.17) ensures a perfect match between the surrogate and the high-fidelity model at the current design $\mathbf{x}^{(i)}$, $\mathbf{s}^{(i)}(\mathbf{x}^{(i)}) = \mathbf{f}(\mathbf{x}^{(i)})$.

All correction terms $\boldsymbol{\Lambda}^{(i)}$, $\boldsymbol{\Delta}_r^{(i)}$ and $\boldsymbol{\delta}^{(i)}$ can be obtained analytically by solving appropriate linear regression problems (Koziel and Leifsson, 2013b).

As an example, let us consider a dielectric resonator antenna (DRA) shown in Fig. 5.6. The DRA is suspended above the ground plane in order to enhance its impedance bandwidth. The high-fidelity model $\mathbf{f}$ is simulated using the CST MWS transient solver (CST, 2016) (∼800,000 mesh cells, evaluation time 20 min). The low-fidelity model $\mathbf{c}$ is also evaluated in CST (∼30,000 mesh cells, evaluation time

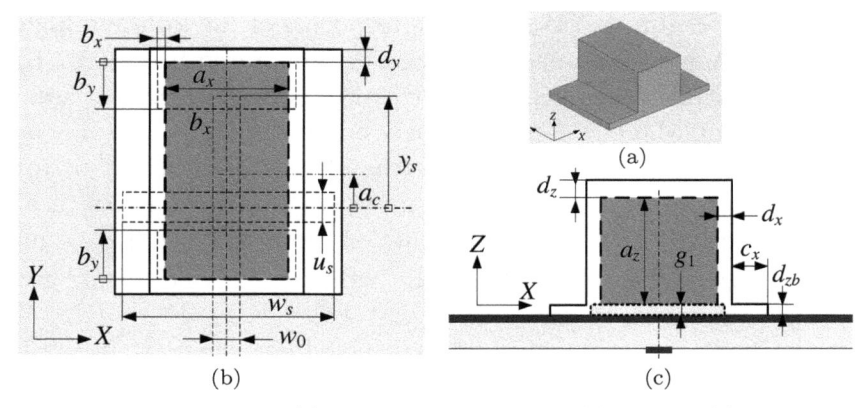

Fundamentals of Surrogate-Based Modeling and Optimization   105

Fig. 5.6. Suspended DRA (a) 3D view of its housing, (b) top and (c) front views.

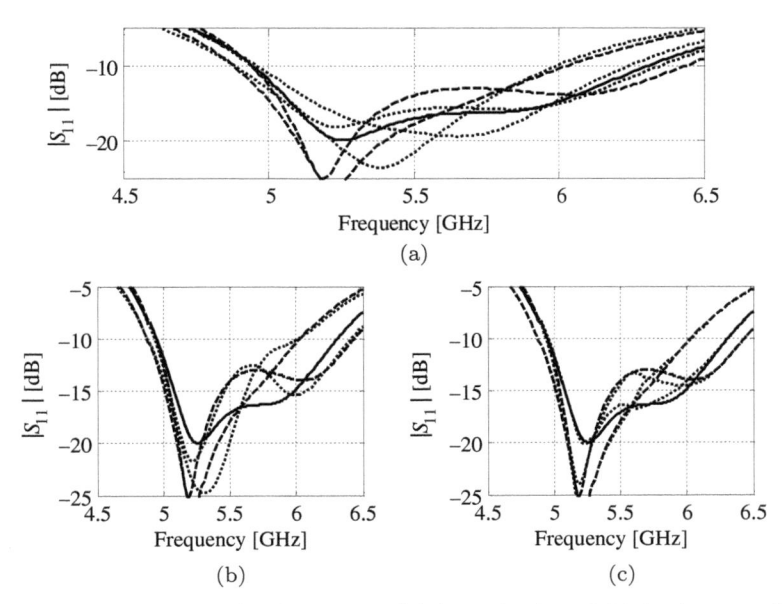

Fig. 5.7. Suspended DRA of Fig. 5.6: (a) low- ($\cdots$) and high- (- - - and —) fidelity model responses at three designs; (b) OSM-corrected low- ($\cdots$) and high- (- - - and —) fidelity model responses at the same designs (OSM correction at the design marked —); (c) multi-point-corrected low- ($\cdots$) and high- (- - - and —) fidelity model responses.

40 seconds). Figure 5.7(a) shows the responses of the low- and high-fidelity DRA model at several designs. As illustrated in Fig. 5.7(b), conventional one-point response correction perfectly aligns **c** and **f** at the design where it is established, but the alignment is not as good for other designs. On the other hand, as shown in Fig. 5.7(c), the multi-point response correction improves model alignment at all designs involved in the model construction (note that Fig. 5.7(c) only shows the global part $\mathbf{\Lambda}^{(i)} \circ \mathbf{c}(\mathbf{x}) + \mathbf{\Delta}_r^{(i)}$ without $\boldsymbol{\delta}^{(i)}$ which would give a perfect alignment at $\mathbf{x}^{(i)}$).

Another approach to low-fidelity model correction is to apply the correction at the level of the model domain. Perhaps the most popular example of such a procedure is input space mapping (ISM) (Bandler et al., 2004a), where the surrogate is created as

$$\mathbf{s}^{(i)}(\mathbf{x}) = \mathbf{c}(\mathbf{x} + \mathbf{q}^{(i)}), \qquad (5.18)$$

where the low-fidelity model and the surrogate are vector-valued.

The model parameters $\mathbf{q}^{(i)}$ are obtained by minimizing the misalignment between the surrogate and the high-fidelity model, $\|\mathbf{f}(\mathbf{x}^{(i)}) - \mathbf{c}(\mathbf{x}^{(i)} + \mathbf{q}^{(i)})\|$; $\mathbf{x}^{(i)}$ is a reference design (e.g., the most recent design encountered during the optimization run) at which the surrogate model is established. The advantages of input space mapping are demonstrated in Fig. 5.8 for microwave filter design (Hong and Lancaster, 2001). The high-fidelity model is evaluated using EM simulation, whereas the low-fidelity model is an equivalent circuit. The response of interest is the reflection coefficient $|S_{11}|$ as a function of frequency. For this particular example, ISM offers both good approximation and generalization capability (cf. Fig. 5.8(d)).

An alternative way of low-fidelity model correction is to exploit parameters that are normally fixed in the high-fidelity model such as microstrip substrate height and/or dielectric permittivity (Koziel et al., 2008a; Bandler et al., 2004b). Because the surrogate model is just an auxiliary tool and it is not supposed to be built or measured, such parameters can be freely adjusted in the low-fidelity model to improve its alignment with the high-fidelity one. Implicit space mapping (Bandler et al., 2003, 2004b; Koziel et al., 2010b) is a

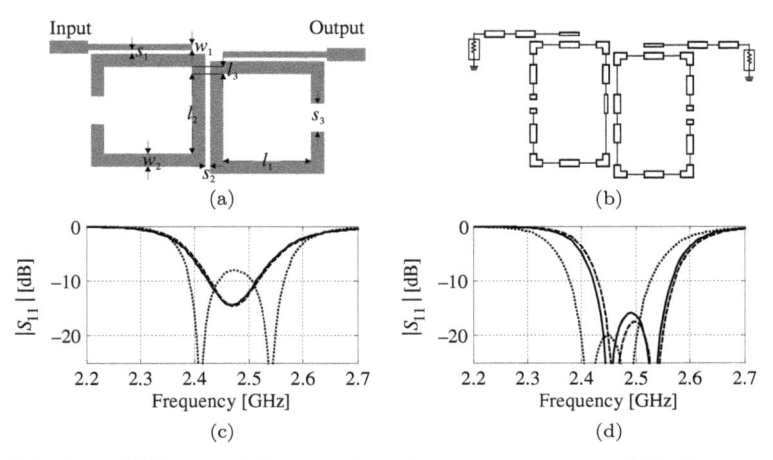

Fig. 5.8. Low-fidelity model correction through parameter shift (input space mapping): (a) microstrip filter geometry (high-fidelity model **f** evaluated using EM simulation); (b) low-fidelity model **c** (equivalent circuit); (c) response of **f** (—) and **c** (· · ·), as well as response of the surrogate model **s** (– –) created using input space mapping; (d) surrogate model verification at a different design (other than that at which the model was created) **f** (—), **c** (· · ·), and **s** (– –). Good alignment indicates excellent generalization of the model.

technique that utilizes this concept. More specifically, the surrogate is obtained as

$$\mathbf{s}^{(i)}(\mathbf{x}) = \mathbf{c}_I(\mathbf{x}, \mathbf{p}^{(i)}). \tag{5.19}$$

The vector **p** in (5.19) denotes a set of implicit space mapping (preassigned) parameters, and $\mathbf{c}_I$ is the low-fidelity model with the explicit dependence on these parameters. When the surrogate is established at the current design $\mathbf{x}^{(i)}$, the vector $\mathbf{p}^{(i)}$ is typically obtained by minimizing the norm-wise discrepancy between the models, $\|\mathbf{f}(\mathbf{x}^{(i)}) - \mathbf{c}_I(\mathbf{x}^{(i)}, \mathbf{p})\|$. Illustration of implicit space mapping for the filter of Figs. 5.8(a) and 5.8(b) is shown in Fig. 5.9. Here, the implicit space mapping parameters are substrate thicknesses of the microstrip-line components (rectangle elements in Fig. 5.8(b)).

Another type of low-fidelity model enhancement is derived from observation of the vector-valued responses of the system which are evaluations of the same design but at different values of certain parameters such as the time, frequency or a specific geometry

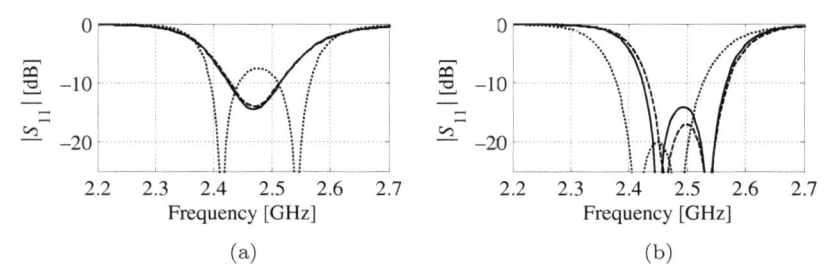

108 *Simulation-Based Optimization of Antenna Arrays*

Fig. 5.9. Low-fidelity model correction through implicit space mapping applied to a microstrip filter of Fig. 5.8: (a) response of **f** (—) and **c** (· · ·), as well as response of the surrogate model **s** (- -) created using implicit space mapping; (b) surrogate model verification at a different design (other than that at which the model was created) **f** (—), **c** (· · ·), and **s** (- -).

parameter. Subsequently, and based on such observations, the low-fidelity model is enhanced by applying a linear or nonlinear scaling to these parameters.

A representative example of such a correction method is frequency scaling. It is popular in electrical engineering where the figures of interest are often responses over frequency, e.g., S-parameters or gain (Koziel et al., 2006; Koziel and Ogurtsov, 2014a).

Let us assume that $\mathbf{f}(\mathbf{x}) = [f(\mathbf{x}, \omega_1) \ f(\mathbf{x}, \omega_2) \ \ldots \ f(\mathbf{x}, \omega_m)]^T$ where $f(\mathbf{x}, \omega_k)$ is the evaluation of the high-fidelity model at a frequency ω_k. Similarly, we have $\mathbf{c}(\mathbf{x}) = [c(\mathbf{x}, \omega_1) \ c(\mathbf{x}, \omega_2) \ \ldots \ c(\mathbf{x}, \omega_m)]^T$. The frequency-scaled surrogate model $\mathbf{s}_F(\mathbf{x})$ is then defined as

$$\mathbf{s}_F(\mathbf{x}, [F_0 \ F_1]) = [c(\mathbf{x}, F_0 + F_1\omega_1) \ldots c(\mathbf{x}, F_0 + F_1\omega_m)]^T. \quad (5.20)$$

Here, F_0 and F_1 are scaling parameters obtained as $[F_0, F_1] = \arg\min\{[F_0, F_1] : \|\mathbf{f}(\mathbf{x}^{(j)}) - \mathbf{s}_F(\mathbf{x}^{(j)}, [F_0 F_1])\|\}$. Figure 5.10 shows an example of frequency scaling applied to the low-fidelity model of a substrate-integrated cavity antenna (Bekasiewicz and Koziel, 2016a). In this example, both the low- and high-fidelity models are evaluated using coarse- and fine-discretization EM simulations.

Interested reader can find more extensive discussion on physics-based surrogates in the literature, e.g., Koziel and Ogurtsov (2014a), Bandler et al. (2004a), Koziel et al. (2013d) and Leifsson and Koziel (2015a).

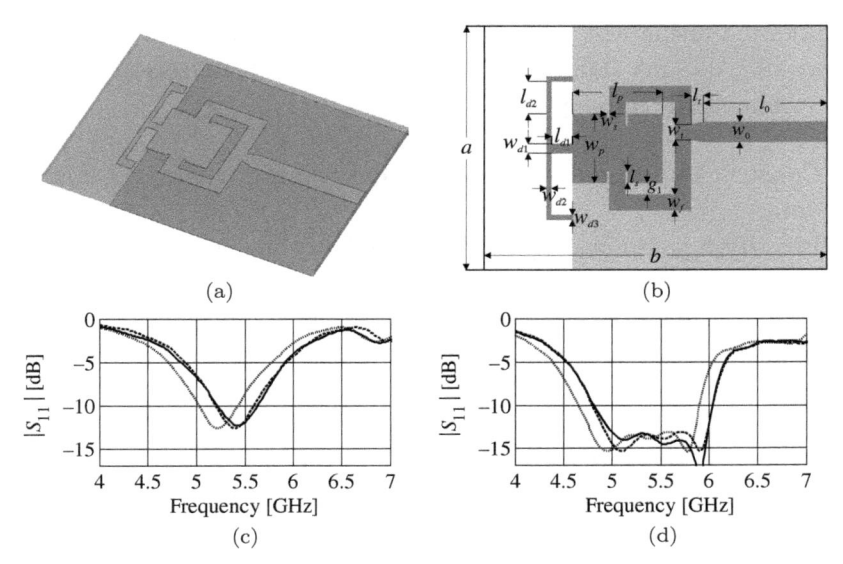

Fig. 5.10. Low-fidelity model correction with frequency scaling: (a) antenna view; (b) antenna geometry (both **f** and **c** evaluated using EM simulation, coarse discretization used for **c**). Response of **f** (—) and **c** (···), as well as response of the surrogate model **s** (- - -) created using frequency scaling at: (c) a certain reference design; and (d) another (test) design. Note that the surrogate properly accounts for the frequency shifts between **c** and **f**.

5.4. Optimization with Data-Driven Surrogates

In this section, we briefly outline the concepts of optimization exploiting data-driven surrogate models. In particular, we discuss a response surface methodology as well as global optimization using kriging surrogates. Furthermore, we highlight a practically important issue of balancing design space exploration and exploitation.

5.4.1. *Optimization by means of response surfaces*

Optimization involving data-driven surrogates, sometimes referred to as response surface approximations (RSAs), consists of the two major stages: (i) construction of the initial surrogate model and (ii) iterative prediction–correction scheme. In the iterative prediction–correction scheme, the surrogate is used to identify the promising locations in the design space; then the surrogate is updated using the high-fidelity model data evaluated at these promising designs (Queipo et al., 2005). The optimization flow can be summarized as follows:

1. Construct an initial surrogate model using pre-sampled training data;
2. Identify a candidate design by optimizing the surrogate model;
3. Evaluate the high-fidelity model at the candidate design;
4. Update the surrogate by incorporating new high-fidelity data;
5. If the termination condition is not satisfied, go to 2.

See also Fig. 5.11 for an illustration of this concept.

It should be emphasized that generating the new sample points by optimizing the surrogate is only one of possible infill criteria. Other criteria may aim at improving global accuracy of the surrogate (see Sec. 5.4.3 for more details). In case of local optimization, the optimization algorithm can be embedded in the trust-region-like framework (Forrester and Keane, 2009) for improved convergence.

A simple yet quite common variation of the above scheme is when the surrogate model is constructed only once and optimized without further refinement. This so-called one-shot approach normally requires considerable amount of training samples to ensure sufficient accuracy, and it is preferred for constructing multiple-use models (Kabir et al., 2008).

5.4.2. *Sequential approximate optimization*

One of the simplest realizations of surrogate-assisted optimization involving data-driven surrogates is sequential approximate optimization (SAO) (Jacobs et al., 2004; Giunta and Eldred, 2000; Pérez et al., 2002; Hosder et al., 2001; Roux et al., 1998; Sobieszczanski-Sobieski and Haftka, 1997; Giunta, 1997). The basic idea of SAO is to restrict the search to a small region of the design and utilize a local approximation model (e.g., low-order polynomial) as a prediction tool. Upon prediction, the search region is modified — upon optimizing the surrogate — using a chosen relocation strategy.

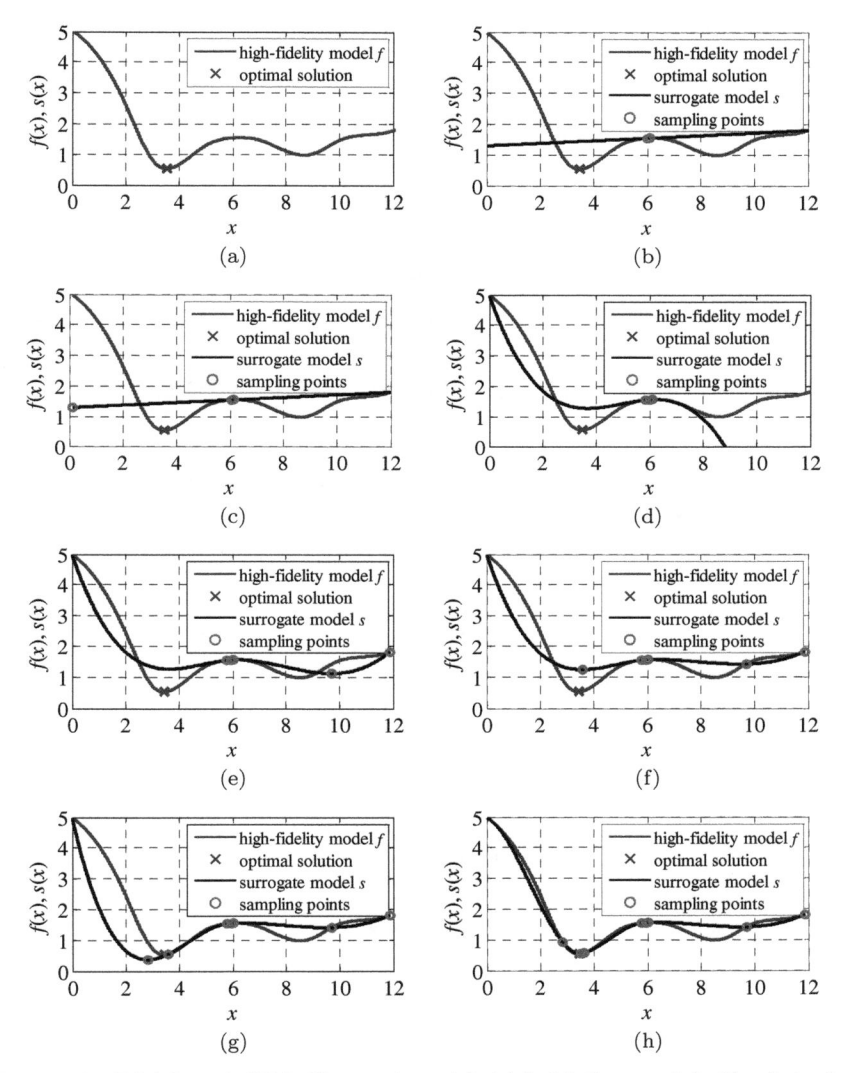

Fundamentals of Surrogate-Based Modeling and Optimization   111

Fig. 5.11. RSA-based SBO illustration: (a) high-fidelity model $f(\cdots)$ to be minimized and the minimum of $f(\times)$; (b) initial surrogate model (here, cubic splines) constructed using two data points; (c) a new data point obtained by optimizing the surrogate (evaluation of f at this point is used to update the surrogate); (d)–(h) after a few iterations, the surrogate model optimum adequately predicts the high-fidelity minimizer; global accuracy of the surrogate is not of primary concern.

The SAO flow is as follows (cf. Fig. 5.12):

1. Set $k = 0$ (iteration count);
2. Evaluate the high-fidelity model f at the design $\mathbf{x}^{(k)}$;
3. Construct a search region $r^{(k)}$ based on a relocation strategy;
4. Sample $r^{(k)}$ using a selected design-of-experiment plan;
5. Acquire high-fidelity data at the designs selected in Step 4;
6. Construct the surrogate $s^{(k)}$ using the data set of Step 5;
7. Find a new design $\mathbf{x}^{(k+1)}$ by optimizing the surrogate $s^{(k)}$ within $r^{(k)}$;
8. Terminate if the stopping criteria are met; otherwise set $k = k+1$ and go to 2.

The number of training samples utilized at each iteration is small because of using simple surrogates. This is an advantage of SAO. On the other hand, only a local search is possible with SAO.

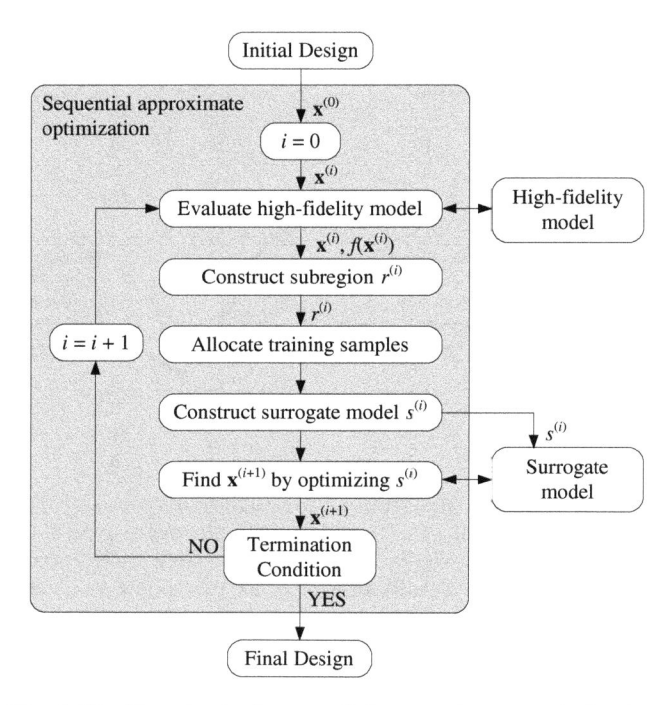

Fig. 5.12. Flowchart of sequential approximate optimization.

An important part of the SAO process is the relocation strategy, that is, the adjustment of size and location of the new region before executing the next iteration (Step 3). A typical strategy is to allocate the center of the new region at the most current optimum (Jacobs et al., 2004). In multi-point strategies, the last iteration optimum may become the corner point for the next search subregion (Toropov et al., 1993). Adjustment of the subregion size may involve trust-region-like frameworks (Giunta and Eldred, 2000; Koziel and Ogurtsov, 2012b).

5.4.3. *SBO with kriging surrogates: Exploration versus exploitation*

Choosing appropriate infill criteria is an important aspect of surrogate-assisted optimization (Forrester and Keane, 2009). The two main goals for selecting the infill points are: (i) reduction of the objective function value, and (ii) improvement of the global accuracy of the surrogate (Couckuyt, 2013).

The simplest infill strategy, suitable for local search, is to allocate a single sample at the surrogate model optimum. This strategy is capable of converging at least to a local minimum of the high-fidelity model assuming that the optimization algorithm is embedded in the trust-region framework (Forrester and Keane, 2009) and the surrogate model of the first-order consistency (Alexandrov et al., 1998). In more general terms, allocation of new training points may be oriented towards global search or constructing a globally accurate surrogate. Clearly, the latter requires a feedback regarding the prediction power of the surrogate model and its generalization capability. In this context, kriging seems to be the most advantageous type of data-driven surrogate because it provides information about the expected model error (Kleijnen, 2008; Jones et al., 1998; Gorissen et al., 2010a).

The following infill criteria based on this feature of kriging models are commonly used:

1. Maximization of the expected improvement, that is, the improvement one expects at an untried point $\mathbf{x}$ (Jones et al., 1998);

2. Minimization of the predicted objective function $\hat{y}(\mathbf{x})$. A reasonable global accuracy of the surrogate has to be assumed (Liu et al., 2012);
3. Minimization of the statistical lower bound, i.e., $LB(\mathbf{x}) = \hat{y}(\mathbf{x}) - As(\mathbf{x})$ (Forrester and Keane, 2009), where $\hat{y}(\mathbf{x})$ is the surrogate model prediction and $s^2(\mathbf{x})$ is the variance; A is a user-defined coefficient;
4. Maximization of the probability of improvement, i.e., identifying locations that give the highest chance of improving the objective function value (Forrester and Keane, 2009);
5. Maximization of the mean square error, i.e., finding locations where the mean square error predicted by the surrogate is the highest (Liu et al., 2012).

It should be emphasized that identifying the new samples according to the above infill criteria requires global optimization, typically realized using population-based metaheuristic algorithms (Couckuyt, 2013). Some comments on constrained SBO processes can be found in Forrester and Keane (2009), Queipo et al. (2005) and Liu et al. (2012).

The aforementioned methods are also referred to as adaptive sampling techniques or sequential sampling. They may either aim at constructing globally accurate surrogates or carrying out global optimization. Putting more focus on design space exploitation usually leads to reduced computational costs. Design space exploration normally results in higher costs and global search capability (Forrester and Keane, 2009). On the other hand, global exploration is often impractical, especially for expensive functions with a medium/large number of optimization variables (more than a few tens). In the optimization context, there should be a balance between exploitation and exploration. Minimization of the statistical lower bound is an example of achieving such a balance controlled by the constant A (from pure exploitation, i.e., $LB(\mathbf{x}) \to \hat{y}(\mathbf{x})$, for $A \to 0$ to pure exploration for $A \to \infty$). However, choosing a good value of A is a non-trivial task (Forrester and Keane, 2009).

Surrogate-assisted optimization methods with adaptive sampling are often referred to as efficient global optimization (EGO) techniques. A detailed discussion of EGO is given in Jones *et al.* (1998) and Forrester and Keane (2009). Kriging-based SBO can also be combined with pattern search methods, e.g., the surrogate management framework (SMF) (Booker *et al.*, 1999). Several EGO-like methods, e.g., the SADEA framework enhanced by memetic operations such as solution pre-screening, have been proposed recently (Liu *et al.*, 2014a).

5.4.4. *Summary*

Surrogate-assisted algorithms utilizing data-driven models work by iterative exploration of the design space so as to improve the model accuracy while searching for the optimum design. One of important problems for this class of methods is finding a balance between exploration of the design space and exploitation of the promising regions. In a majority of antenna-related design problems, the computational cost of high-fidelity model evaluation is rather high; thus, the cost of training data acquisition is typically high. Therefore, approximation-based surrogates may be appropriate for problems with small or medium numbers of designable parameters. For more complex problems, the computational cost of creating the approximation-based surrogate may be prohibitive.

5.5. SBO Using Physics-Based Surrogates

Good generalization capability belongs to the most attractive features of physics-based surrogates. It is a result of the knowledge about the system of interest embedded in the underlying low-fidelity models. As a consequence, a reliable physics-based surrogate can be constructed with a significantly smaller number of high-fidelity training points than approximation-based models (Koziel *et al.*, 2011a). In this section, we outline several selected physics-based SBO methods, including some variations of space mapping (Bandler *et al.*, 2004a; Koziel *et al.*, 2006, 2008a), approximation model management optimization (AMMO, Alexandrov and Lewis, 2001), manifold mapping (Echeverría and Hemker, 2005), shape-preserving response prediction (Koziel, 2010a), adaptively adjusted design specifications

116 *Simulation-Based Optimization of Antenna Arrays*

(Koziel, 2010b), as well as feature-based optimization (Koziel, 2015; Koziel and Bandler, 2015).

5.5.1. *Space mapping*

Space mapping (SM) (Bandler *et al.*, 1994, 2004a; Koziel *et al.*, 2006) refers to a family of design optimization techniques originally developed for solving expensive problems in computational electromagnetics, in particular, microwave engineering. Nowadays, the popularity of SM has spread across many engineering disciplines (Redhe and Nilsson, 2004; Bandler *et al.*, 2004a; Priess *et al.*, 2011; Marheineke *et al.*, 2012; Koziel and Leifsson, 2012a; Tu *et al.*, 2013).

In its first versions, SM was exclusively based on transformation of the low-fidelity model domain (aggressive SM, input SM; Bandler *et al.*, 1994, 1995), which was sufficient for handling many engineering problems, especially in electrical engineering, where the ranges of both the low- and high-fidelity model responses are similar. Other SM variations were developed to handle situations where the low- and high-fidelity models are severely misaligned in terms of the response levels (output SM; Bandler *et al.*, 2003, 2004b; Koziel *et al.*, 2006, 2008a, 2014c; Robinson *et al.*, 2008).

Original SM assumes existence of a mapping $\mathbf{P}$ between the high- and low-fidelity model domains (Bandler *et al.*, 1994), such that $\mathbf{x}_c = \mathbf{P}(\mathbf{x}_f)$ and $\mathbf{c}(\mathbf{P}(\mathbf{x}_f)) \approx \mathbf{f}(\mathbf{x}_f)$. Given $\mathbf{P}$, the direct solution of the original problem (3.1) can be replaced by finding $\mathbf{x}_f^{\#} = \mathbf{P}^{-1}(\mathbf{x}_c^*)$. Here, $\mathbf{x}_c^*$ is the optimal design of $\mathbf{c}$ defined as $\mathbf{x}_c^* = \arg\min\{\mathbf{x}_c : U(\mathbf{c}(\mathbf{x}_c))\}$; $\mathbf{x}_f^{\#}$ can be considered as a reasonable estimate of $\mathbf{x}_f^*$. In other words, the problem (3.1) can be reformulated as

$$\mathbf{x}_f^{\#} = \arg\min_{\mathbf{x}_f} U(\mathbf{c}(\mathbf{P}(\mathbf{x}_f))), \qquad (5.21)$$

where $\mathbf{c}(\mathbf{P}(\mathbf{x}_f))$ is a surrogate model. However, $\mathbf{P}$ is not given explicitly: it can only be evaluated at any $\mathbf{x}_f$ by means of the parameter extraction (PE) procedure, $\mathbf{P}(\mathbf{x}_f) = \arg\min\{\mathbf{x}_c : \|\mathbf{f}(\mathbf{x}_f) - \mathbf{c}(\mathbf{x}_c)\|$. One of practical issues is possible non-uniqueness of the solution to

(5.21) (Bandler et al., 1995). Another issue is the assumption on the similarity of high- and low-fidelity model ranges, which is a very strong assumption (Alexandrov and Lewis, 2001). These and other issues led to numerous improvements, including parametric SM which is outlined later in this section.

Aggressive SM (ASM) (Bandler et al., 1995) is one of the first versions of SM, and it is still popular in microwave engineering (Sans et al., 2014, Rayas-Sanchez, 2016). Assuming uniqueness of $\mathbf{x}_c^*$, the solution to (5.21) is equivalent to reducing the residual vector $\mathbf{f} = \mathbf{f}(\mathbf{x}_f) = \mathbf{P}(\mathbf{x}_f) - \mathbf{x}_c^*$ to zero. The first step of the ASM algorithm is to find $\mathbf{x}_c^*$. Next, ASM iteratively solves the nonlinear system $\mathbf{f}(\mathbf{x}_f) = 0$ for $\mathbf{x}_f$. At the jth iteration, the calculation of the error vector $\mathbf{f}^{(j)}$ requires an evaluation of $\mathbf{P}^{(j)}(\mathbf{x}_f^{(j)}) = \mathbf{P}(\mathbf{x}_f^{(j)}) = \arg\min\{\mathbf{x}_c : \mathbf{f}(\mathbf{x}_f^{(j)}) - \mathbf{c}(\mathbf{x}_c)\|\}$. The quasi-Newton step in the high-fidelity model space is given by

$$\mathbf{B}^{(j)}\mathbf{h}^{(j)} = -\mathbf{f}^{(j)}, \qquad (5.22)$$

where $\mathbf{B}^{(j)}$ is the approximation of the space mapping Jacobian $\mathbf{J}_P = \mathbf{J}_P(\mathbf{x}_f) = [\partial \mathbf{P}^T/\partial \mathbf{x}_f]^T = [\partial(\mathbf{x}_c^T)/\partial \mathbf{x}_f]^T$. Solving (5.22) for $\mathbf{h}^{(j)}$ gives the next iterate $\mathbf{x}_f^{(j+1)} = \mathbf{x}_f^{(j)} + \mathbf{h}^{(j)}$. The algorithm terminates if $\|\mathbf{f}^{(j)}\|$ is sufficiently small. The output of the algorithm is an approximation to $\mathbf{x}_f^{\#} = \mathbf{P}^{-1}(\mathbf{x}_c^*)$. A popular way of obtaining the matrix $\mathbf{B}$ is through a rank one Broyden update (Broyden, 1965) of the form

$$\mathbf{B}^{(j+1)} = \mathbf{B}^{(j)} + \frac{[\mathbf{f}^{(j+1)} - \mathbf{f}^{(j)} - \mathbf{B}^{(j)}\mathbf{h}^{(j)}]\mathbf{h}^{(j)T}}{\|\mathbf{h}^{(j)}\|^2}. \qquad (5.23)$$

Several improvements of the ASM algorithm have been proposed (Bakr et al., 1998, 1999). ASM has been the most popular SM approach in microwave engineering till now (Rayas-Sanchez, 2016).

Parametric SM is another type of space mapping. It is more generic and characterized by explicit analytical form of the surrogate model. In parametric SM, the optimization algorithm is an iterative process (5.1). The two simple examples of input and implicit SM were shown in Sec. 5.3. In general, the input SM surrogate model can take the form (Koziel et al., 2006)

$$\mathbf{s}^{(i)}(\mathbf{x}) = \mathbf{c}(\mathbf{B}^{(i)} \cdot \mathbf{x} + \mathbf{q}^{(i)}). \qquad (5.24)$$

Here, $\mathbf{B}^{(i)}$ and $\mathbf{q}^{(i)}$ are matrices obtained by minimizing misalignment between the surrogate and the high-fidelity model as

$$[\mathbf{B}^{(i)}, \mathbf{q}^{(i)}] = \arg\min_{[\mathbf{B},\mathbf{q}]} \sum_{k=0}^{i} w_{i.k} \|\mathbf{f}(\mathbf{x}^{(k)}) - \mathbf{c}(\mathbf{B} \cdot \mathbf{x}^{(k)} + \mathbf{q})\|. \qquad (5.25)$$

Equation (5.21) is, in fact, a nonlinear regression problem with $w_{i.k}$ being the weighting factors. A common choice of $w_{i.k}$ is $w_{i.k} = 1$ for all i and all k (all previous designs contribute equally) or $w_{i.1} = 1$ and $w_{i.k} = 0$ for $k < i$ (the surrogate depends on the most recent design only).

In general, the SM surrogate model is constructed as follows:

$$\mathbf{s}^{(i)}(\mathbf{x}) = \bar{\mathbf{s}}(\mathbf{x}, \mathbf{p}^{(i)}), \qquad (5.26)$$

where $\bar{\mathbf{s}}$ is a generic SM surrogate model, i.e., the low-fidelity model $\mathbf{c}$ composed with suitable (usually linear) transformations. The parameters $\mathbf{p}$ are obtained in the extraction process similar to (5.25). More information about specific SM surrogates can be found in the literature Bandler et al. (2004a) and Koziel et al. (2006).

Despite its simplicity and demonstrated efficiency, there are some practical issues that are still preventing SM from being widely accepted by industry designers. These issues include (i) finding an appropriate balance between the quality and the speed of the underlying low-fidelity model as well as (ii) selection of appropriate SM transformations (Koziel and Bandler, 2007a,b). Resolving the above listed issues is not a trivial task (Koziel et al., 2008b). As a matter of fact, similar issues are actually common to the majority of physics-based SBO methods. On the other hand, SM has been found to be a very efficient tool for yielding satisfactory designs in various engineering disciplines (Redhe and Nilsson, 2004; Bandler et al., 2004a; Marheineke et al., 2012). A number of enhancements of SM algorithms have been suggested to alleviate some of the difficulties such as potential convergence problems (Koziel et al., 2010a,c).

5.5.2. *Approximation model management optimization*

Approximation model management optimization (AMMO) (Alexandrov and Lewis, 2001) is a simple algorithmic framework with the surrogate model constructed through response correction. It exploits sensitivity data to ensure first-order consistency conditions, in particular, $s^{(i)}(\mathbf{x}^{(i)}) = f(\mathbf{x}^{(i)})$, and $\nabla s^{(i)}(\mathbf{x}^{(i)}) = \nabla f(\mathbf{x}^{(i)})$. Additionally, AMMO utilizes the trust-region methodology (Conn et al., 2000) to guarantee convergence of the optimization process to the high-fidelity model optimum. Assuming that $\beta(\mathbf{x}) = f(\mathbf{x})/c(\mathbf{x})$ is the correction function and

$$\beta_i(\mathbf{x}) = \beta(\mathbf{x}^{(i)}) + \nabla\beta(\mathbf{x}^{(i)})^T(\mathbf{x} - \mathbf{x}^{(i)}), \tag{5.27}$$

the surrogate model is defined as

$$s^{(i)}(\mathbf{x}) = \beta_i(\mathbf{x})c(\mathbf{x}). \tag{5.28}$$

Equation (5.28) can be shown to satisfy the aforementioned consistency conditions. Obviously, (5.28) requires both the low- and high-fidelity derivatives.

5.5.3. *Manifold mapping*

Manifold mapping (MM) can be considered as a variation of output space mapping. It is essentially the response correction technique (Echeverría and Hemker, 2005, 2008) with the surrogate model set up using multiple high-fidelity data points as follows:

$$\mathbf{s}^{(i)}(\mathbf{x}) = \mathbf{f}(\mathbf{x}^{(i)}) + \mathbf{S}^{(i)}(\mathbf{c}(\mathbf{x}) - \mathbf{c}(\mathbf{x}^{(i)})). \tag{5.29}$$

Here, $\mathbf{S}^{(i)} = \mathbf{\Delta F} \cdot \mathbf{\Delta C}^\dagger$ is the $m \times m$ correction matrix, where

$$\mathbf{\Delta F} = [\mathbf{f}(\mathbf{x}^{(i)}) - \mathbf{f}(\mathbf{x}^{(i-1)}) \ldots \mathbf{f}(\mathbf{x}^{(i)}) - \mathbf{f}(\mathbf{x}^{(\max\{i-n,0\})})], \tag{5.30}$$

$$\mathbf{\Delta C} = [\mathbf{c}(\mathbf{x}^{(i)}) - \mathbf{c}(\mathbf{x}^{(i-1)}) \ldots \mathbf{c}(\mathbf{x}^{(i)}) - \mathbf{c}(\mathbf{x}^{(\max\{i-n,0\})})]. \tag{5.31}$$

The pseudoinverse, denoted by †, is defined as $\mathbf{\Delta C}^\dagger = \mathbf{V}_{\mathbf{\Delta C}} \mathbf{\Sigma}^\dagger_{\mathbf{\Delta C}} \mathbf{U}^T_{\mathbf{\Delta C}}$, where $\mathbf{U}_{\mathbf{\Delta C}}$, $\mathbf{\Sigma}_{\mathbf{\Delta C}}$, and $\mathbf{V}_{\mathbf{\Delta C}}$ are the factors in the singular value decomposition of the matrix $\mathbf{\Delta C}$. The matrix $\mathbf{\Sigma}^\dagger_{\mathbf{\Delta C}}$ is the result of inverting the non-zero entries in $\mathbf{\Sigma}_{\mathbf{\Delta C}}$, leaving the zeroes invariant.

Upon convergence, the linear correction $\mathbf{S}^*$ (being the limit of $\mathbf{S}^{(i)}$ with $i \to \infty$) maps the point $\mathbf{c}(\mathbf{x}^*)$ to $\mathbf{f}(\mathbf{x}^*)$, and the tangent plane for $\mathbf{c}(\mathbf{x})$ at $\mathbf{c}(\mathbf{x}^*)$ to the tangent plane for $\mathbf{f}(\mathbf{x})$ at $\mathbf{f}(\mathbf{x}^*)$ (Echeverría and Hemker, 2008). If the sensitivity data is available for both the low- and high-fidelity model, the correction matrix $\mathbf{S}^{(i)}$ can be defined using the exact Jacobians of the models.

5.5.4. Shape preserving response prediction

Shape-preserving response prediction (SPRP) is a parameterless approach (Koziel, 2010a, 2012) where the surrogate model is constructed assuming that the change of the high-fidelity model response due to the adjustment of the design variables can be predicted using the actual changes of the low-fidelity model response. It is critically important for SPRP that the low-fidelity model is physics-based, which ensures that the effect of the design parameter variations on the model response is similar for both models. The change of the low-fidelity model response is described by the translation vectors corresponding to a certain (finite) number of characteristic points of the model's response. These translation vectors are subsequently used to predict the change of the high-fidelity model response with the actual response of $\mathbf{f}$ at the current iteration point, $\mathbf{f}(\mathbf{x}^{(i)})$, treated as a reference.

Figure 5.13 explains the concept of SPRP using an example of a microstrip bandstop filter (Koziel, 2010c). Figure 5.13(a) shows the example of the low-fidelity model response $|S_{21}|$ (transmission coefficient) in the frequency range from 8 to 18 GHz, at the design $\mathbf{x}^{(i)}$, as well as the low-fidelity model response at some other design $\mathbf{x}$. Circles denote five characteristic points of $\mathbf{c}(\mathbf{x}^{(i)})$, here, selected to represent $|S_{21}| = -3$ dB, $|S_{21}| = -20$ dB, and the local $|S_{21}|$ maximum (at about 13 GHz). Squares denote corresponding characteristic points for $\mathbf{c}(\mathbf{x})$, while small line segments represent the translation vectors that determine the "shift" of the characteristic points of $\mathbf{c}$ when changing the design variables from $\mathbf{x}^{(i)}$ to $\mathbf{x}$. Because the low-fidelity model is physics-based, the high-fidelity model response at the given design, here, $\mathbf{x}$, can be predicted using the same translation vectors applied to the corresponding characteristic points of the fine model response at $\mathbf{x}^{(i)}$, $\mathbf{f}(\mathbf{x}^{(i)})$, cf. Fig. 5.13(b).

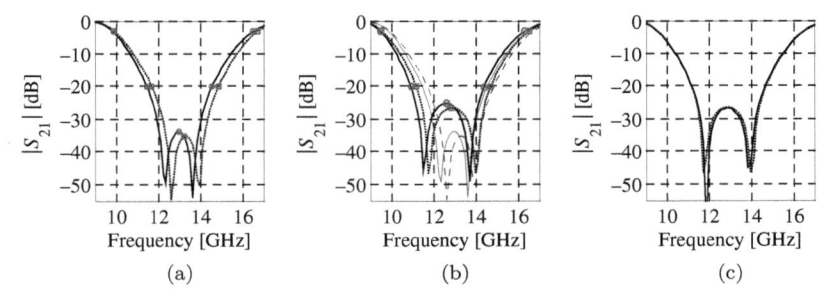

Fundamentals of Surrogate-Based Modeling and Optimization 121

Fig. 5.13. SPRP concept (Koziel, 2010a): (a) Example low-fidelity model response at the design $\mathbf{x}^{(i)}$, $\mathbf{c}(\mathbf{x}^{(i)})$ (—), and at another design $\mathbf{x}$, $\mathbf{c}(\mathbf{x})$ ($\cdots$), characteristic points of $\mathbf{c}(\mathbf{x}^{(i)})$ (o) and $\mathbf{c}(\mathbf{x})$($\square$), and the translation vectors (short lines); (b) High-fidelity model response at $\mathbf{x}^{(i)}$, $\mathbf{f}(\mathbf{x}^{(i)})$ (—) and the predicted fine model response at $\mathbf{x}$($\cdots$) obtained using SPRP based on characteristic points of Fig. 5.11(a); characteristic points of $\mathbf{f}(\mathbf{x}^{(i)})$ (o) and the translation vectors (short lines) were used to find the characteristic points ($\square$) of the predicted high-fidelity model response; low-fidelity model responses $\mathbf{c}(\mathbf{x}^{(i)})$ and $\mathbf{c}(\mathbf{x})$ are plotted using thin solid and dotted line, respectively; (c) predicted ($\cdots$) and actual (—) high-fidelity model response at $\mathbf{x}$.

As indicated in Fig. 5.13(c), the predicted high-fidelity model response at the design $\mathbf{x}$ is in a very good agreement with the actual response, $\mathbf{f}(\mathbf{x})$.

A rigorous formulation of SPRP is available in the literature Koziel (2010a,c). It should be mentioned that an important assumption of SPRP is that the shapes of the high- and low-fidelity model responses are similar in overall. This assumption means that the characteristic points of the coarse model $\mathbf{c}$ and the fine model $\mathbf{f}$ responses are in one-to-one correspondence. If this assumption is not satisfied, the SPRP surrogate cannot be evaluated because the translation vectors are not well defined. Generalizations of SPRP that relax the above listed assumption can be found in some cases (Koziel, 2010c). In the context of antenna optimization, handling reflection characteristics of narrow band and multi-band structures as well as array radiation pattern optimization (e.g., for sidelobe reduction) are examples of problems SPRP is well suited for.

5.5.5. *Adaptively adjusted design specifications*

The physics-based surrogate-assisted algorithms discussed so far in this chapter exploit the idea of low-fidelity model correction in order to improve low-fidelity model alignment with the high-fidelity one. Adaptively adjusted design specifications (AADS) (Koziel, 2010b) proposes an alternative way of exploiting the system-specific knowledge embedded in the low-fidelity model by modifying the design specifications so as to account for the model discrepancies. AADS is not universally applicable, but it is extremely simple to implement as no changes of the low-fidelity model are required. AADS consists of the two basic steps:

1. Modify the design specifications of the original problem to account for the differences between the responses of the high-fidelity model **f** and the low-fidelity model **c** at their characteristic points.
2. Obtain a new design by optimizing the low-fidelity model with respect to the modified specifications.

The way AADS is formulated allows for handling minimax-type of specifications (Koziel, 2010b) so that the characteristic points of the responses should correspond to the relevant design specification levels. The characteristic points may also include local maxima/minima of the respective responses at which the specifications may not be satisfied. Figure 5.14(a) shows the high- and low-fidelity model responses at the optimal design of **c**, corresponding to a bandstop filter example (Koziel, 2010d); design specifications are indicated using horizontal lines; here, $|S_{21}| \leq -30$ dB for 12–14 GHz, $|S_{21}| \geq -3$ dB for 8–9 GHz, and 17–18 GHz. Figure 5.14(b) shows the characteristic points of **f** and **c**, i.e., the points corresponding to -3 and -30 dB levels as well as to the local maxima of the responses. In the first step of the AADS optimization procedure, the design specifications are modified so that the level of satisfying/violating the modified specifications by the low-fidelity model response corresponds to the satisfaction/violation levels of the original specifications by the high-fidelity model response.

In the example of Fig. 5.14, for each edge of the specification line, the edge frequency is shifted by the difference of the frequencies of the corresponding characteristic points, e.g., the left edge of the -30 dB specification line is moved 0.7 GHz to the right. This shift

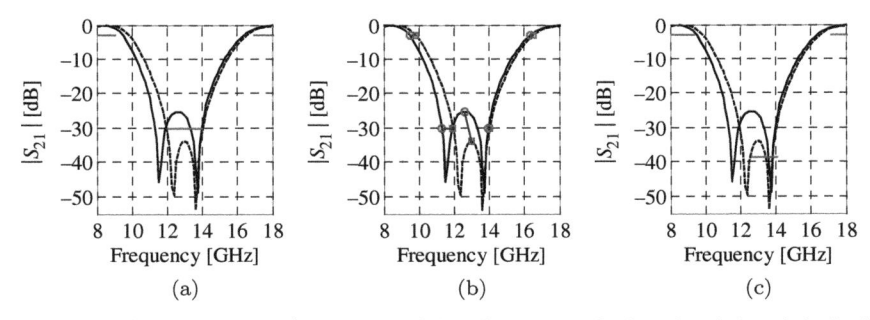

Fig. 5.14. AADS concept (responses of **f** and **c** are marked with solid and dashed line, respectively): (a) high- and low-fidelity model responses at the initial design (optimum of **c**) as well as the original design specifications, (b) characteristic points of the responses corresponding to the specification levels (here, -3 and $-30\,\text{dB}$) and to the local response maxima, (c) high- and low-fidelity model responses at the initial design and the modified design specifications.

is equal to the length of the line connecting the corresponding characteristic points in Fig. 5.14(b). Similarly, the specification levels are shifted by the difference between the local maxima/minima values for the respective points, e.g., the $-30\,\text{dB}$ level is shifted $8.5\,\text{dB}$ down because of the difference of the local maxima of the corresponding characteristic points of **f** and **c**. Modified design specifications are shown in Fig. 5.14(c).

Subsequently, the low-fidelity model is optimized with respect to the modified specifications and the new design obtained this way is treated as an approximated solution to the original design problem. Steps 1 and 2 can be iterated if necessary. If the correlation between the low- and high-fidelity models is good, a substantial design improvement is typically observed after the first iteration; however, additional iterations may bring further design improvements (Koziel, 2010b).

Figure 5.15 illustrates an AADS iteration applied to microstrip-to-SIW transition design (Ogurtsov and Koziel, 2011). Note that optimizing the low-fidelity model with respect to the modified specifications results in improving the high-fidelity model design with respect to the original specifications. Because the model discrepancies may change somehow from one design to another, a few iterations may be necessary to find an optimal high-fidelity design.

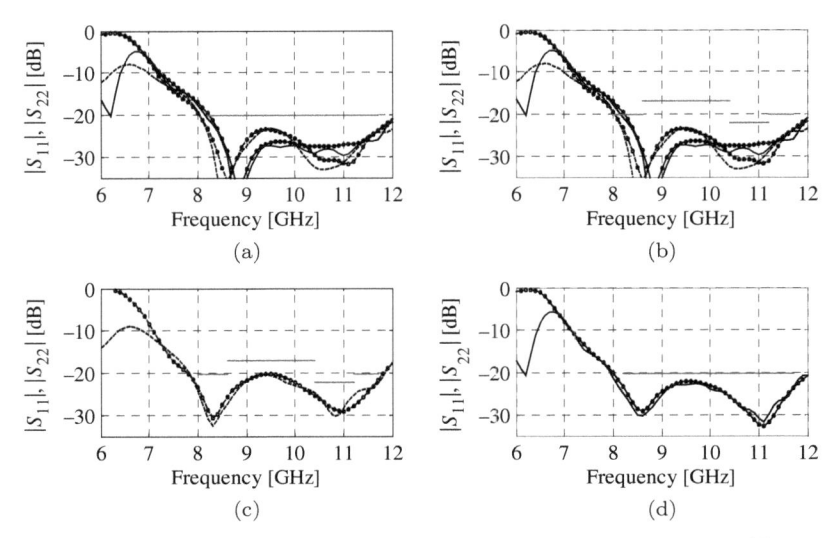

124 *Simulation-Based Optimization of Antenna Arrays*

Fig. 5.15. AADS for optimization of microstrip-to-SIW transition (Ogurtsov and Koziel, 2011): high- and low-fidelity model responses denoted as solid and dashed lines, respectively. $|S_{22}|$ distinguished from $|S_{11}|$ using circles. Design specifications denoted by thick horizontal lines: (a) model responses at the beginning of the iteration and original design specifications; (b) model responses and modified design specifications that reflect the differences between the responses; (c) low-fidelity model optimized with respect to the modified specifications; (d) high-fidelity model at the low-fidelity model optimum shown versus original specifications.

5.5.6. *Feature-based optimization*

The last physics-based approach to discuss in this chapter is feature-based optimization (FBO). FBO is one of the most recent techniques of this class (Glubokov and Koziel, 2014a,b; Koziel and Bekasiewicz, 2015b). Instead of handling complete (often highly nonlinear) responses, FBO processes only selected response features. Dependence of these selected features on the designable parameters is usually less nonlinear than that for the original responses; and lower nonlinearity of the features speeds up the optimization process.

A concept of response features is explained in Fig. 5.16 showing a family of reflection responses of an example antenna considered in Koziel (2015). The response is evaluated along a line segment in the design space, $\mathbf{x} = t\mathbf{x}_a + (1-t)\mathbf{x}_b$, $0 \leq t \leq 1$, where $\mathbf{x}_a$

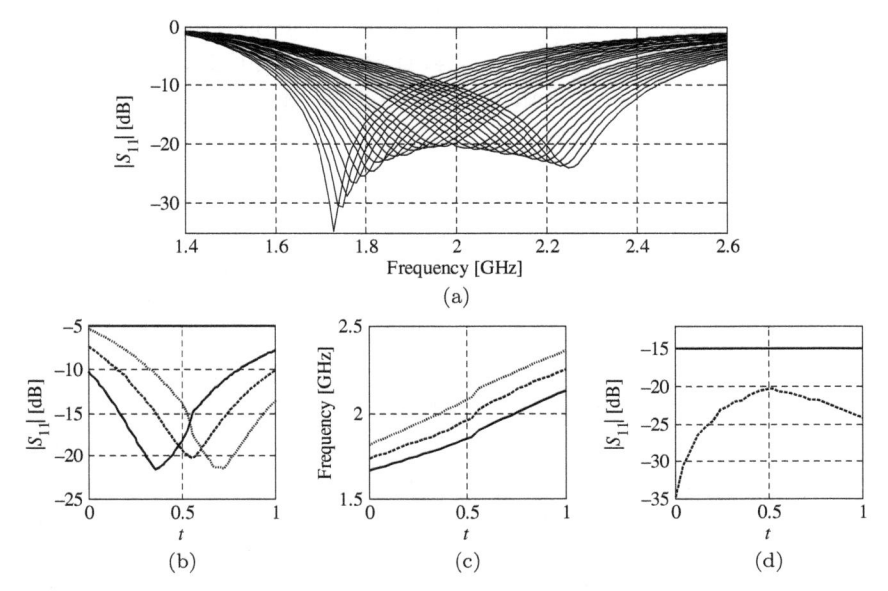

Fundamentals of Surrogate-Based Modeling and Optimization 125

Fig. 5.16. Response features: (a) family of reflection responses of the example antenna structure, evaluated along certain line segment in the design space, example features corresponding to −15 dB levels and center frequency; (b) response variability at selected frequencies, 1.9 GHz (—), 2.0 GHz (- - -), and 2.1 GHz (· · ·), indicating their high nonlinearity; (c) variability of the response features (frequency components), here, corresponding to −15 dB levels as well as the center frequency; (d) variability of the level components of the response features of (c).

and $\mathbf{x}_b$ are arbitrarily selected vectors. Figure 5.16(a) shows the plots of the corresponding features as a function of the parameter t. Figure 5.16(b) shows the reflection characteristic $|S_{11}|$ versus the line segment parameter t for various frequencies, here, 1.9, 2.0, and 2.1 GHz.

As indicated in the plots, the original responses are highly nonlinear whereas the behavior of the response features is much less nonlinear, and, consequently, easier to handle. This results in faster convergence of the optimization algorithm working at the level of features.

We use $f_k(\mathbf{x})$ and $l_k(\mathbf{x})$ to denote the frequency and the level of the kth feature, $k = 1, \ldots, p$, where p is the number of features

126 Simulation-Based Optimization of Antenna Arrays

for a given problem. The response features should be selected in accordance to given specifications so that the original formulation of the design problem (3.1) can be replaced by an equivalent problem formulated for the features

$$\mathbf{x}_f^* = \arg\min_{\mathbf{x}} U_F(\mathbf{f}_F(\mathbf{x})), \quad (5.32)$$

where $\mathbf{f}_F \in R^{2p}$ denotes the feature-based model whereas U_F is the objective function at the feature level. According to our notation, we have $\mathbf{f}_F(\mathbf{x}) = [f_1(\mathbf{x})\ f_2(\mathbf{x})\ \ldots\ f_p(\mathbf{x})\ l_1(\mathbf{x})\ l_2(\mathbf{x})\ \ldots\ l_p(\mathbf{x})]^T$.

Figure 5.17 shows a typical situation for a narrow-band antenna. In this specific example, the goal is to maximize the fractional bandwidth at the $-15\,\text{dB}$ level. For illustration, three response features are utilized corresponding to the $-15\,\text{dB}$ level and to the response minimum. It can be observed that one has a very good prediction of the feature location obtained using the feature model gradients, $\nabla f_k(\mathbf{x})$ and $\nabla l_k(\mathbf{x})$ (i.e., $f_k(\mathbf{x} + d\mathbf{x}) \approx f_k(\mathbf{x}) + \nabla f_k(\mathbf{x})d\mathbf{x}$, and $l_k(\mathbf{x} + d\mathbf{x}) \approx l_k(\mathbf{x}) + \nabla l_k(\mathbf{x})d\mathbf{x}$) for a specific search step size $d\mathbf{x}$, versus the actual feature location (verified by EM simulation). This

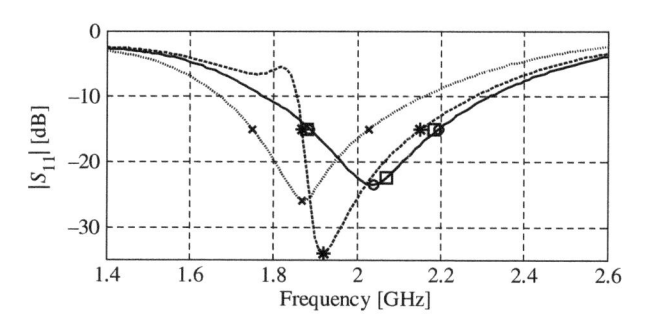

Fig. 5.17. Feature prediction using estimated feature gradients: high-fidelity model at the reference design $\mathbf{x}$ (........) and corresponding feature points (×), high-fidelity model at the perturbed design $\mathbf{x} + d\mathbf{x}$ (——) and the corresponding feature points (o), linear expansion model constructed at the frequency-based response level and evaluated at $\mathbf{x} + d\mathbf{x}$ (........) together with the corresponding feature points (∗), as well as the high-fidelity feature points predicted using feature gradients (□). Feature point prediction obtained using linear expansion (at the frequency-based response level) is inaccurate; feature prediction using feature gradients is much more reliable.

is not the case when the same search step is utilized for the first-order Taylor model constructed for the frequency-based response, i.e., $\mathbf{f}(\mathbf{x} + d\mathbf{x}) \approx \mathbf{f}(\mathbf{x}) + \mathbf{J_f}(\mathbf{x})d\mathbf{x}$, where $\mathbf{J_f}$ denotes the model Jacobian.

The FBO algorithm is formulated as an iterative process generating a series $\mathbf{x}^{(i)}$ of approximations to $\mathbf{x}^*$ (solution to the original problem (3.1)) as follows:

$$\mathbf{x}^{(i+1)} = \arg \min_{\mathbf{x}} U_F(\mathbf{l}^{(i)}(\mathbf{x})), \qquad (5.33)$$

where $\mathbf{l}^{(i)}$ is a linear model of the response features set up at the current design $\mathbf{x}^{(i)}$ using finite differentiation. The model $\mathbf{l}^{(i)}$ is defined as

$$\mathbf{l}^{(i)}(\mathbf{x}) = \begin{bmatrix} \begin{bmatrix} f_1(\mathbf{x}^{(i)}) \\ \vdots \\ f_p(\mathbf{x}^{(i)}) \end{bmatrix} + \begin{bmatrix} \frac{f_1(\mathbf{x}_{d.1}^{(i)}) - f_1(\mathbf{x}^{(i)})}{d_1^{(i)}} \\ \vdots \\ \frac{f_p(\mathbf{x}_{d.p}^{(i)}) - f_p(\mathbf{x}^{(i)})}{d_p^{(i)}} \end{bmatrix} \circ (\mathbf{x} - \mathbf{x}^{(i)}) \\ \begin{bmatrix} l_1(\mathbf{x}^{(i)}) \\ \vdots \\ l_p(\mathbf{x}^{(i)}) \end{bmatrix} + \begin{bmatrix} \frac{l_1(\mathbf{x}_{d.1}^{(i)}) - l_1(\mathbf{x}^{(i)})}{d_1^{(i)}} \\ \vdots \\ \frac{l_p(\mathbf{x}_{d.p}^{(i)}) - l_p(\mathbf{x}^{(i)})}{d_p^{(i)}} \end{bmatrix} \circ (\mathbf{x} - \mathbf{x}^{(i)}) \end{bmatrix}, \qquad (5.34)$$

where $\mathbf{x}^{(i)} = [x_1^{(i)} \ x_2^{(i)} \ \ldots \ x_n^{(i)}]^T$ is a current design, whereas $\mathbf{x}_{d.k}^{(i)} = [x_1^{(i)} \ \ldots \ x_k^{(i)} + d_k^{(i)} \ \ldots \ x_n^{(i)}]^T$ are the perturbed designs. Here, $\mathbf{d}^{(i)} = [d_1^{(i)} \ d_2^{(i)} \ \ldots \ d_n^{(i)}]^T$ is the perturbation size. The symbol $\circ$ denotes component-wise multiplication. Frequency and level coordinates of the feature points are modeled independently. To ensure convergence, the algorithm is embedded in the trust-region framework (Conn et al., 2000), i.e., we have

$$\mathbf{x}^{(i+1)} = \arg \min_{\|\mathbf{x} - \mathbf{x}^{(i)}\| \le \delta^{(i)}} U_F(\mathbf{l}^{(i)}(\mathbf{x})), \qquad (5.35)$$

where $\mathbf{l}^{(i)}$ is optimized within the trust region defined as $\|\mathbf{x} - \mathbf{x}^{(i)}\| \le \delta^{(i)}$; $\delta^{(i)}$ is the trust-region size updated using conventional rules (Conn et al., 2000). The new design $\mathbf{x}^{(i+1)}$ produced by (5.35) is only

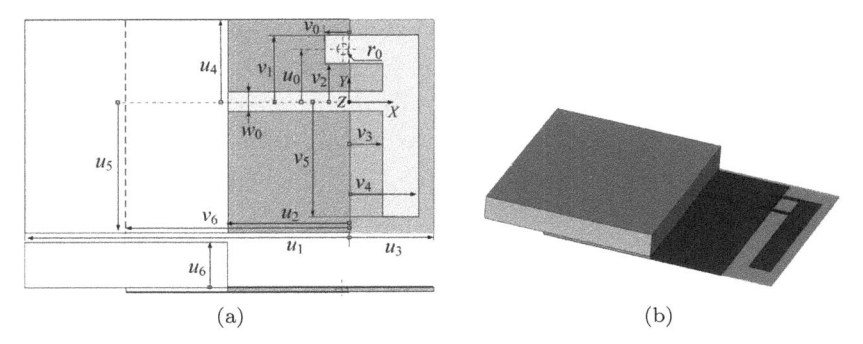

Fig. 5.18. PIFA (Koziel, 2015): (a) top and side views; (b) perspective view. Substrate are shown transparent.

accepted if it leads to the improvement of $\mathbf{f}_F(\mathbf{x})$, i.e., $\mathbf{f}_F(\mathbf{x}^{(i+1)}) < \mathbf{f}_F(\mathbf{x}^{(i)})$.

In order to speed up the optimization process, the gradients of the features can be estimated (cf. (5.34)) using the low-fidelity model evaluations rather than the high-fidelity ones. This is justified as long as the low- and high-fidelity models are sufficiently well correlated; then their absolute discrepancies are not critical (Koziel, 2015).

For the sake of illustration, consider the planar inverted-F antenna (PIFA) (Koziel, 2015) shown in Fig. 5.18. The design variables are $\mathbf{x} = [v_0 \ v_1 \ v_2 \ v_3 \ v_4 \ v_5 \ v_6]^T$. The initial design is $\mathbf{x}^{(0)} = [2 \ 8 \ 5 \ 2 \ 10 \ 25 \ 45]^T$ mm. Fixed parameters are $[u_0 \ u_1 \ u_2 \ u_3 \ u_4 \ u_5 \ u_6 \ u_7 \ w_0 \ r_0]^T = [6.15 \ 50.0 \ 15.0 \ 10.5 \ 29.35 \ 11.65 \ 5.0 \ 1.0 \ 0.5]^T$ mm; 0.508 mm substrate (Fig. 5.18(b) on the right) and the box (on the left) are of Rogers TMM4 and TMM6. The high-fidelity model $\mathbf{f}$ is evaluated in CST Microwave Studio ($\sim$650,000 mesh cells, simulation time 13 min). The low-fidelity model $\mathbf{c}$ is also evaluated in CST Microwave Studio ($\sim$22,000 mesh cells, simulation time 50 s).

The antenna was optimized using the FBO algorithm to obtain as wide fractional bandwidth as possible at $|S_{11}| = -15$ dB (symmetrically) around 2.0 GHz. The final design $\mathbf{x}^* = [2.028 \ 8.226 \ 6.587 \ 2.618 \ 9.134 \ 22.87 \ 45.24]^T$ mm was obtained in 6 iterations with the total cost corresponding to about 11 evaluations of the high-fidelity model ($8 \times \mathbf{f}$ and $50 \times \mathbf{c}$).

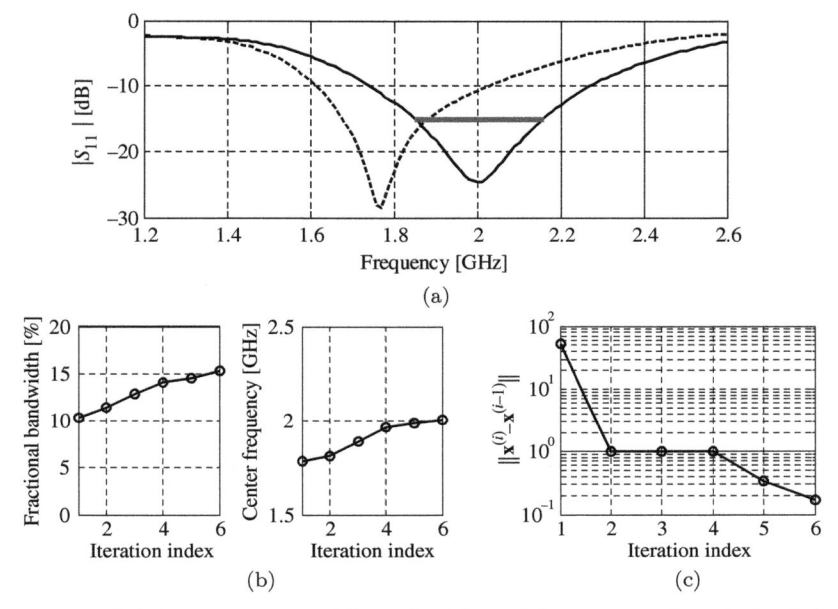

Fig. 5.19. PIFA optimization (Case 1): (a) initial (- - -) and optimized (—) response (fractional bandwidth around 2 GHz of 15 percent); (b) evolution of the fractional bandwidth and the center frequency vs. iteration index; (c) convergence plot. The thick horizontal line in (a) marks the −15 dB antenna bandwidth upon optimization.

Figure 5.19 shows the high-fidelity model responses at the initial and the final designs, as well as the evolution of the fractional bandwidth and the convergence plot. It should be noted that the presented approach is also efficient in terms of low-fidelity model evaluations, which is important when the time evaluation ratio between the **f** and **c** is low as in the present example.

For the sake of comparison the antenna was also optimized using the two benchmark methods: (i) space mapping (SM) (Koziel et al., 2008a), and (ii) the pattern search algorithm (Kolda et al., 2003). The SM algorithm utilized here exploits the low-fidelity model **c** as the underlying coarse model, and two types of model correction, specifically, frequency scaling (Koziel and Ogurtsov, 2014a) and additive response correction (Koziel and Leifsson, 2012b). The SM surrogate is reset at every iteration and re-optimized using pattern

Table 5.1. Planar inverted-F antenna — design optimization cost.

Algorithm	Algorithm component	Number of model evaluations	CPU cost Absolute	CPU cost Relative to $\mathbf{f}$
Direct $\mathbf{f}$ optimization (pattern search)	Evaluating $\mathbf{f}$	$205 \times \mathbf{f}$	2665 min	205.0
Space mapping	Evaluating $\mathbf{c}$	$345 \times \mathbf{c}$	288 min	22.2
	Evaluating $\mathbf{f}$	$5 \times \mathbf{f}$	65 min	5.0
	Total cost	N/A	353 min	27.2
Feature-based optimization	Evaluating $\mathbf{c}$	$50 \times \mathbf{c}$	42 min	4.0
	Evaluating $\mathbf{f}$	$8 \times \mathbf{f}$	104 min	9.0
	Total cost	N/A	146 min	13.0

search. The pattern search algorithm is used in both direct search and space mapping procedure is based on the implementation described in Koziel (2010e).

The final designs obtained using all three methods are comparable, however, the design costs differ considerably (cf. Table 5.1).

An important prerequisite of the FBO process is that the feature point set is consistent along the optimization path. In particular, it is necessary to establish the linear model (5.34) as well as to compare the quality of the designs at subsequent iterations. For certain problems, the above consistency may be difficult to maintain due to considerable changes of the system response during the optimization process. In some situations, it is possible to work around this issue, e.g., by neglecting the feature points that are not present in one (or more) of the designs that are being compared (Glubokov and Koziel, 2014b).

5.5.7. *Summary*

Physics-based SBO algorithms offer improved computational efficiency due to exploiting the problem-specific knowledge embedded in the low-fidelity model. At the same time, they are more difficult in implementation and less transferrable between various classes of problems. Nevertheless, the physics-based SBO methods seem to be the most promising approaches for handling complex design problems in antenna engineering, particularly, problems that involve expensive EM-simulations. It should be mentioned that this chapter only presented selected methods. Discussion of other techniques can be found in the literature, e.g., adaptive response correction (Koziel *et al.*, 2009), adaptive response prediction (Koziel and Leifsson, 2012c), nested space mapping (Koziel *et al.*, 2014c), corrected space mapping (Robinson *et al.*, 2008), or tuning space mapping (Koziel *et al.*, 2011b).

Chapter 6

Antenna Models for Simulation-Based Design

In this chapter, we discuss low-fidelity EM-simulated models as fundamental components of antenna and antenna array design processes which are performed by means of simulation-based surrogate-assisted optimization. These low-fidelity models are also referred as coarse models in the broader context of surrogate-based microwave circuit design optimization (Bandler *et al.*, 1994, 1995, 2003, 2004a; Koziel and Ogurtsov, 2014; Leifsson and Koziel, 2015). We list and discuss general requirements the low-fidelity models need to satisfy as well as overview typical low-fidelity model setups. Numerical examples emphasize the trade-off between the low-fidelity model's accuracy and its computational cost.

6.1. Low-Fidelity Antenna Models in Simulation-Based Optimization

Design methods discussed in this book in a significant part exploit physics-based surrogate models. As explained in Chapter 5, this type of surrogates utilizes underlying low-fidelity (coarse) EM models which are typically simulated with discrete full-wave EM solvers, custom or commercial, e.g., HFSS (2016), CST (2016), and

FEKO (2015). The list of antenna structures, for which discrete full-wave EM simulation is the only reliable modeling and evaluation possibility, includes but is not limited to printed-circuit ultra-wideband (UWB) antennas (Schantz, 2005), dielectric resonator antennas (DRAs) (Petosa, 2007), and antenna arrays with strong element coupling (Balanis, 2005; Mailloux, 2005).

It is worth to note that in an SBO process the low-fidelity model is repeatedly simulated many times, either to provide data points to configure an response surface surrogate (Koziel and Ogurtsov, 2011a), or to yield a prediction of the high-fidelity model optimum (Bandler et al., 2004a). As a result such repeated simulations, the net computational time of low-fidelity simulations can substantially contribute to the total cost of the antenna optimization process. Therefore, one would set up the low-fidelity model of an antenna structure of interest to be as fast as possible. At the same time, one expects that the low-fidelity model of the antenna adequately represents — qualitatively and quantitatively — the original high-fidelity model at different points of the design space over the frequencies of interest as well as for different radiation directions. Setting up the low-fidelity model one trades its accuracy for computational speed. Subsequently, the tolerated inaccuracy of the low-fidelity model is corrected with the surrogate model update.

In practical problems a number of antenna characteristics of radiation and reflection responses, e.g., sidelobe level and reflection coefficient, should simultaneously satisfy certain design requirements. The radiation and reflection characteristics, however, are of different sensitivity to coarseness of the simulated model, e.g., the far-field quantities such as peak directivity or sidelobe level are integral figures, and therefore, they are less sensitive. On the other hand, the input impedance or the reflection coefficient is more sensitive to the model discretization density, in particular, to the quality of the mesh of the feeding elements. Thus, simplifications of the low-fidelity model should be made with respect to the antenna characteristics which are to be utilized in the optimization loop. In some cases, certain

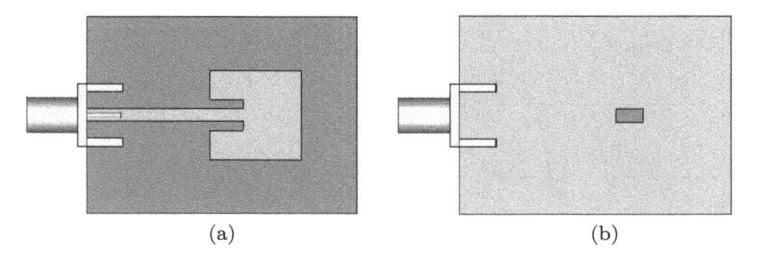

Fig. 6.1. Microstrip patch antenna model: front (a) and back (b).

characteristics are disregarded or considered at separate stage of the antenna design process (Koziel and Ogurtsov, 2014a).

An example of a 5.8-GHz microstrip patch antenna, shown in Fig. 6.1, illustrates differences of its responses evaluated with models of different fidelities. Both of the models are defined, discretized, and simulated using CST MWS (CST, 2016). The reference, fine-model, is discretized with about 1,585,840 mesh cells and with the minimal mesh step of 0.10 mm; it is evaluated in 10 min 6 s with the CST MWS transient solver on a 2.00-GHz 6-core CPU 64 GB RAM computer. A quite dense discretization of the fine-model ensures that no noticeable changes of the response occur with increasing the discretization density. At the same time, the low-fidelity model is discretized with about 151,620 mesh cells and with the minimal mesh step of 0.3 mm; it is simulated on the same computer in only 2 min 15 s.

For this antenna — supposed to operate around 5.8 GHz — the reflection coefficient response of the low-fidelity model, shown in Fig 6.2(a), is substantially misaligned with that of the high-fidelity model. Thus, tuning dimensions of the antenna using the low-fidelity model can result in an unreliable design in terms of reflection coefficient. At the same time, the direct utilization of the high-fidelity model for design optimization will be associated with a substantial total design time.

On the other hand, the model fidelity is less critical for the realized gain value about 5.8 GHz though the realized gain responses are also misaligned in frequency, as shown in Fig. 6.2(b). Finally,

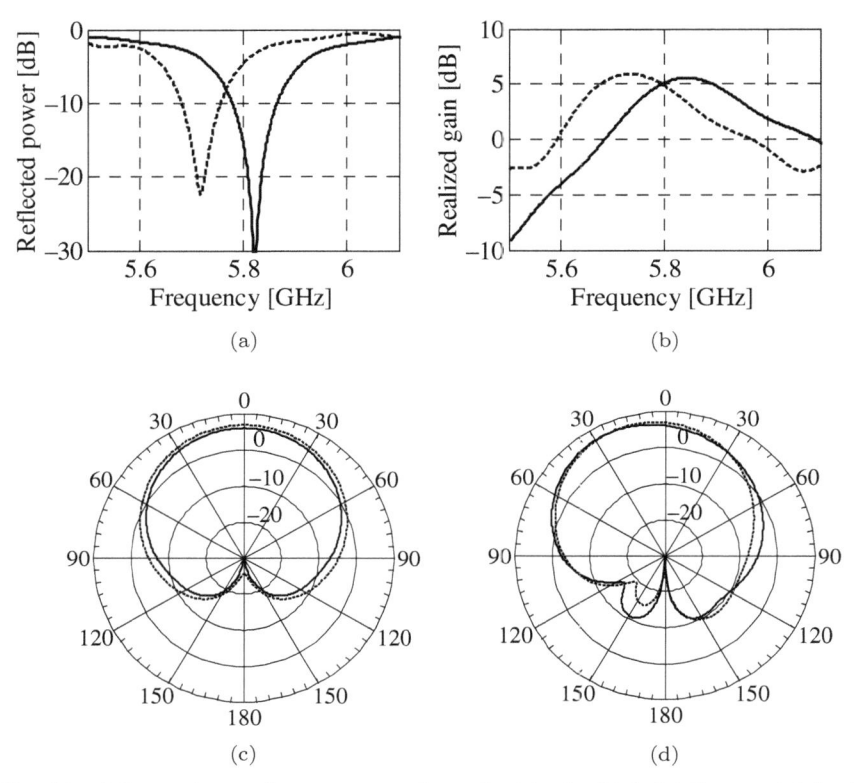

136 *Simulation-Based Optimization of Antenna Arrays*

Fig. 6.2. Microstrip patch antenna evaluated with the high-fidelity model (—) and low-fidelity model (- - -): (a) reflection coefficient; (b) realized gain toward zenith; (c) H-plane directivity pattern at 5.8 GHz; (d) E-plane directivity pattern at 5.8 GHz.

the two models have very similar directivity patterns, as shown with Figs. 6.2(c) and 6.2(d).

6.2. Coarse-Discretization as a Basis of Low-Fidelity Models

Discrete EM simulators, in particular commercial software packages, e.g., HFSS (2016), CST (2016), and FEKO (2015), are extensively used for contemporary antenna and microwave design in industry and academia. Not long ago discrete EM simulators were used mostly for component analysis and design verification. Nowadays, due to the progress in computing hardware as well as development of computational methods, discrete EM simulators turn to be indispensable for the entire design process. Discrete full-wave simulations are also allow for obtaining reliable antenna reflection and radiation responses with respect to installation environment.

Discretization density, local as well as overall, has the strongest effect on accuracy and computational time of the models evaluated

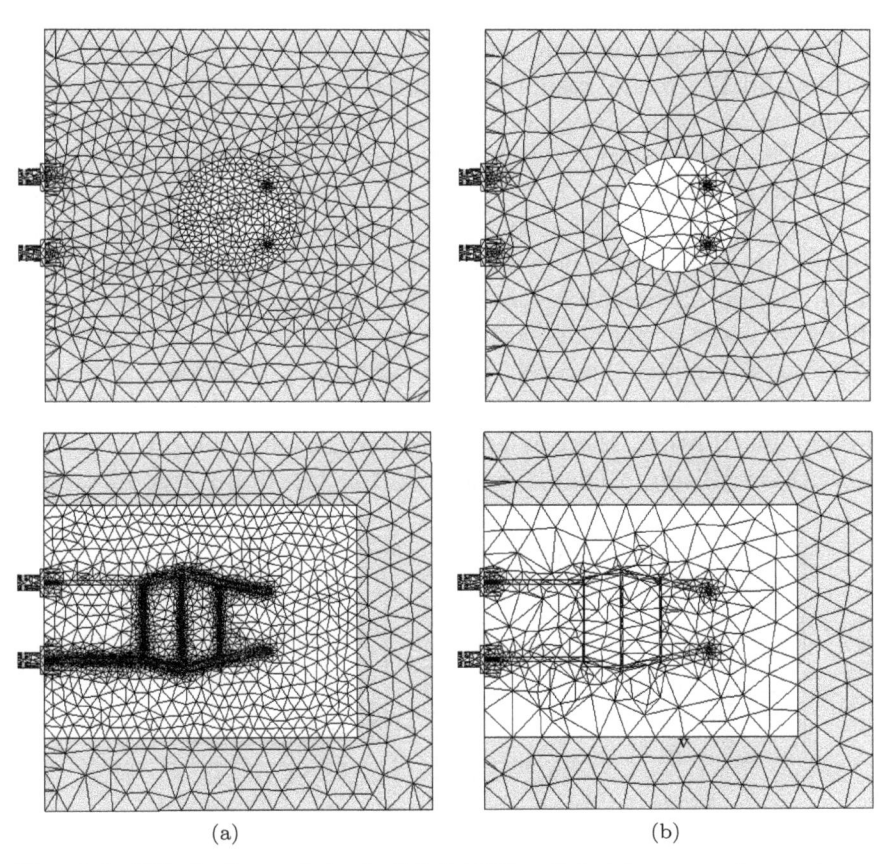

Fig. 6.3. Elliptically polarized 2.0–2.4 GHz dielectric resonator antenna (Ogurtsov et al., 2017) with the dielectric resonator core excited through a microstrip branchline coupler: (a) tetrahedral mesh of the high-fidelity model; (b) coarse mesh of the low-fidelity model. The upper views are for the antenna front, and the lower views are for the back.

by the EM solver. At the same discretization density is the most efficient parameter allowing to trade accuracy for speed. Thus, a straightforward way to create a low-fidelity model of an antenna is to relax mesh settings compared to those of the high-fidelity model, e.g., as illustrated in Fig. 6.3. The low-fidelity model **c** is faster than model **f** due to introduced simplifications. Typically, in EM simulation model **c** can be made 5 to 50 times faster than model **f**. However, model **c** is obviously not as accurate as model **f**. Thus, the low-fidelity model cannot simply replace the high-fidelity model in the design optimization process. Figure 6.4(a) shows the high- and low-fidelity model responses simulated with different hexahedral meshes at the same design of the antenna of Fig. 6.3. Figure 6.4(b) shows the relationship between overall mesh coarseness and antenna simulation time.

Both local and global mesh coarseness strongly affect model simulation time, model accuracy and, thus, the overall performance of the design optimization process. The coarser model is faster and allows a lower numerical cost per iteration of the SBO process. The coarser model, however, is less accurate; therefore, it results in a larger number of iterations necessary to find a satisfactory design. In addition, there is an increased risk for the optimization algorithm utilizing a coarser model to fail to obtain a good design

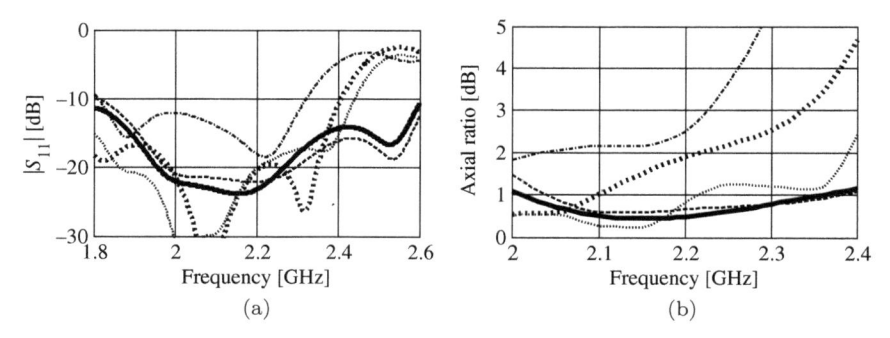

Fig. 6.4. Antenna of Fig. 6.3 at a certain design simulated using the CST MWS transient solver (CST, 2016): (a) reflection coefficient and (b) axial ratio toward zenith. Simulations with hexahedral meshes comprising different numbers of cells: 793,728 ($\cdot - \cdot$); 1,975,974 (• • •); 3,667,700 ($\cdots$); 5,636,160 (- - -); and 15,857,400 (—).

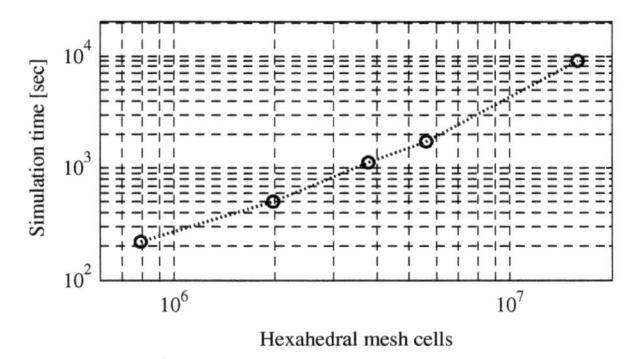

Fig. 6.5. Antenna of Fig. 6.3 at a certain design simulated using the CST MWS transient solver (CST, 2016) on a 2.00-GHz 6-core CPU 64 GB RAM computer: model simulation time versus mesh size. Data points (o) correspond to meshes listed in Fig. 6.4 caption.

(Koziel and Ogurtsov, 2012a). On the other hand, finer models are more expensive but they are more likely to produce an improved and reliable design with a smaller number of iteration. One can infer from Fig. 6.4 that the "finest" coarse-discretization model (5,636,160 cells mesh cells) represents the high-fidelity model responses (shown with thick solid lines) very closely in terms of reflection coefficient and axial ratio. The model with 3,667,700 cells can be considered as a borderline in terms of reflection coefficient and axial ratio. The two remaining models could be considered inaccurate; in particular, the model with 793,728 cells is essentially unreliable (Fig. 6.5).

6.3. Other Simplifications of Low-Fidelity Models

In addition to coarse discretization other simplifications can be made in the simulation setup of low-fidelity models. Possible computational simplifications include:

(a) shrinking the computational domain and applying simple absorbing boundaries with the finite-volume methods (HFSS, 2015; CST, 2016; Taflove and Hagness, 2005);
(b) using low-order basis functions with the finite-element and moment method solvers (Lin, 2008, 2014; Harrington, 1993; Makarov, 2002);

(c) using more relaxed simulation termination criteria such as the S-parameter error for the frequency domain methods with adaptive meshing, e.g., HFSS (2015) and CST (2016), and residue energy for time-domain solvers (CST, 2016).

Simplification of physics of low-fidelity antenna models can include:

(a) ignoring dielectric and metal losses as well as material dispersion;
(b) using evanescent thickness of metallization for printed antennas;
(c) ignoring moderate anisotropy of substrates;
(d) energizing antennas with discrete sources rather than waveguide ports (HFSS, 2015; CST, 2016; FEKO, 2015; Taflove and Hagness, 2005).

In fact the computational and physics simplifications are closely related — they are listed in two groups mostly for classification. For example, ignoring dielectric losses and material dispersion in the time-domain finite-difference method turn into a simpler formulation of the computational problem, fewer unknowns and, shorter simulation time for the same mesh (Taflove and Hagness, 2005). The following example illustrates the effect of the above listed simplifications on accuracy of the low-fidelity model and its simulation time.

Consider a printed Yagi antenna shown in Fig. 6.6. The substrate is a 0.635 mm Rogers RT6010 material (RT/duroid 6006/6010, 2011) with lateral dimensions of 24.65 mm by 17.5 mm. The ground plane is 11.3 mm by 17.5 mm. The input to the antenna is a 50-ohm microstrip. Metallization is with 50 μm copper.

Antenna models are defined in the CST MWS environment and simulated with the CST MWS transient solver (CST, 2016). The high-fidelity model $\mathbf{f}$ is discretized with a fine mesh assuring that no noticeable changes in the reflection and radiation responses are observed with refining the mesh. This requirement results in 45 mesh cells per wavelength at the center frequency of the simulation bandwidth for a design with dimensions $\mathbf{x} = [v_1\ v_2\ v_3\ w_1\ w_2\ w_3\ w_4\ u_1\ u_2\ u_3\ u_4\ u_5\ u_6\ u_7\ u_8\ u_9]^T = [8.9\ 4.2\ 3.0\ 0.6\ 1.2\ 0.3\ 0.3\ 4.0\ 1.5\ 4.8\ 1.8\ 1.5\ 4.0\ 3.0\ 3.35\ 3.0]^T$ mm. As a result, the model $\mathbf{f}$ contains

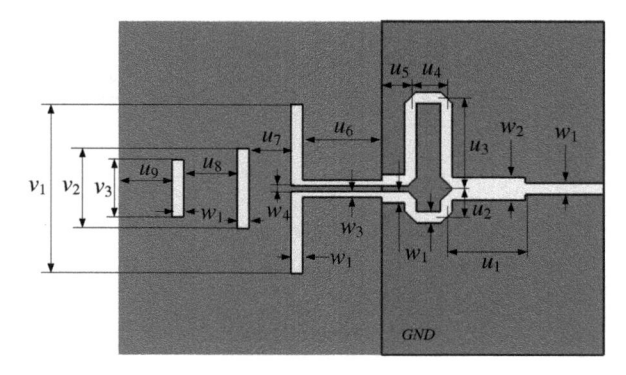

Fig. 6.6. Printed Yagi antenna (Koziel and Ogurtsov, 2014a).

1,611,624 mesh cells, and it is simulated in 12 min 13 s. on a 2.33-GHz 8-core CPU with 8 GB RAM computer. A low-fidelity model c_1 is different from model f only in the mesh density and, thus, in the number of mesh cells. In particular, the model c_1 is discretized with only 15 cells per wavelength resulting in 96,000 mesh cells. Its simulation time is only 8% of that of model f. Responses of the models are shown in Fig. 6.7.

Another low-fidelity model c_2 has discretization density of the model c_1. Additional simplifications of c_2 include: the dielectric substrate and copper are modeled as lossless; copper thickness is evanescent; the number of absorbing boundary layers has been reduced from 6 to 4; the distances to the absorbing boundaries have been made 40 percent of those of the models f and c_1; the termination condition has been set to -25 dB of the residual energy (CST, 2016) versus -40 dB of the models f and c_1. With these simplifications, the number of mesh cell in the model c_2 has been reduced not much compared to that of c_1, to 90,000; however, its evaluation time is only 5% of that of model f.

By visual inspection of the reflection coefficient responses in Fig. 6.7(a) one concludes that the low-fidelity models c_1 and c_2 have similar quality in overall when compared to the reflection response of the high-fidelity model. Further, the gain responses of all models shown in Fig. 6.7(b) are quite close and consistent. The difference of gain of the model c_2 from gain of f and c_1 is due to its lossless

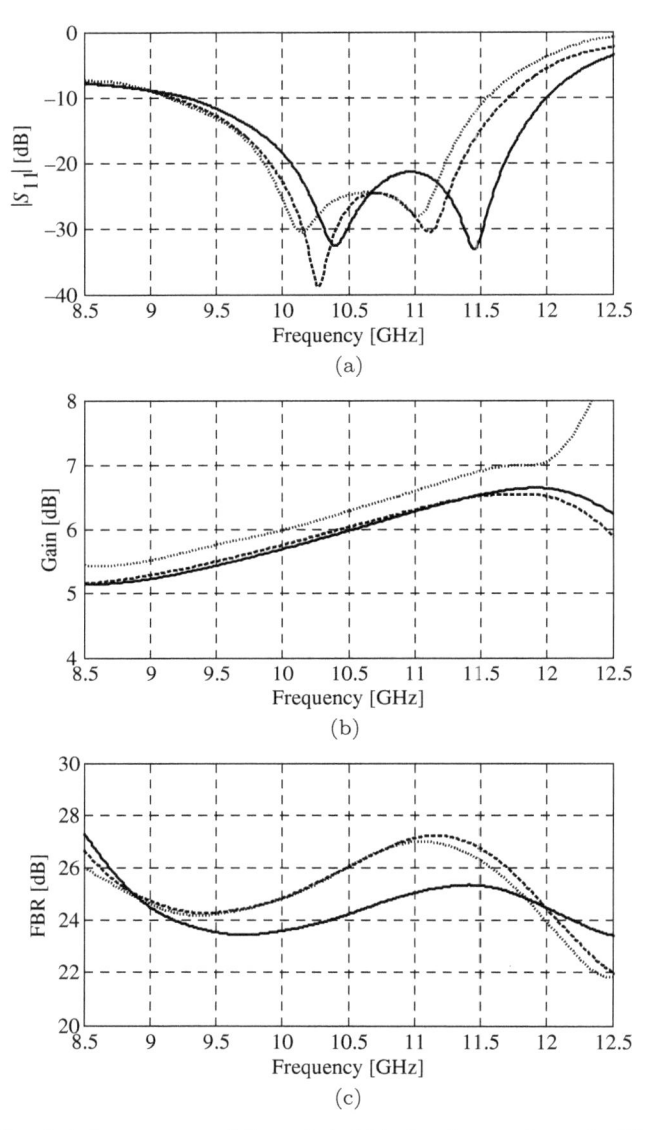

142 *Simulation-Based Optimization of Antenna Arrays*

Fig. 6.7. Simulated responses of the printed Yagi antenna shown in Fig. 6.5 with the high-fidelity model **f** (—), low-fidelity models **c**1 (- - -) and **c**2 ($\cdots$) at a certain design: (a) reflection coefficient; (b) IEEE gain; (c) front-to-back ratio (Koziel and Ogurtsov, 2014a).

description, and it can be easily taken into account up to 12 GHz (the upper useable frequency of this design) through the space mapping correction or adaptive response correction (Koziel *et al.*, 2008b, 2009; Koziel and Ogurtsov, 2011a). Front-to-back ratios of the low-fidelity models are essentially the same and equally different from that figure of model **f**.

A possible computational effort, which would be required to correct low-fidelity responses in the SBO processes, is expected to be about the same for both models c_1 and c_2 because the models' responses are of similar quality; however, the use of model c_2 in the SBO process may result is a lower computational cost of the final design because c_2 is almost two times faster than c_1.

6.4. Automated Selection of Model Fidelity

It is worth to mention that a visual inspection of the model response and the relationship between the high- and low-fidelity models is an important part of the model selection process. It is essential that the low-fidelity model captures all important features of the high-fidelity model response. Nevertheless, because the low-fidelity model subjects two conflicting requirements of accuracy and speed, the optimal choice is unclear *a priori*. In particular, both accuracy and speed of the low-fidelity model depend, in general, on a location in the design space. Therefore, the problem of model fidelity selection should be quantitatively addressed within execution of a particular SBO algorithm. A more detailed discussion about the model fidelity selection and model fidelity effect on the performance of the SBO process, including both quality of the final design and the overall optimization cost, is available in Koziel and Ogurtsov (2012a).

Chapter 7

Element Design: Case Studies

In this chapter, we demonstrate the use of surrogate-assisted optimization techniques for the design of individual printed-circuit antenna structures including those used as elements of antenna arrays. In all cases, we utilize different variations of surrogate modeling methods, variable-fidelity EM simulation models, response correction techniques, some of which have been discussed in the previous parts of this book, particularly, in Chapter 5. We also emphasize a point that depending on particular design objectives, generic optimization methods need to be tailored for a specific design task, e.g., when dealing with explicit reduction of the antenna size, or especially when finding design trade-offs between conflicting antenna figures. Some of the case studies are provided with experimental verification although — from the optimization viewpoint — the ultimate validating model is the high-fidelity EM simulation one.

7.1. EM-Driven Design of a Planar UWB Dipole Antenna with Integrated Balun

In this section, we present the design optimization of a UWB dipole antenna with an integrated balun using space mapping technology specifically, we use frequency scaling and additive response correction. The final design is validated with high-fidelity EM simulation and measurement of the fabricated prototype.

7.1.1. *Antenna geometry*

The antenna layout is shown in Fig. 7.1 (Koziel *et al.*, 2015). The design objective is the lowest as possible reflection coefficient within the 3.1–10.6 GHz band. The antenna includes a balun which is a microstrip-to-coplanar strip (CPS) transition having a ground edge of a linear profile and an open radial microstrip stub. The antenna is implemented on a 0.76-mm thick Taconic RF-35 substrate (RF-35, 2018). Dimensional parameters of the dipole, CPS section, and the balun are the design variables: $\mathbf{x} = [r_x\ r_y\ x_3\ y_2\ y_3\ y_4\ y_5\ r_s\ \alpha\ s_0]^T$. Parameters $w_0 = 1.73, x_1 = 5, x_2 = 1.73$, and $y_1 = 11$, all in mm, are fixed.

The structure is modeled with CST MWS and simulated using the CST MWS transient solver (CST, 2016). We utilize two discrete EM models: the high-fidelity model $\mathbf{f}$ ($\sim$11,000,000 mesh cells, simulation time $\sim$53 min) and the low-fidelity model $\mathbf{c}$ ($\sim$600,000 cells, simulation time $\sim$100 s). It is assumed that, in actual applications, there should be no connector in the antenna proximity. Therefore, the discrete models to be used in optimization were defined with waveguide ports (CST, 2016). A high-fidelity model equipped with an SMA edge-mount connector had been also defined to compare the simulated final design to its measured prototype.

7.1.2. *Optimization procedure*

The design task is formulated as a nonlinear minimization problem of the form (3.1): $\mathbf{x}^* = \mathrm{argmin}\{\mathbf{x} : U(\mathbf{f}(\mathbf{x}))\}$. The objective function is defined as a maximum in-band reflection coefficient. The high-fidelity model is computationally expensive, and the number of geometry parameters is large. Still, we aim at simultaneous adjustment of both the parameters of the dipole itself and those of the balun. Therefore, we utilize the iterative SBO process (5.1), recalled here for the convenience of the reader. The SBO algorithm yields a sequence

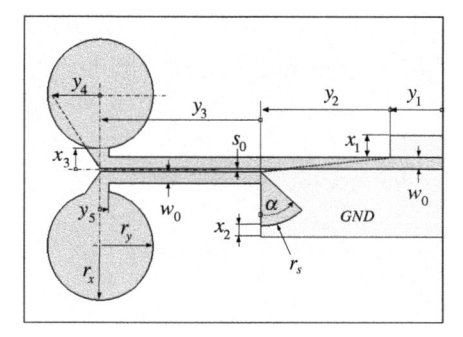

Fig. 7.1. UWB dipole antenna with an integrated balun: layout (Koziel et al., 2015a).

of approximations $\mathbf{x}^{(i)}, i = 0, 1, \ldots,$ of the optimum $\mathbf{x}^*$

$$\mathbf{x}^{(i+1)} = \arg\min_{\mathbf{x}} U(\mathbf{s}^{(i)}(\mathbf{x})), \tag{7.1}$$

where $\mathbf{s}^{(i)}$ is the surrogate model at iteration i. The surrogate model $\mathbf{s}$ is constructed by correcting the low-fidelity model $\mathbf{c}$. There are the following major kinds of misalignment between the models $\mathbf{c}$ and $\mathbf{f}$: shifts in frequency and level. Therefore, two types of correction are used: frequency scaling and additive response correction.

The surrogate model is defined as

$$\mathbf{s}^{(i)}(\mathbf{x}) = \mathbf{c}(\mathbf{x}, \alpha^{(i)} F) + \mathbf{d}^{(i)}, \tag{7.2}$$

where $\mathbf{c}(\mathbf{x}, F)$ denotes explicit dependency of $\mathbf{c}$ on frequency. Further, we have

$$\mathbf{d}^{(i)} = \mathbf{f}(\mathbf{x}^{(i)}) - \mathbf{c}(\mathbf{x}^{(i)}, \alpha^{(i)} F), \tag{7.3}$$

$$\alpha^{(i)} F = \alpha_0^{(i)} + \alpha_1^{(i)} F, \tag{7.4}$$

where (7.4) describes the affine frequency scaling (Bandler et al., 2004a). The frequency scaling parameters are obtained as

$$[\alpha_0^{(i)}\ \alpha_1^{(i)}] = \arg\min_{\mathbf{x}} ||\mathbf{f}(\mathbf{x}^{(i)}) - \mathbf{c}(\mathbf{x}^{(i)}, \alpha_0^{(i)} + \alpha_1^{(i)} F)||. \tag{7.5}$$

The response correction term (7.3) ensures perfect alignment of the surrogate and the high-fidelity model at the current iteration point $\mathbf{x}^{(i)}$. Solving (7.5) is a nonlinear regression problem but, in practice, it does not require extra EM simulations as the response at the scaled frequencies can be obtained with interpolation of the response evaluated at the original sweep. The surrogate model optimization (7.1) is realized using the pattern search algorithm (Koziel, 2010e).

7.1.3. Numerical results

The initial design is $\mathbf{x}^{\text{ini}} = [r_x \ r_y \ x_3 \ y_2 \ y_3 \ y_4 \ y_5 \ r_s \ \alpha \ s_0]^T = [13 \ 13 \ 3 \ 25 \ 25 \ 10 \ 2 \ 10 \ 45 \ 0.4]^T$, where all variables but α are in mm; α is in degrees. The optimum, $\mathbf{x}^* = [12.43 \ 12.33 \ 2.40 \ 27.34 \ 34.07 \ 10.50 \ 1.71 \ 6.13 \ 44.9 \ 0.76]^T$, has been found in four iterations of the optimization procedure described in Sec. 7.1.2. Each iteration required about 120 simulations of the low-fidelity model $\mathbf{c}$ and one simulation of the high-fidelity model $\mathbf{f}$. The reflection coefficient of the final design is shown in Fig. 7.2, where $|S_{11}|$ is below -17 dB for the entire UWB band. The total numerical cost of this design corresponds to about 19 simulations of the UWB antenna high-fidelity model.

7.1.4. Experimental validation

A photograph of the manufactured design prototype is shown in Fig. 7.3(a). The antenna under test was excited through an edge-mount connector (SMA connector, 2012). Simulated and measured reflection coefficients are shown in Fig. 7.3(b). Simulated and measured radiation patterns in the plane orthogonal to the antenna substrate at selected frequencies are shown in Fig. 7.4. The radiation of the final design stays essentially omnidirectional in this plane up to 7 GHz.

7.2. Design of Compact UWB Slot Antenna

In this section, we demonstrate size reduction of a UWB slot antenna using surrogate-assisted optimization. In case of a highly dimensional parameter space, which we deal with here, it is neither possible to adjust all relevant parameters nor simultaneously control the

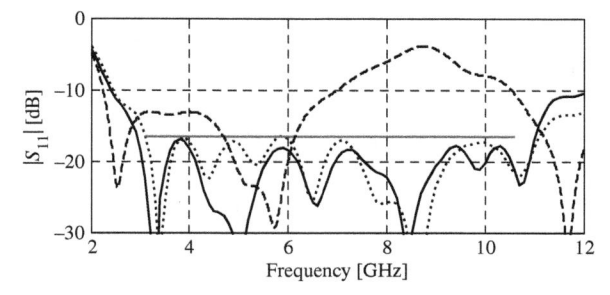

Fig. 7.2. Simulated reflection coefficient: the initial design (- - -); the final design without connector (···) and with connector (—). The horizontal line marks the maximum in-band reflection level of the optimized antenna.

(a)

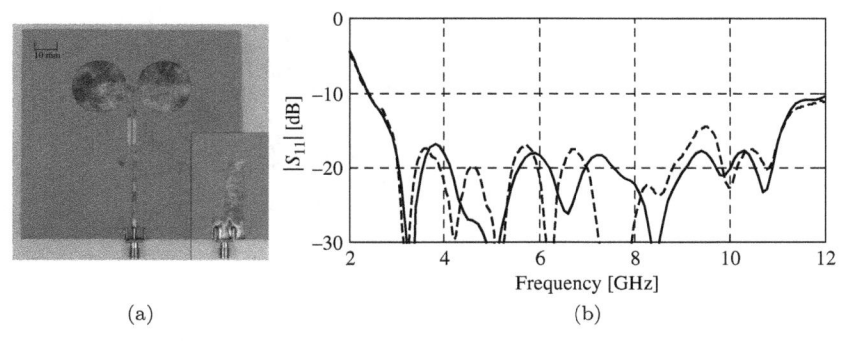

(b)

Fig. 7.3. UWB-dipole antenna with balun: (a) manufactured final design, (b) reflection response of the final design: simulated with connector (—) and measured (- - -).

antenna footprint and performance parameters without numerical optimization.

7.2.1. *Antenna structure*

We consider a compact UWB slot antenna of Fig. 7.5. The structure is implemented on a 0.762-mm thick Taconic RF-35 substrate. The design variables are $\mathbf{x} = [l_0\ l_{g1r}\ l_{g2}\ l_{s1}\ l_{s2}\ l_{s3}\ l_{s4}\ l_{s5}\ l_{f1r}\ l_{f2r}\ w_{s1}\ w_{s2}\ w_{s3}\ w_{s4}\ w_{f2}\ o_{f1r}\ o_{s2r}\ o_{s3r}\ o_{s4r}]^T$. The parameter $w_{f1} = 1.7$ is fixed to ensure 50-ohm input impedance. Moreover, the dimensions $l_{g1} = l_{g1r}(l_0 - w_{s1})$, $l_{f1} = l_{f1r} \cdot l_0(1 - l_{f2r})$, $l_{f2} = l_0 \cdot l_{f1r} \cdot l_{f2r}$, $o_{f1} = o_{f1r}(l_{s1} + l_{s2} + l_{s3} + l_{s4} + w_{s4} + l_{g2} - w_{f1})$, $o_{s2} =$

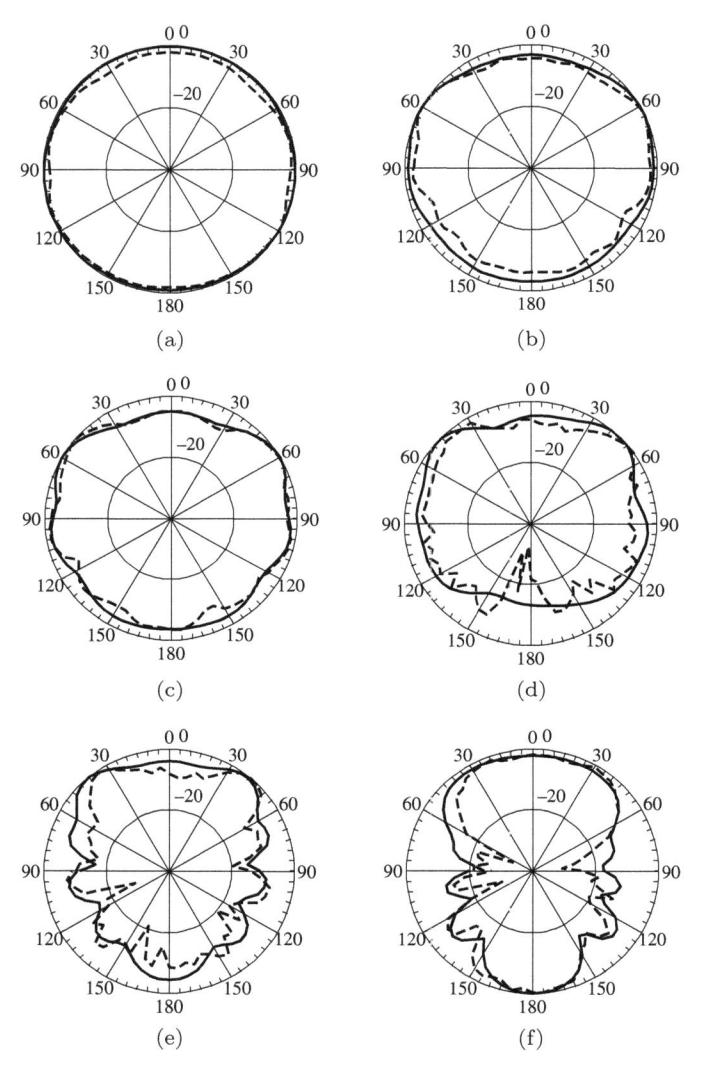

150   Simulation-Based Optimization of Antenna Arrays

Fig. 7.4. Normalized power pattern of the final design: simulated with connector (—) and measured (- - -) at selected frequencies: (a) 3 GHz, (b) 5 GHz, (c) 6 GHz, (d) 7 GHz, (e) 9 GHz, and (f) 10 GHz. 90 degrees on the left correspond to the normal direction above the antenna. 180 degrees correspond to "to-the-connector".

(a) (b)

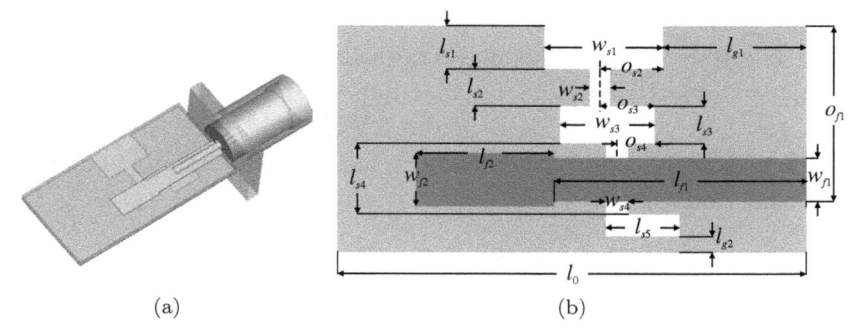

Fig. 7.5. Compact UWB slot antenna (Bekasiewicz and Koziel, 2016b): (a) 3D view with the SMA connector; (b) geometrical details of the structure with highlighted design variables.

$0.5(o_{s2r}(w_{s1} - w_{s2}) - w_{s2})$, $o_{s3} = 0.5(o_{s3r}(w_{s2} - w_{s3}) - w_{s3})$ and $o_{s4} = 0.5(o_{s4r}(w_{s3} - w_{s4}) - w_{s4})$ are set as relative to ensure geometrical consistency of the design during the optimization process.

The simulation models of the antenna are implemented in CST MWS: the high-fidelity model $\mathbf{f}$ ($\sim$3,800,000 mesh cells, simulation time 22 min), and the low-fidelity model $\mathbf{c}$ ($\sim$420,000 cells, simulation time 2 min). As the antenna structure is electrically small, the SMA connector is included in the EM models.

The design objective is to minimize the antenna footprint while ensuring that the reflection coefficient is below -10 dB over the frequency range from 3.1 to 10.6 GHz. The initial design, $\mathbf{x}_0 = [22.0\ 0.52\ 1.4\ 2.3\ 2.1\ 1.6\ 0.7\ 2.7\ 0.68\ 0.35\ 4.6\ 0.8\ 3.2\ 0.4\ 1.7\ 0.72\ 0\ 0\ 0]^T$, is extracted from dimensions of the reference structure of (Chu et al., 2013).

7.2.2. Optimization algorithm

The design task is formulated as in (3.1). The primary objective is the minimal antenna footprint while having $|S_{11}| \leq -10$ dB for 3.1–10.6 GHz. The objective function is defined as (Bekasiewicz and Koziel, 2016b)

$$U(\mathbf{f}(\mathbf{x})) = S(\mathbf{x}) + \beta \cdot g(\mathbf{f}(\mathbf{x}))^2, \qquad (7.6)$$

where $S(\mathbf{x})$ is the antenna footprint. The penalty factor β is set to 1000. The reflection coefficient is controlled with a penalty function $g(\mathbf{f}(\mathbf{x})) = \max\{(\max\{|S_{11}|_{3.1-10.6\,\text{GHz}}\} + 10)/10,\ 0\}$. Notice that $g(\mathbf{f}(\mathbf{x})) = 0$ if $|S_{11}| \leq -10\,\text{dB}$ in the UWB band. The SBO process (5.1) is used to speed up design where the surrogate model is defined as (Koziel and Ogurtsov, 2014a)

$$\mathbf{s}^{(i)}(\mathbf{x}) = \mathbf{s}_{\text{RSA}}^{(i)}(\mathbf{x}) + [\mathbf{f}(\mathbf{x}^{(i)}) - \mathbf{s}_{\text{RSA}}^{(i)}(\mathbf{x}^{(i)})]. \tag{7.7}$$

The surrogate model (7.7) is obtained as a local response surface approximation (RSA) model $\mathbf{s}_{\text{RSA}}^{(i)}(\mathbf{x})$ which is subsequently corrected using output space mapping (Bandler et al., 2004b).

The RSA model here is a second-order polynomial without mixed terms:

$$\mathbf{s}_{\text{RSA}}^{(i)}(\mathbf{x}) = \mathbf{s}_{\text{RSA}}^{(i)}([x_1 \ldots x_n]^T) = \lambda_0 + \sum_{l=1}^{n} \lambda_l x_l + \sum_{l=1}^{n} \lambda_{n+l} x_l^2. \tag{7.8}$$

The RSA model (7.8) is constructed from the training data acquired at the design $\mathbf{x}^{(i)}$ and $2n$ perturbations around it allocated the using star-distribution design of experiment (Koziel et al., 2011a). The coefficients $\lambda_0, \ldots, \lambda_{2n+1}$ are found solving a linear regression problem $\mathbf{s}_{\text{RSA}}^{(i)}(\mathbf{x}_B^{(k)}) = \mathbf{f}(\mathbf{x}_B^{(k)})$, $k = 1, \ldots, 2n+1$, where $\mathbf{x}_B^{(k)}$ are the training points. A concept of the local RSA model construction is depicted in Fig. 7.6.

7.2.3. *Numerical results and experimental validation*

The final design $\mathbf{x}^* = [17.92\ 0.41\ 1.5\ 2.8\ 1.82\ 1.48\ 0.45\ 2.85\ 0.65\ 0.35\ 4.57\ 0.78\ 3.66\ 0.87\ 2.08\ 0.72\ -0.07\ -0.78\ 0.91]^T$ is obtained in five iterations of the optimization algorithm. The antenna size at the final design is $8.93\,\text{mm} \times 17.92\,\text{mm}$, i.e., the antenna footprint is only $1.6\,\text{cm}^2$. The initial and final designs are compared in Fig. 7.7.

The antenna has been manufactured (cf. Fig. 7.8(a)) and measured. The simulated and measured reflection characteristics of the antenna are compared in Fig. 7.8(b). The characteristics are well aligned at the edges of the operating bandwidth. The measured

Element Design: Case Studies   153

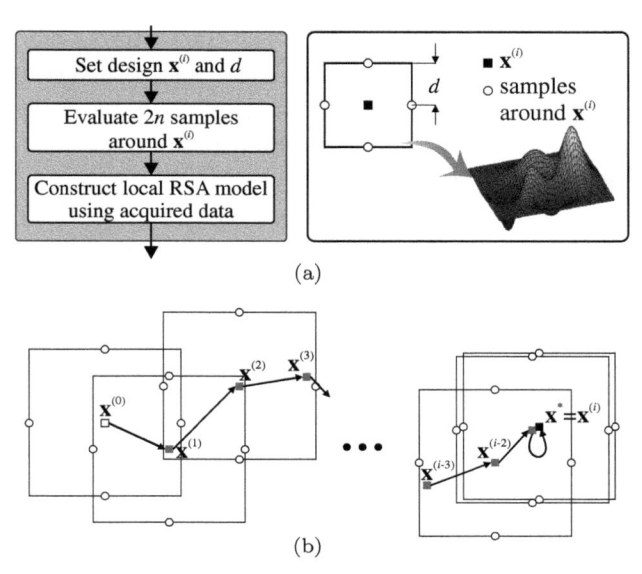

Fig. 7.6. Conceptual illustration of the surrogate-assisted scheme for UWB slot antenna miniaturization (Bekasiewicz and Koziel, 2016b): (a) construction of the local RSA model where the perturbation size **d** is determined using sensitivity analysis of the structure; (b) the exemplary path between the initial and the final design in the two-dimensional design space.

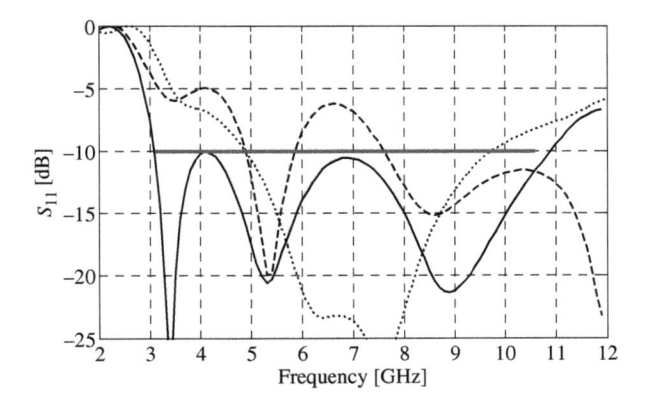

Fig. 7.7. Reflection coefficient of the compact UWB slot antenna: the antenna model with the SMA connector simulated at the initial design (- - -); the antenna model with the SMA connector simulated at the final design (—); the antenna model without the SMA connector simulated at the final design (· · ·). One can observe that the SMA connector affects the electrical length of the antenna. The horizontal line marks the design specifications for antenna reflection.

154 *Simulation-Based Optimization of Antenna Arrays*

results exhibit additional resonance at the center frequency; however, the response is below $-10\,\mathrm{dB}$ within the entire UWB band. The visible discrepancy between the characteristics is introduced by fabrication (etching, milling, etc.) and assembly imperfections (allocation of the connector, soldering, etc.).

A comparison of the simulated and measured realized gain in a direction which is normal to the antenna front surface over the UWB band is shown in Fig. 7.9 where the realized gain responses agree well.

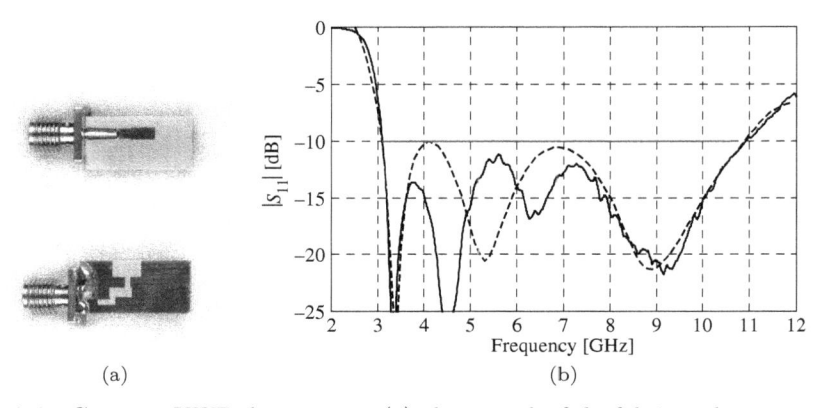

Fig. 7.8. Compact UWB slot antenna: (a) photograph of the fabricated prototype (front and back); (b) simulated (- -) and measured (—) reflection coefficient. The horizontal line marks the design specifications for antenna reflection.

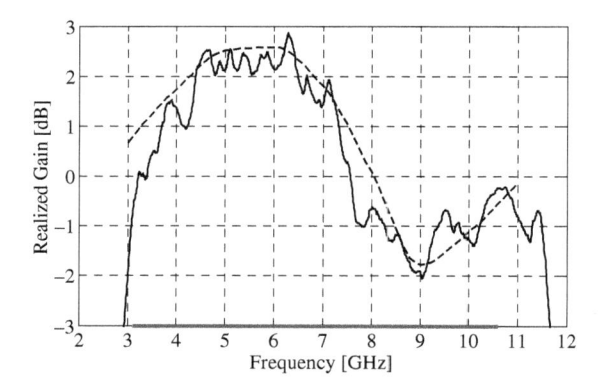

Fig. 7.9. Compact UWB slot antenna: simulated (- -) and measured (—) realized gain in the direction normal to the antenna's front surface. The horizontal line marks the UWB frequency band.

The responses are relatively stable within 4–7 GHz. Above 7 GHz the simulated and measured gain in the measured direction weaken by about 2 dB and remains more or less stable within the 8–11 GHz range.

The simulated and measured H-plane radiation patterns at 3.1, 5, 7, and 9 GHz are shown in Fig. 7.10 from where one sees that the antenna provides wideband omnidirectional radiation in the H-plane.

7.3. Optimization of Slot-Ring Coupled Patch Antenna

Here, we optimize a slot-ring coupled patch antenna. We use the feature-based optimization algorithm (FBO) (Sec. 5.5.6 in

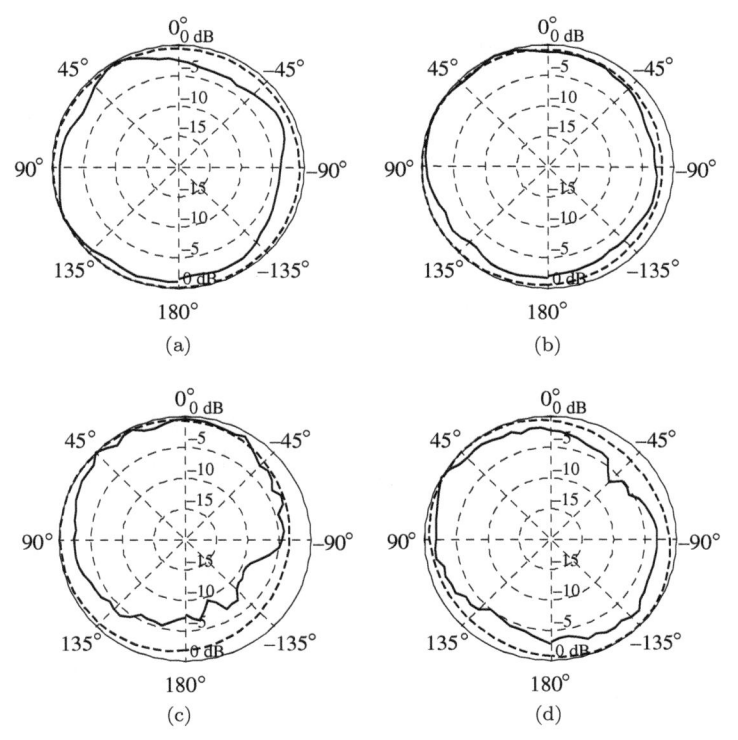

Fig. 7.10. Simulated (− −) and measured (—) H-plane patterns of the compact UWB slot antenna at: (a) 3.1 GHz; (b) 5 GHz; (c) 7 GHz; and (d) 9 GHz.

156 Simulation-Based Optimization of Antenna Arrays

Chapter 5). FBO is well suited for adjusting narrowband structures, in particular, this antenna. We also discuss an important issue pertinent to variable-fidelity optimization methods, which is selection of the low-fidelity simulation model.

7.3.1. Antenna structure

An antenna shown in Fig. 7.11 (Chen, 2007) contains a rectangular patch coupled to a rectangle-shaped ground-plane slot. The slot is coupled to a coplanar waveguide (CPW) line which is the antenna input. The antenna is implemented on a 0.762-mm thick Taconic RF-35 substrate ($\varepsilon_r = 3.5$, $\tan\delta = 0.0018$). Design variables are $\mathbf{x} = [d\ d_1\ d_2\ l_1\ s_1\ s_2\ s_3\ w_1\ o_1\ o_2]^T$ mm. Fixed dimensions are $w_0 = 3, l_0 = 30$ and $s_0 = 0.15$, all in mm.

The design goal is to minimize the antenna reflection coefficient at the operation frequency $f_0 = 2.45\,\text{GHz}$ (and, at the same time, allocating the minimum at the operation frequency).

The antenna EM models are implemented in CST Microwave Studio (CST, 2016). The high-fidelity model is with 1,200,000 mesh cells and simulated in about 12 min.

7.3.2. Low-fidelity model selection

The reliability and the computational cost of the FBO procedure (Sec. 5.5.6 in Chapter 5) very much depend on the correlation

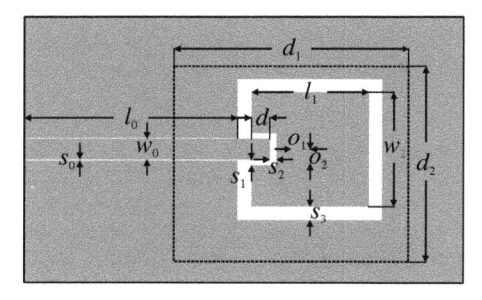

Fig. 7.11. CPW-fed slot-ring coupled patch antenna: geometry with highlighted design parameters. Note that dashed line represents the patch located on the opposite side of the ground plane shown with gray color. White color represents the ground plane slots.

between the low- and high-fidelity models. In order to find the optimal discretization level, e.g., the number of mesh-cells per characteristic wavelength (CST, 2016), for the low-fidelity model we utilize statistical analysis which is briefly explained below (Koziel and Bekasiewicz, 2016a).

Let vectors $\mathbf{x}^{(1)}$ and $\mathbf{x}^{(2)}$ stand for the reference designs. Having the fine model and the coarsely discretized model both simulated at $\mathbf{x}^{(1)}$ and $\mathbf{x}^{(2)}$, the following vectors are defined:

$$\Delta \mathbf{F} = \mathbf{F}(\mathbf{x}^{(2)}) - \mathbf{F}(\mathbf{x}^{(1)}), \tag{7.9}$$

$$\Delta \mathbf{F}_c = \mathbf{F}_c(\mathbf{x}^{(2)}) - \mathbf{F}_c(\mathbf{x}^{(1)}). \tag{7.10}$$

Vectors (7.9) and (7.10) represent variations of the frequency components of the feature coordinates for the high- and low-fidelity models of the structure at hand, namely, $\mathbf{F}(\mathbf{x}^{(k)}) = [f_1(\mathbf{x}^{(k)}) \ldots f_p(\mathbf{x}^{(k)})]^T$ of $\mathbf{x}^{(k)}, k = 1, 2$ (see Sec. 5.5.6 in Chapter 5 for explanation of the notation used here). The definition of $\mathbf{F}_c$ is analogous.

For illustration, let us consider an example with five feature points that correspond to the -10 dB and -15 dB levels of reflection coefficient, as well as the minimum value of the reflection coefficient. Here we only consider frequency coordinates of the feature points for the sake of brevity. After defining the feature points, the data sets $\{\Delta \mathbf{F}, \Delta \mathbf{F}_c\}$ and $\{\Delta \mathbf{L}, \Delta \mathbf{L}_c\}$ (representing corresponding variations of the level components) are utilized to construct the linear regression models. Then, the determination coefficient r^2 (Draper and Smith, 1998) is computed. The coefficient r^2 reflects the level of correlation between the fine- and coarse-discretization data sets. In other words, r^2 represents the reliability of the gradient estimation (cf. (5.34)) when the low-fidelity model is utilized for sensitivity analysis. In particular, when $r^2 = 1.0$, the correlation between the models is ideal. It is assumed here that the low-fidelity model is sufficiently accurate if the determination coefficient is $r^2 = 0.9$ or higher. To find an appropriate discretization level, the correlation analysis has to be performed for a family of the low-fidelity models of increasing mesh density. For the computational models implemented in CST Microwave Studio, the mesh density is adjusted by setting the

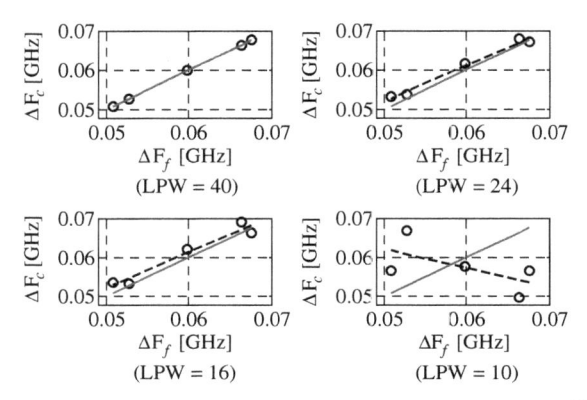

158   Simulation-Based Optimization of Antenna Arrays

Fig. 7.12. Scatter plots $\Delta \mathbf{F}_c$ versus $\Delta \mathbf{F}$ obtained for two designs $\mathbf{x}^{(1)}$ and $\mathbf{x}^{(2)}$ for the LPW parameter set to 40 (fine model), 24, 16, and 10. Solid lines represent regression functions for the reference data sets $\{\Delta \mathbf{F}, \Delta \mathbf{F}\}$ and $\{\Delta \mathbf{L}, \Delta \mathbf{L}\}$, whereas the linear regression function is marked using dashed lines.

value of the (mesh cell) lines per (characteristic) wavelength (LPW). The LPW parameter of the high-fidelity model is usually set to 40 or higher. The analysis of correlation is performed for the low-fidelity model LPW parameter of 8 and higher.

Figure 7.12 shows the selected scatter plots of $\Delta \mathbf{F}_c$ versus $\Delta \mathbf{F}$ and their corresponding regression lines obtained for an exemplary antenna structure (Koziel and Bekasiewicz, 2016a). The LPW values for which the plots have been obtained were set to 10, 16, 24, and 40. Based on these results one can conclude that the mesh with LPW = 10 is too coarse, whereas the model alignment for LPW = 16 is significantly improved. At the same time, the model with LPW = 24 can be considered sufficient for estimating the gradients. The correlations for LPW = 40 correspond to the high-fidelity model.

It should be emphasized that the specific values of the determination coefficient depend on a particular design for which they are calculated. Also, some fluctuations may be present for low LPW values that are result of numerical noise. Here, this problem is mitigated through analysis of correlation at K reference designs so that $(K-1)K/2$ independent vectors $\Delta \mathbf{F}$ and $\Delta \mathbf{F}_c$ are obtained for each selected LPW value.

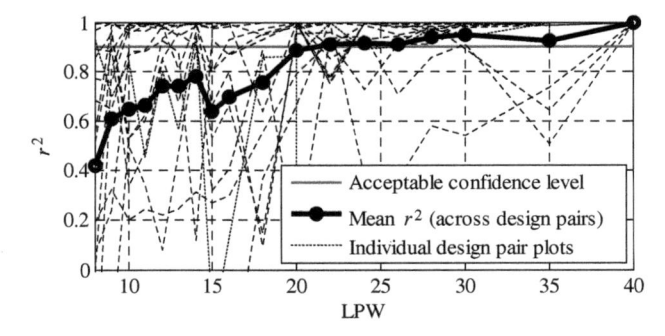

Fig. 7.13. Slot-ring coupled patch antenna: the determination coefficient plots for the $\{\Delta \mathbf{F}, \Delta \mathbf{F}_c\}$ data sets versus model discretization level (here, determined by the LPW parameter). Original data are plotted using dotted lines for all design pairs $\{\mathbf{x}^{(k)}, \mathbf{x}^{(l)}\}, k \neq l$. Acceptable confidence level (here set to 0.9) is marked with the horizontal line.

The r^2 plots in Fig. 7.11 corresponding to LPW parameters in a range from 8 to 40 are shown in Fig. 7.13. Based on this data, we select the low-fidelity model with LPW = 20 (simulation time of 140 s).

7.3.3. Results, benchmarking, and experimental validation

Figure 7.14 shows the antenna reflection responses at the initial design, $\mathbf{x}^{\text{init}} = [4\ 5\ 5\ 16\ 0.3\ 1\ 1\ 16\ 0\ 0]^T$ mm, and the FBO design (LPW = 20), $\mathbf{x}^* = [2.1\ 3.67\ 4.94\ 15.95\ 0.81\ 1.48\ 0.20\ 16.38\ -0.29\ -0.11]^T$ mm. For comparison, the antenna was also optimized with the low-fidelity models of different LPW such as 10 (simulation time of 30 s) and 28 (simulation time of 5 min). The algorithm failed to find a satisfactory design when worked with the very coarse low-fidelity model (LPW = 10). Optimization with LPW = 28 yielded a similar design; however, the computational cost of the process is higher as shown in Table 7.1.

The FBO design $\mathbf{x}^*$ has been fabricated and measured. Figure 7.15 shows the photographs of the prototype. Figure 7.16 shows that the simulated and measured reflection responses agree quite well.

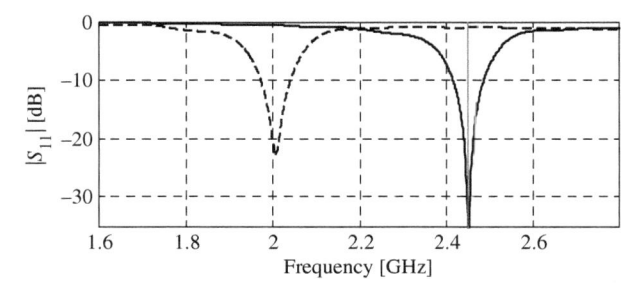

160 Simulation-Based Optimization of Antenna Arrays

Fig. 7.14. Slot-ring coupled patch antenna: reflection response at the initial design (- - -) and the design obtained with FBO and the low-fidelity model with LPW = 20 (—). The vertical line indicates the intended operating frequency of the antenna.

Table 7.1. Slot-ring coupled patch antenna design cost.

	Number of model simulations		Optimization cost	
Algorithm (Low-fidelity model LPW)	Low fidelity	High fidelity	Absolute (hours)	Relative (high-fidelity simulations)
Single-fidelity FBO (40[#])	80	10	18	90
Variable-fidelity FBO (28)	80	11	8.9	44.3
Variable-fidelity FBO (20)	60	11	4.5	22.4
Variable-fidelity FBO (10)		N/A (algorithm is divergent)		
Pattern search	–	356	71.2	356

[#] Here, the low- and high-fidelity models are the same.

7.4. Low-Cost Modeling and Optimization of Ring Slot Antenna

As mentioned in Chapter 5, data-driven modeling of antennas is an expensive process due to the fact that a large number of training data samples is normally needed to construct a reliable model. In this section, a technique for constrained sampling is presented that allows us to focus the modeling process on the region that is relevant from the antenna design optimization standpoint. Such a modeling approach results in considerable savings in terms of necessary data samples and thus in much lower design optimization numerical costs.

Element Design: Case Studies   161

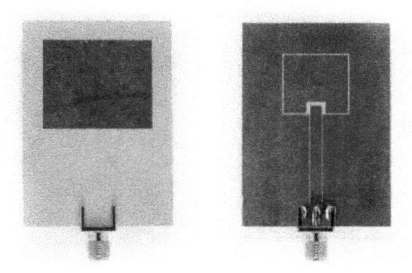

Fig. 7.15. Photographs of the prototype of the optimized slot-ring coupled patch antenna: front (left) and back (right).

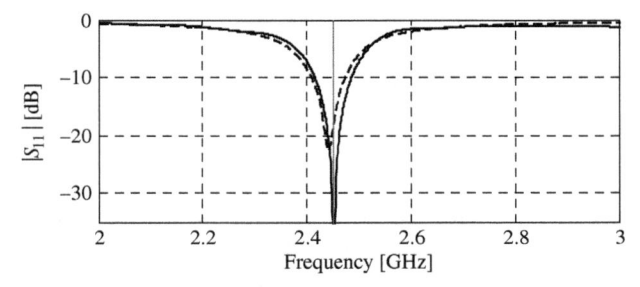

Fig. 7.16. Slot-ring coupled patch antenna: simulated (—) and measured (---) reflection response for the optimized design $\mathbf{x}^*$. The vertical line indicates the intended operating frequency of the antenna.

7.4.1. Modeling methodology

A critical component of a modeling process is definition of the region of surrogate model validity. Such definition is based on reference designs corresponding to a set of operating conditions and material parameters of the antenna of interest. For the purpose of presentation, operating frequency f and relative dielectric permittivity ε_r of the substrate are considered as the operating condition and material parameter of interest, respectively. The surrogate should to be reliable within the ranges of operating frequencies $f_{\min} \leq f \leq f_{\max}$ and permittivities $\varepsilon_{\min} \leq \varepsilon_r \leq \varepsilon_{\max}$. Let $\mathbf{f}(\mathbf{x})$ be a response of an EM-simulated antenna model. Here, $\mathbf{x}$ is a vector of antenna parameters (in general, both of geometry and materials). Symbol $\mathbf{x}^*(f, \varepsilon_r)$ will denote the design optimized for the operating frequency

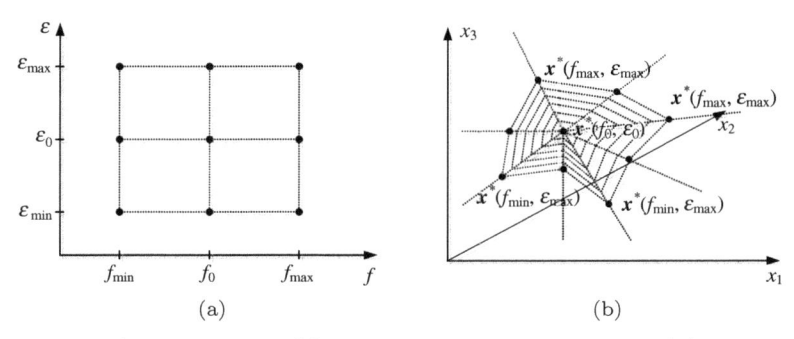

162 Simulation-Based Optimization of Antenna Arrays

Fig. 7.17. Reference designs: (a) distributed on the f-ε plane; (b) allocated in a 3D space where the manifold (shaded) determines the region of the surrogate model construction.

f and substrate dielectric permittivity ε_r. The region of model validity is defined as a vicinity of the manifold spanned by nine reference designs covering the aforementioned ranges of operating frequency and permittivity, $f_{\min} \leq f \leq f_{\max}$ and $\varepsilon_{\min} \leq \varepsilon_r \leq \varepsilon_{\max}$. These are $\mathbf{x}^*(f^\#, \varepsilon_r^\#)$, for all combinations of $f^\# \in \{f_{\min}, f_0, f_{\max}\}$ and $\varepsilon_r^\# \in \{\varepsilon_{\min}, \varepsilon_{r0}, \varepsilon_{\max}\}$, cf. Fig. 7.17.

We define the following eight vectors: $\mathbf{v}_1 = \mathbf{x}^*(f_{\min}, \varepsilon_{\min}) - \mathbf{x}^*(f_0, \varepsilon_{r0})$; $\mathbf{v}_2 = \mathbf{x}^*(f_{min}, \varepsilon_{r0}) - \mathbf{x}^*(f_0, \varepsilon_{r0})$; $\mathbf{v}_3 = \mathbf{x}^*(f_{\min}, \varepsilon_{\max}) - \mathbf{x}^*(f_0, \varepsilon_{r0})$; $\mathbf{v}_4 = \mathbf{x}^*(f_0, \varepsilon_{\max}) - \mathbf{x}^*(f_0, \varepsilon_{r0})$; $\mathbf{v}_5 = \mathbf{x}^*(f_{\max}, \varepsilon_{\max}) - \mathbf{x}^*(f_0, \varepsilon_{r0})$; $\mathbf{v}_6 = \mathbf{x}^*(f_{\max}, \varepsilon_{r0}) - \mathbf{x}^*(f_0, \varepsilon_{r0})$; $\mathbf{v}_7 = \mathbf{x}^*(f_{\max}, \varepsilon_{\min}) - \mathbf{x}^*(f_0, \varepsilon_{r0})$; and $\mathbf{v}_8 = \mathbf{x}^*(f_0, \varepsilon_{\min}) - \mathbf{x}^*(f_0, \varepsilon_{r0})$. See Fig. 7.18(a) for illustration. We also define a manifold M, which is spanned by eight pairs of vectors $[\mathbf{v}_1, \mathbf{v}_2], [\mathbf{v}_2, \mathbf{v}_3], \ldots, [\mathbf{v}_8, \mathbf{v}_1]$, as

$$M = \bigcup_{k=1}^{8} M_k = \bigcup_{k=1}^{8} \{\mathbf{y} = \mathbf{x}^*(f_0, \varepsilon_0) + \alpha \mathbf{v}_k + \beta \mathbf{v}_{k+1} : \alpha, \beta \geq 0, \ \alpha + \beta \leq 1\}. \quad (7.11)$$

Here, for consistency of notation we define $\mathbf{v}_9 = \mathbf{v}_1$. Figure 7.18(b) shows a point $\mathbf{z}$ and its projection $P_k(\mathbf{z})$ onto the hyperplane containing M_k. Point P_k is defined in a conventional sense, i.e., as the point on the hyperplane that is the closest to $\mathbf{z}$. The projection corresponds to the following expansion coefficients with respect to

Element Design: Case Studies   163

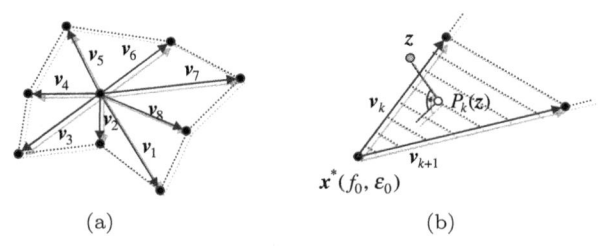

(a)   (b)

Fig. 7.18. Auxiliary components of the region of validity of the surrogate model: (a) the manifold of Fig. 7.17(b) with the spanning vectors $\mathbf{v}_k$ marked; we have $\mathbf{v}_1 = \mathbf{x}^*(f_{\min}, \varepsilon_{\min}) - \mathbf{x}^*(f_0, \varepsilon_{r0})$, $\mathbf{v}_2 = \mathbf{x}^*(f_{\min}, \varepsilon_{r0}) - \mathbf{x}^*(f_0, \varepsilon_{r0})$, $\mathbf{v}_3 = \mathbf{x}^*(f_{\min}, \varepsilon_{\max}) - \mathbf{x}^*(f_0, \varepsilon_{r0}), \ldots, \mathbf{v}_8 = \mathbf{x}^*(f_0, \varepsilon_{\min}) - \mathbf{x}^*(f_0, \varepsilon_{r0})$; (b) manifold M_k with its spanning vectors as well as a point $\mathbf{z}$ and its projection onto the hyperplane containing M_k.

$\mathbf{v}_k$ and $\mathbf{v}_{k+1}$:

$$\arg\min_{\bar{\alpha},\bar{\beta}} \|\mathbf{z} - [\mathbf{x}^*(f_0, \varepsilon_{r0}) + \bar{\alpha}\mathbf{v}_k + \bar{\beta}\mathbf{v}_{k+1}^{\#}]\|^2, \qquad (7.12)$$

where $\mathbf{v}_{k+1}^{\#} = \mathbf{v}_{k+1} - p_k \mathbf{v}_k$ with $p_k = v_k^T v_{k+1}(v_k^T v_k)$. Thus, $\mathbf{v}_{k+1}^{\#}$ is a component of $\mathbf{v}_{k+1}$ that is orthogonal to $\mathbf{v}_k$. Note that (7.12) is equivalent to

$$[\mathbf{v}_k \ \mathbf{v}_{k+1}^{\#}] \begin{bmatrix} \bar{\alpha} \\ \bar{\beta} \end{bmatrix} = \mathbf{z} - \mathbf{x}^*(f_0, \varepsilon_{r0}). \qquad (7.13)$$

The least-square solution to (7.13) is

$$\begin{bmatrix} \bar{\alpha} \\ \bar{\beta} \end{bmatrix} = (\mathbf{V}_k^T \mathbf{V}_k) \mathbf{V}_k (\mathbf{z} - \mathbf{x}^*(f_0, \varepsilon_{r0})), \qquad (7.14)$$

where $\mathbf{V}_k = [\mathbf{v}_k \mathbf{v}_{k+1}^{\#}]$. For practical reasons, we are interested in the expansion coefficients with respect to $\mathbf{v}_k$ and $\mathbf{v}_{k+1}$ given as $\alpha = \bar{\alpha} - p_k \bar{\beta}$, $\beta = \bar{\beta}$. Note that $P_k(\mathbf{z}) \in M_k$ if and only if $\alpha \geq 0$, $\beta \geq 0$, and $\alpha + \beta \leq 1$.

We define $\mathbf{x}_{\max} = \max\{\mathbf{x}^*(f_0, \varepsilon_{r0}) + \mathbf{v}_1, \ldots, \mathbf{x}^*(f_0, \varepsilon_{r0}) + \mathbf{v}_8\}$ and $\mathbf{x}_{\min} = \min\{\mathbf{x}^*(f_0, \varepsilon_{r0}) + \mathbf{v}_1, \ldots, \mathbf{x}^*(f_0, \varepsilon_{r0}) + \mathbf{v}_8\}$. Vector $\mathbf{dx} = \mathbf{x}_{\max} - \mathbf{x}_{\min}$ is the range of variation of antenna geometry parameters within the manifold M. The surrogate model domain X_S is defined as follows: a vector $\mathbf{y} \in X_S$ if and only if: (i) the set $K(\mathbf{y}) = \{k \in \{1, \ldots, 8\}: P_k(\mathbf{y}) \in M_k\}$ is not empty, and (ii) $\min\{||(\mathbf{y} - P_k(\mathbf{y}))||\mathbf{dx}|| : k \in K(\mathbf{y})\} \le d_{\max}$, where || denotes component-wise division ($d_{\max}$ is a user-defined parameter).

The first condition ensures that $\mathbf{y}$ is sufficiently close to M in a "horizontal" sense. In (ii), the normalized distance between $\mathbf{y}$ and its projection onto that M_k to which the distance is the shortest is compared to the user-defined $d_{\max}$. Due to normalization with respect to the parameter ranges $\mathbf{dx}$, $d_{\max}$ determines the "perpendicular" size of the surrogate model domain (as compared to the "tangential" size given by $\mathbf{dx}$). Therefore, a typical value of $d_{\max}$ would be 0.2 or so.

By definition, all reference designs as well as the entire manifold M belong to X_S. At the same time, the size of X_S is dramatically smaller (volume-wise) than the size of the hypercube containing the reference designs (i.e., $\mathbf{x}$ such that $\mathbf{x}_{\min} \le \mathbf{x} \le \mathbf{x}_{\max}$).

The surrogate model is constructed using kriging interpolation (Queipo et al., 2005) based on the training data sampled within X_S. The design of experiments technique is random sampling within the interval $[\mathbf{x}_{\min}, \mathbf{x}_{\max}]$ assuming uniform probability distribution. The samples allocated outside $[\mathbf{x}_{\min}, \mathbf{x}_{\max}]$ are rejected.

7.4.2. Ring slot antenna structure

Consider an antenna shown in Fig. 7.19(a) (Sim et al., 2014). It comprises a microstrip line that feeds a circular ground plane slot with a defected ground structure (DGS). The thickness and loss tangent of the substrate are 0.762 mm and 0.0018. The set of parameters is $\mathbf{x} = [l_f l_d w_d r_s s_d o g \varepsilon_r]^T \cdot \varepsilon_r$ represents the relative permittivity of the substrate. To insure the 50-ohm line impedance, the width w_f is computed for each value of ε_r. The antenna high-fidelity model $\mathbf{f}$ is implemented in CST MWS (CST, 2016) with about 300,000 mesh cells and 90 s of simulation time.

The modeling problem is already challenging due to a large number of parameters. To make it even more challenging, we assume wide ranges of operating frequencies, $f_{\min} = 2.5$ GHz to $f_{\max} = 6.5$ GHz, and substrate permittivities, $\varepsilon_{\min} = 2.0$ to

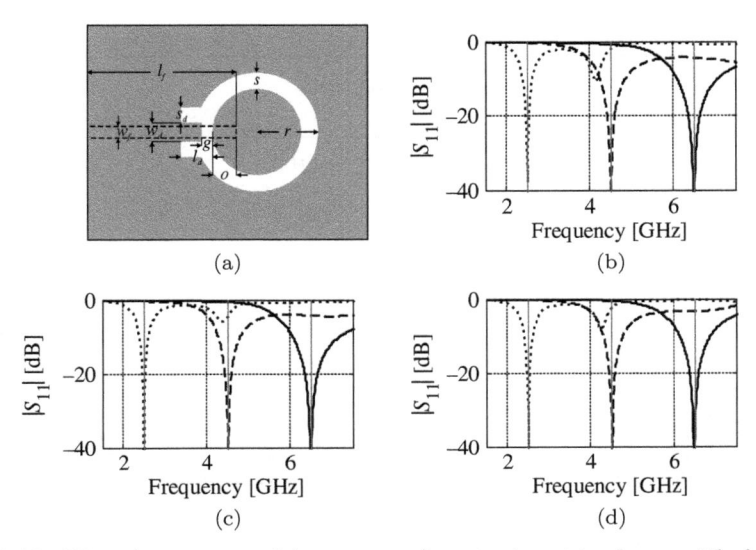

Fig. 7.19. Ring slot antenna: (a) geometry (input microstrip shown with dashed lines); reflection coefficient of the antenna for nine reference designs: (b) $\varepsilon_r = 2.0$, (c) $\varepsilon_r = 3.5$, (d) $\varepsilon_r = 5.0$; ($\cdots$) $f_0 = 2.5$ GHz, (---) $f_0 = 4.5$ GHz, (—) $f_0 = 6.5$ GHz. The vertical lines indicate the intended operating frequencies of the antenna.

$\varepsilon_{\max} = 5.0$. The reference designs have been obtained by optimizing the antenna for all combinations of $f \in \{2.5, 4.5, 6.5\}$ GHz and $\varepsilon_r \in \{2.0, 3.5, 5.0\}$ using feature-based optimization (Sec. 5.5.6 in Chapter 5). The optimization costs are 40–50 antenna simulations per design. Antenna reflection responses of all nine reference designs are shown in Figs. 7.19(b)–(d).

The modeling approach has been verified for $d_{\max} = 0.2$ by setting up the kriging interpolation surrogate with 100, 200, 500, and 1000 random samples. The test set contained 100 random points. For benchmarking, the kriging model was also constructed using 1000 training points allocated in a conventional unconstrained manner. Table 7.2 shows the average RMS errors for all considered models. Selected two-dimensional projections of the training sets for uniform and constrained sampling are shown in Fig. 7.20. It can be observed that constrained sampling allows for 3.5-fold improvement of the predictive power of the surrogate. At the same time, comparable

Table 7.2. Ring slot antenna: Modeling results.

Design space sampling and surrogate modeling technique*	Average RMS error (%)
Uniform sampling in the original space, $N = 1000$	7.3
Constrained sampling, $N = 100$	7.8
Constrained sampling, $N = 200$	5.5
Constrained sampling, $N = 500$	3.3
Constrained sampling, $N = 1000$	2.1

*In all cases, the surrogate model was constructed using kriging interpolation.

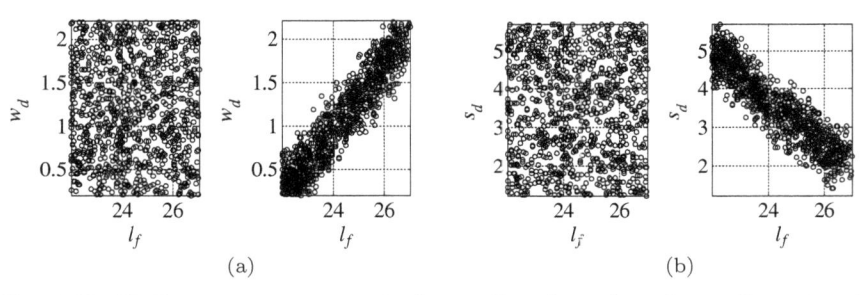

Fig. 7.20. Uniform versus constrained sampling for selected two-dimensional projections onto (a) $l_f - w_d$ plane, (b) $l_f - s_d$ plane.

modeling error is achieved with ten-fold reduction of the number of training samples. Figure 7.21 shows the surrogate and EM model responses at the selected test designs.

7.4.3. *Application examples and experimental validation*

To demonstrate a practical application of the discussed modeling approach, the antenna of Fig. 7.19 was designed — by optimizing the surrogate — for various substrate permittivities and operating frequencies. Figure 7.22 shows the optimization results for designs listed in Table 7.3. Excellent agreement between the surrogate and EM simulated reflection coefficient responses can be observed. In addition, the antenna responses are well centered at the operating frequencies.

Element Design: Case Studies   167

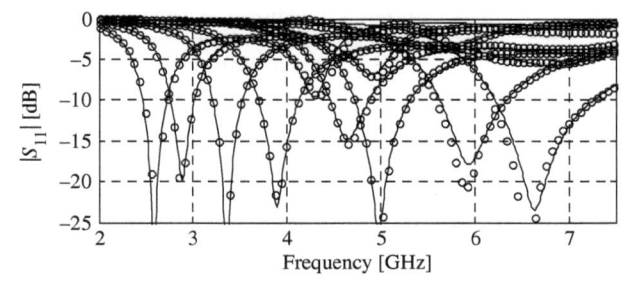

Fig. 7.21. Reflection coefficient of the ring slot antenna of Fig. 7.19(a) at the selected test designs for $N = 1000$: high-fidelity EM model (—); surrogate model (o).

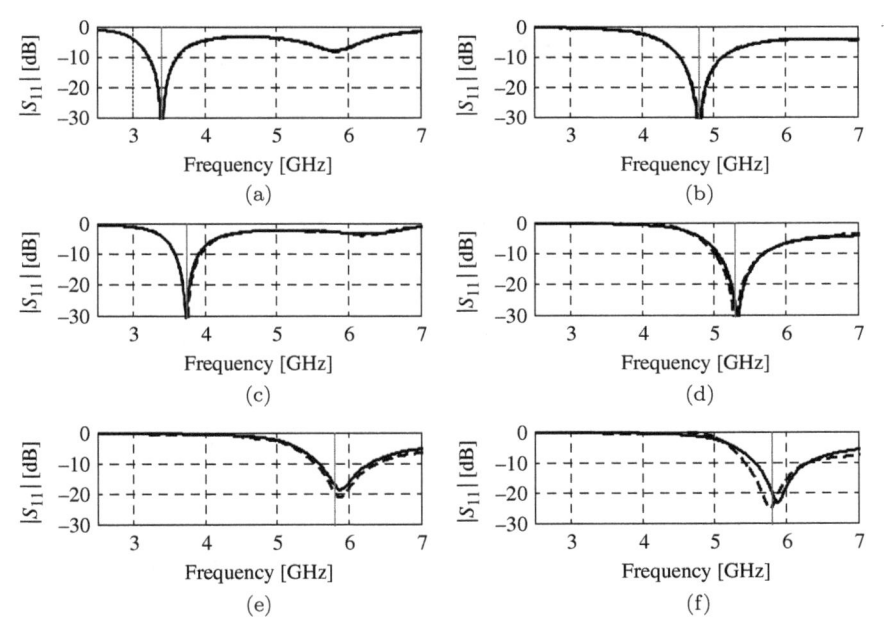

Fig. 7.22. Surrogate (- - -) and EM-simulated responses (—) of the ring slot antenna of Fig. 7.19(a) at the designs obtained by optimizing the proposed surrogate model for (a) $\varepsilon_r = 2.2$ and $f_0 = 3.4\,\text{GHz}$, (b) $\varepsilon_r = 2.6$ and $f_0 = 4.8\,\text{GHz}$, (c) $\varepsilon_r = 4.3$ and $f_0 = 3.75\,\text{GHz}$, (d) $\varepsilon_r = 4.1$ and $f_0 = 5.3\,\text{GHz}$, (e) $\varepsilon_r = 2.6$ and $f_0 = 5.8\,\text{GHz}$, (f) $\varepsilon_r = 3.5$ and $f_0 = 5.8\,\text{GHz}$. Requested operating frequencies marked using vertical lines.

168 Simulation-Based Optimization of Antenna Arrays

Table 7.3. Ring slot antenna: Optimized verification designs.

ε_r	f_0 [GHz]	l_f	l_d	w_d	r	s	s_d	o	g
2.2	3.4	23.55	5.70	1.08	12.85	5.21	3.33	5.47	0.96
2.6	4.8	22.04	5.04	0.23	9.93	3.34	4.75	5.80	1.20
4.3	3.75	24.06	5.33	0.60	10.97	3.91	4.12	5.42	0.87
4.1	5.3	22.31	4.58	0.23	8.58	3.53	5.04	5.30	2.00
2.6	5.8	22.02	4.39	0.51	7.66	3.31	4.39	5.18	1.93
3.5	5.8	22.14	4.19	0.40	7.66	3.06	4.39	4.78	2.03

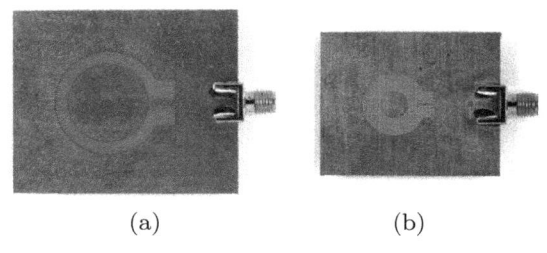

(a) (b)

Fig. 7.23. Prototypes of designs: (a) with $\varepsilon_r = 2.2$, $f_0 = 3.4$ GHz; (b) with $\varepsilon_r = 2.6$, $f_0 = 5.8$ GHz.

Antenna designs with $\varepsilon_r = 2.2$ and $f_0 = 3.4$ GHz, as well as with $\varepsilon_r = 3.5$ and $f_0 = 5.8$ GHz have been fabricated and measured. Figure 7.23 shows the photographs of the prototypes. Agreement between the simulations and measurements is good for both radiation and reflection characteristics as indicated in Fig. 7.24.

7.5. Multi-objective Design of Planar Yagi Antenna

In antenna practice it is necessary to consider several often conflicting performance requirements. The simplest way of handling various specifications is to select a primary one (e.g., size reduction in case of compact antennas) and handle the remaining ones through appropriately defined constraints. On the other hand, available trade-offs can be identified with multi-objective design optimization which finds a so-called Pareto set representing the best possible trade-offs between conflicting goals. The most popular approaches for

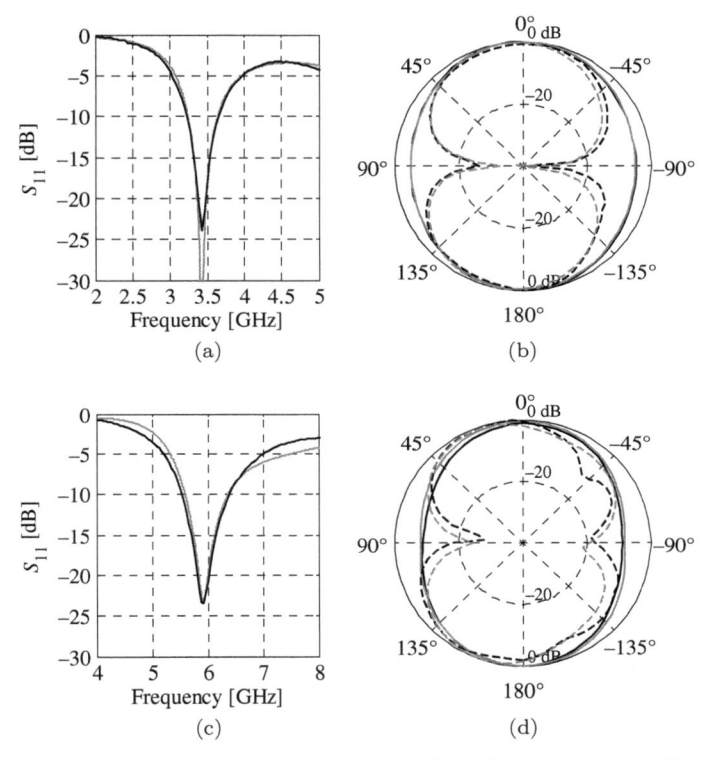

Fig. 7.24. Simulated (gray) and measured (black) reflection coefficient and radiation pattern characteristics of the antennas of Fig. 7.23: (a) and (b) design with $\varepsilon_r = 2.2$, $f_0 = 3.4\,\text{GHz}$; (c) and (d) design with $\varepsilon_r = 2.6$, $f_0 = 5.8\,\text{GHz}$. Radiation patterns are simulated and measured at the design operating frequencies, (b) 3.4 GHz and (d) 5.8 GHz. H-plane patterns are shown with solid lines and E-plane patterns are shown with dashed lines.

multi-objective optimization are population-based metaheuristics, e.g., genetic algorithms (Kerkhoff and Ling, 2007), particle swarm optimizers (Lim and Isa, 2014), etc.

In case of antennas, utilization of metaheuristics for multi-objective optimization can be often unfeasible due to high computational costs of repeated simulation of realistic EM models. Reduced-cost antenna design can be realized by means of surrogate-assisted procedures. In this section, two-objective optimization of a

170 Simulation-Based Optimization of Antenna Arrays

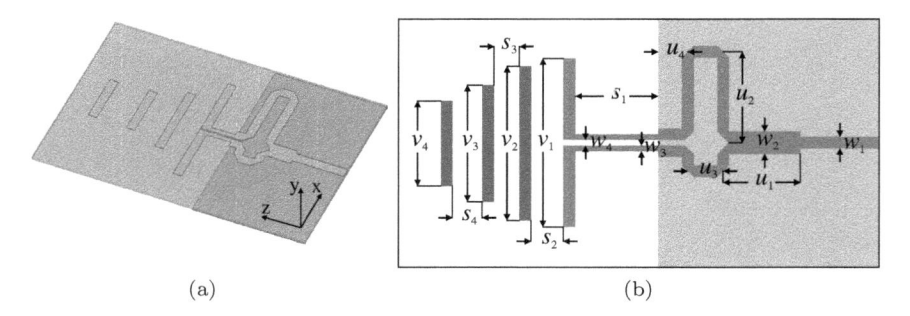

Fig. 7.25. A planar Yagi antenna: (a) view; (b) dimensional parameters.

planar Yagi antenna is discussed. More information about surrogate-based multi-objective antenna optimization is available in Koziel and Bekasiewicz (2016d).

7.5.1. *Antenna structure and problem statement*

Consider a Yagi antenna which is shown in Fig. 7.25. The 50-ohm microstrip input is connected to the dipole element through a balun, a microstrip-to-coplanar strip transition which includes a quarter-wave transformer. The antenna is implemented on a 0.762-mm Taconic RF-35 substrate (RF-35, 2018). The design variables are $\mathbf{x} = [s_1\ s_2\ s_3\ s_4\ v_1\ v_2\ v_3\ v_4\ u_1\ u_2\ u_3\ u_4]^T$ mm. Fixed parameters are $w_1 = 1.7$, $w_2 = 3$, $w_3 = 0.85$, and $w_4 = 0.85$, all in mm. The original design space is determined by the following lower/upper bounds $\mathbf{l} = [2\ 2\ 2\ 2\ 18\ 7\ 7\ 7\ 3\ 7\ 2\ 1]^T$ and $\mathbf{u} = [10\ 10\ 10\ 10\ 30\ 15\ 15\ 15\ 12\ 16\ 6\ 3]^T$. We consider the following two objectives: F_1 — minimization of voltage standing wave ratio (VSWR), and F_2 — maximization of the end-fire gain. Both objectives are considered within the 5.2–5.8 GHz frequency range. The maximum allowed VSWR is 2 for the entire frequency range. The design process utilizes two EM models, both implemented in CST Microwave Studio (CST, 2016): the high-fidelity model $\mathbf{f}$ (∼600,000 mesh cells, 25 min per simulation), and the low-fidelity model $\mathbf{c}$ (∼110,000 mesh cells, 2.5 min per simulation).

7.5.2. Design procedure and numerical results

The first step of the optimization process is to reduce the ranges of geometry parameters. Typically, the Pareto fronts for antenna structures are quite regular; therefore, it is reasonable to assume that the majority of the Pareto front is contained in the hypercube spanned by the extreme Pareto-optimal designs obtained by solving

$$\mathbf{x}^{*(k)} = \arg \min_{\mathbf{l} \leq \mathbf{x} \leq \mathbf{u}} F_k(\mathbf{c}(\mathbf{x})), \qquad (7.15)$$

where k = 1, 2, i.e., the single-objective optima. To lower the cost of the process, the single-objective optima are found using the low-fidelity EM model. The reduced bounds are $\mathbf{l}^* = \min\{\mathbf{x}^{*(1)}, \mathbf{x}^{*(2)}\}$ and $\mathbf{u}^* = \max\{\mathbf{x}^{*(1)}, \mathbf{x}^{*(2)}\}$ where min and max are understood in a component-wise sense. We have $\mathbf{l}^* = [4.05\ 3.75\ 2.93\ 2\ 22.89\ 13\ 14.6\ 8\ 4.93\ 12.34\ 4.2\ 1.96]^T$ mm, and $\mathbf{u}^* = [7.39\ 9.75\ 8.93\ 10\ 24.22\ 15\ 14.6\ 15\ 8.93\ 13.01\ 4.2\ 2.62]^T$ mm. The reduced space is five orders of magnitude (volume-wise) smaller than the initial one. The response surface approximation model was subsequently constructed using 1300 low-fidelity training samples allocated by means of Latin hypercube sampling (Beachkofski and Grandhi, 2002). The generalization error of the model is estimated using cross-validation technique (Queipo et al., 2005) and it is only 1% for VSVR and 0.1% for gain. It should be emphasized that design space reduction is essential for constructing the accurate data-driven surrogate model using a reasonable amount of training data.

The initial Pareto optimal set has been identified using a multi-objective evolutionary algorithm applied to the surrogate model **s** (setup: the population size of 500; the maximum number of iterations of 50). At the last stage, a set of ten designs has been selected from the initial Pareto set and refined using output space mapping (Koziel and Bekasiewicz, 2016d). The results are shown in Table 7.4, which includes detailed antenna dimensions for the selected designs, and Fig. 7.26, which includes the initial and refined Pareto sets.

It can be observed that the minimum VSVR is 1.177 (with the corresponding average gain of 6.47 dB). The maximum average gain possible for this antenna is 8 dB while still maintaining the VSVR

172 Simulation-Based Optimization of Antenna Arrays

Table 7.4. Yagi antenna: Dimensions of selected high-fidelity Pareto-optimal designs.

		Designs								
		$\mathbf{x}_f^{(1)}$	$\mathbf{x}_f^{(2)}$	$\mathbf{x}_f^{(3)}$	$\mathbf{x}_f^{(4)}$	$\mathbf{x}_f^{(5)}$	$\mathbf{x}_f^{(6)}$	$\mathbf{x}_f^{(7)}$	$\mathbf{x}_f^{(8)}$	$\mathbf{x}_f^{(9)}$	$\mathbf{x}_f^{(10)}$
F_1 (VSWR)		2.00	1.86	1.80	1.62	1.57	1.47	1.24	1.23	1.21	1.18
F_2 (Gain [dB])		−8.04	−7.98	−7.96	−7.93	−7.77	−7.64	−7.42	−7.41	−6.59	−6.47
Antenna	s_1	4.30	4.59	4.87	5.45	6.03	5.04	6.19	6.19	7.28	7.38
parameters	s_2	9.75	9.40	9.46	8.99	7.57	6.56	3.82	3.82	3.75	3.79
	s_3	8.44	7.43	8.10	8.28	6.94	8.05	7.24	7.24	3.73	3.08
	s_4	9.92	9.99	9.98	9.75	9.75	9.49	9.61	9.64	9.77	9.41
	v_1	22.96	23.06	23.03	23.10	23.00	23.09	23.96	23.96	23.92	24.03
	v_2	14.99	14.96	14.92	14.93	14.97	14.89	14.95	14.95	14.71	14.65
	v_3	14.60	14.60	14.60	14.60	14.60	14.60	14.60	14.60	14.60	14.60
	v_4	15.00	15.00	15.00	15.00	14.78	15.00	14.81	14.75	11.70	11.52
	u_1	5.11	5.38	5.44	5.41	5.49	8.60	7.36	7.36	7.90	8.08
	u_2	12.34	12.39	12.35	12.40	12.40	12.79	12.39	12.39	12.36	12.36
	u_3	4.20	4.20	4.20	4.20	4.20	4.20	4.20	4.20	4.20	4.20
	u_4	2.35	2.23	2.01	2.02	1.98	1.98	2.59	2.59	2.39	2.24

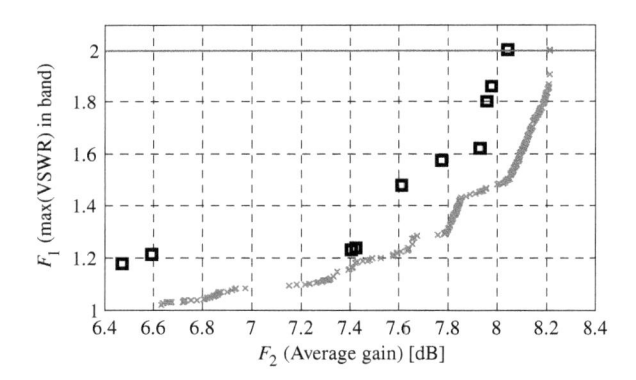

Fig. 7.26. Comparison of the low- (×) and high-fidelity (□) representations of the Pareto front (Bekasiewicz et al., 2014a). The horizontal line indicates the maximum acceptable VSWR level.

level of 2.0 within the entire frequency band. Figures 7.27 and 7.28 show reflection coefficient and gain responses of the antenna for a few designs selected along the high-fidelity Pareto-optimal set).

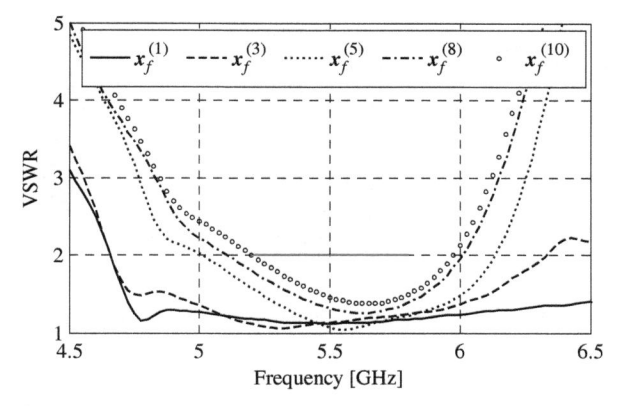

Fig. 7.27. VSWR of the Yagi antenna for designs selected from Table 7.4 (Bekasiewicz et al., 2014a). The horizontal red line marks the 5.2–5.8 GHz band.

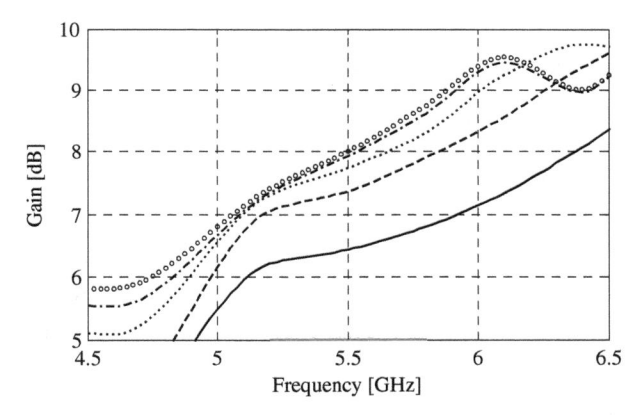

Fig. 7.28. Gain of the Yagi antenna for designs selected from Table 7.4. See Fig. 7.27 for the legend (Bekasiewicz et al., 2014a). The horizontal red line denotes 5.2–5.8 GHz.

The cost of the optimization process comprises: 334 simulations of the low-fidelity model for design space reduction, 1300 simulations of the low-fidelity model for surrogate model construction, and 30 simulations of the high-fidelity model for design refinement. Multi-objective optimization of the surrogate is very fast: the cost of which corresponds to less than one simulation of the high-fidelity model. The total cost of the Yagi antenna optimization is about 194 simulations of the high-fidelity model (∼81 hours of CPU time) that

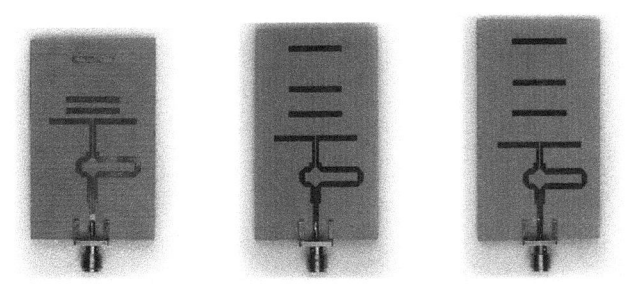

Fig. 7.29. Fabricated Yagi antenna designs: left-to-right (a) $\mathbf{x}_f^{(1)}$, (b) $\mathbf{x}_f^{(5)}$, (c) $\mathbf{x}_f^{(10)}$.

is very low in comparison to direct multi-objective optimization using population-based metaheuristics which typically needs at least a few thousands of model evaluations.

7.5.3. *Experimental validation*

Selected designs, $\mathbf{x}_f^{(1)}, \mathbf{x}_f^{(5)}, \mathbf{x}_f^{(10)}$ of Table 7.4, had been fabricated (cf. Fig. 7.29). A comparison of simulations and measurements is shown with Fig. 7.30.

Although the measurements agree with the simulations, some discrepancies can be observed. These discrepancies are mostly due to the SMA connectors which were not included in the high-fidelity models of the Yagi antenna. This resulted not only in the frequency shifts between the responses but also slightly different shapes of the characteristics.

7.6. Design of Microstrip Patch Antennas

Microstrip patch antennas (MPAs) are low-profile, mechanically robust, low-weight, integrable with printed circuits and inexpensive to manufacture with printed-circuit technology. Broadside radiation of MPAs is a convenient feature for antenna subarray and arrays. Functional disadvantages of MPAs include modest efficiency, low power handling capability, narrow impedance bandwidth. Methods to widen the MPA impedance bandwidth include the use of thicker substrates, stacking and suspending MPA radiators.

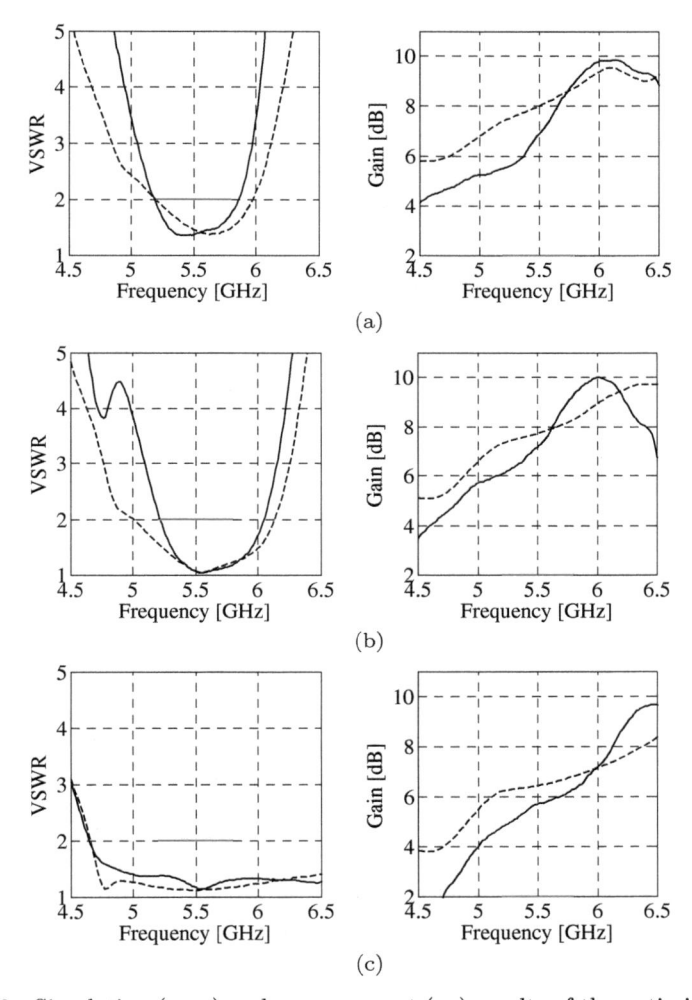

Fig. 7.30. Simulation (- -) and measurement (—) results of the optimized Yagi antennas for designs: (a) $\mathbf{x}_f^{(1)}$; (b) $\mathbf{x}_f^{(5)}$; (c) $\mathbf{x}_f^{(10)}$ (Bekasiewicz et al., 2014a). The right-hand plots represent the endfire gain. The horizontal red line denotes 5.2–5.8 GHz.

Initial MPA dimensioning in case of simple patch shapes (e.g., square, rectangular, or circular) can be easily done using the transmission line and/or cavity models (Garg et al., 2000; Waterhouse, 2003; Balanis, 2005; Jackson, 2007). However, accurate evaluation of MPAs' performance characteristics (e.g., radiation pattern, total efficiency, axial ratio, etc.) and, especially, adjustment of MPA characteristics requires repeated high-fidelity EM simulations. In this situation, surrogate-based optimization allows reliable MPA designs at reasonable time/effort costs.

In this section, we demonstrate and validate two MPA designs of configurations whose are quite often used in printed-circuit antenna arrays. One is a single-layer MPA comprising a rectangular patch fed with recessed microstrip line. Another is a double layer slot-energized MPA. Both MPAs are dimensioned for 5.8 GHz operation.

7.6.1. *Recessed microstrip line fed MPA: Geometry and model setup*

An MPA fed through a recessed 50 ohms microstrip line is outlined in Fig. 7.31. This MPA should be designed on a single layer 0.762-mm Taconic RF-35 (RF-35, 2018) for operation at and about 5.8 GHz. Metallization of the ground, input microstrip, and patch is with 18 μm copper. The vector of design variables $\mathbf{x} = [x_1\ y_1\ x_2\ y_2\ x_3\ y_3\ y_4\ y_5]^T$ includes: dimensions of the patch, x_1 and y_1; dimensions of the recession, x_2 and y_2; size and location of a rectangular aperture of the ground, $x_3\ y_3$, and y_4; and patch center-to-connector distance, y_5. The width of the input microstrip w_0 and dimension of the substrate s are fixed to 1.85 mm and 15 mm, respectively. The non-radiating aperture of the ground is introduced to enhance matching. The MPA is supposed to operate at the TM_{010} mode. The reference direction for the mode definition is the zenith. The following figures of interest are considered for formal handling: reflection coefficient in dB at the connector input with the frequency bandwidth centered at 5.8 GHz; radiation pattern in the E-plane.

The MPA is modeled and evaluated using CST MWS (CST, 2016). An SMA connector is included in the MPA models. Connector dimensions are according to the manufacturer data (SMA, 2012). Two discrete EM models are utilized: a high-fidelity model $\mathbf{f}$ comprising 1,430,000 mesh-cells and simulated in about

25 min at the initial design; a low-fidelity model **c** comprising 178,000 mesh-cells and simulated in 2 min at the initial design. The horizontal dash-dot line depicts the E-plane of the far-field that is also a magnetic wall of the EM model.

7.6.2. *Recessed microstrip line fed MPA: Optimization procedure*

Feature-based optimization (FBO) (Koziel and Bekasiewicz, 2016a) is adopted for adjustment of the MPA dimensions. The choice of FBO is motivated by typical discrepancies between the high-fidelity and low-fidelity discrete EM models of the MPA and by the narrowband and substantially resonant behaviors of the reflection coefficient and the realized gain over frequency.

7.6.3. *Recessed microstrip line fed MPA: Numerical results and experimental validation*

Dimensions of initial design, $\mathbf{x}^{ini} = [13.60\ 13.60\ 1.40\ 5.00\ 2.00\ 4.00\ 0.00\ 50.00]^T$ mm, were set up using the resonant frequency formula of the cavity model of the TM_{010} mode (Balanis, 2005) and subsequent parameter sweeps using the low-fidelity model of the MPA. The optimum $\mathbf{x}^* = [13.63\ 13.63\ 1.36\ 5.08\ 2.00\ 4.01\ 0.02\ 50.00]^T$ mm has been found in just a few iterations of the FBO algorithm.

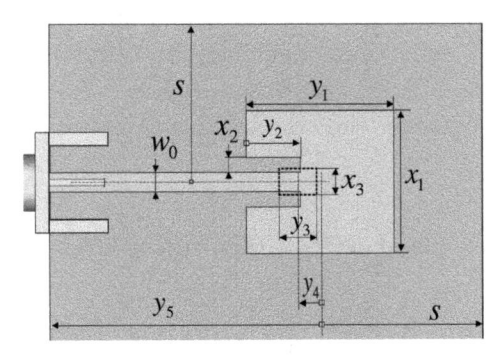

Fig. 7.31. Geometry of the recessed microstrip line fed MPA model: top view. The patch and substrate are shown transparent. The dash-line rectangle depicts the rectangular aperture of the ground.

178   *Simulation-Based Optimization of Antenna Arrays*

Fig. 7.32. Photographs of the fabricated MPA: front (a) and back (b).

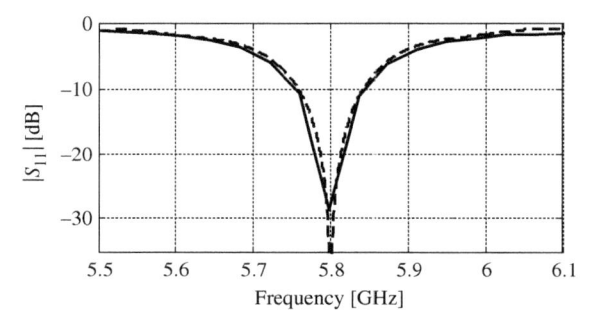

Fig. 7.33. Reflection coefficient of the optimal design: simulated MPA model **f** (- - -) and measured MPA (—).

The optimal design had been manufactured (cf. Fig. 7.32) and tested. Measured reflection coefficient and radiation patterns are compared with simulated responses in Figs. 7.33 and 7.34 where close agreement between the simulated and measured responses can be observed.

7.6.4. *Slot-energized MPA: Geometry and model setup*

This 5.8 GHz MPA is with two 0.762 mm Taconic RF-35 layers (RF-35, 2018) and shown in Fig. 7.35. The MPA patch is fed through a ground-plane rectangular slot with a 50-ohm open-end microstrip residing on the back side of the MPA. Metallization of the ground,

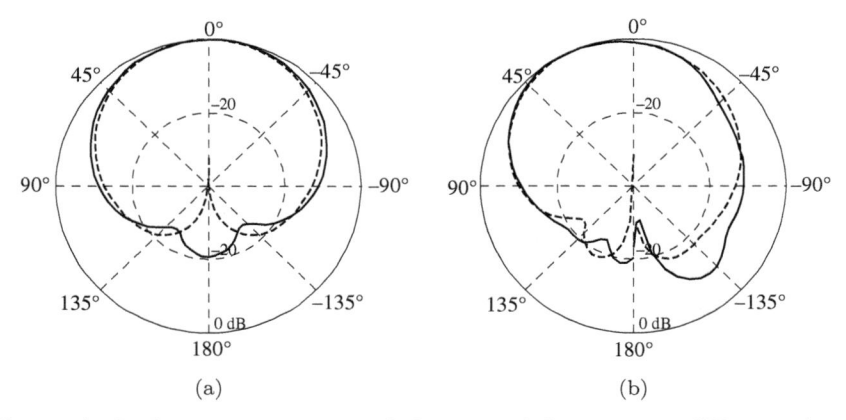

Fig. 7.34. Radiation pattern cuts of the optimal design at 5.8 GHz, simulated MPA model **f** (- - -) and measured MPA (——): (a) H-plane and (b) E-plane.

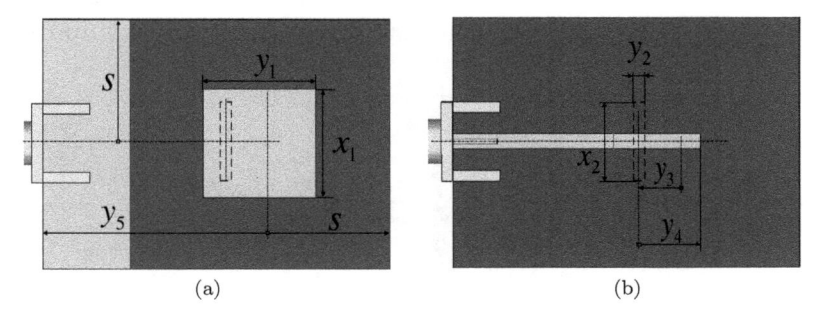

Fig. 7.35. Geometry of the slot-energized MPA model: top (a) and view (b).

input microstrip, and patch is with $18\,\mu$m copper. The vector of design variables $\mathbf{x} = [x_1\ y_1\ x_2\ y_2\ y_3\ y_4\ y_5]^T$ includes: dimensions of the patch, x_1 and y_1; dimensions and location of the slot, x_2, y_2, and y_3; length of the microstrip open-end stub (relative to the slot center); patch center-to-connector distance, y_5. The width of the input microstrip w_0 and dimension of the substrate s are fixed to 1.85 mm and 15 mm, respectively. The MPA is to operate at the TM_{010} mode where the reference direction for the mode definition is the zenith. The horizontal dash-dot lines in Fig. 7.35 depict the E-plane of the far-field and a magnetic symmetry wall of the simulated EM models. The figures of interest include: magnitude of the reflection coefficient versus frequency and radiation pattern at 5.8 GHz.

The MPA is modeled and evaluated using CST MWS (CST, 2016). An SMA connector is included in the MPA models. Connector dimensions are according to the manufacturer data (SMA, 2012). Two discrete EM models are utilized: a high-fidelity model **f** comprising 1,604,000 mesh-cells and simulated in about 13 min and 40 s at the initial design; a low-fidelity model **c** comprising 116,365 mesh-cells and simulated in 1.5 min at the initial design.

7.6.5. *Slot-energized MPA: Design with optimization, results, and validation*

Dimensions of this MPA are adjusted using feature-based optimization (FBO) (Koziel and Bekasiewicz, 2016a). Initial dimensions, $\mathbf{x}^{ini} = [13.00\ 13.05\ 9.00\ 1.00\ 5.50\ 9.00\ 50.00]^T$ mm, were set up using

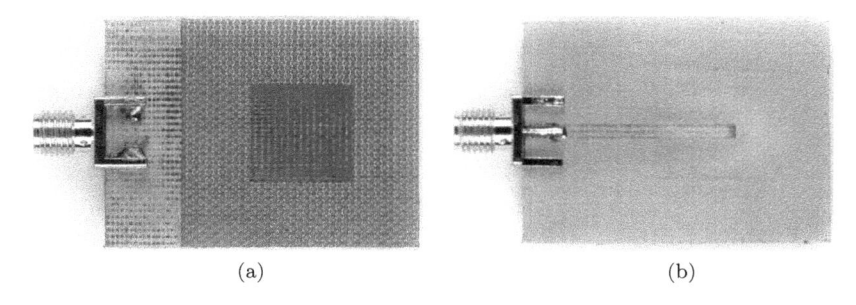

Fig. 7.36. Photographs of the fabricated MPA: front (a) and back (b).

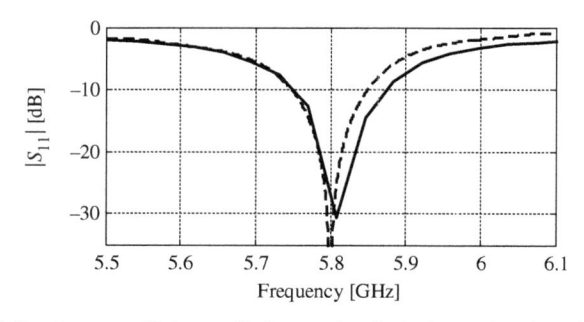

Fig. 7.37. Reflection coefficient of the optimal design: simulated MPA model **f** (- - -) and measured MPA (—).

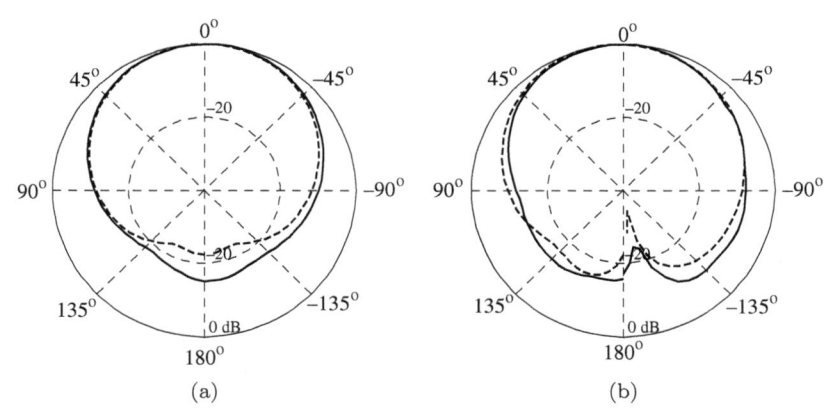

Fig. 7.38. Radiation pattern cuts of the optimal design at 5.8 GHz, simulated MPA model **f** (---) and measured MPA (——): (a) H-plane and (b) E-plane.

the resonant frequency formula of the TM_{010} mode (Balanis, 2005) and subsequent parameter sweeps using the low-fidelity model **c**. The optimum $\mathbf{x}^* = [12.90\ 13.13\ 9.01\ 1.04\ 5.68\ 8.52\ 49.84]^T$ mm has been found in a few iterations of the FBO algorithm.

The optimal design had been manufactured (cf. Fig. 7.36) and tested. Manufacturing tolerances of metallization are ±4 um. Measured reflection coefficient response and radiation patterns are compared with those simulated in Figs. 7.37 and 7.38 where close agreement between the responses can be observed.

Chapter 8

Microstrip Antenna Subarray Design Using Simulation-Based Optimization

Subarrays of microstrip antennas are important elements of contemporary low-profile arrays (Mailloux, 2005; Haupt, 2010). The low profile and low fabrication costs of microstrip subarrays justify their utility. Moreover, they can also be produced with the same fabrication techniques that are used to realize elements of amplitude and phase control, amplification, as well as other necessary antenna arrays elements, all on the same substrate (Garg *et al.*, 2000; Waterhouse, 2003). At the same time, microstrip patch antennas are characterized by narrow fractional bandwidth when built on thin substrates and in their basic configurations (Jackson, 2007).

One of the most popular configurations of microstrip patch antenna subarrays (MPASs) is the one with the corporate feed residing on the same side of the substrate as the patches (Munson, 1974; Hall and Hall, 1988; Balanis, 2005). For this case, currents of the feed can significantly contribute to certain sensitive radiation figures, e.g., to the sidelobe level (SLL) (Hall and Hall, 1988; Levine *et al.*, 1989). Therefore, as far as low SLLs and well-matched designs are of interest, the electromagnetic (EM) models of the subarrays should include both the radiating elements and the feed. Reliable EM modeling also turns to be essential for another MPAS configuration which comprises patch radiators and a microstrip feed built on two substrates separated by a common ground plane. This configuration allows a more compact footprint (Secmen et al., 2006; Jackson, 2007) and alleviates the issue of parasitic radiation of the feed.

Optimal designs of subarrays of the aforementioned types cannot be found by simple means, e.g., using manual parameter sweeps, due to complex EM interactions within the structure. Thus, optimization of the subarrays, with respect to both radiation and reflection figures, and the subarrays' dimensions considered as variables, is necessary for obtaining optimal designs meeting given performance requirements. On the other hand, a large number of designable parameters and the corresponding CPU costs of repeated EM simulations (performed within the optimization loop) make such optimization task challenging.

In this chapter, we discuss and demonstrate design of microstrip subarrays involving a customized surrogate-based optimization (SBO) algorithm and variable-fidelity EM simulations (Koziel and Ogurtsov, 2014a). Numerical examples of four-patch subarray illustrate the SBO algorithm in terms of quality of the final designs and computational costs. Simulated and measured responses of three selected optimal designs of the one-side MPAS configuration are provided to illustrate reliability of the discussed approach.

8.1. Design Method: Optimization Algorithm

MPAS design is formulated as a nonlinear minimization problem (Koziel and Ogurtsov, 2014a)

$$\mathbf{x}^* = \arg\min_x U(\mathbf{f}(\mathbf{x})), \qquad (8.1)$$

where $\mathbf{f}$ is a structure's response, a radiation pattern cut at a frequency of interest or the reflection coefficient over a frequency range of interest, whereas $\mathbf{x}$ is a vector of designable parameters of the subarray of interest. The objective function U encodes the design specifications.

Antenna design is a multi-objective problem in general. Here, we handle objectives by means of scalarization (Nocedal and Wright, 2006). Specifically, only one objective, the SLL is optimized directly. Sufficient levels for other objectives are secured by means of suitable penalty functions (Nocedal and Wright, 2006). For example, the objective function, defined to minimize the SLL while maintaining the reflection coefficient at frequency f_0 at -20 dB level (or lower), is defined as

$$U(f(\mathbf{x})) = U_{\text{SLL}} + \beta \cdot \max(|S_{11}|(f_0) + 20, 0)^2. \qquad (8.2)$$

Here, U_{SLL}, the SLL for a given angular sector, is our primary objective. The second term of (8.2) is a penalty controlling the reflection coefficient at the -20 dB level at the frequency of interest f_0. The penalty factor β is normally set to a large value, here, 1000. In case of several objectives, the function U would be defined as

$$U(f(\mathbf{x})) = U_{\text{SLL}} + \sum_{k=1}^{K} \beta_k \cdot p_k(f(\mathbf{x}))^2, \qquad (8.3)$$

where β_k and $p_k(\mathbf{f}(\mathbf{x}))$ are the kth penalty coefficient and penalty function, respectively. The penalty functions $p_k(\mathbf{f}(\mathbf{x}))$ are normally defined so that $p_k(\mathbf{f}(\mathbf{x})) = 0$ for designs at which a given objective satisfies its design specifications and $p_k(\mathbf{f}(\mathbf{x})) > 0$ if its design specifications are violated.

Solving problem (8.1) through direct optimization of the high-fidelity model $\mathbf{f}$ can be very challenging in a case of many variables (here, 15) and simulation time of model $\mathbf{f}$ (20 min and more per design in this work). To alleviate the challenge of direct optimization of the high-fidelity model, a series of approximate solutions to (8.1) is generated as

$$\mathbf{x}^{(i+1)} = \arg\min_x U(\mathbf{s}^{(i)}(\mathbf{x})), \qquad (8.4)$$

where $\mathbf{s}^{(i)}$ denotes the surrogate model at iteration i. The surrogate model is constructed from the low-fidelity model $\mathbf{c}$ which is a coarse-mesh version of the high-fidelity model $\mathbf{f}$. Model $\mathbf{c}$ is simulated with the same EM solver as model $\mathbf{f}$. Other simplifications of model $\mathbf{c}$

include: relaxed convergence criteria; modeling metals as perfect conductors; evanescent thickness of metallization; lossless dielectrics; and frequency independent materials.

A typical ratio of the simulation times of models **f** and **c** is between 10 and 50. The reduced simulation time of model **c** is obtained at the expense of its degraded accuracy. To serve as a reliable surrogate in the design process, model **c** has to be corrected. Two types of corrections to align responses of model **s** with those of model **f** are used: (i) frequency scaling which is applied to the reflection coefficient of **c**; and (ii) additive response correction which is applied to both reflection coefficient and realized gain patterns.

Frequency scaling is an affine transformation defined as $F(\omega) = f_0 + f_1\omega$ (Bandler et al., 2004a), where f_0 and f_1 are unknown parameters to be determined. Let $\mathbf{c}(\mathbf{x}) = [c(\mathbf{x}, \omega_1), \ldots, c(\mathbf{x}, \omega_m)]^T$ represent the reflection coefficient of the low-fidelity model **c** evaluated at frequencies $\omega_j, j = 1, \ldots, m$. The frequency-scaled model is then defined as (Bandler et al., 2004a)

$$\mathbf{s}_F(\mathbf{x}) = [c(\mathbf{x}, F(\omega_1)), \ldots, c(\mathbf{x}, F(\omega_m))]^T, \quad (8.5)$$

where

$$[f_0^{(i)}, f_1^{(i)}] = \arg \min_{[f_0, f_1]} \sum_{k=1}^{m} [\mathbf{f}(\mathbf{x}^{(i)}, \omega_k) - \mathbf{s}_F(\mathbf{x}^{(i)}, f_0 + f_1\omega_k)]^2. \quad (8.6)$$

Frequency-scaled parameters are obtained without referring to an EM simulation because all the necessary responses $\mathbf{s}_F(\mathbf{x}^{(i)}, f_0 + f_1\omega_k)$ are evaluated with interpolating/extrapolating the known values $\mathbf{c}(\mathbf{x}^{(i)}, \omega_k), k = 1, \ldots, m$.

An additive response correction is implemented as

$$\mathbf{s}^{(i)}(\mathbf{x}) = \mathbf{c}(\mathbf{x}) + \mathbf{f}(\mathbf{x}^{(i)}) - \mathbf{c}(\mathbf{x}^{(i)}). \quad (8.7)$$

Correction (8.7) ensures zeroth-order consistency (Alexandrov and Lewis, 2001) between the surrogate and the high-fidelity model at $\mathbf{x}^{(i)}$. The correction of the reflection coefficient is subsequently applied to the frequency-scaled model (8.5).

Figure 8.1 shows the effect of frequency scaling on the reflection coefficient, as well as the necessity of response correction due to

Microstrip Antenna Subarray Design 187

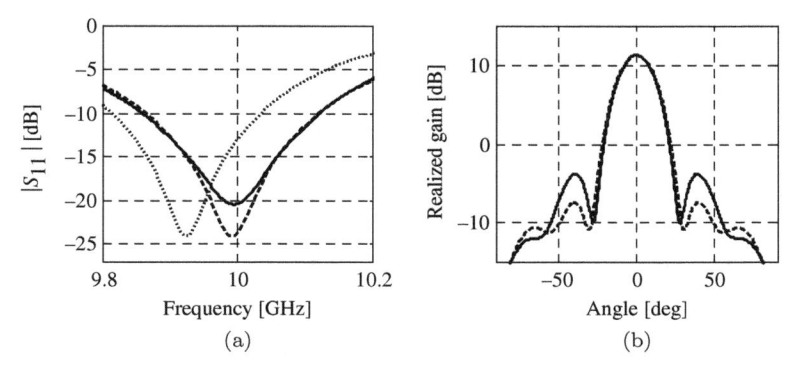

Fig. 8.1. Example of simulated responses of the models utilized in the SBO algorithm: (a) reflection coefficient of the low- (···) and high- (—) fidelity models, and the frequency-scaled (- - -) low-fidelity model at an example design; (b) realized gain pattern cut of the low- (- - -) and high- (—) fidelity models at the same design indicating the necessity of response correction (8.7) to adequately represent the gain pattern by the surrogate model.

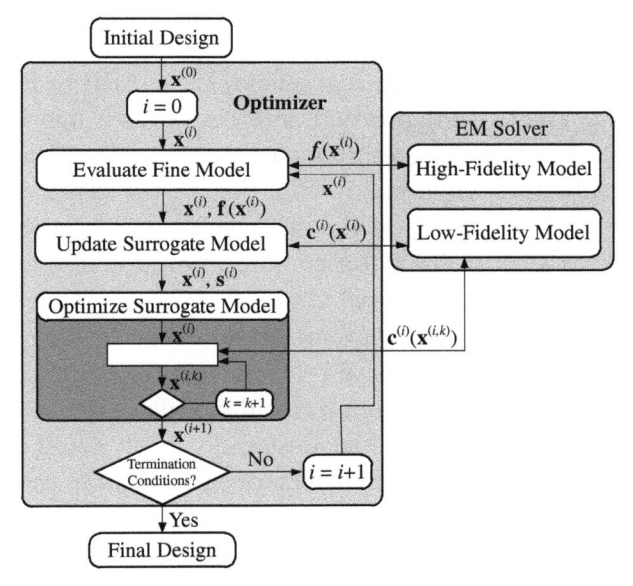

Fig. 8.2. Simulation-driven design: an SBO approach. The original (high-fidelity) model **f** is evaluated once per iteration. The new approximation of the optimum design $\mathbf{x}^{(i)}$ is obtained by optimizing the surrogate model **s**. For a well-working SBO algorithm, the number of iterations (thus, the number of high-fidelity EM simulations) is low, which dramatically reduces the design cost compared to conventional optimization techniques.

misalignment between the low- and high-fidelity patterns beyond the main-beam, particularly for low SLLs. Correction terms are generally design-dependent. Therefore, a few iterations of the algorithm (8.2) are necessary to arrive at the optimum design. The flow of the SBO algorithm is outlined in Fig. 8.2.

8.2. Design Formulation and EM Models of Microstrip Antenna Subarrays

The following task is considered: design a microstrip fed four patch linear MPAS (cf. Fig. 8.3) residing on a 0.5-mm thick Taconic RT-35 substrate and operating at 10 GHz. Cases targeting designs with different beamwidths and SLLs in the H-plane are considered: Design I with the 60° beamwidth; Design II with the 50° beamwidth; and Design III with the 40° beamwidth. Beamwidth is defined so that the pattern is minimized beyond the chosen range of angles. For all designs, the realized gain toward the zenith should not be less than 10 dB and the reflection coefficient should be −15 dB or lower, all at 10 GHz. Two MPAS configurations are considered: one is with the corporate feed and patches residing on the same side of the grounded microstrip substrate as shown in Fig. 8.3(a); the other one is with the corporate feed residing on the other substrate than the patches, shown in Fig. 8.3(b). In the latter configuration, the feed and patches are interfaced with through-hole vias.

It is worth to mention that inclusion of the feeding structures into the EM model is essential for adequate evaluation of the subarrays. To illustrate that point, results of numerical experiments are presented below for selected designs of both configurations of MPASs. The results show the effect of the feed on the radiation response, in particular, on the SLL.

For both configurations, the initial settings are obtained as follows. Given the permittivity and thickness of the substrate, the resonant dimensions of a single patch are estimated as 8 mm × 8 mm. The inset type of the feed is adopted for a one-side configuration shown in Fig. 8.3(a). Dimensions of the patch, width of the input microstrip, and the inset section are found using optimization of the

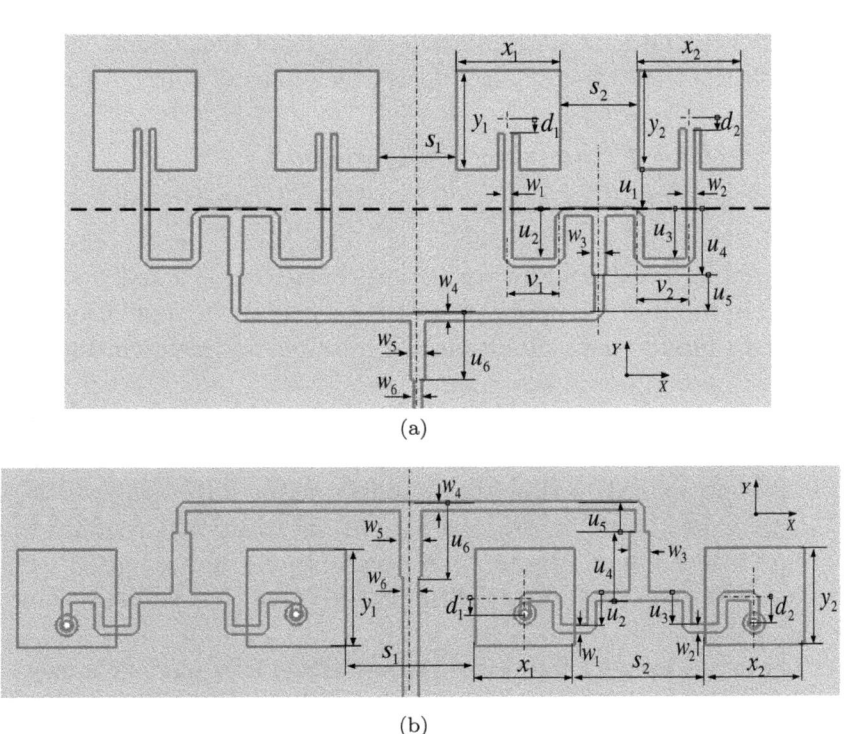

Fig. 8.3. Microstrip patch antenna subarray (MPAS) configurations, layouts: (a) MPAS I with one substrate and one metallization layer (in addition to the ground plane); (b) MPAS II with two substrates and two metallization layers (in addition to the ground plane). The symmetry plane is marked with the vertical dash-dot line at the center.

EM model of a single patch antenna. The single patch antenna is defined and simulated with the CST MWS software (CST, 2016).

Similarly for the other MPAS configuration, dimensions of the patch and location of the probe are found using simulation-based optimization of the corresponding EM model of a single antenna. Subsequently, these dimensions are used as the initial values of the design variables of the MPAS models. All EM models are defined and simulated with infinite lateral dimensions of the grounded substrate.

Numerical experiments are conducted as follows. First, MPAS models without the corporate feed are optimized. Then, having the optimal dimensions and spacings of the subarrays, the corporate feeds are added to the models. Radiation patterns of the models without and with the feed are compared for every subarray configuration.

8.2.1. MPAS I (one-side configuration)

The first numerical experiment compares normalized radiation patterns of the following two models. One is an MPAS model optimized for the reduced SLL and low reflection coefficient; it is fed with four microstrip inputs (the upper part of Fig. 8.3(a)). In simulations the microstrip inputs were simultaneously excited by identical incident waves.

The design variables are $\mathbf{x} = [s_1\ s_2\ y_1\ y_2\ d_1\ d_2\ w_1\ w_2]^T$. The patches are assumed to be square ($x_1 = y_1$ and $x_2 = y_2$). The SLL was optimized having the nominal major lobe beamwidth of 50 degrees as shown in Fig. 8.4(a) and an additional requirement for the realized gain toward the zenith to be above 10 dB. The H-plane radiation pattern with SSL of -16 dB of that model is shown in Fig. 8.4(a).

The second model comprises the optimized first model connected to a corporate feed. The corporate feed was designed with an ordinary mean of quarter wave transformers (Pozar, 2011). The H-plane SLLs of the two models are noticeably different as shown in Fig. 8.4(a). The E-plane patterns of the models also differ noticeably in Fig. 8.4(b). The structural difference of the models is only in their feeds, and consequently, in the simulated currents. This difference is illustrated in Fig. 8.5 which shows the current amplitude distribution in the substrate at 10 GHz. A standing wave pattern at the input of the corporate feed model shown in Fig. 8.5(b) as well as the minimum of the simulated reflection coefficient located at 9.91 GHz in Fig. 8.4(c) both indicate a need of further simulation-based tuning of the MPAS designs.

8.2.2. MPAS II (two-side configuration)

Two models of the two-side configuration are shown in Fig. 8.3(b). The first one is fed with probes, and optimized for reduced SLL and

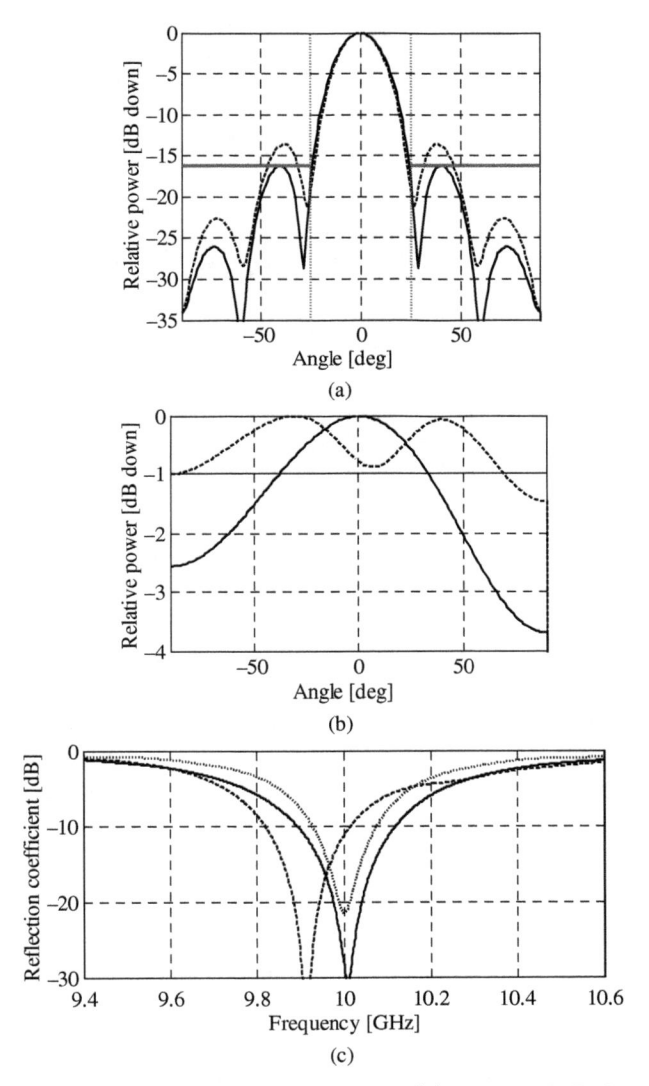

Fig. 8.4. MPAS I 50-degree beamwidth design: (a) simulated H-plane radiation pattern with the specifications imposed on the SLL and the major lobe beamwidth and (b) E-plane normalized radiation pattern at 10 GHz, with four microstrip inputs (—) and a corporate feed (– –); (c) simulated reflection coefficients at four microstrip inputs (—) and (···), and at the input of the corporate feed (– –).

192 Simulation-Based Optimization of Antenna Arrays

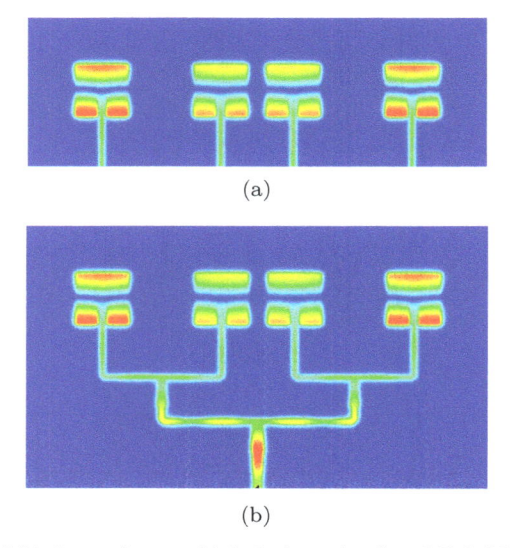

Fig. 8.5. MPAS I 50-degree beamwidth design, simulated E-field amplitude in the substrate at 10 GHz: (a) according to the model without the feed; (b) according to the model with the feed.

low reflection. Design variables are $\mathbf{x} = [s_1\ s_2\ y_1\ y_2\ d_1\ d_2\ R_0]^\mathrm{T}$, where R_0 is the impedance of the probe sources to be cast into the width, w_1 and w_2 (here $w_1 = w_2$) of the input microstrips at the feed design step. The patches are set to be square. The SLL was optimized with the nominal major lobe beamwidth of 40 degrees as shown in Fig. 8.6(a) and with an additional requirement for the realized gain toward zenith to be above 10 dB. The second model comprises the first one, the optimized probe-fed patches, and a feed. The radiators and the feed are connected through vias as depicted in Fig. 8.3(b).

Simulated patterns and reflection coefficients of the two models are shown in Fig. 8.6. The radiation patterns and the reflection coefficients indicate that simulation-based tuning is required. It can be concluded, that as far as a low SSL and good matching are simultaneously of interest, EM models of the MPAS need to include the feed with its dimensional parameters being a part of the design variable vector. Such models will be repeatedly simulated and adjusted through the SBO process as described further.

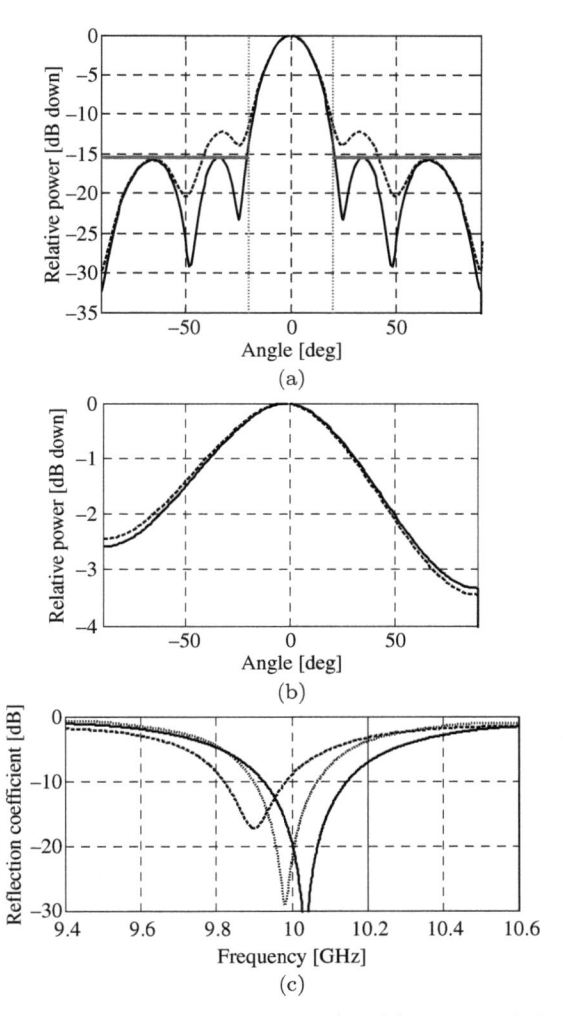

Fig. 8.6. MPAS II 40-degree beamwidth design: (a) simulated H-plane pattern and (b) E-plane pattern, both at 10 GHz, with four probe inputs (—) and a corporate feed (– –); (c) simulated reflection coefficients at the probes (—) and (· · ·), and at the input of the corporate feed (– –).

8.3. Optimization Results

The SBO algorithm described in Sec. 8.1 is used for both configurations, MPAS I and MPAS II. Optimization results outlined are presented per configuration.

8.3.1. MPAS I (one-side configuration)

We use two simulated EM models per subarray: the original (high-fidelity) model **f**, and the coarse-discretization (low-fidelity) model **c**. Both models are defined with CST MWS and simulated using the CST transient solver on 8-core Intel Xeon 2.00 GHz processor with 64 GB RAM. The model **f** comprises 852,000 mesh cells at the initial design and simulates in 18 min. The model **c**, a coarse-mesh version of **f**, comprises only 72,500 mesh cells at the initial design and simulated in less than 1 min. Other simplifications in the low-fidelity model **c** are the following: lossless materials; a coarser residual energy termination condition (−25 dB for model **c** versus −40 dB for model **f**).

The vector of design variables includes dimensions of the patches, spacings of the patches, insets, and dimensions of feed: $\mathbf{x} = [s_1 \ s_2 \ y_1 \ y_2 \ d_1 \ d_2 \ w_1 \ w_2 \ u_3 \ u_4 \ u_6 \ w_3 \ w_4 \ w_5]^T$, as shown in Fig. 8.3. Other parameters are either fixed or dependent. Lateral extensions of the ground and substrate are modeled and simulated as infinite. The MPASs are fed with 50 ohms inputs (the width of the input microstrip $w_6 = 1.15$ mm). Metallization is with 50 µm thick copper. The meander sections are introduced in the corporate feed of the designs as shown in Fig. 8.3(a). Final dimensions of the sections are to be determined through optimization.

Cases, corresponding to 60-, 50-, and 40-degree beamwidths in the H-plane, referred as Design I, II, and III, have been considered. For each case the H-plane SLL and reflection coefficient were minimized at 10 GHz. The initial design $\mathbf{x}^{(0)} = [8.0 \ 8.0 \ 7.85 \ 7.85 \ -1.0 \ -1.0 \ 0.65 \ 0.65 \ 8.0 \ 8.0 \ 5.5 \ 6.0 \ 1.1 \ 0.65 \ 1.35]^T$ mm shows SLL = −10.7 dB, $|S11|$ = −15 dB at 10 GHz, and the minimum of the reflection coefficient at 9.94 GHz.

For Design I, the optimum, $\mathbf{x}^{(3)} = [8.1 \ 10.0 \ 7.865 \ 7.870 \ -1.0 \ -0.67 \ 0.63 \ 0.67 \ 8.0 \ 8.2 \ 5.7 \ 6.1 \ 1.1 \ 0.65 \ 1.35]^T$ mm with SLL = −20 dB and $|S11|$ = −30 dB at 10 GHz, has been obtained in three iterations of the algorithm (8.4) at the total cost corresponding to only about 25 evaluations of the high-fidelity model **f**. It took ∼150 × **c** and 1 × **f**

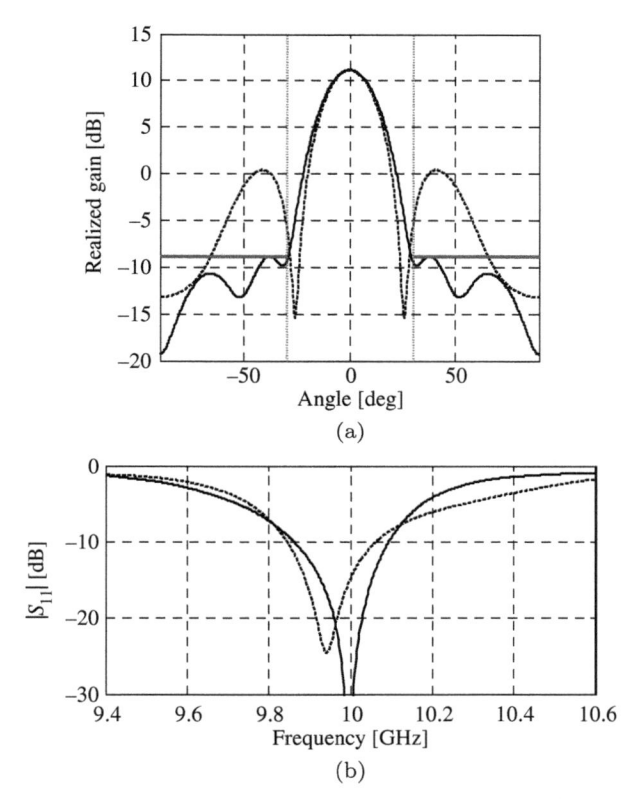

Fig. 8.7. Design I (MPAS I with 60-degree H-plane beamwidth) optimized (—) and initial (- - -): (a) H-plane realized gain; (b) reflection coefficient.

simulations per iteration. Figure 8.7 shows the subarray's responses at the initial $\mathbf{x}^{(0)}$ and optimized $\mathbf{x}^{(3)}$ designs.

The optimum of Design II, $\mathbf{x}^{(4)} = [10.01\ 12.60\ 7.855\ 7.86\ -2.0\ -0.67\ 0.68\ 0.72\ 8.5\ 8.0\ 6.3\ 5.12\ 0.95\ 0.75\ 1.15]^T$ mm with SLL $= -18.2$ dB and $|S11| = -32$ dB at 10 GHz, has been obtained in four iterations of the algorithm (8.4) at the total cost corresponding to only 33 simulations of model $\mathbf{f}$. Figure 8.8 shows subarray's responses at the initial $\mathbf{x}^{(0)}$ and optimized $\mathbf{x}^{(4)}$ designs.

The optimum for the Design III, $\mathbf{x}^{(4)} = [8.5\ 12.8\ 7.865\ 7.86\ -1.5\ -0.87\ 0.63\ 0.72\ 8.5\ 8.0\ 6.5\ 6.12\ 1.1\ 0.7\ 1.15]^T$ mm with

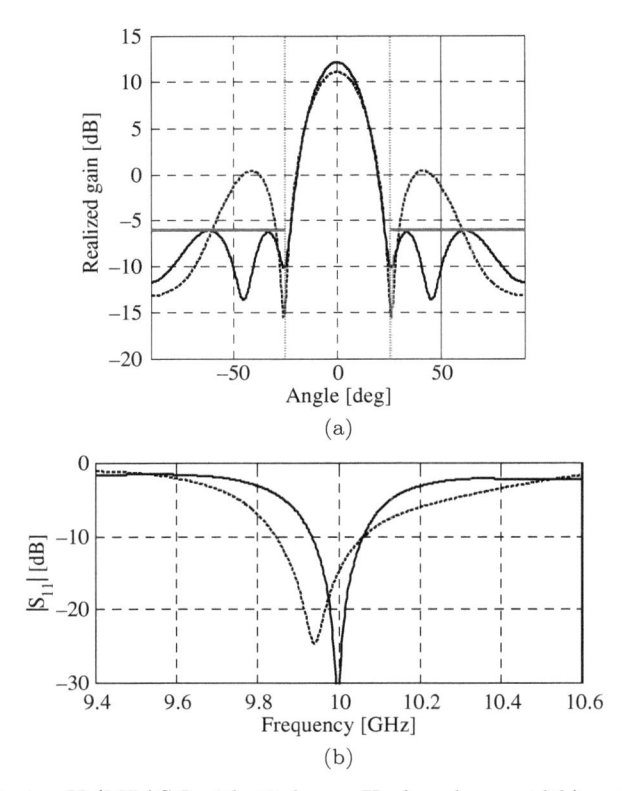

196 *Simulation-Based Optimization of Antenna Arrays*

Fig. 8.8. Design II (MPAS I with 50-degree H-plane beamwidth) optimized (—) and initial (- - -): (a) H-plane realized gain; (b) reflection coefficient.

SLL = −16 dB and $|S11|$ = −26 dB at 10 GHz, has been found in four iterations of the algorithm (8.4) at the total cost corresponding to 33 evaluations of **f**. The initial and optimized responses of the Design III are shown in Fig. 8.9. The impedance transformer section with dimensions w_5 and u_6 has been eliminated in Designs II and III with optimization by bringing the optimal value of w_5 to 1.15 mm.

The peak realized gain of the optimal designs, 11.2 dB of Design I, ∼12.2 dB of Design II, and 11.7 of Design III, are close to the maximal attainable value of 12.2 dB which is estimated assuming uncoupled lossless antennas and with the single antenna peak directivity of 6 dBi.

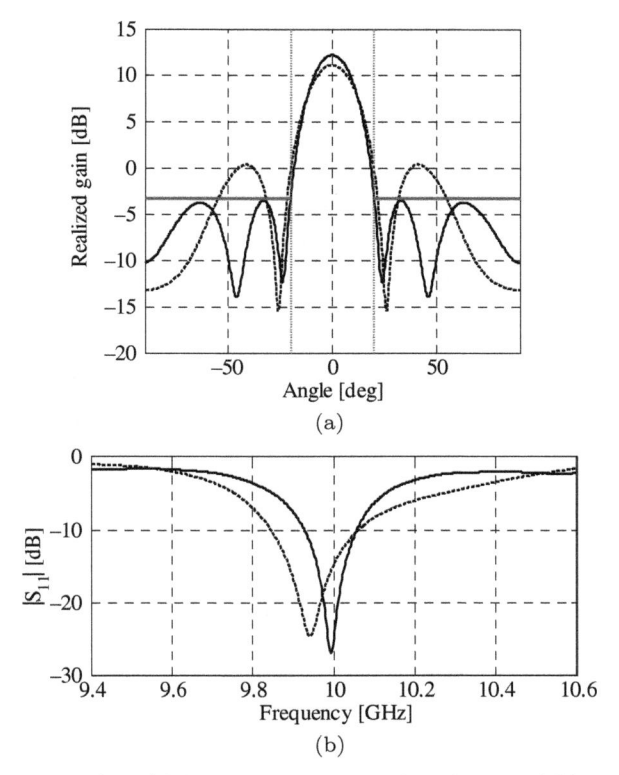

Fig. 8.9. Design III (MPAS I with 40-degree H-plane beamwidth) optimized (—) and initial (- - -): (a) H-plane realized gain; (b) reflection coefficient.

8.3.2. *MPAS II (two-side configuration)*

The computational savings can be even more substantial when the SBO methodology applies to the two-side subarray configuration (Fig. 8.3(b)). We consider two cases of the 60- and 50-degree H-plane beamwidth. The design variables are $\mathbf{x} = [s_1\ s_2\ d_1\ d_2\ y_1\ y_2\ u_2\ u_3\ u_4\ u_6\ w_1\ w_3\ w_5]^T$. Trace widths are set to $w_1 = w_2 = w_4$. The initial design for both cases is $\mathbf{x}^{(0)} = [10.5\ 10.5\ 7.55\ 7.55\ -1.5\ -1.5\ 0.0\ 0.0\ 5.0\ 6.0\ 0.65\ 1.2\ 1.4]^T$ mm.

Discrete EM models of this subarray configuration are more graded and comprise larger numbers of mesh cells. In particular, model $\mathbf{f}$ comprises 3,100,000 mesh cells and needs 45 min for simulation, whereas the coarse-mesh model **c** comprises 265,000 cells and needs 3 min, both at the initial design.

For Design I, the optimum, $\mathbf{x}^{(2)} = [9.9\ 9.83\ 7.55\ 7.54\ -2.1\ -1.3\ 0.17\ 0.0\ 5\ 5.47\ 0.67\ 1.48\ 1.5]^T$ mm with SLL $= -20$ dB, has been found in two iterations of the SBO algorithm at the total cost corresponding to about 19 evaluations of the high-fidelity model **f**. Responses of the optimum are shown in Fig. 8.10. Realized gain toward the zenith is 10 dB.

For Design II, the optimum, $\mathbf{x}^{(3)} = [9.9\ 9.9\ 7.55\ 7.55\ -2.0\ -1.4\ 0.0\ 0.0\ 5.0\ 5.33\ 0.67\ 1.50\ 1.5]^T$ mm with SLL $= -18$ dB, has been found in three iterations at the total cost corresponding to about 27 evaluations of the high-fidelity model **f**. Responses of this optimum are shown in Fig. 8.11.

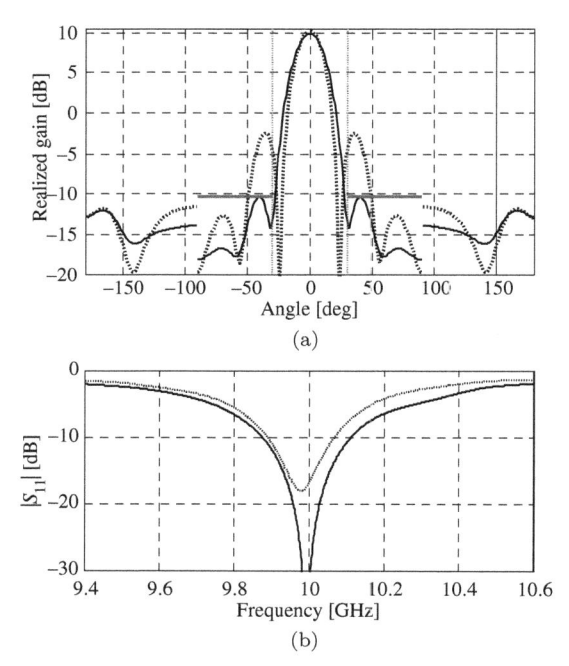

Fig. 8.10. Design I (MPAS II with 60-degree H-plane beamwidth) optimized (—) and initial (· · ·): (a) H-plane realized gain; (b) reflection coefficient.

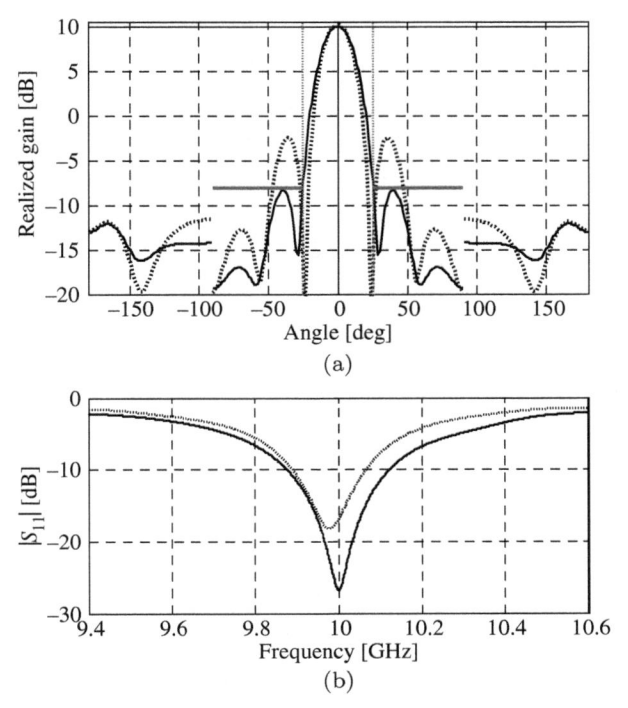

Fig. 8.11. Design II (MPAS II with 50-degree H-plane beamwidth) optimized (—) and initial (···): (a) H-plane realized gain; (b) reflection coefficient.

8.4. Validation by Measurements

The optimized designs of the one-side configuration have been manufactured on the 96 × 68 mm RF-35 microstrip substrate and equipped with 50 ohms SMA connectors as shown in Fig. 8.12. Radiation and reflection responses of the fabricated designs have been measured in the Anechoic Chamber of Gdansk University of Technology using a setup with a dual polarized horn antenna (QPar, 2012) and E5071C ENA Network Analyzer (Agilent, 2013). Measured and simulated responses are in close agreement for Designs I–III as shown in Figs. 8.13–8.15.

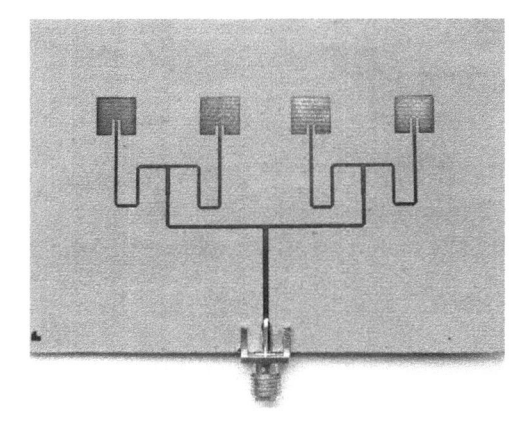

200 *Simulation-Based Optimization of Antenna Arrays*

Fig. 8.12. Photograph of a fabricated subarray: Configuration MPAS I Design II.

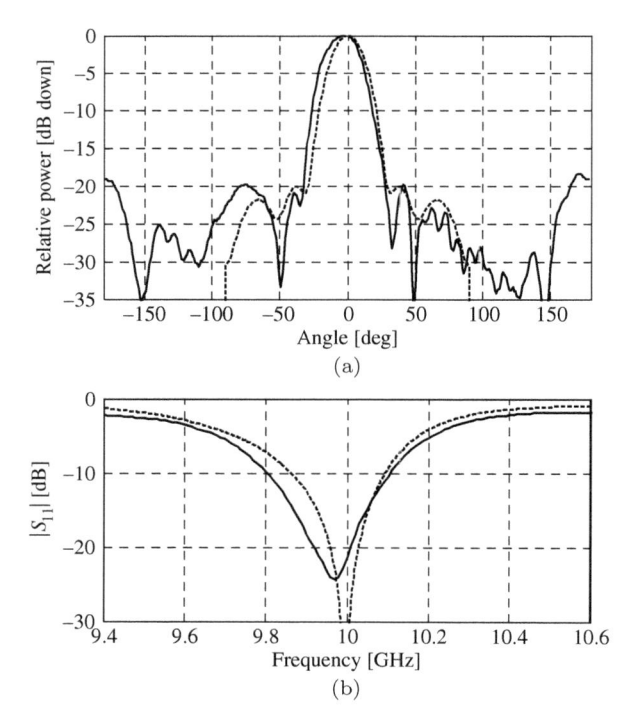

Fig. 8.13. MPAS I Design I measured (—) and simulated (− −): (a) H-plane radiation pattern at 10 GHz; (b) reflection coefficient.

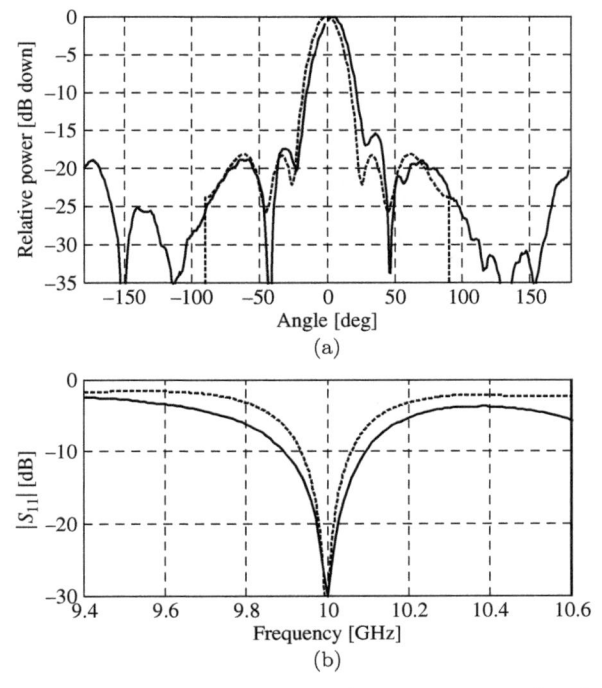

Fig. 8.14. MPAS I Design II measured (—) and simulated (– –): (a) H-plane radiation pattern at 10 GHz; (b) reflection coefficient.

8.5. Summary

In this chapter, numerical efficiency and reliability of the surrogate-based optimization methodology for simulation-driven design of printed-circuit subarrays has been demonstrated by conducting design of microstrip patch antenna subarrays energized with microstrip corporate feeds. In the used approach, the effect of the feed on the sidelobe level is taken into account by simultaneous adjustment of all relevant geometry parameters of the antennas and the feed. That allows us to control radiation and reflection responses of the subarray in a single optimization process.

It is also demonstrated that including the corporate feed in the simulated model of the subarray is essential to obtain reliable designs. The design process is realized as surrogate-based optimization involving coarse-discretization simulations as the underlying low-fidelity

202 Simulation-Based Optimization of Antenna Arrays

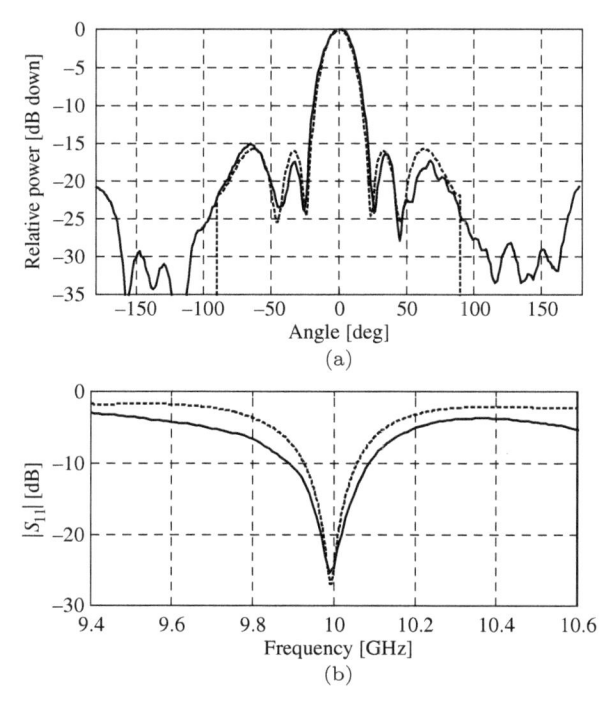

Fig. 8.15. MPAS I Design III measured (—) and simulated (– –): (a) H-plane radiation pattern at 10 GHz; (b) reflection coefficient.

model. The latter, upon suitable correction, serves as a reliable predictor leading to an improved design. The optimization process aims at reducing the sidelobe level, the primary objective, while ensuring low reflection coefficient at the operating frequency with penalty functions. The final designs are obtained at a low computational cost corresponding to about 30 evaluations of the high-fidelity models of the designed subarray. Comparison to measured responses of the fabricated designs illustrates reliability of the utilized models and the applied optimization approach.

Chapter 9

Antenna Array Models for Simulation-Based Design and Optimization

The optimization-based design of antenna arrays (AAs) is justified, first, by a large number of designable parameters describing composition and geometries of the aperture, elements, and feeding circuit. Second, array radiation and reflection response figures typically exhibit nonlinear behavior as function of the designable parameters. Accurate evaluations of the response figures can only be achieved through full-wave electromagnetic simulations of the discrete models of array circuits, which can be quite computationally expensive. Further, AA design problems are inherently multi-objective ones with several performance requirements to be controlled at the same time.

Metaheuristic optimization approaches, e.g., genetic algorithms (Aljibouri *et al.*, 2000, 2008; Ares-Pena *et al.*, 1999; Deb *et al.*, 2014; Haupt, 1995; Karamalis *et al.*, 2009; Petko and Werner, 2007; Rodriguez and Ares, 2000; Wen *et al.*, 2016), particle swarm optimizers (Chamaani *et al.*, 2010; Datta and Misra, 2009; Deb *et al.*, 2014; Elragal *et al.*, 2011; Fong *et al.*, 2016; Goudos *et al.*, 2009; Jang *et al.*, 2016; Jin and Rahmat-Samii, 2006, 2007; Liang *et al.*, 2017; Shelokar *et al.*, 2007), and other population-based methods (Liang *et al.*, 2017; Rajo-Iglesias and Quevedo-Teruel, 2007; Liu *et al.*, 2016;

Rocca et al., 2011), have become popular for AA design problems. Utilization of metaheuristic algorithms is definitely beneficial in the presence of multiple local optima (Yang, 2010a); however, population-based methods require hundreds, thousands and even tens of thousands of objective function calls. Thus, metaheuristics are generally applicable to problems in which numerical costs of the objective function evaluations are not of concern, i.e., to problems described by either simple or preconfigured models. Simple models of AAs such as analytical array factors as models of AA apertures or network models for AA feeds are fast; however, they can be quite approximate, e.g., incapable to account for certain full-wave effects such as parasitic emission, excitation of higher-order modes, coupling, and proximity effects.

An important component of an AA CAD process is an accurate, realistic, yet a handleable model of a structure under design. In other words, it is the availability of an accurate representation of the structure of interest which can be an array element, radiation aperture, feed, even the entire aperture-feed circuit. Given a reliable model, potential benefits of an optimization process can be exercised to the full extend, so that the effort spent on setting up and handling the optimization problem is justified.

Parameterization of the model using all relevant parameters is a necessary prerequisite. Furthermore, simulated responses should allow connection of AA subcircuits so that the net radiation and reflection responses of a superseding structure can be evaluated. Reusability of the simulated data in a fast way is another important point such models need to provide. Generic models of this kind can be configured and evaluated only with discrete EM modelers/solvers (CST, 2016; HFSS, 2016; Sonnet EM, 2016; FEKO, 2015; XFDTD, 2014; Burke et al., 1981), which are based on computational electromagnetic methods (Taflove and Hagness, 2005; Jin, 2014; Rautio and Harrington, 1987; Harrington, 1993; Pan, 2003). Discrete EM simulations of AA apertures, and especially apertures with feeds, can be quite expensive when accurate, e.g., if the structure is simulated with fine discretization. Therefore, an option to trade accuracy for speed, with subsequent correction, is highly desirable for simulation-based methods. In this chapter, we overview options (or types) of EM models for AA design and outline their application areas.

9.1. Array Factor-Based Models of Antenna Array Apertures

The fastest and the simplest model of an antenna array far-field is possible with the array factor of the array aperture,

$$\mathbf{E}(\theta, \varphi) = \mathbf{E}_e(\theta, \varphi) \sum_{n=1,\ldots,N} a_n e^{j\psi_n}, \qquad (9.1)$$

where a_n is the amplitude, complex in general, n is the index of the element, ψ_n is the phase shift of the nth element contribution to the far-field, given with (2.18)–(2.22) for the free-space propagation and for various apertures, whereas $\mathbf{E}_e(\theta, \varphi)$ is the far-field of the element, and the sum of

$$\mathrm{AF}(\theta, \varphi) = \sum_{n=1,\ldots,N} a_n e^{j\psi_n} \qquad (9.2)$$

is referred as the array factor. Using (9.2), the array factor can be written as

$$\mathbf{E}(\theta, \varphi) = \mathbf{E}_e(\theta, \varphi) \mathrm{AF}(\theta, \varphi) \qquad (9.3)$$

for the electric far-field, and

$$U(\theta, \varphi) = U_e(\theta, \varphi) |\mathrm{AF}(\theta, \varphi)|^2 \qquad (9.4)$$

for the radiation intensity.

The formulas (9.1) and (9.2) assume identical element patterns. Typically (9.3) or (9.4) can serve as predictors of radiation patterns for fixed-beam apertures and apertures with relatively weak interactions of the elements.

The element pattern can be acquired either by analytical (Elliott, 2003; Balanis, 2005; Jackson, 2007) or computational means (Makarov, 2002; Maloney et al., 2005; Jin, 2008; Kogure et al., 2011) as a pattern of a stand-alone antenna. With the latter method, boundary conditions of the computational model can be of the open or periodic type. The periodic boundary conditions with the period equal to the array spacing can approximately account for the presence of other antennas of the array aperture. Descriptive accuracy of the array factor models, based on (9.1)–(9.4), depends on a configuration of a particular array aperture, the type of the array element, and interactions between the elements.

An obvious benefit from the design point of view, implied with (9.3) and (9.4), is that the tasks of element design and aperture design, assuming weak interactions of the array elements, can be conducted separately from each other.

It is worth to clarify the physical meaning of weights a_n in the array factor (9.2) in relation to modeling and design of antenna arrays. Notice that formula (9.1) implies linearity and superposition properties for the total field. Therefore, under the weight a_n in summation of (9.1) or (9.2) we understand the complex amplitude of the incident wave at the input of the nth embedded element of the array aperture. This way, a clear connection between estimations with simple models (9.1)–(9.4), on one hand, and responses computed with full-wave discrete EM solvers and, especially, measured S-parameters, on the other hand, are established for microwave frequencies.

Mutual interactions of the elements including their active reflection coefficients, contributions of the aperture periphery, e.g., contributions of the ground plane edges to the radiation pattern, the effects of parasitic surface waves, as well as radiation from the feed are beyond capability of the array factor models. Nevertheless, array factor-based models can be considered as antenna array radiation coarse models which are indispensable for quick estimations of the radiation patterns, including peak directivity values, sidelobe levels, beamwidths, etc., at the prototyping stages. Moreover, upon a proper correction, they can be utilized in the design process, e.g., for pattern optimization. Description of array factor-based model correction and utilization of the array factor-based models in simulation-based design processes are provided in Chapters 10–17.

Basic utility options of the array factor in the context of simulation-based surrogate-assisted antenna array design can be classified as follows:

(i) Estimating radiation responses, e.g., major lobes' shape, sidelobes'/grating lobes' location and magnitudes, axial ratio, for fixed-beam and phased array apertures prior implementation as a function of the excitation taper and aperture dimensions (cf. Chapter 10). This is a standard yet extremely useful application of the array factor model.

(ii) Providing a coarse model of the radiation response for subsequent surrogate-assisted optimization of fixed-beam apertures as well as of phased array apertures. The element pattern can be included in (9.3) and (9.4) either as a pre-simulated pattern of a stand-alone aperture element or as an effective aperture-embedded element pattern relating the array factor (9.2) to radiation characteristics of an EM-simulated high-fidelity model of the aperture. This book, in a significant part (Chapters 10–17), describes and illustrates how to set up, update, and apply the array factor-based surrogate models for various tasks related to simulation-based optimization of antenna array apertures.

(iii) Being a component of fast models of array feeds as described in Chapter 14. This application of the array factor allows searching for optimal feed architectures (before implementation) under various constrains (from those imposed on the radiation pattern and aperture spacings up to those imposed on power splits within the feed). Because this application precedes the implementation (e.g., with microstrip technology) of the array circuit, no correction of the coarse model with respect to simulated responses are involved. This is a rather new and generic (from an implementation technology viewpoint) application of array factor-based models; it is demonstrated for microstrip corporate feeds in Chapter 14.

(iv) Providing a coarse model of the radiation pattern of an array with aperture and feed integrated as presented in Sec. 14.2 of Chapter 14. This is a combination of options (iii) and (iv) elaborated to realize a required sidelobe level by redesign of the corporate feed. It is worth to mention that this application allows for modeling of the array excitation taper within an integrated array circuit.

9.2. Computational EM Models of Antenna Arrays

The most accurate and complete description of antenna array components, array radiation apertures and aperture-feed modules is available with full-wave electromagnetic simulations of the corresponding discrete electromagnetic models. Contemporary electromagnetic simulations of antennas, antenna array components and structures are performed by antenna designers typically using commercially available software packages, e.g., CST (2016), HFSS (2016), Sonnet EM (2016), FEKO (2015), XFDTD (2014). These packages incorporate the graphical user interface, scripting capabilities, material libraries, transmission line calculators, built-in optimizers, post-processing features as well as features for import and export of CAD projects in various formats. Hardware acceleration, e.g., with FPGA and graphics processor units (GPUs), becomes available for certain type of solvers (CST, 2016). Parameterization of dimensions, material parameters, sources, and excitations is crucial for design, and therefore, has become a standard feature.

Full-wave discrete solvers of theses software packages implement a number of numerical methods of computational electromagnetics (CEM) which have been developed for the last decades. These CEM methods include: the finite-difference time-domain (FDTD) method (Taflove and Hagness, 2005), the finite-element method (FEM) (Jin, 2014), and the method of moments (MoM) (Harrington, 1993; Rao, et al., 1982). The FDTD is considered to be the most versatile and robust among the above techniques. It solves the Maxwell's equations directly in time domain. Thus, FDTD solvers can handle nonlinear materials and evaluate responses of the circuits over wide frequency bands in one run. A strong feature of the explicit formulation of the FDTD (Taflove and Hagness, 2005) is that it does not involve any preconditioning or matrix inversion. In addition, hardware-accelerated FDTD computations are realizable on mass-produced GPUs (Taflove and Hagness, 2005; GPU Computing Guide, 2016).

The FEM (Jin, 2014) and MoM (Harrington, 1993; Rao et al., 1982) solve the EM problems in the frequency domain and at every frequency of interest. The strongest features of the FEM algorithms include the automated mesh adaptation solution process and the sparse system matrices. The later feature of the FEM allows handling computational problems with relatively large number of discrete elements in contrast to the basic MoM algorithms operating with densely filled matrices (Harrington, 1993; Rao et al., 1982; Makarov, 2002). On the other hand, developments of iterative CEM solvers, in particular, the multi-level fast multipole method (MLFMM), can offer drastic improvements in speed and memory for the MoM formulation (Chew et al., 2000). The MLFMM solution option becomes available in commercial EM packages, e.g., CST (2016) and FEKO (2015). However, the computational benefits of the MLFMM are not always available for antenna array problems (Haupt, 2010, Sec. 6.61).

FDTD and FEM-based solvers discretize and solve computational domains of volumes while MoM-based solvers discretize and solves for physical and/or equivalent surface currents. These CEM foundations define the problems from the application point of view in which every solver can be more efficient than the others. For instance, the MoM can be more suitable for solving scattering problems and for problems where interacting components, e.g., antennas or scatterers, are spaced by distances of few and more wavelengths so that the uniformly filled medium between components is excluded from the discretization and solution processes. The FDTD and FEM are more suitable for solving volumes of complex content for which the MoM description in terms of currents turns to be much more computationally expensive. Hybrid solvers, simultaneously utilizing the strongest features of the FEM and MOM for more numerically efficient simulations, e.g., for elimination of large voids (e.g., between antennas or between an antenna and ground/reflector) and numerical absorbing boundaries, are available, e.g., FEKO (2015).

The progress in development of the computing hardware makes it possible to simulate realistic models of antenna array circuits with the software packages on personal computers and laptops in reasonable time. On the other hand, modern technology imposes more and more stringent requirements on accuracy of simulated responses of microwave components and antenna modules. The simulated structures are becoming of more integrated composition, and therefore, need to be described with realistic, and consequently, with computationally expensive models which are dependent on a large number of design variables. Such models when used in the simulation-based design processes should be evaluated multiple times at different points of the design space to obtain the optimal design.

Typical trade-offs between the antenna array model accuracy and its computational costs are similar to the single antenna trade-offs which have been already discussed in Chapter 6. The challenges of computational costs associated to repeated simulations of antenna arrays, within the simulation-driven optimization processes, are even more substantial due to their size, composition, and accuracy requirement which are imposed on the sensitive antenna array performance figures such as sidelobes, distribution of power within the feed, excitation amplitudes and phase shifts at the elements' inputs.

Similar to high-fidelity models of stand-alone antennas (Chapter 6), high-fidelity computational models of the array components up to entire array circuits are defined with fine meshes (obtained with automated adaptive meshing or mesh convergence supervised studies) and account for many influencing factors such as material losses and dispersion, finite thickness of metallization, etc. In the examples of this book, such models are primarily used as references to set up the surrogate models originated from simplified analytical or computational description of the modeled structures, correct the surrogate models at different points of the design space within the SBO process, and verify SBO-obtained intermediate and final designs. In certain cases, when the designed component is described using just a few design variables and the EM model of the structure contains a limited number of discrete elements, i.e., fast, the high-fidelity model can be used directly, e.g., to set up a response surface model for (Sec. 14.2 in Chapter 14), optimization (modeling and optimization of unequal-split microstrip junctions).

Coarse (or low-fidelity) EM simulated models, also referred as coarsely discretized models in this book, in addition to simplified physics (infinitesimal thickness of metallization, lossless materials, simplified absorbing boundaries and sources), as the name implies, are defined with coarse meshes. Basic requirements to coarsely discretized models in the context of array design are similar to those already listed for stand-alone antennas in Chapter 6. In addition, the model coarseness is an even more critical factor affecting the results of SBO numerical costs and quality of obtained designs of array circuits. In particular, different subcircuits within the same array circuit can be designed with quite different model coarseness not only due to structural differences but also due to the different sensitivity of the simulated characteristics and accuracies of the used EM solvers to the mesh quality.

Applications of coarsely discretized models in the context of simulation-based surrogate-assisted antenna array design can be summarized as follows:

(i) Inexpensive approximate simulation of radiation and reflection responses' characteristics of the designed components and entire array circuits.
(ii) Providing a coarse model of the radiation and reflection responses for subsequent surrogate-assisted optimization at different stages of the SBO-assisted design process, from optimization of a single component, stand-alone aperture element (antenna), radiating aperture, feed, and in certain cases (Sec. 14.1 in Chapter 14) up to the entire integrated array circuit.

Being physics-based models, coarsely discretized simulated models, are very versatile in terms of modeled structures and evaluated figures and indispensable components of SBO methods at implementation steps. Their use is discussed and illustrated throughout the case studies of this book.

9.3. Simulation-Based Superposition Models

Reusable models of array radiating apertures, which are capable of returning characteristics of radiation and reflection responses evaluated at the high-fidelity EM level of description as function of applied excitations, are required for many tasks of aperture design and performance evaluation. Simulation-based superposition models are configured from the simulated responses of the computational EM models of array apertures (Kowalski and Jin, 2000; Koziel and Ogurtsov, 2014b, 2015a,d). Simulation-based superposition models are numerical realization of early listed relations (2.2), for fields in general, or (2.3), for far-fields, and (2.36) or (2.37), for reflection response (Koziel and Ogurtsov, 2014b, 2015a,d).

Once the fields and S-parameters of the aperture of interest with fixed dimensions are acquired (by means of N simulations of the aperture with only one element being active at a time where N is the number of the aperture radiators in general) and stored, the response of the aperture to any applied excitation can be evaluated very fast with vectorized computations (Matlab, 2010). Accuracy of the evaluated responses is the accuracy of the processed EM simulations up to numerical round offs. Computational costs of acquisition of simulation-based superposition models can be cut down significantly using the surrogate-assisted approaches involving coarsely discretized models (Chapters 10 (Sec. 10.4) and 12). In this book, we describe and demonstrate the use of simulation-based superposition models in optimization of linear and planar arrays (Chapters 10 (Sec. 10.4) and 12).

Simulation-based superposition models are not the fastest option for optimization of fixed-beam apertures; however, such models are definitely indispensable for accurate evaluation of phased array apertures and excitation tapers (including those designed with various means, e.g., as in Chapters 10 and 15) in terms of different characteristics (peak realized gain, beam width, sidelobes, grating lobes, active reflection coefficients, total reflected power, etc.) all versus the scan angle(s).

Chapter 10

Design of Linear Antenna Array Apertures Using Surrogate-Assisted Optimization

This chapter addresses design of linear antenna array apertures using conventional and surrogate-based optimization techniques. A number of specific methods and design cases are considered. We start with several cases of end-fire and broadside linear array apertures where radiation patterns are evaluated using the array factor. It is demonstrated, as opposed to common practice, the design optimization process does not need to use population-based metaheuristics with the array factor-based models. Instead, a random-search-based initialization and the conventional gradient search are sufficient for the considered problems.

Full-wave EM simulation models of linear antenna array apertures are introduced in later sections of this chapter. In particular, we discuss response correction techniques as well as utilization of simulation-based superposition models. In all considered cases minimum sidelobe designs are obtained at the cost corresponding to a few high-fidelity simulations of the respective array structures.

Throughout this chapter, under the term of linear array, we will understand linear array apertures. Thus, the outcome of the design processes will include excitation tapers, amplitudes and/or phases of antenna array radiating elements. For the simulation-based designs, i.e., when the optimizers use simulated responses of the full-wave EM models, these amplitudes and/or phases are of the incident waves at the antenna inputs.

10.1. Optimum Design of Array Factor Models Using Smart Random Search and Gradient-Based Optimization

In this section, we describe a simple technique for reliable optimization of array factor models and provide several examples of linear array pattern synthesis.

10.1.1. *Problem formulation*

The array factor of a linear array comprising N elements, depicted in Fig. 10.1, can be written as

$$F(\theta) = \sum_{n=1}^{N} a_n \exp\left\{j\left(\alpha_n + 2\pi z_n \cos(\theta)\right)\right\}, \qquad (10.1)$$

where a_n are weights of the array elements, their excitation amplitudes (here we assume all $a_n = 1$, i.e., the uniform amplitude excitation), z_n (in wavelengths) are coordinates of the elements' locations, and α_n are phases of the nth element; also

$$z_n = \sum_{k=1}^{n} d_k, \qquad (10.2)$$

$$\alpha_n = \sum_{k=1}^{n} \Delta\alpha_k, \qquad (10.3)$$

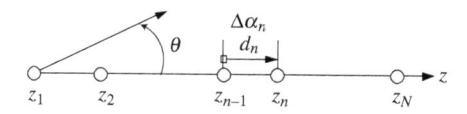

Fig. 10.1. A linear end-fire array with notation used throughout the section.

where d_k and $\Delta\alpha_k$ stand for the spacing and phase shift, respectively. Here the references, d_1 and $\Delta\alpha_1$, are set to zero.

Directivity of the array is (Balanis, 2005)

$$D(\theta) = \frac{2\,|F(\theta)|^2}{\int_0^\pi |F(\theta)|^2 \sin\theta\, d\theta}. \tag{10.4}$$

The design specifications can be formulated as $D(\theta) \leq D_{\max}$ or $D(\theta) \geq D_{\min}$ for specific ranges of angle θ.

We consider the following major optimization cases:

1. SLL reduction with two design variables: d ($d_n = d, n = 2, \ldots, N$) and $\Delta\alpha$ ($\Delta\alpha_n = \Delta\alpha, n = 2, \ldots, N$);
2. directivity maximization with two design variables;
3. SLL reduction with $N-1$ design variables: $\Delta\alpha_n, n = 2, \ldots, N$;
4. SLL suppression with $2N - 2$ design variables: d_n and $\Delta\alpha_n$, $n = 2, \ldots, N$.

10.1.2. *Optimization methodology*

Here, we exploit gradient search combined (optionally) with smart random initialization. We demonstrate that this is both an efficient and computationally cheap way of solving this array synthesis problem.

The main optimization engine is the standard sequential-quadratic programming (SQP) algorithm (Nocedal and Wright, 2006) implemented in Matlab *fminimax* (Matlab, 2010) optimization routine. This routine is very convenient as the array synthesis problem can be formulated as a minimax optimization task with upper and lower specifications. Although the synthesis problem has, in general, many design variables, analytical derivatives can easily be calculated for the array factor (10.1) and array directivity (10.4). Analytical derivatives greatly reduce the optimization cost because with gradient optimization a typical number of objective function evaluations is under few hundreds instead of thousands or tens of thousands which are normally required by population-based methods such as GAs or PSO.

As the synthesis problem may be multi-modal with many local minima, we also include an optional random search as an initial synthesis step. Our smart random search algorithm can be formulated as follows:

$$\mathbf{x}^{(i+1)} = \lambda^{(i)} \mathbf{x}^{\text{rand}} + (1 - \lambda^{(i)}) \mathbf{x}^{\text{best}}, \qquad (10.5)$$

where $\mathbf{x}^{\text{rand}}$ is a randomly generated point, $\mathbf{x}^{\text{best}}$ is the best design found so far, whereas $\lambda^{(i)}$ is a scalar coefficient that decreases towards the end of the search process. In particular, we have $\lambda^{(i)} = i/i_{\max}$, where $i_{\max}$ is the maximum allowed number of function calls for the random search stage. The procedure (10.5) biases the search towards the best design found so far and turns to a local search when i gets closer to $i_{\max}$. This kind of random search is more efficient than pure random search and nearly as efficient as most of population-based search techniques.

In practice, random initialization is only necessary for cases where both phases and spacings were considered simultaneously as design variables for $N = 20$ or 40. For other cases, direct gradient optimization starting from the reference design was sufficient.

10.1.3. Case study 1: Linear end-fire array optimization

Linear arrays comprising a different number of elements have been optimized using the procedure described in Sec. 10.1.2. In each case the initial design was the end-fire designs with the half-wavelength (HW) spacings (Balanis, 2005), to be referred as HW designs, and a lower bound of one-eighth of the wavelength is imposed for the spacings.

10.1.3.1. Sidelobe reduction with two variables

In this case, the optimization objective is the SLL minimized in the minimax sense for $\theta_0 \leq \theta \leq 180°$ where θ_0 is the first null angle of the corresponding HW design (Balanis, 2005).

For $N = 10$ the initial designs are $\mathbf{x}^{\text{init}} = [d \Delta \alpha]^T = [0.225 \ -97.73°]^T$, and the final design $\mathbf{x}^{(*)} = [0.424 \ -150°]^T$.

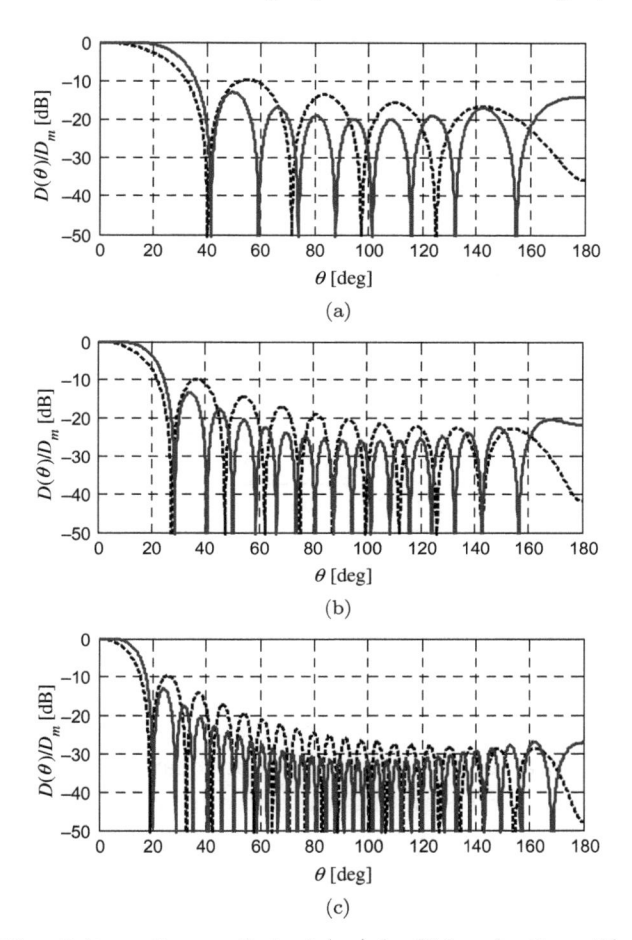

Fig. 10.2. Directivity pattern optimized (—) for SLL reduction with two design variables, uniform spacing and progressive phase shift for different numbers of elements, N: (a) $N = 10$; (b) $N = 20$; (c) $N = 40$. The references are the HW arrays (– –). D_m is the peak directivity.

For $N = 20$ the initial designs are $\mathbf{x}^{\text{init}} = [0.2375 \ -93.87°]^T$, and the final design $\mathbf{x}^{(*)} = [0.418 \ -150°]^T$. For $N = 40$ the initial designs are $\mathbf{x}^{\text{init}} = [0.2438 \ -91.93°]^T$, and the final design $\mathbf{x}^{(*)} = [0.417 \ -149.9°]^T$. The optimization results, directivity patterns, are shown in Fig. 10.2. The peak directivity and SLL are listed in Table 10.1.

218 Simulation-Based Optimization of Antenna Arrays

Table 10.1. Optimization results for SLL reduction with uniform spacing and uniform phase shift.

	Peak directivity (dBi)		SLL (dB)	
Number of elements	Initial (HW)	Optimized	Initial (HW)	Optimized
10	12.16	11.52	−9.66	−12.97
20	15.36	14.90	−9.86	−13.19
40	18.47	17.94	−9.91	−13.25

Table 10.2. Optimization results for maximum directivity with uniform spacing and uniform phase shift.

	Peak directivity (dBi)		SLL (dB)	
Number of elements	Initial (HW)	Optimized	Initial (HW)	Optimized
10	12.16	13.89	−9.66	−10.50
20	15.36	17.38	−9.86	−10.68
40	18.47	20.73	−9.91	−10.50

It is worth mentioning that no substantial improvement in the SLL can be obtained if the major lobe beamwidth is kept from broadening, e.g., by restricting the directivity by the value of the initial design at a particular angle.

10.1.3.2. Peak directivity maximization with two variables

In this case, the optimization objective is the peak directivity for the zero angle. The initial designs are the same as in the cases before. The final design for $N = 10$ is $\mathbf{x}^{(*)} = [0.394 \ -156.4°]^T$. The final design for $N = 20$ is $\mathbf{x}^{(*)} = [0.445 \ -167.4°]^T$. The final design for $N = 40$ is $\mathbf{x}^{(*)} = [0.460 \ -169.3°]^T$. The directivity patterns are shown in Fig. 10.3. The peak directivity and SLL are listed in Table 10.2. Here the optimized designs have closely the same SLL of about −10.5 dB, major beams narrower than the corresponding HW designs and directivity about 2 dB higher than the HW designs.

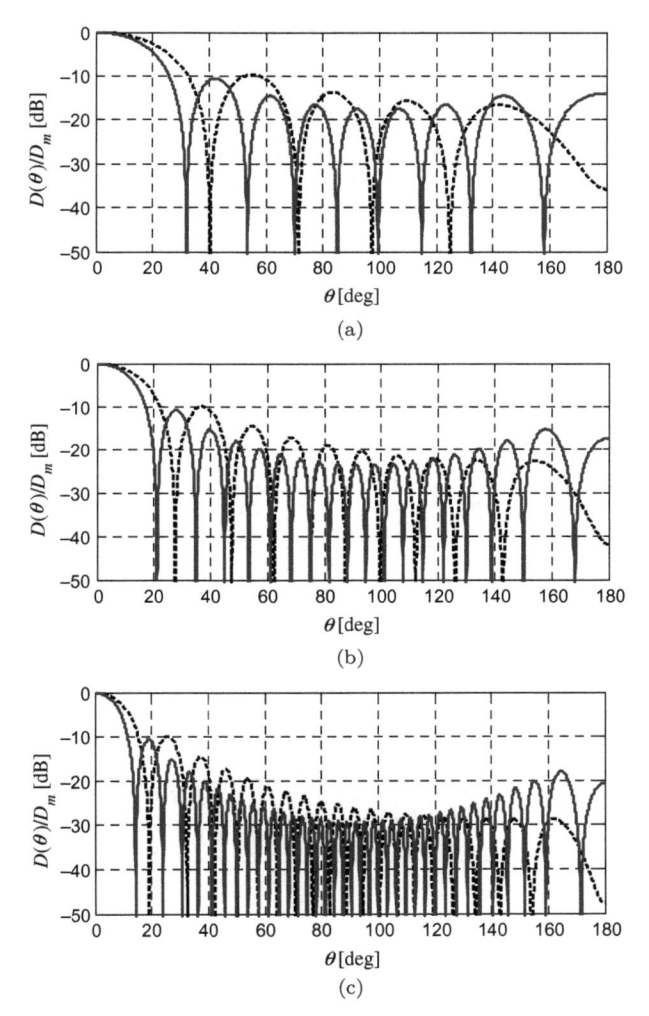

Fig. 10.3. Directivity pattern of the end-fire array optimized (—) for maximum peak directivity, D_m, with two design variables, uniform spacing and progressive phase shift for different numbers of elements, N: (a) $N = 10$; (b) $N = 20$; (c) $N = 40$. The references are the HW arrays (- -).

10.1.3.3. *Sidelobe reduction with different phase shifts*

In this case, the optimization objective is to minimize the SLL. The initial designs are the same HW designs as in the two previous cases.

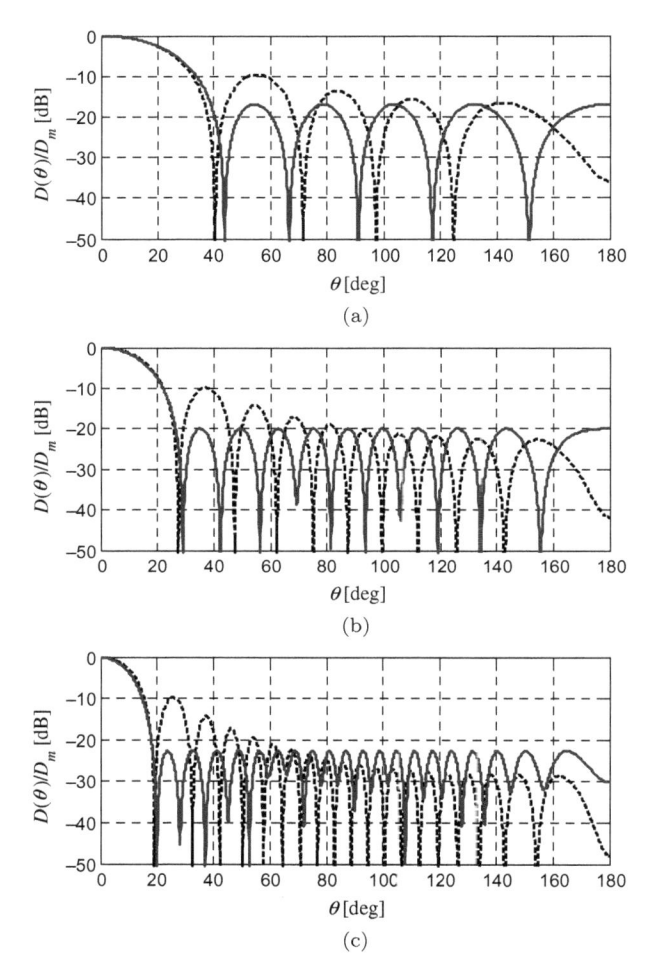

220 *Simulation-Based Optimization of Antenna Arrays*

Fig. 10.4. Directivity pattern optimized (—) for SLL reduction with $N-1$ design variables, phase shifts, for different numbers of elements, N: (a) $N = 10$; (b) $N = 20$; (c) $N = 40$. The references are the HW arrays (− −). D_m is peak directivity. Spacing is fixed to half-wavelength.

For every N (the number of array elements), the spacing is fixed to that of the corresponding HW design. There are $N - 1$ design variables, phase shifts $\Delta\alpha_n, n = 2, \ldots, N$. The directivity patterns of the final designs are shown in Fig. 10.4. The peak directivity and SLL are listed in Table 10.3. The phase shifts of the optimal designs are plotted in Fig. 10.5.

Table 10.3. Optimization results for SLL reduction with $N-1$ different phase shifts and the HW spacing.

	Peak directivity (dBi)		SLL (dB)	
Number of elements	Initial (HW)	Optimized	Initial (HW)	Optimized
10	12.16	13.06	−9.66	−16.96
20	15.36	16.62	−9.86	−20.02
40	18.47	19.84	−9.91	−22.70

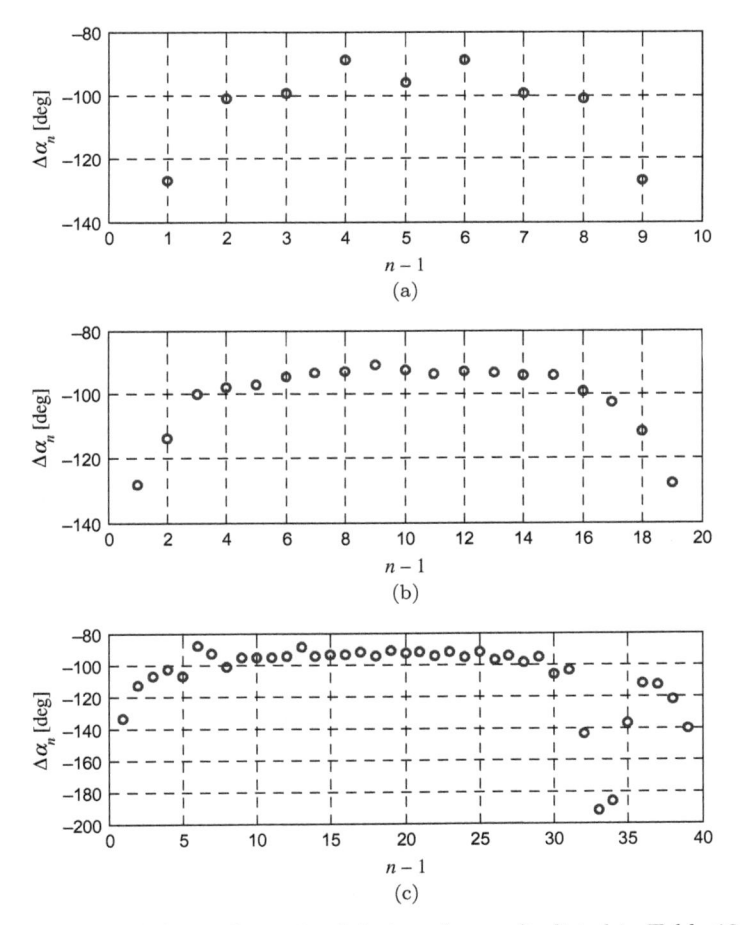

Fig. 10.5. Phase shifts at the optimal designs for results listed in Table 10.3: (a) $N=10$; (b) $N=20$; (c) $N=40$. In each case, the spacing is fixed to that of the corresponding HW design. End-fire radiation points to the right.

10.1.3.4. Sidelobe reduction with different phase shifts and spacing

The initial designs are the HW designs. We consider $2N - 2$ design variables, phase shifts $\Delta \alpha_n$, and spacing $d_n, n = 2, \ldots, N$. The final designs have much lower SLLs and higher peak directivities than in the previous cases, the values are listed in Table 10.4. The phase shifts and spacing of the optimal designs for 20 elements are shown in Fig. 10.6.

Table 10.4. Optimization results for SLL reduction with $2N - 2$ design variables, different phase shifts and spacing.

Number of elements	Peak directivity (dBi) Initial (HW)	Optimized	SLL (dB) Initial (HW)	Optimized
10	12.16	14.14	-9.66	-19.70
20	15.36	18.01	-9.86	-24.28
40	18.47	21.63	-9.91	-28.27

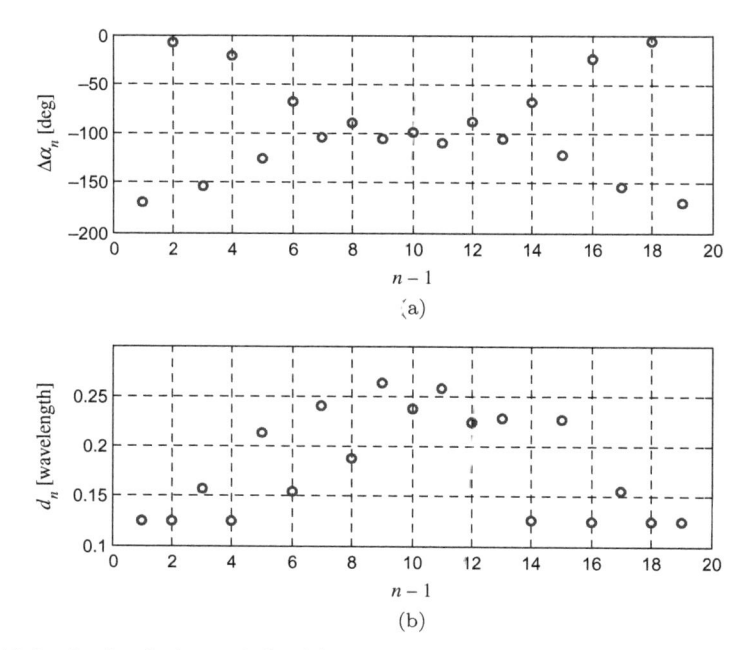

Fig. 10.6. Optimal phase shifts (a) and spacing (b) at the final design for SLL reduction with $N = 20$ elements for results listed in Table 10.4; index n stands for the element number according to Fig. 10.1.

10.2. Null Controlled Pattern Design

Here, linear arrays comprising of 10, 20, and 40 elements have been optimized using the procedure described in Sec. 10.2 with the objective being SLL minimization and obtaining pattern nulls. In each case, the initial design was the HW design corresponding to the particular number of array elements. For the element spacing, the

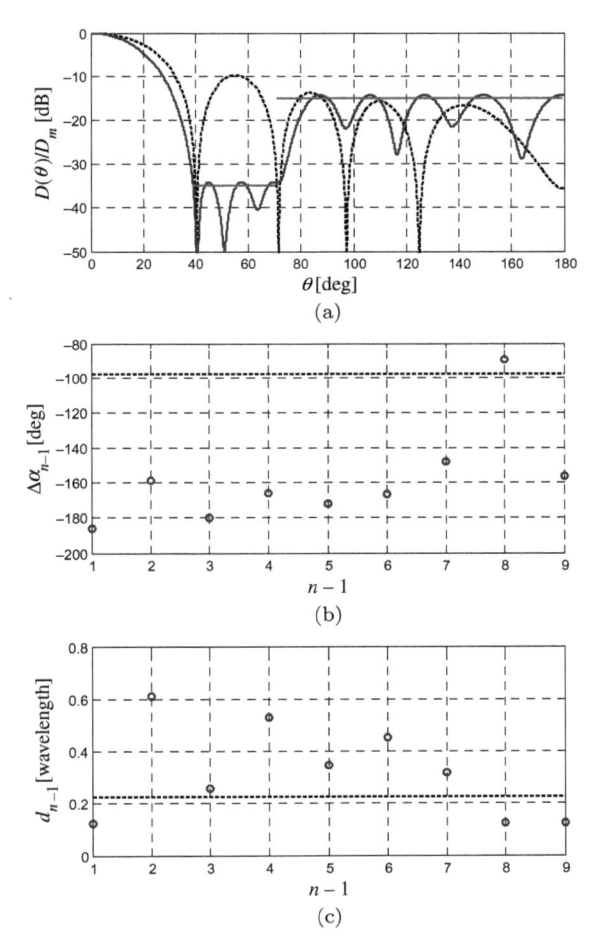

Fig. 10.7. Array of 10 elements, with the lower bound on element spacing of 0.125, optimization results: (a) normalized pattern; (b) progressive phase shift; and (c) spacing where n stands for a particular element number. The HW array (- - -) is the reference.

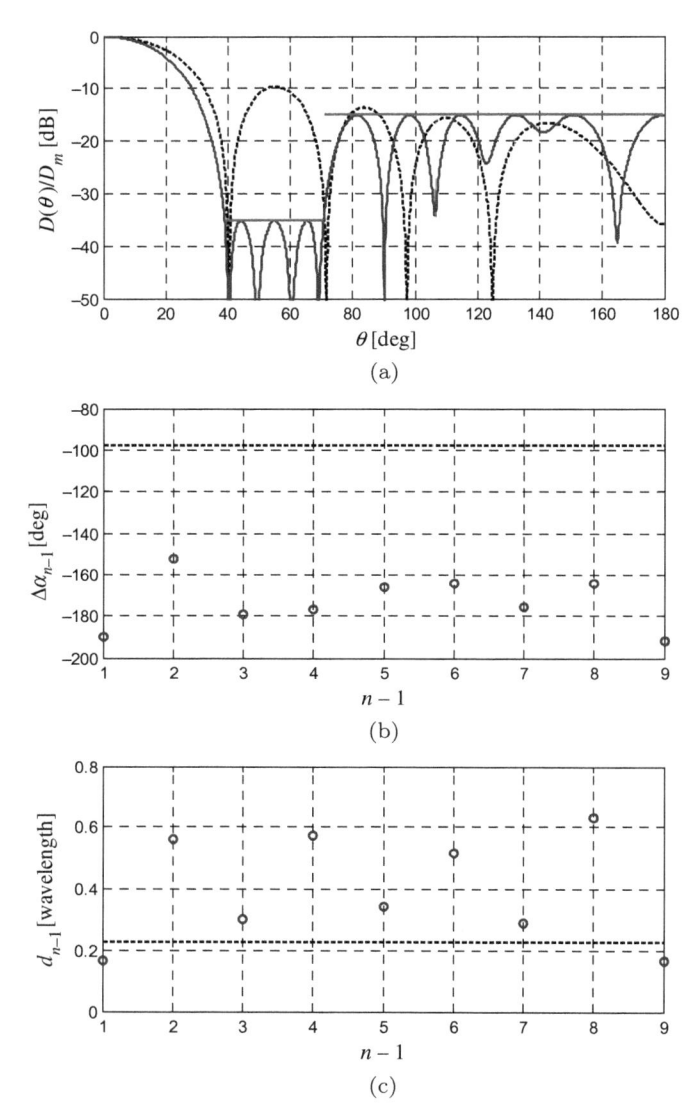

224 Simulation-Based Optimization of Antenna Arrays

Fig. 10.8. Array of 10 elements, with the lower bound on element spacing of 0.167, optimization results: (a) normalized pattern; (b) progressive phase shift; and (c) spacing where n stands for a particular element number. The HW array (- - -) is the reference.

Design of Linear Antenna Array Apertures   225

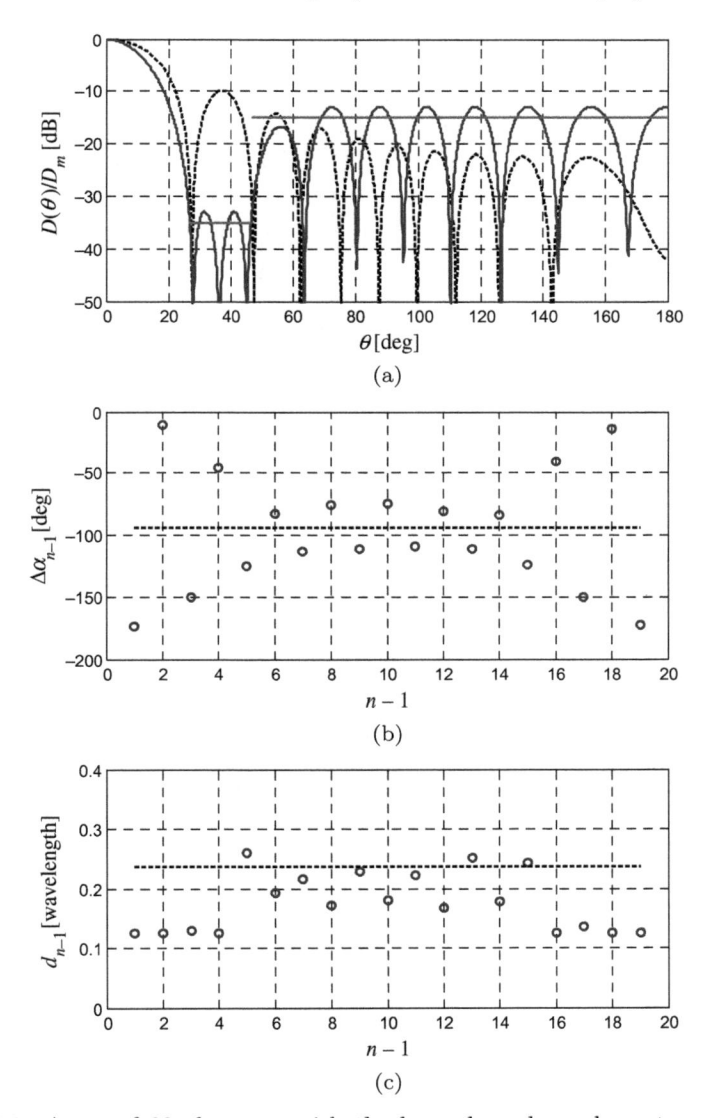

Fig. 10.9. Array of 20 elements, with the lower bound on element spacing of 0.125, optimization results: (a) normalized pattern; (b) progressive phase shift; and (c) spacing where n stands for a particular element number. The HW array (- - -) is the reference.

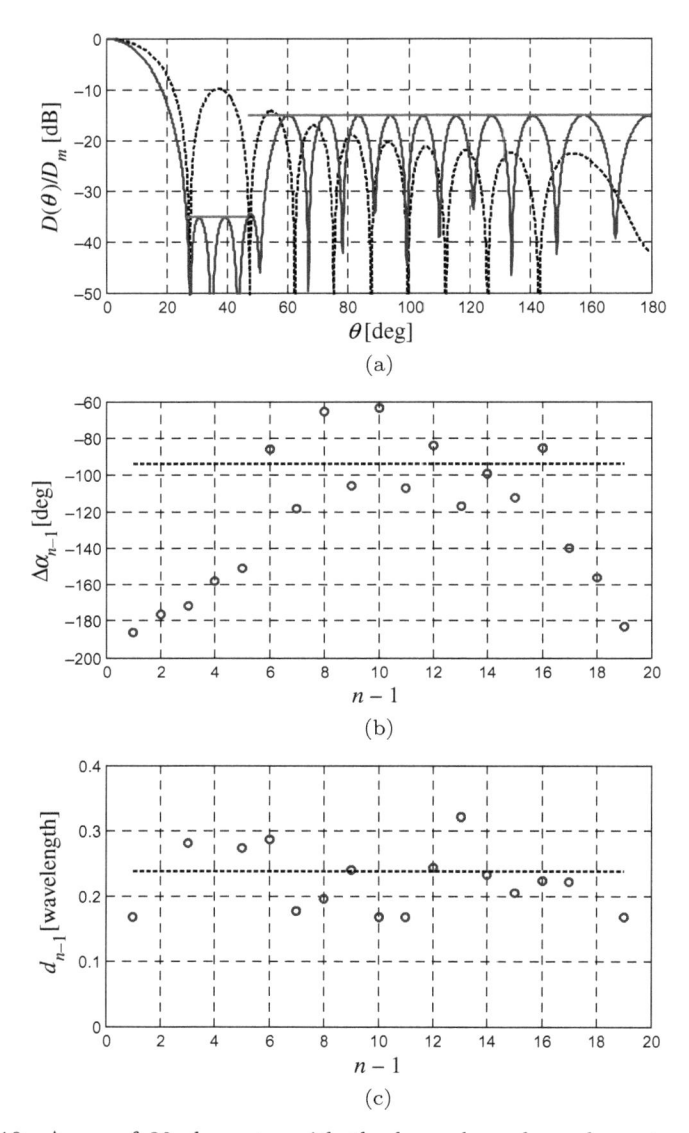

226 *Simulation-Based Optimization of Antenna Arrays*

Fig. 10.10. Array of 20 elements, with the lower bound on element spacing of 0.167, optimization results: (a) normalized pattern; (b) progressive phase shift; and (c) spacing where n stands for a particular element number. The HW array (- - -) is the reference.

two cases of the lower bound of 0.125 (one eights of the wavelength) and 0.167 are considered in the examples. Figures 10.7, 10.9, and 10.11 show the initial and optimized patterns for the first case, whereas Figs. 10.8, 10.10, and 10.12 show the patterns for the second case. Tables 10.5 and 10.6 indicate the values of peak directivity and SLL for all considered test cases as well as their comparison with the

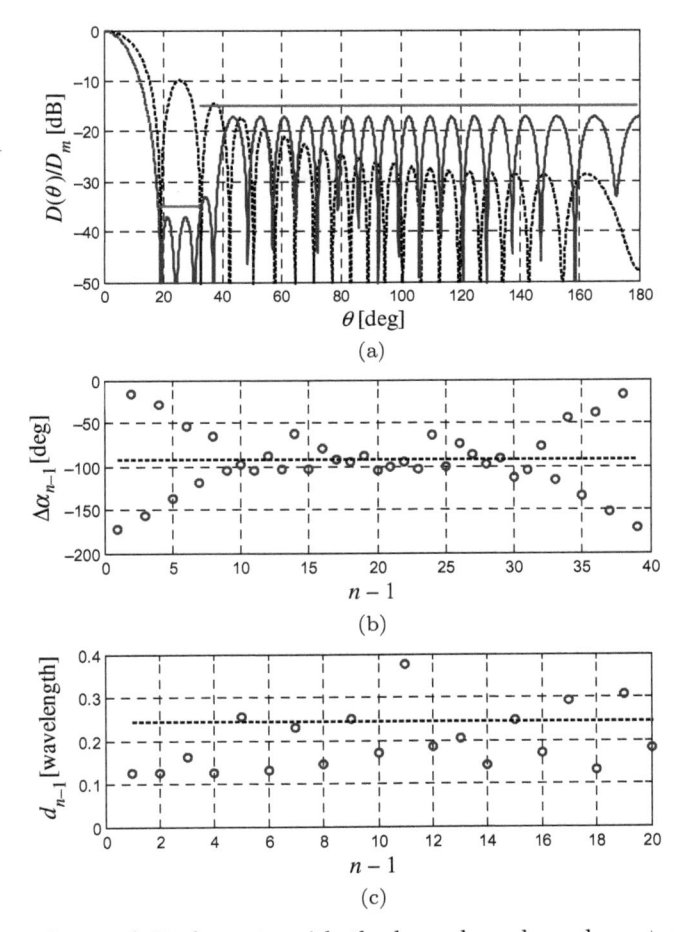

Fig. 10.11. Array of 40 elements, with the lower bound on element spacing of 0.125, optimization results: (a) normalized pattern; (b) progressive phase shift; and (c) spacing where n stands for a particular element number. The HW array (- - -) is the reference.

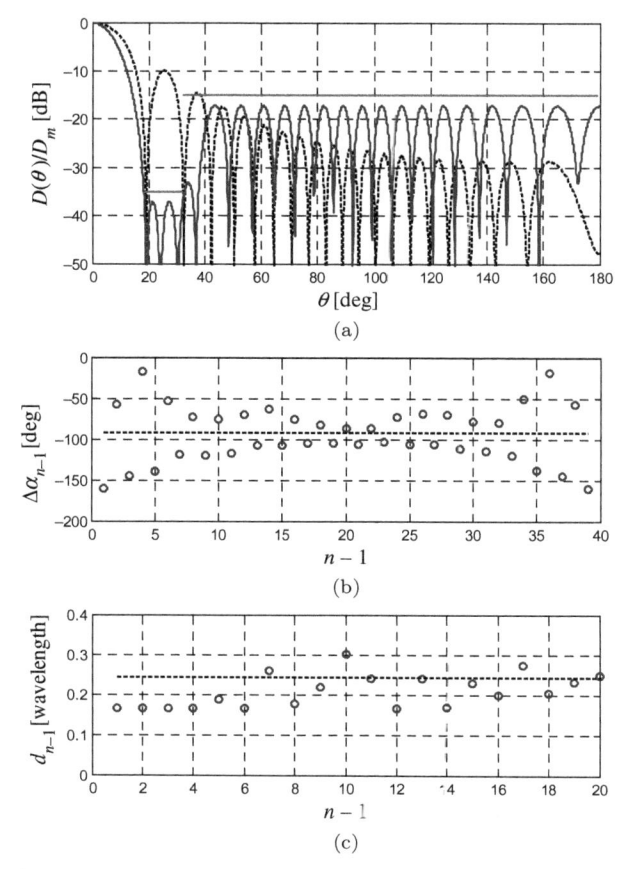

228 Simulation-Based Optimization of Antenna Arrays

Fig. 10.12. Array of 40 elements, with the lower bound on element spacing of 0.167, optimization results: (a) normalized pattern; (b) progressive phase shift; and (c) spacing where n stands for a particular element number. The HW array $(--)$ is the reference.

Table 10.5. Design results with the lower bound on element spacing of 0.167.

Number of elements	Peak directivity (dBi) Initial (HW)	Peak directivity (dBi) Optimized	SLL (dB) Initial (HW)	SLL (dB) Optimized[#]
10	12.16	14.0	−9.66	−15.0
20	15.36	16.3	−9.86	−15.0
40	18.47	17.6	−9.91	−17.5

[#]The first sidelobe is around 20 dB lower than indicated in the table.

Table 10.6. Design results with the lower bound on element spacing of 0.125.

	Peak directivity (dBi)		SLL (dB)	
Number of elements	Initial (HW)	Optimized	Initial (HW)	Optimized[#]
10	12.16	14.0	−9.66	−14.5
20	15.36	15.1	−9.86	−13.0
40	18.47	18.6	−9.91	−17.5

[#]The first sidelobe is around 20 dB lower than indicated in the table.

reference HW designs. The results obtained with the lower bound of 0.167 are usually slightly better than the results for the 0.125 bound, even though the latter corresponds to a larger design space.

10.3. 20-Element Broadside Array Design for Pattern Nulls and Sector Beam

Consider the following specifications: $|F(\theta)| \leq -60$ dB for $50° \leq \theta \leq 60°$ and $120° \leq \theta \leq 130°$ (pattern nulls), $|F(\theta)| \leq -40$ dB for $0° \leq \theta \leq 80°$ and $100° \leq \theta \leq 180°$ (SLL suppression) and $|F(\theta)| \geq -3$ dB for $87° \leq \theta \leq 93°$ (beam). Only excitation amplitudes a_n and uniform spacing d are considered as design variables. In-phase excitation is assumed, i.e., α_n are zeros.

The pattern was obtained using gradient-search with analytical derivatives, and it is shown in Fig. 10.13. It should be noted that the gradient-search allows very accurate nullification in the sector of interest (50°–60° and 120°–130°), which is difficult to achieve using other methods (see, e.g., Jin and Rahmat-Samii, 2006). The design cost is only 300 function evaluations. No random-search pre-optimization was used in this case.

The second design case concerns the following pattern specifications: $|F(\theta)| \leq -25$ dB for $0° \leq \theta \leq 70°$ and $110° \leq \theta \leq 180°$ (SLL suppression) and $|F(\theta)| \geq -0.5$ dB for $80° \leq \theta \leq 100°$ (sector beam). Excitation amplitudes a_n, phases α_n, and spacing d are the design variables. The obtained pattern is shown in Fig. 10.14. The design cost is 1500 function evaluations out of which 1000 evaluations were used for the smart random search and 500 were used for the

230 *Simulation-Based Optimization of Antenna Arrays*

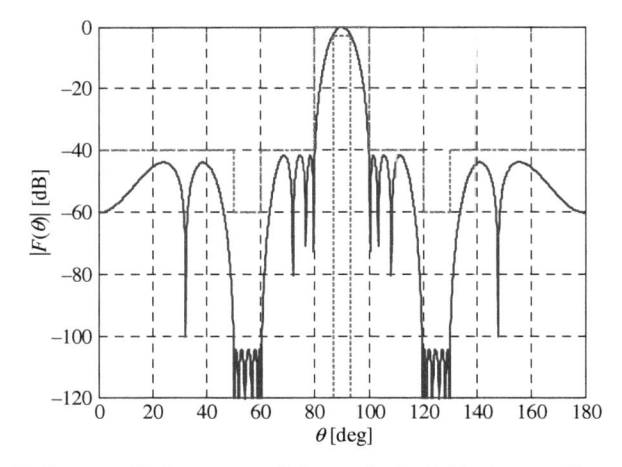

Fig. 10.13. Null controlled pattern of the optimized 20-element linear array. The dashed lines denote the pattern specifications.

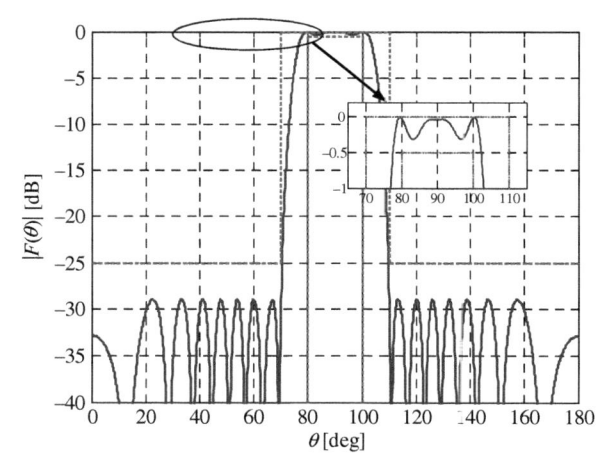

Fig. 10.14. Sector beam pattern of the optimized 20-element linear array. The dashed lines denote the pattern specifications.

gradient search with analytical derivatives. Notice that the pattern satisfying the same specifications had been obtained in Jin and Rahmat-Samii (2006) using Taguchi's method at the cost of almost 5000 function evaluations. The synthesis using PSO required around 15,000 function evaluations (Jin and Rahmat-Samii, 2006).

10.4. Phase–Spacing Optimization of Linear Arrays Using Simulation-Based Surrogate Superposition Models

In this section, a technique for simulation-driven optimization of the phase excitation tapers and spacings for linear array of microstrip patch antennas is discussed. It exploits two models of the array under optimization: an analytical model which is based on the array factor and an electromagnetic (EM) simulation-based surrogate model of the entire array aperture. The former is used to provide initial designs which meet the design requirements imposed on the radiation response. The latter is used for tuning of the array radiation response while controlling the active reflection coefficient at the antenna inputs as well as for validation of the final design.

Furthermore, the simulation-based surrogate model allows for subsequent evaluation of the array radiation and active reflection responses in the beam scanning operation at negligible computational costs. The simulation-based surrogate model is constructed with a superposition of simulated radiation and reflection responses, complex far-fields and complex S-parameters, respectively, of the designed array aperture with only one array element being excited at a time. Moreover, the low computational cost of the surrogate model is ensured by the coarse-mesh simulation of the array aperture EM model. Reliability of such a model is ensured by means of a suitable correction which is carried out with respect to the array aperture EM high-fidelity model. The correction is performed iteratively throughout the optimization process. Performance, numerical efficiency, and accuracy of this technique are demonstrated with radiation pattern synthesis of linear arrays comprising 32 microstrip patch antennas by phase–spacing optimization.

10.4.1. *Array aperture geometry and design problem outline*

An important design problem for arrays radiating broadside, e.g., as shown in Fig. 10.15, is to find array element excitations and spacings resulting in reduced sidelobes while maintaining other figures,

232 *Simulation-Based Optimization of Antenna Arrays*

e.g., peak realized gain and maximal active reflection coefficient at predefined levels at the operational frequency. For an array of interest we will search for phases of the signals incident at the elements' inputs and element spacings resulting in the reduced sidelobe level (SLL). Cases of uniform and non-uniformly spaced elements will be considered.

Consider an array of identical probe-fed microstrip patch antennas (MPAs) residing on a finite substrate as shown in Fig. 10.15(a). The MPA patches and the substrate are coplanar to the YOZ-plane of the coordinate system shown in Fig. 10.15(a). The XOZ plane is the E-plane for the considered array by locations of the probes exciting the dominant TM_{001}^x mode of the MPAs. Such location of the MPA feeds on the array line (z-axis in Fig. 10.15(a)) makes the element pattern in the upper E-plane more omnidirectional, that is a potential benefit for a phased array operation. On the other hand,

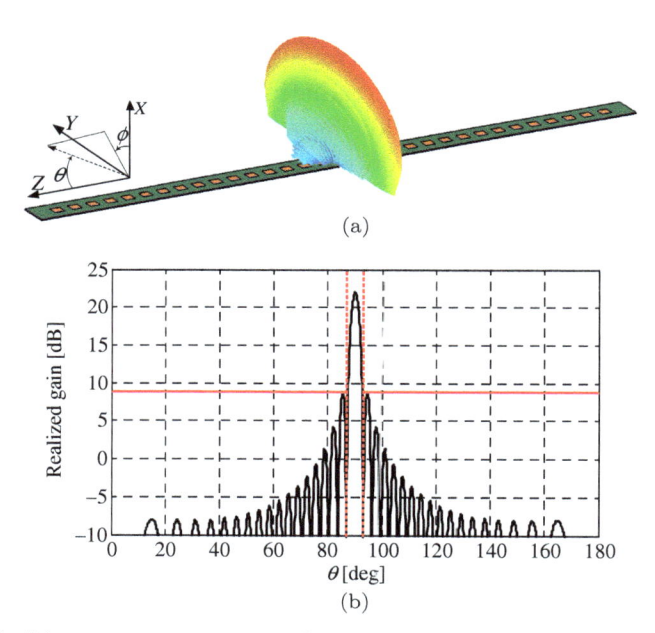

Fig. 10.15. Linear array aperture with 32 microstrip patch antennas on a finite substrate: (a) 3D-view (the ground plane is not shown) and the simulated radiation pattern (50 dB range); (b) realized gain in the E-plane of the uniform amplitude-phase-spacing design.

it increases undesirable effects of near-field interactions between the elements so that to get a reliable evaluation of the array reflection and radiation responses discrete EM simulations become necessary.

We use two types of array models, analytical model which is essentially the array factor and a simulation-based model. Phase–spacing synthesis is conducted as numerical optimization. Optimization of the analytical model is fast yet it locates an improved design approximately. The simulation-based model of the array aperture is built with superposition. The simulation-based model provides the detailed description and produces sufficiently accurate radiation and reflection responses. Further adjustment of the design is performed as optimization of the superposition model which is configured from coarse-mesh simulations and, therefore, the model should be iteratively corrected.

The superposition model also allows us to control the reflection response, validate the final design, and evaluate the final design for different scan angles. Moreover, the superposition model can be used for adjustment of the excitation phases at different scan angles without extra EM simulations. In particular, the data utilized for superposition model construction are sufficient for evaluation of radiation and reflection responses of the antenna array excited with any complex taper.

10.4.2. Array factor model for the radiation response estimation

Radiation responses of arrays can be estimated using the array factor assuming that the array elements produce the same far-fields. Using this concept, the electric far-field in the E-plane of the antenna array can be written as

$$\mathbf{E}(\theta) = \mathbf{E}_e(\theta, \phi = 0) \cdot \mathrm{AF}(\theta), \tag{10.6}$$

where $\mathbf{E}_e(\theta, \phi = 0)$ is the element far-field in the E-plane and $\mathrm{AF}(\theta)$ is the array factor,

$$\mathrm{AF}(\theta) = \sum_{n=1}^{N} a_n e^{j 2\pi z_n \cos\theta}, \tag{10.7}$$

where a_n and z_n are, respectively, the excitation coefficient and the n-th element's z-coordinate normalized to the operational wavelength.

Assuming that the elements produce the same far-fields, omnidirectional element radiation in the plane of interest and no change of the active reflection coefficient with excitation, the realized gain of the array at the operational frequency can be estimated as follows:

$$G_r(\theta) \propto |AF(\theta)|^2. \qquad (10.8)$$

Equation (10.8) is adopted as a fast estimator $\mathbf{R}_a$ of the radiation pattern at different points of the design space. The simulation-based model of the array aperture is an accurate evaluator of the radiation and reflection responses.

10.4.3. *Design using optimization of simulation-based surrogates*

In this section, we provide the details of the design technique. One of the objectives was to reduce the computational cost of array optimization.

10.4.3.1. *Problem formulation: Objective function*

The high-fidelity model $\mathbf{f}(\mathbf{x})$ is an accurate EM-simulated representation of the antenna array, where $\mathbf{x}$ is a vector of designable parameters (e.g., excitation amplitudes). More specifically, $\mathbf{f}(\mathbf{x})$ is a vector-valued function of $\mathbf{x}$ that represents the active reflection coefficients versus frequency for all ports of the array, and realized gain versus the θ angle. The design problem can be, therefore, formulated as an optimization task

$$\mathbf{x}^* = \arg\min_x H(\mathbf{f}(\mathbf{x})), \qquad (10.9)$$

where the objective function H encodes the design specifications.

Here the radiation and reflection responses are simultaneously handled by optimizing the SLL directly while ensuring low reflection using a suitable penalty function (Nocedal and Wright, 2006). The objective function takes the form (similar to the one used in

Koziel et al. (2014d)):

$$H(\mathbf{f}(\mathbf{x})) = U_{\text{SLL}} + \beta \cdot \max_{n=1,\ldots,N}\{|S_n(f_0)| + 10.0\}^2. \qquad (10.10)$$

In (10.12), U_{SLL} stands for the SLL, the primary objective. The penalty function controls the active reflection coefficients $|S_n|$ at the level at most -10 dB at the center frequency f_0. The penalty factor β is set to a large value, here, 1000, so that a relative violation of the constraint beyond a few percent leads to noticeable increase of the entire objective function value.

10.4.3.2. *Optimization of the array factor model*

The array factor is utilized for two purposes: (i) to search for the optimized values of element spacing because this task cannot be done using the superposition model as the latter is constructed for fixed dimensions; and (ii) to provide a starting point for optimizing the superposition models with respect to excitation phases.

As a main optimization engine, we exploit the standard sequential-quadratic programming (SQP) algorithm (Nocedal and Wright, 2006) implemented in Matlab *fminimax* optimization routine. This routine is very convenient as the array synthesis problem can be formulated as a minimax optimization task with upper and lower specifications. To reduce the computational cost of the process, we utilize analytical derivatives which can be easily calculated for the array factor model. A typical number of objective function evaluations is a few hundred instead of thousands or tens of thousands normally required by population-based methods such as GAs or PSO.

Antenna array design problems are normally multi-modal with many local minima; therefore, we also include an optional random search (described in Sec. 10.1.2) as an initial step.

10.4.3.3. *Correction of the simulation-based low-fidelity model*

In evaluation of the array responses to different applied excitations, the configured superposition model $\mathbf{f}_s$ is fast compared to a direct simulation of the array; however, the CPU cost of acquiring the

236 *Simulation-Based Optimization of Antenna Arrays*

data necessary for creation of $\mathbf{f}_s$ is high, namely $N \times \mathbf{f}$ where N is the number of high-fidelity simulations of the entire array aperture comprising N antennas. Therefore, here, we only use the model $\mathbf{f}_s$ for comparison.

For the sake of computational savings, our design methodology is based on the superposition model $\mathbf{c}_s$, which is constructed using N evaluations of the coarse-mesh model $\mathbf{c}$. The time evaluation ratio between $\mathbf{f}$ and $\mathbf{c}$ is 15–50. The starting point for the optimization process is obtained by optimizing the array factor.

The radiation pattern of $\mathbf{c}_s$ is a relatively reliable representation of that of the high-fidelity model $\mathbf{f}$ (simulated with simultaneous excitation of all elements). Possible discrepancies are compensated using additive response correction which is described in Sec. 10.5.3.4 (Eq. (10.14)).

In the same time the misalignments between corresponding active reflection coefficients can be quite significant as illustrated in Fig. 10.16. In order to use $\mathbf{c}_s$ as a reliable representation of the

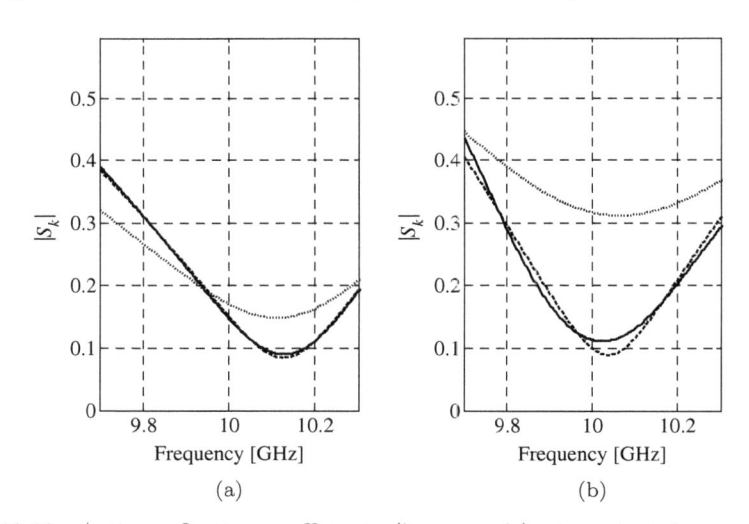

Fig. 10.16. Active reflection coefficients (linear scale) at a selected port of the antenna array of Fig. 10.15(a): (a) the low-fidelity model $\mathbf{c}_s$ ($\cdots$), the high-fidelity model $\mathbf{f}$ (—), and the frequency/response scaled low-fidelity (- - -) responses at the same design. It is seen that the scaling procedure effectively reduces the misalignment between the model responses; (b) $\mathbf{f}$, $\mathbf{c}_s$ and scaled $\mathbf{c}_s$ responses at another design, showing good prediction capability of the surrogate model.

high-fidelity model **f** in terms of reflection, the following scaling procedure is utilized (Bandler et al., 2004a):

$$S_{s.k}(f) = \alpha_0^k \cdot S_{c.k}(\alpha_1^k + f\alpha_2^k), \qquad (10.11)$$

where f stands for frequency; $S_{c.k}$ is the kth active reflection coefficient of model $\mathbf{c}_s$, whereas $S_{s.k}$ is the scaled response. We use an affine frequency scaling (Yelten et al., 2012) with the parameters α_1^k and α_2^k, as well as multiplicative response correction implemented through α_0^k.

The scaling parameters are obtained independently for each k by solving the nonlinear regression problem

$$[\alpha_0^k, \alpha_1^k, \alpha_2^k] = \arg \min_{[\bar{\alpha}_0^k, \bar{\alpha}_1^k, \bar{\alpha}_2^k]}$$

$$\times \int_{f_{\min}}^{f_{\max}} [S_{f.k}(f) - \bar{\alpha}_0^k \cdot S_{c.k}(\bar{\alpha}_1^k + f\bar{\alpha}_2^k)]^2 df. \qquad (10.12)$$

Notice that $S_{f.k}$ stands for the kth active reflection coefficient of the high-fidelity model **f** with simultaneous excitation of all elements.

Both $f_{\min}$ and $f_{\max}$ are normally selected relatively close to the operation frequency f_0 because we are only interested in matching the models around the operation frequency. For example, if $f_0 = 10$ GHz then $f_{\min}$ and $f_{\max}$ can be chosen as $f_{\min} = 9.7$ GHz and $f_{\max} = 10.3$ GHz. On the other hand, the particular choice of $f_{\min}$ and $f_{\max}$ is not critical; the only requirement is that the minima of the reflection coefficients should be between $f_{\min}$ and $f_{\max}$.

The frequency-scaled model serves as a surrogate model of the reflection response of the array aperture under design.

10.4.3.4. *Optimization algorithm*

As mentioned before, the element spacings are adjusted by optimizing the array factor model, and then they are kept fixed at the next optimization stage. The next optimization stage is carried out using the corrected simulation-based superposition model where the correction of the reflection coefficients is implemented as described in Sec. 10.4.3.3.

The radiation pattern of the array is optimized using the following iterative scheme (Koziel and Ogurtsov, 2014a):

$$\mathbf{x}^{(i+1)} = \arg\min_{\mathbf{x};||\mathbf{x}-\mathbf{x}^{(i)}||\leq \delta^{(i)}} H(\mathbf{s}^{(i)}(\mathbf{x})), \qquad (10.13)$$

where $\mathbf{s}^{(i)}$ is the surrogate model at iteration i. The radiation pattern of $\mathbf{s}^{(i)}$ is obtained as (Koziel and Ogurtsov, 2014a)

$$r_s^{(i)}(\mathbf{x}) = r_c^{(i)}(\mathbf{x}) + [r_f^{(i)}(\mathbf{x}) - r_c^{(i)}(\mathbf{x})], \qquad (10.14)$$

where r_c and r_f are radiation patterns of $\mathbf{c}_s$ and $\mathbf{f}$, respectively. Equation (10.14) ensures perfect agreement between $r_s^{(i)}$ and $r_f^{(i)}$ at $\mathbf{x}^{(i)}$.

To ensure convergence, the algorithm (10.13) is embedded in the trust-region (TR) framework (Conn et al., 2000), i.e., the surrogate model optimization (10.15) is constrained to the vicinity of the current design $\mathbf{x}^{(i)}$. Here $\mathbf{x}^{(i)}$ is defined as $||\mathbf{x} - \mathbf{x}^{(i)}|| \leq \delta^{(i)}$, where $\delta^{(i)}$ is the TR radius at iteration i. The TR radius is adjusted using the standard rules (Conn et al., 2000).

The flow diagram of the entire design optimization process is shown in Fig. 10.17. The process is automated using an in-house interface between the EM solver, here CST MWS (CST, 2016), and the optimization algorithm implemented in Matlab.

10.4.4. *Optimization results*

In this section, we describe a design process applied to an array of 32 MPAs. The process comprises three stages. The techniques of each stage are described in the previous sections. We consider two cases which are different in the number of design variables: (i) non-uniform spacings and non-uniform phases (NSNP); and (ii) uniform spacings and non-uniform phases (USNP). All excitation amplitudes are equal. In all cases, spacings and phases are imposed to be symmetrical on the z-axis (depicted in Fig. 10.15(a)) relative to the array center. In the same time, the aperture geometry is not symmetrical on the z-axis because of the feeding probes.

The probes, exciting the TM_{001}^x dominant mode of the microstrip patches, have the same offset relatively the patch centers, 2.9 mm in the z-direction of Fig. 10.15(a). The microstrip substrate is a 1.575-mm-thick Rogers RT5880 with lateral extends (beyond the patch edges) of 18.4 mm in the z-direction and 9.2 mm in the y-direction. The metal ground plane is modeled as infinite. The patches are 9.2 × 9.2 mm. Patch metallization is with 70 µm copper.

For a single MPA, the peak directivity is 7.3 dBi, peak realized gain is 7.2 dB, E-plane half-power beamwidth is 95°, all at 10 GHz. The reflection coefficient at the input of a single MPA is below −35 dB at 10 GHz and has the 6% bandwidth at the −10 dB level.

The NSNP case contains 32 design variables: 16 spacings and 16 phases. The USNP case contains 17 design variables, one spacing and 16 phases. In both cases, NSNP and USNP, designs with various constraints with respect to the maximum spacing are considered, leading to corresponding optimum designs. In this section, we only demonstrate operation and report numerical costs for selected designs; comparison of the designs including their performance in the beam-scanning operation mode will be given in Sec. 10.4.5.

The primary objective function is the sidelobe level (SSL) in the E-plane. The SLL should be minimized for 0–87° and 93–180° angles. Peak realized gain should be maintained closely to that of the initial design, $\mathbf{x}^{\text{init}}$, which is a uniform amplitude/phase/spacing array with 17 mm spacing, namely to 22 dB. This requirement is taken into account through a properly defined penalty function. Active reflection coefficients should be under −10 dB at 10 GHz; this requirement is also taken into account with appropriate penalty functions.

10.4.4.1. *Optimization with non-uniform spacing and phases*

The design variables for the NSNP case were $\mathbf{x} = [s_1 \; s_2 \; \ldots \; s_{16} \; \varphi_1 \; \varphi_2 \; \ldots \; \varphi_{16}]^T$ where s_n and φ_n stand for center-to-center element spacings normalized to the wavelength and excitation phases. Spacings s_n ($n = 1, \ldots, 16$) were constrained, $s_{\min} \leq s_n \leq s_{\max}$, and $s_n = s_{\{32-n\}}$ for $n = 17, \ldots, 31$. Various maximum allowed spacings $s_{\max}$ have been considered, whereas the minimum spacing $s_{\min}$ has been fixed to 0.37. Results for $s_{\max} = 1.0$ are presented.

240 *Simulation-Based Optimization of Antenna Arrays*

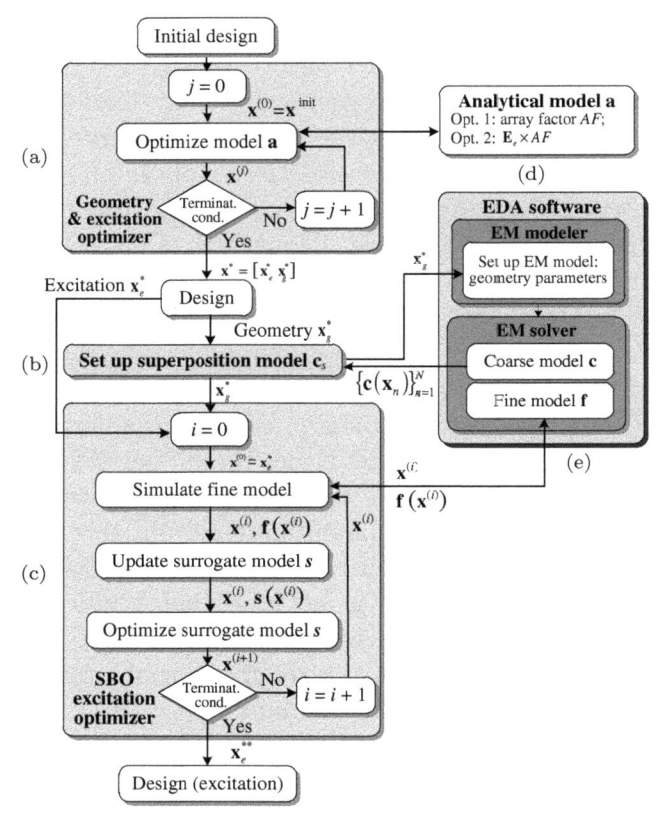

Fig. 10.17. A flow-chart of the three-stage SBO approach with a simulation-based superposition model for phase–spacing optimization: (a) optimization of the analytical array model **a** for excitation denoted as $\mathbf{x}_e$ and spacings denoted as $\mathbf{x}_g$; (b) setting up the coarse-mesh simulation-based superposition model $\mathbf{c}_s$ at the optimum of the analytical model **a** denoted as $\mathbf{x}^* = [\mathbf{x}_e^* \mathbf{x}_g^*]$; (c) surrogate model optimization for excitation; (d) the analytical model **a** of aperture radiation, implemented in Matlab, in this work. $\mathbf{E}_e$ stands for the pre-simulated far-field of the isolated element; (e) EM modeler and solver. In this work, the vector $\mathbf{x}_e$ contains excitation phases, and vector $\mathbf{x}_g$ contains center-to-center element spacings normalized to the wavelength.

The optimization starts from the random initial design $\mathbf{x}^{\text{init}}$. At the first stage, outlined in Fig. 10.17(a), the array factor model **a**, (10.7) and (10.8) has been optimized for the reduced SLL using the approach outlined in Sec. 10.5.3. The computational cost of this stage is a few hundred of model evaluations; that is about 1 min of CPU

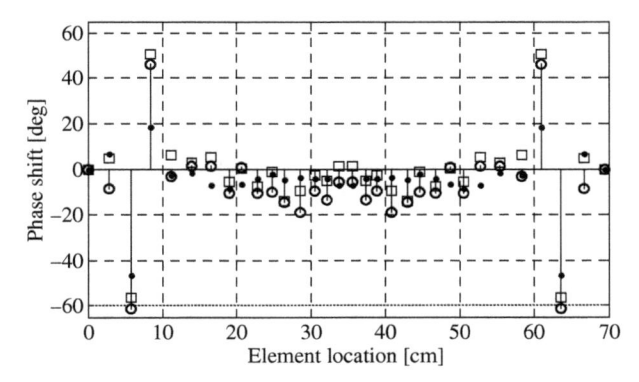

Fig. 10.18. Excitation phases versus location of the NSNP array elements: at the optimum of the fast model **a**, $\mathbf{x}^{\text{opt.1}}$ (·); at the optimum of the simulation-based surrogate model $\mathbf{s}_s$ at 10 GHz, $\mathbf{x}^{\text{opt.2}}$ (o) shown with stem lines; and at the optimum of the high-fidelity model $\mathbf{f}_s$ at 10 GHz, $\mathbf{x}_f^{\text{opt.2}}$ (□).

time in total including analytical gradient calculation. This time can be neglected if compared to the array simulation time. The maximum and minimum spacing of the first stage optimum $\mathbf{x}^{\text{opt.1}}$ are 1.0 and 0.53, respectively. Phases and location of the array elements of the optimum $\mathbf{x}^{\text{opt.1}}$ are shown in Fig. 10.18 using dots where the first element is a reference.

At the second stage depicted in Fig. 10.17(b) the coarse-mesh simulation-based superposition model $\mathbf{c}_s$ as well as high-fidelity superposition model $\mathbf{f}_s$ (used only for comparison) have been set up at the optimum $\mathbf{x}^{\text{opt.1}}$ with 32 simulations of the coarse-mesh model **c** and with 32 simulations of the high-fidelity model **f**, respectively. In addition to coarser meshes, model **c** was defined with lossless materials to speed up its simulations further. Both models **c** and **f** were defined with the CST MWS environment (CST, 2016) and simulated using the CST MWS transient solver on a 2-GHz 12 core dual CPU 64 GB RAM computer.

Discrete model **f** comprising 4,213,620 mesh-cells required 17 min and 29 s for every simulation at $\mathbf{x}^{\text{opt.1}}$ while discrete model **c** comprising 226,800 mesh-cells required only 58 s for every simulation. Thus, the entire model $\mathbf{c}_s$ had been acquired at the time cost of about 1.77 simulations of model **f**.

242 *Simulation-Based Optimization of Antenna Arrays*

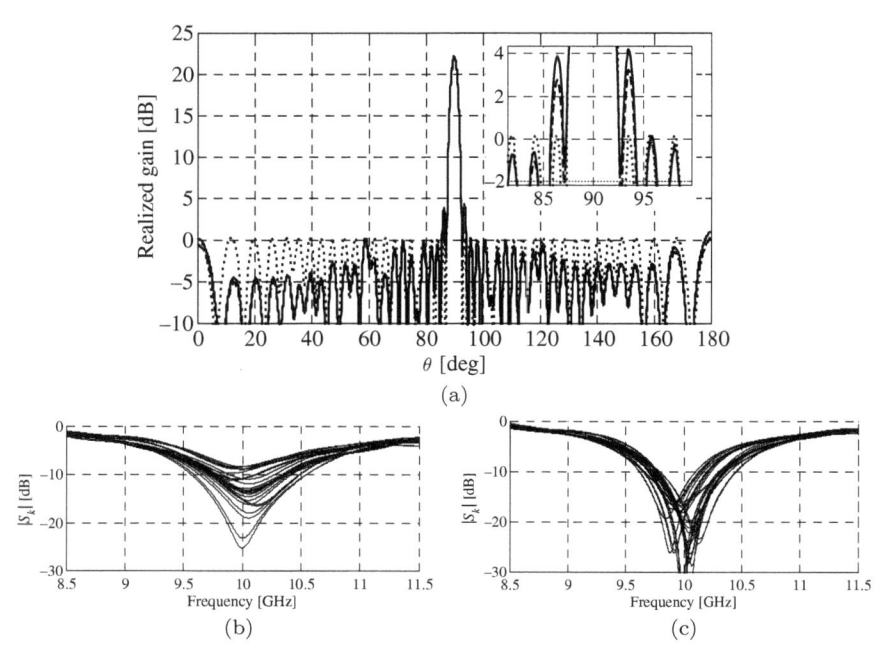

Fig. 10.19. NSNP array responses at the model **a** optimum $\mathbf{x}^{\text{opt.1}}$: (a) realized gain in the E-plane of $\mathbf{f}_s$ (—) with the $-18\,\text{dB}$, $\mathbf{c}_s$ (- -) with the $-19\,\text{dB}$, and **a** ($\cdots$) with the $-22\,\text{dB}$ SLL; (b) active reflection coefficients of $\mathbf{c}_s$; (c) active reflection coefficients of $\mathbf{f}_s$.

Responses of models **a**, $\mathbf{c}_s$, and $\mathbf{f}_s$ at design $\mathbf{x}^{\text{opt.1}}$ are shown in Fig. 10.19. The radiation pattern of the array factor model **a** has been scaled in Fig. 10.19(a) to the peak realized gain of $\mathbf{f}_s$. Figures 10.19(a)–(c) also illustrate different sensitivity of the radiation and, in especially, the active reflection coefficients to fidelity of the discrete model. Figure 10.19(a) shows two spikes on the radiation responses of the simulation-based models next to the major lobe. This issue is fixed with phase optimization at the third stage of the process (depicted in Fig. 10.17(c)) using the simulation-based surrogate model $\mathbf{s}_s$.

The SBO optimizer has found the surrogate model optimum $\mathbf{x}^{\text{opt.2}}$ in two iterations with only two simulations of $\mathbf{f}$ which were needed to correct the surrogate $\mathbf{s}_s$. The correction has been done for the active reflection response using scaling (10.11), (10.12), and for

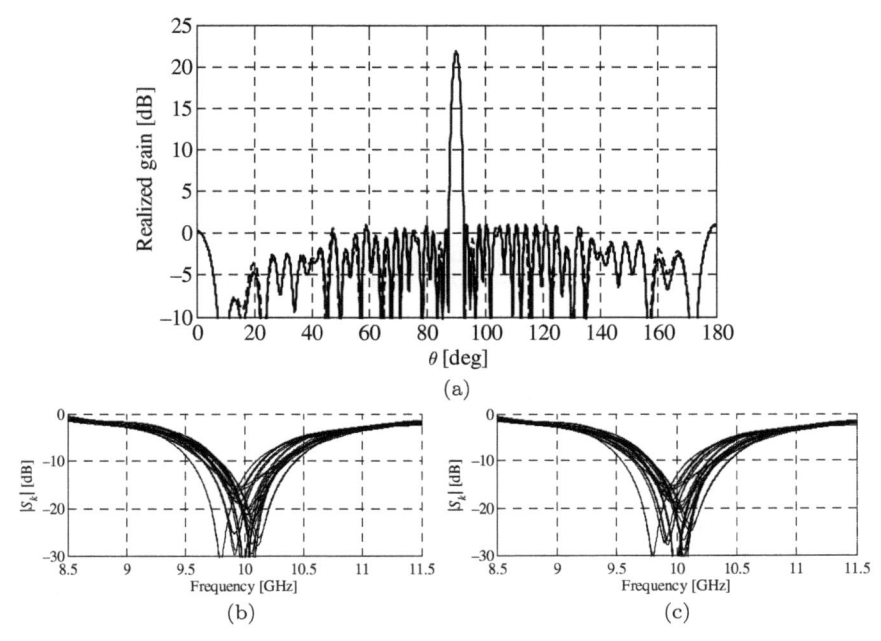

Fig. 10.20. NSNP array evaluated with the high-fidelity model $\mathbf{f}$: (a) realized gain in the E-plane with the -21 dB SLL at the surrogate model's optimum $\mathbf{x}^{\text{opt.}2}$ (- -) and with the same -21 dB SLL at the high-fidelity model's optimum $\mathbf{x}_f^{\text{opt.}2}$ (—); (b) active reflection coefficients at the surrogate model's optimum $\mathbf{x}^{\text{opt.}2}$; (c) active reflection coefficients at the high-fidelity model's optimum $\mathbf{x}_f^{\text{opt.}2}$.

radiation using response correction (10.14). The correction accounted for inaccuracies of the low-fidelity radiation and reflection responses. Phases of the optimum $\mathbf{x}^{\text{opt.}2}$ are shown in Fig. 10.18.

Response of the high-fidelity model $\mathbf{f}$ at the final design $\mathbf{x}^{\text{opt.}2}$ is shown in Fig. 10.20. Model $\mathbf{f}_s$ had been also optimized for comparison. The phases of its optimum design $\mathbf{x}_f^{\text{opt.}2}$ are shown in Fig. 10.18. Notice that radiation and reflection responses of two high-fidelity models $\mathbf{f}_s$ and $\mathbf{f}$ at the same design are identical up to numerical noise. Therefore, Figs. 10.20(a) and 10.20(c) also show responses of model $\mathbf{f}_s$ at $\mathbf{x}_f^{\text{opt}}$ with the same solid lines.

Figure 10.20 shows that optimization of the surrogate $\mathbf{s}_s$ and superposition model $\mathbf{f}_s$ yield designs of quite similar quality; however, the design of the surrogate model has been obtained at the total costs of less than three evaluations of **f**. Thus, the total design cost using the applied techniques is less than five high-fidelity simulations of the array aperture. This total cost includes the cost of the surrogate model setup.

10.4.4.2. *Optimization with uniform spacing and non-uniform phases*

In this study, the design variables are the uniform spacing and excitation phases. The design variable vector becomes $\mathbf{x} = [s_1 \; \varphi_1 \; \varphi_2 \; \ldots \; \varphi_{16}]^T$ where s_1 is the constrained center-to-center spacing and $\varphi_1, \ldots, \varphi_{16}$ are the phases. Cases of different upper bounds for spacing, $s_{\max}$, had been also considered. The lower bound for spacing, $s_{\min}$, has been fixed to the same value as in the NSNP case. Results presented in this section are for $s_{\max} = 0.8$.

Optimization starts from the initial design $\mathbf{x}^{\text{init}}$. At the first stage (outlined in Fig. 10.17(a)), the array factor model **a**, (10.7), namely the normalized power pattern, has been optimized for the reduced SLL using the approach described in Sec. 10.5.3.

Phases of the optimum $\mathbf{x}^{\text{opt.1}}$ are shown in Fig. 10.21 using dots where the first element is a reference. Spacing of the optimum $\mathbf{x}^{\text{opt.1}}$ was obtained as $s_1 = 0.649$. It took about 1 min to optimize the array

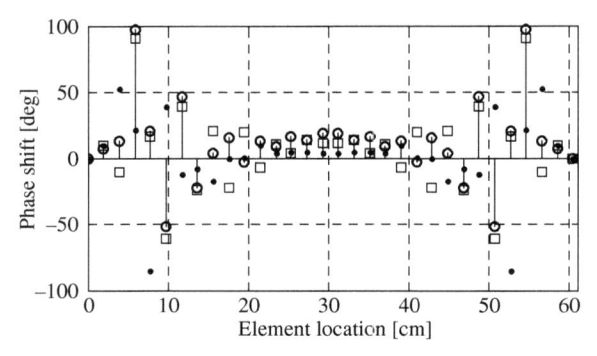

Fig. 10.21. Excitation phases of the USNP array: at the optimum of the fast model **a**, $\mathbf{x}^{\text{opt.1}}$ (·); at the optimum of the simulation-based surrogate model $\mathbf{s}_s$ at 10 GHz, $\mathbf{x}^{\text{opt.2}}$ (o) shown with stem lines; and at the optimum of the high-fidelity model $\mathbf{f}_s$ at 10 GHz, $\mathbf{x}_f^{\text{opt.2}}$ (□).

Design of Linear Antenna Array Apertures 245

factor model; this time is negligible in comparison to those spent for other design stages.

The high-fidelity discrete model **f** comprised 3,795,012 mesh-cells and required 39 min and 23 s for simulation at $\mathbf{x}^{\text{opt.1}}$. The low-fidelity discrete model **c** comprised 205,000 mesh-cells and needed only 53 s for simulation at $\mathbf{x}^{\text{opt.1}}$. Thus, the low-fidelity superposition model $\mathbf{c}_s$ had been set up at the cost of about 0.72 simulations of **f**. Both models **f** and **c** were evaluated using the CST MWS transient solver on a 2-GHz 12 core dual CPU 64 GB RAM computer.

Simulation time of the high-fidelity model **f** of this uniformly spaced antenna array is more than two times longer than that of the high-fidelity model of the non-uniformly spaced array considered earlier. Such an increase of the simulation time is associated with the periodic, thus, more resonant, structure of the uniformly spaced

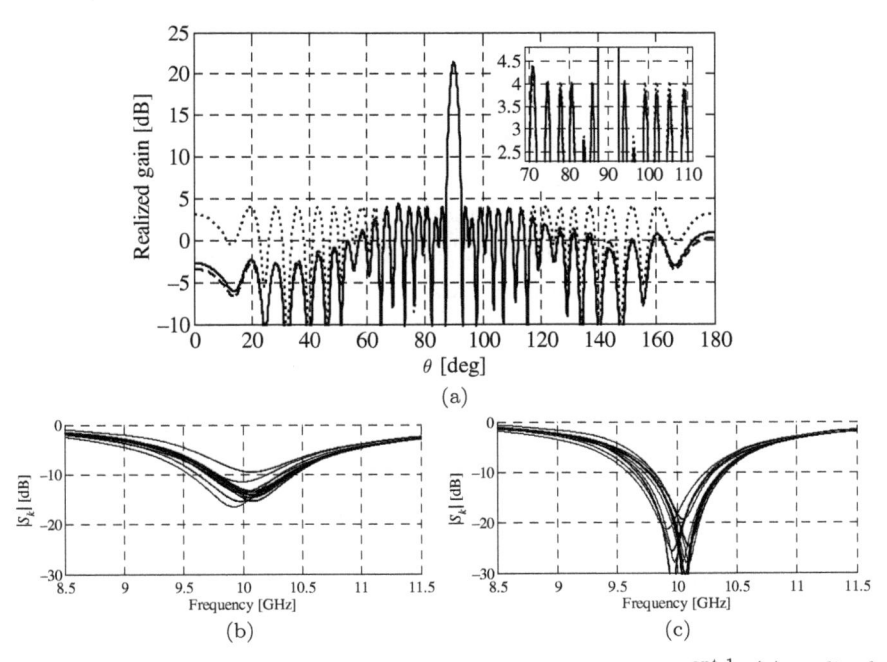

Fig. 10.22. USNP array responses at the model **a** optimum $\mathbf{x}^{\text{opt.1}}$: (a) realized gain in the E-plane of $\mathbf{f}_s$ (—) with the -17.0 dB, $\mathbf{c}_s$ (- -) with the -17.0 dB, and **a** ($\cdots$) with the -17.4 dB SLL; (b) active reflection coefficients of $\mathbf{c}_s$; (c) active reflection coefficients of $\mathbf{f}_s$.

array. Consequently, the transient solver needs substantially more time steps to evolve the time-domain signal before the signals decays to the same threshold value.

Responses evaluated with the array factor **a**, low-fidelity superposition $\mathbf{c}_s$, and high-fidelity superposition model $\mathbf{f}_s$ at design $\mathbf{x}^{\text{opt}.1}$ are shown in Fig. 10.22. The radiation pattern of the array factor is scaled in Fig. 10.22(a) to the peak of $\mathbf{f}_s$.

The simulation-based superposition models have been set up at $\mathbf{x}^{\text{opt}.1}$ technically in the same way as in the NSNP case. The third stage of the design process, phase optimization, starts from the phase values of $\mathbf{x}^{\text{opt}.1}$ and uses the surrogate model $\mathbf{s}_s$ defined with geometry and material parameters of $\mathbf{x}^{\text{opt}.1}$. The SBO optimizer had found the surrogate model optimum $\mathbf{x}^{\text{opt}.2}$ in three iterations with three evaluations of $\mathbf{f}$ needed to correct the surrogate $\mathbf{s}_s$. Phases of the optimum $\mathbf{x}^{\text{opt}.2}$ are plotted in Fig. 10.21. Response of the high-fidelity model $\mathbf{f}$ at the final design $\mathbf{x}^{\text{opt}.2}$ is shown in Figs. 10.23(a) and 10.23(b).

The surrogate model has been optimized using the objective function similar to (10.10), with the penalty function enforcing all active reflection coefficients to be no larger than $-10\,\text{dB}$ at $10\,\text{GHz}$. Responses of the high-fidelity model $\mathbf{f}_s$ at its optimum $\mathbf{x}_f^{\text{opt}.2}$ are shown in Figs. 10.23(a) and 10.23(c). Simulated responses at $\mathbf{x}^{\text{opt}.2}$ are quite similar to those at $\mathbf{x}_f^{\text{opt}.2}$; however, the total cost of obtaining $\mathbf{x}^{\text{opt}.2}$ is low: only four simulations of the high-fidelity model of the array aperture.

10.4.5. *Optimized designs as phased array apertures*

In this section, we provide a comparative performance analysis of the designs obtained in the previous section and operating as apertures of phased array antennas. Specifically, the NSNP and USNP designs $\mathbf{x}^{\text{opt}.2}$ obtained through surrogate-based optimization are compared to uniform spacing–amplitude–phase (USAP) designs which will be referred as $\mathbf{x}_{1,u}$ and as $\mathbf{x}_{2,u}$. The USAP design $\mathbf{x}_{1,u}$ has the same aperture length (the sum of the center-to-center spacings) as the NSNP design $\mathbf{x}^{\text{opt}.2}$. The USAP design $\mathbf{x}_{2,u}$ has the center-to-center spacing as the SBO USNP design $\mathbf{x}^{\text{opt}.2}$, 0.6494.

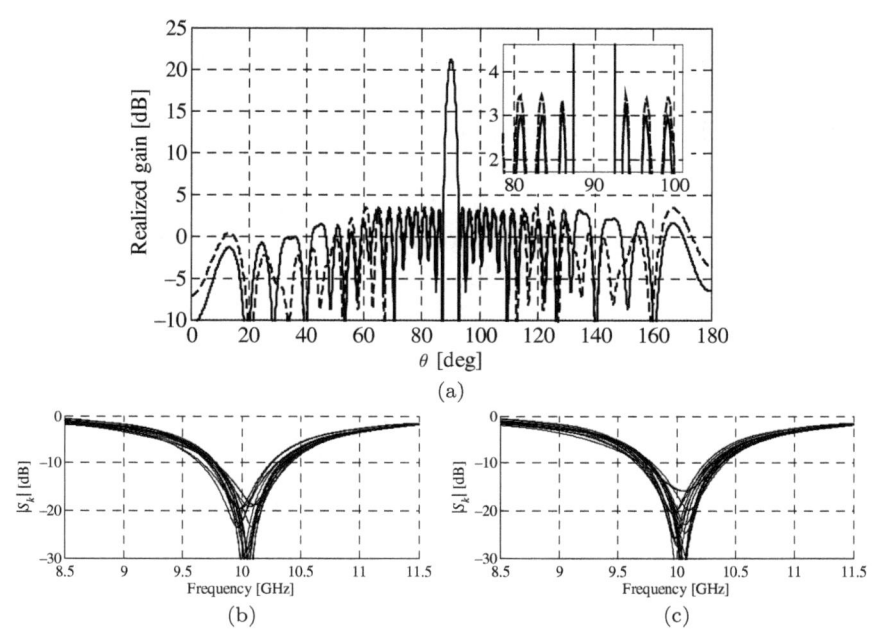

Fig. 10.23. USNP array responses evaluated with the high-fidelity model $\mathbf{f}$: (a) realized gain in the E-plane with the -17.8 dB SLL at the surrogate model's optimum $\mathbf{x}^{\text{opt.2}}$ ($--$) and with the -17.8 dB SLL at the high-fidelity model's optimum $\mathbf{x}_f^{\text{opt.2}}$ (—); (b) active reflection coefficients at the surrogate model's optimum $\mathbf{x}^{\text{opt.2}}$; (c) active reflection coefficients at the high-fidelity model's optimum $\mathbf{x}_f^{\text{opt.2}}$.

Scanning of the main beam is obtained by adding extra phase shifts (Mailloux, 2005) to the excitation phase-tapers of the designs. For the NSNP design $\mathbf{x}^{\text{opt.2}}$, the phase shift accounts for location of a particular element. Figures of interest of the above designs, the realized gain scan loss, SLL, grating lobe level (GLL), and the half-power beamwidth (HPBW), are evaluated, at the high-fidelity level of description using the surrogate superposition models $\mathbf{s}_s$.

Notice that because of applied corrections, the model $\mathbf{s}_s$ yields responses almost identical to the simulated responses of the corresponding high-fidelity model $\mathbf{f}$ for array apertures of identical geometry and with identical applied excitation.

The surrogate superposition model for the USAP design $\mathbf{x}_{1,u}$ has been acquired with 32 coarse-mesh EM simulations, i.e., at the total

248 Simulation-Based Optimization of Antenna Arrays

cost of two high-fidelity EM simulations that include the surrogate model correction. Notice that the surrogate model $\mathbf{s}_s$ is already available for the USAP design $\mathbf{x}_{2,u}$: it is that of the NSNP design $\mathbf{x}^{opt.2}$ with the zero phase taper applied. The figures of interest for the designs under study are evaluated for different scan angles with essentially no extra computational cost. All designs were also evaluated using array factor model $\mathbf{a}$, Eq. (10.6), with the simulated far-field of the isolated MPA element.

Behavior of the peak realized gain versus the scan angle for the above described designs is shown in Fig. 10.24. It can be observed

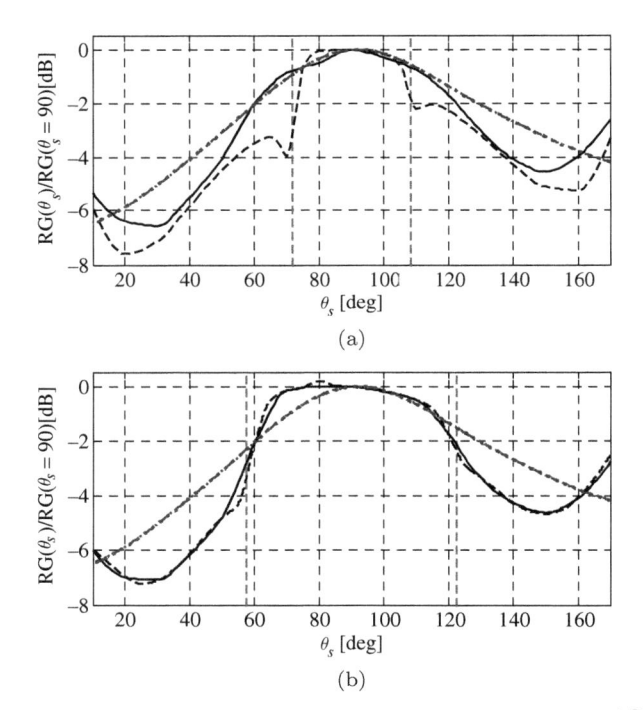

Fig. 10.24. Peak realized gain scan loss: (a) NSNP optimum $\mathbf{x}^{opt.2}$ evaluated with its surrogate model $\mathbf{s}_s$ (—) and with its array factor model $\mathbf{a}$ ($\cdots$); uniform design $\mathbf{x}_{1,u}$ with its surrogate model $\mathbf{s}_s$ (– –) and with its array factor model $\mathbf{a}$ ($\cdot$ – $\cdot$); (b) USNP optimum $\mathbf{x}^{opt.2}$ evaluated with its surrogate model $\mathbf{s}_s$ (—) and with its array factor model $\mathbf{a}$ ($\cdots$); uniform design $\mathbf{x}_{2,u}$ (– –) with its surrogate model $\mathbf{s}_s$ (– –) and with its array factor model $\mathbf{a}$ ($\cdot$ – $\cdot$). Vertical dash-lines show the maximal allowed scan angle with no grating lobes in case (a) for design $\mathbf{x}_{1,u}$ and in case (b) for all designs. $\theta_s = 90°$ is for the boresight direction.

from Fig. 10.24(a) that the scan losses of the NSNP design are substantially smaller than scan losses of the design $\mathbf{x}_{1,u}$. The peak the realized gain of the latter noticeably decreases when approaching the critical scan angle.

The USNP design $\mathbf{x}^{opt.2}$ and USAP design $\mathbf{x}_{2,u}$ show quite similar behavior of the realized gain scan loss versus the scan angle, as shown in Fig. 10.24(b). Notice that the array factor models $\mathbf{a}$ produce curves of realized gain scan loss which are substantially different from those

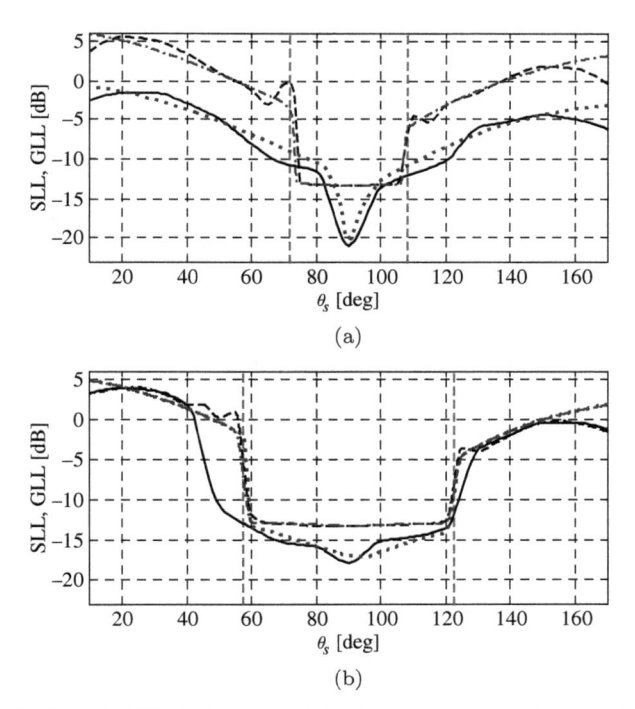

Fig. 10.25. SLL and GLL (whichever is higher at a particular angle) versus the scan angle: (a) NSNP optimum $\mathbf{x}^{opt.2}$ evaluated with its surrogate model $\mathbf{s}_s$ (—) and with its array factor model $\mathbf{a}$ (···); uniform design $\mathbf{x}_{1,u}$ with its surrogate model $\mathbf{s}_s$ (– –) and with its array factor model $\mathbf{a}$ (·– ·); (b) USNP optimum $\mathbf{x}^{opt.2}$ evaluated with its surrogate model $\mathbf{s}_s$ (—) and with its array factor model $\mathbf{a}$ (···); uniform design $\mathbf{x}_{2,u}$ (– –) with its surrogate model $\mathbf{s}_s$ (– –) and with its array factor model $\mathbf{a}$ (·–·). Vertical dash-lines show the maximal allowed scan angle with no grating lobes for case (a) for design $\mathbf{x}_{1,u}$ and for (b) for all designs. $\theta_s = 90°$ is for the boresight direction.

evaluated with the simulation-based superposition models $\mathbf{s}_s$. Thus, the predictions of the scan loss using the array factor model are unreliable for these designs.

The SLL and GLL versus the scan angle for the NSNP as well as USNP cases are shown in Figs. 10.25(a) and 10.25(b), respectively. In overall, the optimized designs show lower SLL and GLL. Even the NSNP shows the lower GLL than the uniform design except for scan angles about 80°. The SLL/GLL versus scan angle curves evaluated with the array factor models are in much better agreement with the corresponding levels evaluated with the simulation-based superposition models.

The HPBW versus the scan angle evaluated for all designs with the surrogate models, shown in Fig. 10.26, approximately follows the HPBW behavior in the large array limit (Mailloux, 2005).

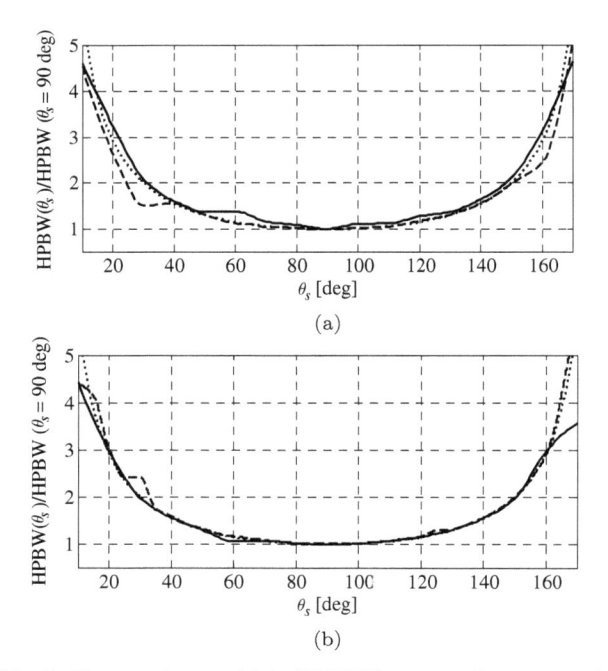

Fig. 10.26. The half power beamwidth (HPBW) versus the scan angle: (a) NSNP optimum $\mathbf{x}^{\text{opt.2}}$ (—), uniform design $\mathbf{x}_{1,\text{u}}$ (– –); (b) USNP optimum $\mathbf{x}^{\text{opt.2}}$ (—), uniform design $\mathbf{x}_{2,\text{u}}$ (– –). HPBW in the large array limit ($\cdots$) (Mailloux, 2005). $\theta_s = 90°$ is for the boresight direction.

10.5. Summary

In this chapter, design of linear array apertures using conventional and surrogate-assisted optimization techniques has been presented. It was demonstrated that utilization of variable-fidelity EM models and various kinds of surrogate modeling techniques (e.g., simulation-based superposition models and response correction techniques) permits dramatic reduction of the design time cost and accurate evaluation of far-field figures, e.g., radiation patterns, as well as reflection responses, e.g., active reflection coefficients.

Chapter 11

Design of Planar Microstrip Antenna Arrays Using Variable-Fidelity EM Models

In this chapter, a technique for rapid simulation-driven design of planar microstrip antenna arrays is discussed. It exploits the surrogate-based optimization (SBO) paradigm and variable-fidelity electromagnetic (EM) simulations. The design process includes radiation pattern optimization and matching in term of active reflection coefficients at the inputs of the elements of the array aperture.

Two low-fidelity models are utilized: a coarse-mesh EM model of the array aperture, and a fast array model that combines the simulated single element radiation response and the array factor. These low-fidelity models, after suitable correction, guide the optimization process towards the optimum of the antenna array high-fidelity EM model. In its own turn, the high-fidelity EM model is used for models' correction and for verification of the improved designs so that referring to it is minimized. The design optimization of 25- and 49-element microstrip antenna array apertures is conducted and described to demonstrate the operation as well as efficiency of the discussed technique. In the considered examples, the computational cost of design optimization is equivalent to a few high-fidelity

254 *Simulation-Based Optimization of Antenna Arrays*

simulations of the entire array aperture despite a large number of adjustable model parameters.

11.1. Planar Antenna Array Design Problem

Consider an array of Fig. 11.1. Discrete EM models of the entire array are necessary here to account for element coupling, contribution of the aperture edges, and reliably evaluate the radiation and reflection responses. The array aperture is required to have a linear polarization and operate at 10 GHz. Each microstrip patch is fed by a 50-ohm probe.

The design tasks are: to maintain the array peak directivity at 20 dBi; to have the direction of maximum radiated power density normal to the plane of the aperture; to suppress the sidelobe level (SLL) down to −20 dB; to keep active reflection coefficients lower than −10 dB, all at 10 GHz.

Initial dimensions of the element, microstrip patch antenna (MPA), are 11 mm × 9 mm. A grounded 1.58-mm thick layer of RT/duroid 5880 is the substrate. The substrate/metal ground

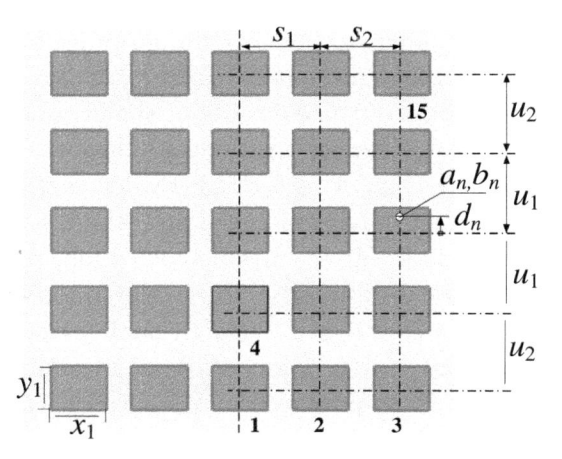

Fig. 11.1. Microstrip antenna array aperture. A symmetry (magnetic) wall is shown with the vertical dash-line at the center. Design variables (51 in total) include the patch sizes (x_1 and y_1), element spacing (horizontal s_1, s_2, and vertical u_1, u_2), probe locations ($d_1, \ldots, d_{15}$), amplitudes ($a_1, \ldots, a_{15}$), and phases ($b_1, \ldots, b_{15}$) of the excitation signals.

extends to a half of the patch size in a particular direction beyond the outermost patch edges. Locations of the feeds at the initial design are at the center of the patch in horizontal direction and 2.9 mm off the center in the vertical direction.

The symmetry wall, shown in Fig. 11.1, is introduced to the EM models, which means that the array dimensions and excitation signals (both amplitudes and phases) are set to be symmetrical with respect to the wall. Thus we will adjust the following dimensions: spacings (s_1, s_2, u_1, u_2), patch size (x_1, y_1), location of probes $(d_1, \ldots, d_{15})$, excitation amplitudes $(a_1, \ldots, a_{15})$, and phases $(b_1, \ldots, b_{15})$.

To evaluate the responses of the array we adopt the following two EM models: (i) a high-fidelity (or fine) discrete EM model of the array aperture, **f**; and (ii) a coarse EM model of the array aperture **c** which is a coarse-mesh version of **f**. These models are utilized in the SBO process as described in the following section.

11.2. Design Optimization Methodology

In this section, we discuss an approach to planar array optimization that reduces computational costs of its simulation-based design. First, the SBO basics which are closely relevant to antenna array optimization are overviewed. Then, we present a combination of techniques which has been developed for handling the design problems with the design tasks highlighted in Sec. 11.1.

11.2.1. *Surrogate-based optimization*

The array design problem is formulated here as a nonlinear minimization problem,

$$\mathbf{x}^* = \arg\min_x U(\mathbf{f}(\mathbf{x})), \qquad (11.1)$$

where $\mathbf{f}(\mathbf{x}) \in R^m$ is the response vector of a high-fidelity EM model, representing all figures of interest, in particular, the radiation response, as well as the active reflection coefficients at all ports; U is the objective function; $\mathbf{x} \in R^n$ is a vector of design variables, here, representing all adjustable parameters as described in Sec. 11.1.

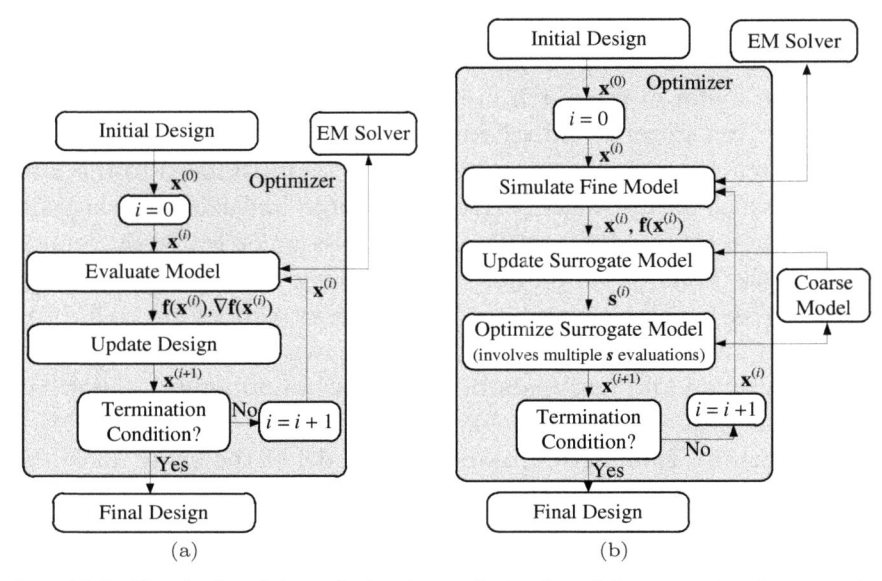

Fig. 11.2. Simulation-driven design by optimization: (a) conventional approach; (b) generic SBO approach.

The objective function U is defined so that a smaller value of U corresponds to a better design.

Direct optimization (Fig. 11.2(a)) of the high-fidelity EM model **f** is replaced by an iterative correction and optimization of its fast surrogate as shown in Fig. 11.2(b). Typically, the model **f** is only evaluated once per iteration, namely, at every new design $\mathbf{x}^{(i+1)}$ after optimizing the surrogate model to update the surrogate. The number of iterations for a well performing SBO algorithm is substantially smaller than for direct optimization methods (Koziel *et al.*, 2011a), hence such an SBO algorithm finds an optimum with a much fewer simulations of the high-fidelity model of the optimized structure and, thus, in a much shorter time.

A generic SBO scheme produces approximations $\mathbf{x}^{(i)}$ of $\mathbf{x}^*$ as follows (Koziel *et al.*, 2011a):

$$\mathbf{x}^{(i+1)} = \arg\min_x U(\mathbf{s}^{(i)}(\mathbf{x})), \qquad (11.2)$$

where $\mathbf{s}^{(i)}$ is the surrogate model at iteration i. In general, the surrogate model is constructed with a suitable correction of the underlying low-fidelity model. In this work, the low-fidelity model $\mathbf{c}$ is a lossless coarse-mesh version of the original high-fidelity model $\mathbf{f}$ (Koziel and Ogurtsov, 2012a). The quality of model $\mathbf{c}$ with a particular discretization and other relaxed simulation settings is determined by a visual inspection of its responses simulated at and about the initial design and with respect to model $\mathbf{f}$, i.e., the quality of $\mathbf{c}$ is inferred from numerical experiments involving the user's judgments. A major requirement is that the responses of the low-fidelity model should capture major features of the high-fidelity model responses. Automatic settings of the low-fidelity model for optimal performance of the SBO process in terms of the design quality and numerical cost are addressed in Koziel and Ogurtsov (2012a).

Quality of the low-fidelity model $\mathbf{c}$ turns into prediction capability of the surrogate (Koziel and Ogurtsov, 2011c). At the same time, $\mathbf{c}$ should be much faster than $\mathbf{f}$ so that the total cost of optimization and surrogate update is reasonably small. For discrete models of antennas, the ratio of the evaluation times of the high- and low-fidelity models is usually from 5 to 50, hence the computational cost of the low-fidelity model is not negligible. Therefore, developing SBO algorithms for antennas, especially for antenna array design, it is also important to reduce the number of low-fidelity model simulations.

11.2.2. *Surrogate-based optimization for array design*

The evaluation time of the high-fidelity model $\mathbf{f}$, comprising 2,531,760 hexahedral mesh-cells at the initial design of the array of Fig. 11.1, is around 20 min that makes its direct optimization impractical. Therefore, we exploit an auxiliary low-fidelity model $\mathbf{c}$ but with a coarser mesh of 83,790 hexahedral cells at the initial design and materials assigned to be lossless so that its evaluation time is only around 1 min. Both models $\mathbf{f}$ and $\mathbf{c}$ are simulated with the transient solver of CST MWS (CST, 2016) on a 2-GHz Intel(R) Xeon(R) CPU 64 GB RAM computer. The model $\mathbf{c}$ represents the radiation pattern quite accurately but its reflection response is not particularly accurate in comparison to that of the model **f**.

One splits the design variable vector **x** into two parts: $\mathbf{x} = [\mathbf{x}_p^T \mathbf{x}_m^T]^T$, where $\mathbf{x}_p = [s_1 \ s_2 \ u_1 \ u_2 \ x_1 \ y_1 \ a_1 \ldots a_{15}]$ are variables used to optimize the array pattern, and $\mathbf{x}_m = [d_{y1} d_{y2} \ldots d_{y15}]$ are variables used to adjust the active reflection coefficients. Having such separation in mind, the following 3-step procedure (also outlined in Fig. 11.3(a)) has been developed:

Step 1: Optimize the directivity pattern of the low-fidelity model **c** using $\mathbf{x}_p$ with fixed at the initial value $\mathbf{x}_m = \mathbf{x}_{m.0}$. The optimized $\mathbf{x}_p$ will be referred to as $\mathbf{x}_p^*$. Optimization is performed using the pattern search algorithm (Kolda et al., 2003) in order to overcome the problem of numerical noise present in the simulated responses of model **c**. Optimization of model **c** at this step is realized using auxiliary first-order response surface models constructed using large-step design perturbations, and the trust-region framework to ensure convergence.

Step 2: Evaluate model **f** at $\mathbf{x} = [(\mathbf{x}_p^*)^T \ (\mathbf{x}_{m.0})^T]^T$. Use **c** to estimate the necessary changes in $\mathbf{x}_m$ to improve reflection responses. Here, it is assumed that a small change of a given $\mathbf{x}_m$ component, i.e., the location of a particular probe, will noticeably affect the reflection coefficient of the corresponding patch antenna and not reflection coefficients of other antennas. It has been verified with numerical experiments that this assumption is satisfied for the structure under design for the used range of the design variables. Necessary changes in $\mathbf{x}_m$ are performed as follows:

(i) evaluate model **c** at $\mathbf{x} = [(\mathbf{x}_p^*)^T (\mathbf{x}_{m.0})^T]^T$ and at the two perturbed designs varied by $\pm \Delta d_k$ corresponding to a reflection response of patch k that does not satisfy matching requirements (cf. Fig. 11.3(b));

(ii) using interpolation of the data obtained in (i), estimate the change of d_k that gives reasonable change of the response (this takes into account the fact that responses

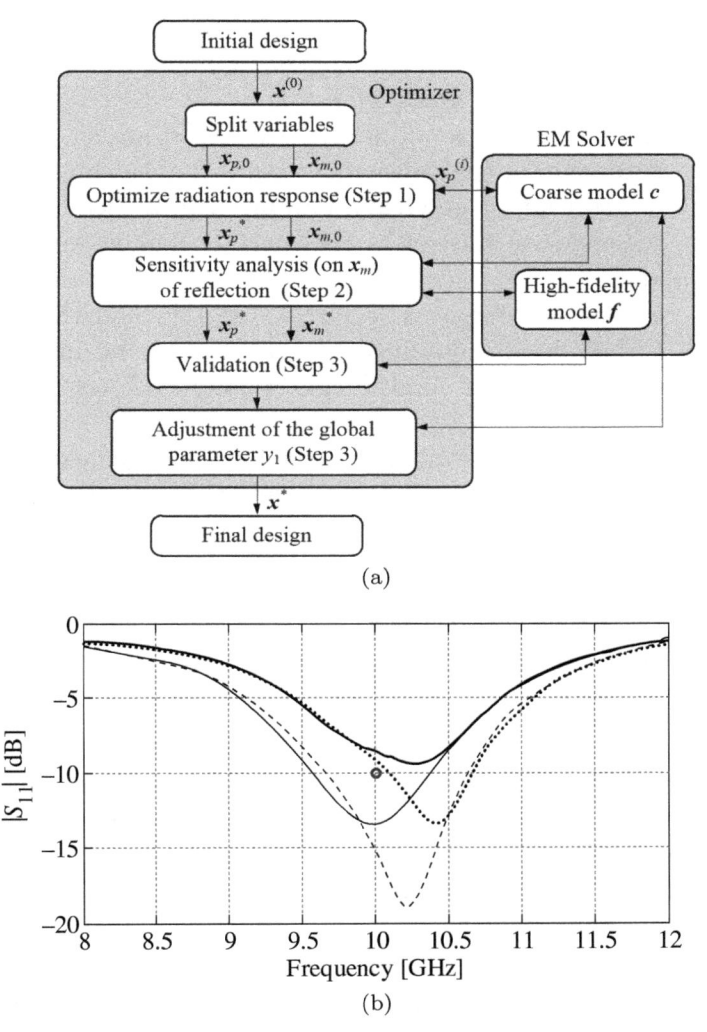

Fig. 11.3. (a) SBO procedure of Sec. 11.2 with two simulated models, coarse-mesh c and high-fidelity f; (b) active reflection coefficient at a selected port of f (—) and of c (—) at $x = [(x_p^*)^T \ (x_{m.0})^T]^T$ and at a design with variable d_k corresponding to port k perturbed by certain Δd_k (thick and thin dotted lines). Based on these responses at $[(x_p^*)^T \ (x_{m.0})^T]^T$ a proper perturbation for d_{yk} is found as described in Step 2. Additional "horizontal" correction of this response may be necessary as described in Step 3. A circle denotes design specifications.

of **f** and **c** are shifted both in frequency and amplitude relative to each other). The modified vector $\mathbf{x}_m$ will be referred to as $\mathbf{x}_m^*$.

Step 3: Evaluate **f** at $\mathbf{x} = [(\mathbf{x}_p^*)^T \ (\mathbf{x}_m^*)^T]^T$. Adjust the global parameter y_1 (patch length) to shift the matching responses in frequency as necessary. Parameter y_1 is used here to shift the response in frequency because the patches are energized at the TM_{010} mode. The change of y_1 is estimated using evaluation of **c** at $\mathbf{x} = [(\mathbf{x}_p^*)^T \ (\mathbf{x}_m^*)^T]^T$ and the two perturbed designs obtained by changing y_1 and interpolating the results. The final design is obtained after this step is referred to as $\mathbf{x}^*$.

It should be noticed that the high-fidelity model **f** is only evaluated in Step 2 (once) and in Step 3 (twice). From the generic SBO scheme (11.2) point of view, the above procedure is a one-iteration SBO approach.

11.3. Implementation and Numerical Results

The initial design is an array with $\mathbf{x}^{(0)} = [s_1 \ s_2 \ u_1 \ u_2 \ x_1 \ y_1 \ a_1 \ldots a_{15} \ d_{y1} \ d_{y2} \ldots d_{y15}] = [16 \ 16 \ 16 \ 16 \ 11 \ 9 \ 1 \ldots 1 \ 2.9 \ldots 2.9]^T$, where geometry dimensions are in mm and excitation amplitudes are normalized to the maximal excitation amplitude. Responses of this design $\mathbf{x}^{(0)}$ are shown in Figs. 11.4(a) and 11.4(b) where in Fig. 11.4(b) the reflection coefficient of the isolated single element is given for reference.

The simulated peak directivity of the element noticeably depends on the environment the element is embedded, e.g., it is about 7.4 dBi with the 22 mm × 18 mm grounded finite substrate for a single isolated element, and about 8 dBi with the 44 mm × 36 mm grounded substrate. Thus, the expected peak directivity of the initial design should be expected to be at least 21.9 dBi. At the same time, according to simulations, the peak directivity of the initial design $\mathbf{x}^{(0)}$ is 19.8 dBi, as shown in Fig. 11.4(a), therefore the effective peak directivity per element is only about 5.8 dBi. The reason for this lower value of simulated peak directivity is that directivities of

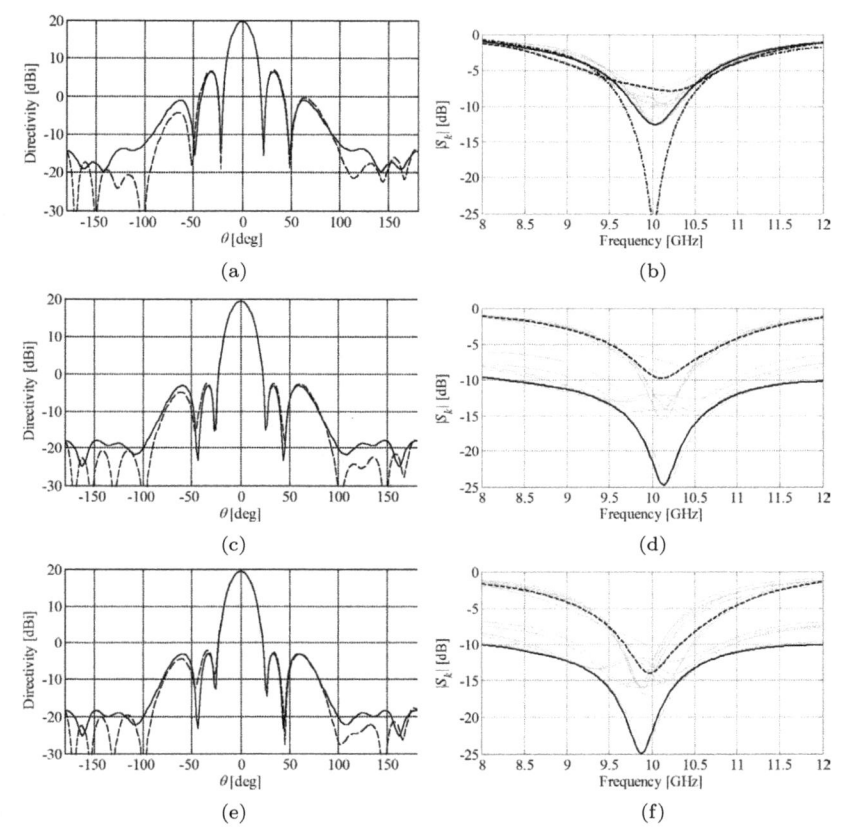

Design of Planar Microstrip Antenna Arrays 261

Fig. 11.4. High-fidelity model **f**: (a) directivity pattern cuts in the E (– –) and H (—) planes at the initial design; (b) active reflection coefficient of element 3 (—) (see Fig. 11.1) and 7 (– –) (see Fig. 11.1), at the initial design, and the reflection coefficient of a single element (· –·); (c) directivity pattern cuts in the E (– –) and H (—) planes after Step 1 (directivity optimization); (d) active reflection coefficient of elements 3 (—) and 7 (– –); (e) directivity pattern cuts in the E (– –) and H (—) planes of the final design (after Step 2 and 3); (f) active reflection coefficient of elements 3 (—) and 7 (– –). Other reflection coefficients are shown with the gray lines in (b), (d), and (f). Reflection coefficients are normalized to the maximal incident amplitude. Directivity patterns are for 10 GHz.

the elements embedded into the array for the particular spacing are noticeably lower than those of the isolated element. For example, the peak directivity of element 7, which is in the center of the array is only 5.6 dBi (the lowest peak directivity of embedded elements) and that of element 3, which is at the corner, is 6.9 dBi (the highest peak directivity of embedded elements). Notice also that the maximum of radiation of element 3 is 6 degrees off the normal direction in the H-plane and 11 degrees off the normal in the E-plane. With these lower values the expected peak directivity of the initial design should be somewhere within 19.6–20.8 dBi, which is consistent with the simulated directivity of 19.8 dBi.

The elements at design $\mathbf{x}^{(0)}$ are closely spaced: the edge-to-edge distance is only 5 mm in the x-direction and 7 mm in the y-directions, that is about 0.17 and 0.23 wavelengths at 10 GHz. Such proximity additionally justifies the use of discrete full-wave models in the design process.

Design specifications for Step 1 (directivity pattern optimization) are the following: minimize sidelobes, namely, minimize H- and E-plane directivity (in the minimax sense) outside the initial design $\mathbf{x}^{(0)}$ major lobe beamwidth, $[-21.5°, 21.5°]$, which is defined with the first null angles in the H- and E-planes. Additionally, we were using a penalty factor to ensure that the maximum directivity of the optimized design is not considerably lower than that of the initial design.

At Step 1, direct optimization of the coarse model $\mathbf{c}$, results in design $\mathbf{x}_p^* = [16.363\ 16.588\ 16.498\ 16.910\ 11.072\ 8.926\ 0.9845\ 0.4529\ 0.3718\ 0.9873\ 0.9748\ 0.4500\ 0.9970\ 0.9754\ 0.9919\ 0.9548\ 0.9369\ 0.5503\ 1.0000\ 0.4671\ 0.3621]^T$. Responses of model $\mathbf{f}$ after Step 1 are shown in Figs. 11.4(c) and 11.4(d). The cost of this Step 1 is 182 evaluations of the coarse-discretization model $\mathbf{c}$.

At Step 2 (matching correction I), we change d_{yk} for ports where matching is not sufficient (i.e., $|S_k| > -10$ dB). For ports 4, 7, 8, and 10, d_{yk} ($k = 4, 7, 8$, and 10) are increased to 3.4 mm. The correction I cost is $(8 \times \mathbf{c} + 1 \times \mathbf{f})$ model simulations.

At Step 3, (matching correction II) the global parameter y_1 is changed to 9.1 mm to shift the reflection responses in frequency to the left. The correction II cost is $(2 \times \mathbf{c} + 2 \times \mathbf{f})$ model simulations.

Responses of the final design with SLL of -21 dB are shown in Figs. 11.4(e) and 11.4(f). The total cost of the whole design optimization process in terms of model simulation is $(192 \times \mathbf{c} + 3 \times \mathbf{f})$ = $12.5 \times \mathbf{f}$, i.e., it is equivalent in time to only 12.5 high-fidelity simulations of the array.

The simulated half-power beamwidth (HPBW) of the final design is 20.4 degrees in the E-plane and 20.2 degrees in the H-plane. HPBWs of a 5 × 5-element Dolph–Chebyshev array, with the same aperture dimensions ($s_1 = s_2 = 0.549$ wavelengths, $u_1 = u_2 = 0.557$ wavelengths), and the same SLL of –21 dB in both principal planes, are estimated as 20.2 degrees in the E-plane and 20.0 degrees in the H-plane according to the analytical model of the array which combines the array factor and the power pattern of the single element. The array factor of this Dolph–Chebyshev design is with SLLs of -18.8 dB and -18.0 dB in E- and H-plane respectively so that the extra reduction of SLLs up to –21 dB is the effect of the element pattern. At the same time, according to discrete simulations, HPBWs of this Dolph–Chebyshev design are slightly wider, namely, 20.8 degrees in E-plane and 20.3 degrees in H-plane. In addition the simulated SLL of the Dolph–Chebyshev design is increased to –19.5 dB.

Other design cases were also considered. For instance, the final design with an additional -10 dB sidelobe suppression in the $[-31.5°, -21.5°]$ and $[21.5°, 31.5°]$ sectors and in-phase excitation with amplitudes as variables is shown in Figs. 11.5(a) and 11.5(b). Its total cost is $13 \times \mathbf{f}$. It should be noticed that additional suppression in the aforementioned sectors compromises the SLL, which is -17 dB for this design. The discussed method is sufficiently flexible to handle various types of design specifications. Responses of the final design with excitation amplitudes and phases as variables are shown in Figs. 11.5(c) and 11.5(d). In this case, the problem is with 51 variables and it is more complex from the optimization point of view. However, the discussed method allows for obtaining the optimized design at the total cost of only about $21 \times \mathbf{f}$ simulations. The additional degrees of freedom allow obtaining the –24 dB SLL. Peak directives of the final designs are within 19.5–20 dBi, i.e., close to 20 dBi required with the design specifications.

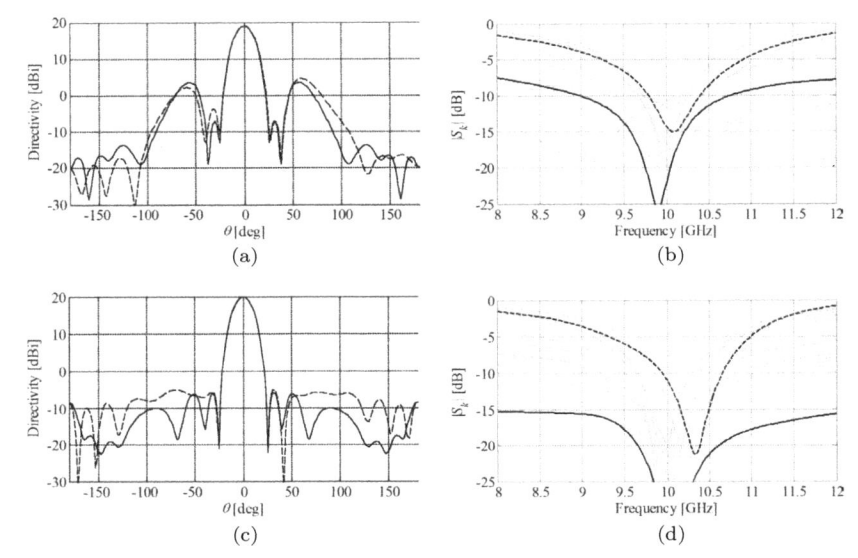

264 *Simulation-Based Optimization of Antenna Arrays*

Fig. 11.5. High-fidelity model **f** at the final designs: (a) directivity pattern cuts in the E (– –) and H (—) planes and (b) active reflection coefficient with additional suppression of SLL and non-uniform amplitude excitation; (c) directivity pattern cuts in the E (– –) and H (—) planes and (d) active reflection coefficient with non-uniform amplitude and phase excitation. Reflection coefficients of element 3 (—) (see Fig. 11.1), element 7 (– –) (see Fig. 11.1), and other elements (shown with the gray lines) are normalized to the maximal incident amplitude. Directivity patterns are for 10 GHz.

11.4. Rapid Optimization of Radiation Response

In this section, a 49-element planar microstrip array is considered. The methodology described in Sec. 11.2 is used to adjust the active reflection coefficients. Optimization of the directivity pattern is performed using a surrogate model configured from an analytical model of the radiation response.

11.4.1. *Design case: 49-element microstrip array*

Consider a planar array aperture shown in Fig. 11.6(a). The aperture comprises 49 microstrip path elements and is for operation at 10 GHz. The microstrip patches are excited at the dominant mode and linearly polarized at 10 GHz. Every patch is fed with a probe. System impedance is 50 ohms. Initial dimensions of the patches are

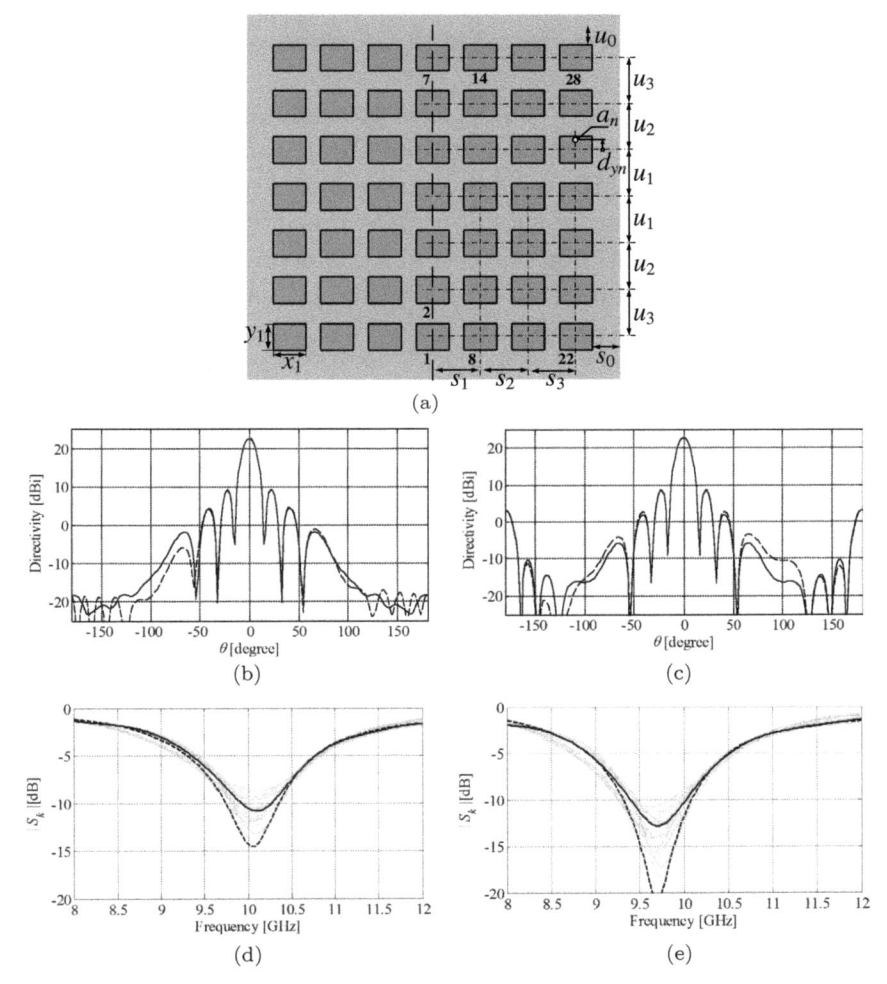

Design of Planar Microstrip Antenna Arrays 265

Fig. 11.6. Array of 49 microstrip patch antennas: (a) front view — a symmetry plane is shown with the vertical dash-line at the center; (b) directivity pattern cuts in the E (− −) and H (—) planes at the initial design evaluated at 10 GHz with the high-fidelity model **f**; (c) directivity pattern cuts in the E (− −) and H (—) planes at the initial design evaluated at 10 GHz with the analytical model **a**; (d) active reflection coefficient of element 4 (—) which is at the center and that of element 22 (− −) which is at the corner with the high-fidelity model **f**; (e) active reflection coefficient of element 4 (—) and that of element 22 (− −) with the coarse-discretization model **c**. One can see a frequency shift as well as vertical misalignment of the reflection coefficients.

11 mm × 9 mm. A 1.58-mm thick layer of RT/duroid 5880 is the microstrip substrate. Extensions of the substrate and ground, s_0 and u_0, are set to 15 mm.

At 10 GHz, the array should meet the following design requirements: (a) the SLL should be under −20 dB; (b) the peak directivity should be about 20 dBi; (c) active reflection coefficients should be under −10 dB. The array should radiate broadside with the major lobe null-to-null beamwidth restricted to 32 degrees.

A starting point for optimization is a uniform array, $\mathbf{x}^{(0)} = [s_1\ s_2\ s_3\ u_1\ u_2\ u_3\ x_1\ y_1\ a_1\ldots a_{28}\ d_{y1}\ldots d_{y28}]^T = [16\ 16\ 16\ 16\ 16\ 11\ 9\ 1\ \ldots\ 1\ 2.9\ \ldots\ 2.9]^T$ where all dimensional parameters are in mm, the excitation amplitudes are normalized, and the phase shifts are in degrees.

Initial values of the spacings (assumed to be all the same) are easily found using the array factor. The feed offset of 2.9 mm (assumed to be the same for all patches), d_{yn} shown in Fig. 11.6(a), is obtained by optimizing the EM model of the single antenna. The SLL of this initial design is about −13 dB as expected, and the peak directivity is 22.7 dBi (cf. Fig. 11.6(b)).

11.4.2. *Utilized models*

The majority of the numerical cost of the procedure described in Sec. 11.2 and illustrated in Sec. 11.3 is related to optimization of the EM-simulated coarse-discretization model **c**. Such a contribution would be even larger for a larger array considered here. This overhead can be reduced by replacing the coarse-discretization model by the analytical model at the stage of radiation response optimization. The resulting design approach is a two-stage process described below.

Stage 1 (directivity pattern optimization): Here, we adopt the analytical model **a** representing directivity $D_a(\theta,\phi) \sim D_e(\theta,\phi) \cdot |A(\theta,\phi)|^2$, which embeds the EM-simulated radiation response of the single microstrip patch antenna $D_e(\theta,\phi)$ and the array factor $A(\theta,\phi)$. The analytical model is extremely fast; however, it is not as accurate as the EM-simulated coarse-discretization model **c**; the directivity patterns of model **a** are shown in Fig. 11.6(b). Therefore, the directivity pattern is optimized iteratively using an SBO scheme which is outlined in Fig. 11.7.

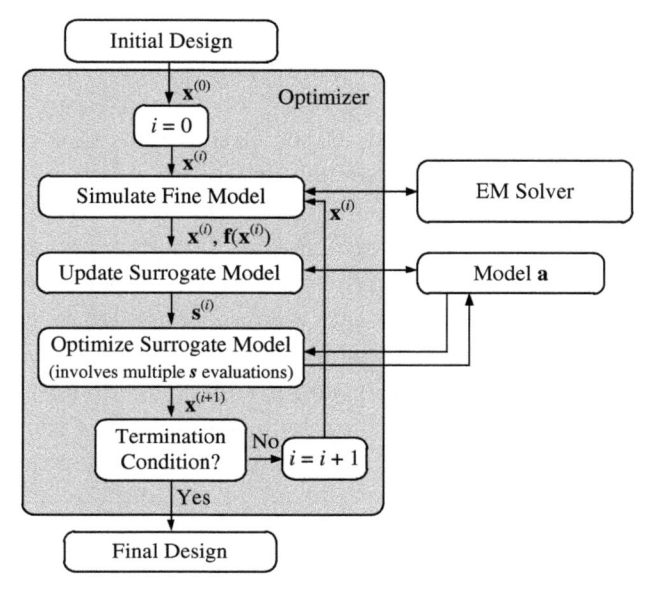

Fig. 11.7. SBO approach for optimization of the radiation response using the analytical model **a** as the low-fidelity model. Surrogate model **s** is constructed according to (11.3).

The surrogate is created by means of the following additive response correction (also referred to as output space mapping (Bandler et al., 2006),

$$\mathbf{s}^{(i)}(\mathbf{x}) = \mathbf{a}(\mathbf{x}) + \mathbf{f}(\mathbf{x}^{(i)}) - \mathbf{a}(\mathbf{x}^{(i)}), \qquad (11.3)$$

where $\mathbf{x}^{(i)}$ is the current design. This kind of correction ensures zeroth-order consistency between the surrogate and the high-fidelity model at the current design $\mathbf{x}^{(i)}$, namely, $\mathbf{s}^{(i)}(\mathbf{x}^{(i)}) = \mathbf{f}(\mathbf{x}^{(i)})$. It should be emphasized that the additive response correction is well suited for constructing the surrogate model in our case of radiation pattern optimization because the major discrepancy between the analytical and the high-fidelity is the vertical difference as indicated in Figs. 11.6(b) and 11.6(c). Usually, two to three iterations of the SBO algorithm (2) with the surrogate model (3) are necessary to yield a satisfactory design in terms of the radiation pattern. One iteration requires only one evaluation of **f**.

Stage 2 (active reflection coefficient adjustment): the coarse-discretization model **c** is used to adjust active reflection coefficients. After completion of Stage 1, the reflection responses are shifted in frequency so that the minima of the active reflection coefficients $|S_k|$ are not exactly at the frequency of 10 GHz. $|S_k|$ can be shifted in frequency individually by adjusting the feed offsets d_n, and collectively by adjusting the patch size, y_1. The amounts of necessary adjustments are estimated using **c**, because the dependencies of $|S_k|$ with respect to design variables are similar for models **c** and **f**. Both models **c** and **f** are simulated using the same discrete EM solver so that their active reflection responses are well correlated as illustrated in Figs. 11.6(d) and 11.6(e) despite the misalignment of the responses in frequency and levels. The computational cost of adjusting the active reflection coefficients is only one **f** simulation and a few **c** simulations. Actual number of **c** simulations depends on how many coefficients $|S_k|$ should be adjusted.

11.4.3. *Optimization with non-uniform amplitude excitation*

Design is carried out with the excitation amplitudes as design variables. Maximal spacings are restricted to 20 mm. The cost of Stage 1, directivity pattern optimization, is only 3 × **f** simulations and the cost of optimizing the analytical model **a** can be neglected. At Stage 2, reflection response adjustment, we change the y-size of the patches, the global parameter y_1, to 9.14 mm in order to move the active reflection coefficients to the left in frequency. The probe offsets d_n are adjusted individually and only for the elements violating the design specifications. The cost of this step is (5 × **c** + 1 × **f**) simulations. The final design is found at $\mathbf{x}^* = [s_1\ s_2\ s_3\ u_1\ u_2\ u_3\ x_1\ y_1\ a_1 \ldots a_{28}]^T =$ [15.97 17.35 20.00 14.38 17.98 19.99 11.00 9.14 0.922 0.787 1.000 0.835 0.953 0.779 0.770 0.958 0.966 1.000 0.810 0.963 0.989 0.925 0.452 0.620 0.832 0.842 0.814 0.631 0.576 0.072 0.752 0.697 0.872 0.821 0.703 0.037$]^T$ where the dimensional parameters are in mm and excitation amplitudes are normalized. Most of the probe offsets d_n are left at the initial design value, 2.9 mm, except four ones adjusted to

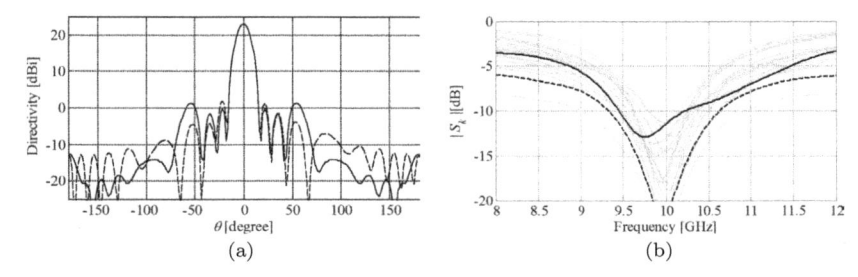

Fig. 11.8. Array of 49 microstrip patches with optimized excitation amplitudes: (a) directivity pattern cuts in the E (– –) and H (—) planes at 10 GHz; (b) active reflection coefficient of element 4 (—) which is at the center, and that of element 22 (– –) which is at the corner. Active reflection coefficients of other elements are shown with the gray solid lines.

$d_4 = d_{11} = d_{18} = 3.9$ mm and $d_{10} = 3.4$ mm. The total cost of $\mathbf{x}^*$ is about $5 \times \mathbf{f}$ simulations. The radiation and reflection responses of the final design $\mathbf{x}^*$ are shown in Fig. 11.8(a). The directivity pattern of $\mathbf{x}^*$ has the SLL under –20 dB and the peak directivity of 22.9 dBi. The HPBWs are 13.4 degrees in both E and H planes.

HPBWs of a 7×7-element Dolph–Chebyshev design with the same aperture dimensions (spacings of $s_1 = s_2 = s_3 = 0.592$ wavelengths and $u_1 = u_2 = u_3 = 0.582$ wavelengths) and the same SLL of –20 dB in both principal planes are estimated as 13.7 degrees in the E-plane and 13.3 degrees in the H-plane according to the analytical model $\mathbf{a}$. The array factor of the Dolph–Chebyshev design is with SLLs of -19.1 dB and -18.9 dB in E- and H-plane, respectively; the extra reduction of SLLs up to –20 dB is due to the element pattern. EM simulations of this Dolph–Chebyshev array, comprising the same elements, show the E-plane HPBW = 13.2 degrees, the H-plane HPBW = 13.5 degrees, and the SLL = –18.3 dB.

11.4.4. Optimization with non-uniform phase excitation

Design is carried out with the excitation phase shifts as design variables. The excitation amplitudes are kept uniform. Maximal spacings are restricted to 20 mm. The final design is found at

270 *Simulation-Based Optimization of Antenna Arrays*

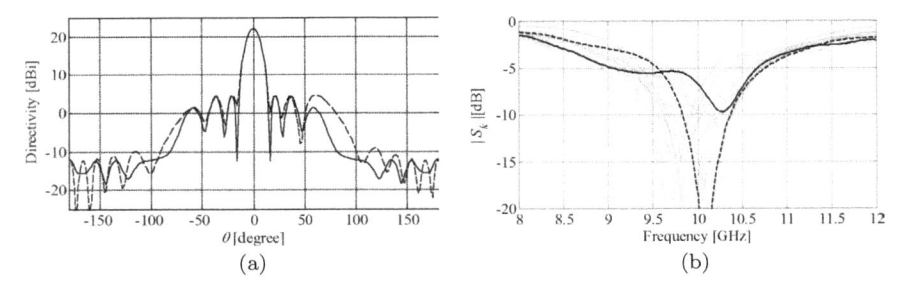

Fig. 11.9. Array of 49 microstrip patches with optimized excitation phases and uniform excitation amplitudes: (a) directivity pattern cuts in the E ($--$) and H (—) planes at 10 GHz; (b) active reflection coefficient of element 4 (—) which is at the center, and that of element 22 ($--$) which is at the corner. Active reflection coefficients of other elements are shown with the gray solid lines.

$\mathbf{x}^* = [s_1\ s_2\ s_3\ u_1\ u_2\ u_3\ x_1\ y_1\ b_1 \ldots b_{28}]^T = [15.00\ 15.00\ 20.00$ 15.15 15.46 19.95 11.00 9.10 08.6 -6.3 1.1 4.3 2.6 3.1 33.3 0.3 11.0 -4.9 5.3 -14.6 45.7 -60.7 17.4 5.8 29.6 -7.0 39.4 -48.9 -17.7 46.5 -13.8 22.5 -1.65 47.9 $-38.9]^T$ where the dimensional parameters are in mm, the phase shifts are in degrees and given relatively the element 1. The total cost of $\mathbf{x}^*$ is about $5 \times \mathbf{f}$ simulations. Radiation and reflection responses of the design $\mathbf{x}^*$ is shown in Fig. 11.9. The directivity pattern of $\mathbf{x}^*$ at 10 GHz has the SLL under -17 dB and the peak directivity of 22.2 dBi. The active reflection coefficients $|S_k|$ are higher than in the previous case as shown in Fig. 11.9(b). Further improvement in the active reflection response should be searched using either extra matching elements or with design of the feed network.

11.5. Summary

In this chapter, a low-cost surrogate-assisted approach for simulation-driven design of planar arrays of microstrip patch antennas has been discussed and illustrated. By utilizing variable-fidelity EM simulations and a suitably corrected quasi-analytical model of the array radiation response, the design goals in radiation and reflection responses can be met at the cost of only a few high-fidelity simulations of the array in the case of excitation amplitudes treated as design variables. It seems that combination of coarse-discretization simulations with surrogate-based optimization techniques (response correction in particular) is a promising way to conduct EM-driven design of realistic antenna array models in a computationally feasible manner. As demonstrated, the use of quasi-analytical models offers further design speed up; however, at the expense of somehow limited control of the reflection response.

Chapter 12

Design of Planar Microstrip Array Antennas Using Simulation-Based Superposition Models

In this chapter, a technique for simulation-driven design of excitation tapers for planar antenna arrays is discussed. The technique exploits models constructed as a superposition of simulated radiation and reflection responses of the array under design. Low computational costs of these models are ensured by using iteratively corrected EM-simulation data computed with coarse-meshes. The technique allows for simultaneous control of the radiation pattern and the reflection coefficients. Numerical efficiency as well as scalability of the technique is demonstrated using the design examples of various sizes and topologies. The examples include 16-element and 100-element microstrip patch antenna arrays of the Cartesian lattice and a 100-element microstrip antenna array of the hexagonal lattice. The technique is versatile as it also can be applied for simulation-based optimization of antenna arrays comprising other types of radiation elements, e.g., wire dipoles or dielectric resonator antennas.

12.1. Design Problem and Array Models

This section outlines the design problem and discusses the simulation-based superposition models utilized throughout the chapter for low-cost design optimization of planar arrays.

12.1.1. *Design problem and antenna array geometry*

A typical design problem or a feasibility study for arrays is sidelobe reduction for a particular planar aperture with already defined geometry parameters (array spacings, dimensions of radiation elements, etc.) by finding an excitation taper (excitation amplitudes and/or phases). Other performance figures (e.g., peak directivity, peak realized gain, major lobe beamwidth, maximal active reflection coefficient) should conform to certain values or levels. To make our discussion concise, in this chapter we consider planar arrays comprising individually fed identical microstrip patches residing on finite microstrip substrates as shown in Fig. 12.1.

12.1.2. *Superposition models and discrete EM models*

The radiation and active reflection responses of the array under design can be obtained for any applied excitation using superposition:

$$\mathbf{E}(\theta, \phi) = \sum_{n=1}^{N} a_n \mathbf{E}_n(\theta, \phi), \tag{12.1}$$

$$S_m = \left(\sum_{n=1}^{N} a_n S_{mn} \right) \Big/ a_m, \tag{12.2}$$

where $\mathbf{E}_n$ and S_{mn} are the EM-simulated far-field and S-parameters, respectively; S_m stands for the active reflection coefficient at the mth port for a particular excitation; a_n is the complex amplitude of the signal incident on the nth port; and N is the number of the array elements. $\mathbf{E}_n$ and S_{mn} are obtained with simulation of the array aperture with only the nth element being active at a time. $\mathbf{E}_n$ in (12.1) is normalized to the complex excitation

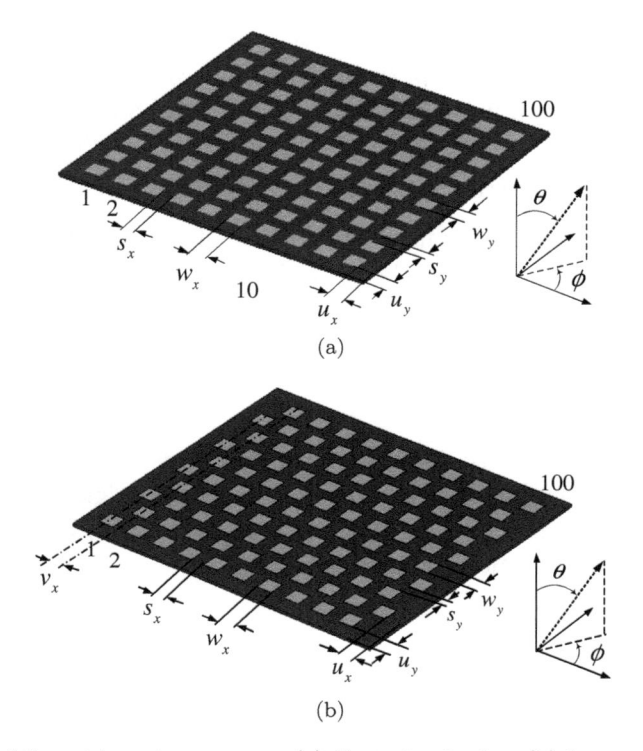

Fig. 12.1. Microstrip antenna array: (a) Cartesian lattice; (b) hexagonal lattice.

amplitude it has been computed with. Directivity and/or realized gain patterns are obtained from the far-field **E** using the standard means (Mailloux, 2005).

The analytical model (cf. (12.1) and (12.2)) combines N pre-simulated far-fields and scattering responses using a particular excitation taper $\{a_n\}_{n=1,...,N}$ so that both radiation and active reflection responses can be obtained for any particular set of excitation $\{a_n\}_{n=1,...,N}$. Two models of this kind are considered in this work: a model $\mathbf{a}_f$ which is configured from N high-fidelity simulations of the discrete array model (denoted as **f**); model $\mathbf{a}_c$ which is configured from N simulations of the coarse-mesh model of the antenna array (denoted as **c**). Thus, $\mathbf{a}_c$ and **c** are low-fidelity versions of $\mathbf{a}_f$ and **f**, respectively. Model $\mathbf{a}_c$, implemented in Matlab, is exploited in the surrogate-based design process as described in Sec. 12.2, whereas the model $\mathbf{a}_f$ is used for comparison only. Full-wave discrete models $\mathbf{f}$ and $\mathbf{c}$ are simulated using the same discrete EM solver.

12.2. Design Optimization Using Surrogates

Here, we provide rigorous formulation of the design optimization problem and describe the optimization methodology involving variable-fidelity models and appropriate correction techniques.

12.2.1. *Design problem formulation: Objective function*

The model $\mathbf{a}_f$ — obtained with (12.1) and (12.2) from array-embedded single-patch responses simulated at high-fidelity — is essentially a perfect representation of the array both in terms of radiation and reflection. The design problem can be formulated as an optimization task of

$$\mathbf{x}^* = \arg\min_{\mathbf{x}} H(\mathbf{a}_f(\mathbf{x})), \qquad (12.3)$$

where $\mathbf{x}$ is a vector of designable parameters (e.g., excitation amplitudes), whereas H is an objective function encoding given performance specifications. Simultaneous handling of radiation and active reflection responses results in two-objective problem, which is scalarized here by optimizing the sidelobe level (SLL) directly and ensuring a safe level of active reflection coefficients using a suitable penalty functions. The objective function takes the form

$$H(\mathbf{a}_f(\mathbf{x})) = U_{\text{SLL}} + \beta \cdot \max_{n=1,\ldots,N} \{|S_n|_{f_0} + 10.0\}^2. \qquad (12.4)$$

Here, U_{SLL}, i.e., the SLL for a given main beam width (a worst case SLL for all considered directivity pattern cuts), is our primary objective function. The penalty function is introduced to control active reflection coefficients $|S_n|$ at the level at most $-10\,\text{dB}$ at the center frequency f_0. The penalty factor β is set to a large value, here, 1000.

12.2.2. Low-fidelity model: Model correction

The analytical model $\mathbf{a}_f$ is fast; however, the CPU cost of acquiring the data necessary for its creation is high, namely $N \times \mathbf{f}$ (N full-wave high-fidelity simulations of the entire array) with N being the number of array elements. For the sake of computational savings, our design methodology is based on the analytical model $\mathbf{a}_c$, which is constructed using N evaluations of the coarse-discretization array model $\mathbf{c}$. The CPU cost of acquiring the data necessary for setting up $\mathbf{a}_c$ normally corresponds to only a few simulations of $\mathbf{f}$; it typically equals to the ratio of simulation times of $\mathbf{f}$ and $\mathbf{c}$ which is from 15 to 50.

The radiation response of $\mathbf{a}_c$ is a relatively accurate representation of the radiation of the high-fidelity model (directly simulated with simultaneous excitation of all elements). Possible discrepancies can be compensated using additive response correction (cf. Sec. 12.2.3). However, misalignment of active reflection responses can be quite significant as illustrated in Fig. 12.2(b). In order to use $\mathbf{a}_c$ as a reliable representation of the high-fidelity model in terms of active reflection coefficients, the following scaling procedure is utilized:

$$S_{s.k}(f) = \alpha_0^k \cdot S_{a.k}(\alpha_1^k + f\alpha_2^k). \tag{12.5}$$

Here, f stands for frequency, $S_{a.k}$ is the kth active reflection response of $\mathbf{a}_c$, whereas $S_{s.k}$ is the scaled response. We use an affine frequency scaling (Bandler et al., 2004a) with parameters α_1^k and α_2^k, as well as multiplicative response correction implemented with α_0^k. The scaling parameters are obtained independently for each k by solving the nonlinear regression problem of

$$[\alpha_0^k, \alpha_1^k, \alpha_2^k] = \arg\min_{[\bar{\alpha}_0^k, \bar{\alpha}_1^k, \bar{\alpha}_2^k]} \int_{f_{\min}}^{f_{\max}} [S_{f.k}(f) - \bar{\alpha}_0^k \cdot S_{a.k}(\bar{\alpha}_1^k + f\bar{\alpha}_2^k)]^2 df, \tag{12.6}$$

where $S_{f.k}$ stands for the kth active reflection coefficient of the model $\mathbf{f}$. Both $f_{\min}$ and $f_{\max}$ are normally selected relatively close to the center frequency f_0 because we are only interested aligning

278 Simulation-Based Optimization of Antenna Arrays

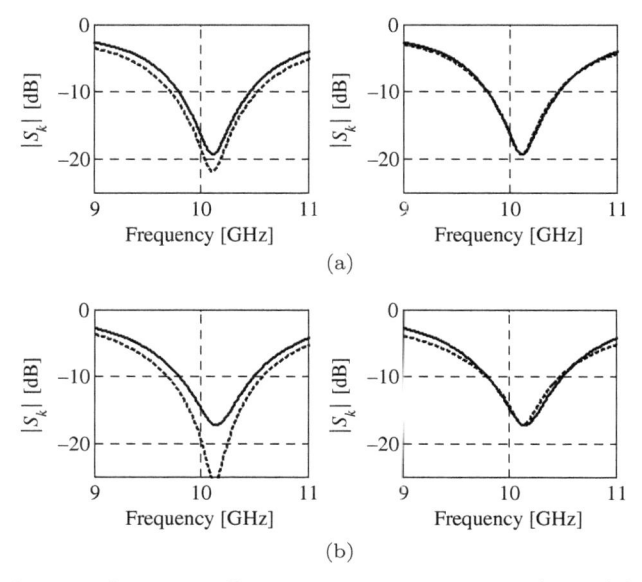

Fig. 12.2. Active reflection coefficients at two selected ports (plots (a) and (b)) of the array of Fig. 12.1(a): the low-fidelity model **c** (- - -) and the high-fidelity model **f** (—) [left-hand side], and frequency/response scaled low- (- - -) and high-fidelity (—) responses at the same design [right-hand side]. It is seen that the scaling procedure effectively reduces the misalignment between the model responses.

the models around f_0. For example, if $f_0 = 10\,\text{GHz}$ then $f_{\min}$ and $f_{\max}$ can be chosen as $f_{\min} = 9.7\,\text{GHz}$ and $f_{\max} = 10.3\,\text{GHz}$.

12.2.3. *Optimization algorithm*

The radiation pattern of the array is optimized using the following iterative scheme (Koziel *et al.*, 2011a; Yelten *et al.*, 2012):

$$\mathbf{x}^{(i+1)} = \arg \min_{\mathbf{x};\,\|\mathbf{x}-\mathbf{x}^{(i)}\| \leq \delta^{(i)}} H(\mathbf{s}^{(i)}(\mathbf{x})), \qquad (12.7)$$

where $\mathbf{s}^{(i)}$ is the surrogate model at iteration i. The radiation response $r_s^{(i)}(\mathbf{x})$ of $\mathbf{s}^{(i)}$ is obtained as $r_s^{(i)}(\mathbf{x}) = r_{a.c}^{(i)}(\mathbf{x}) + [r_{a.f}(\mathbf{x}^{(i)}) - r_{a.c}(\mathbf{x}^{(i)})]$, where $r_{a.c}$ and $r_{a.f}$ are radiation responses of $\mathbf{a}_c$ and $\mathbf{a}_f$, respectively. The radiation response $r_s^{(i)}$ of $\mathbf{s}^{(i)}$ is of zeroth-order consistency with that of $\mathbf{a}_f$. The active reflection response is obtained using the scaling procedure (12.5) and (12.6). The surrogate model is

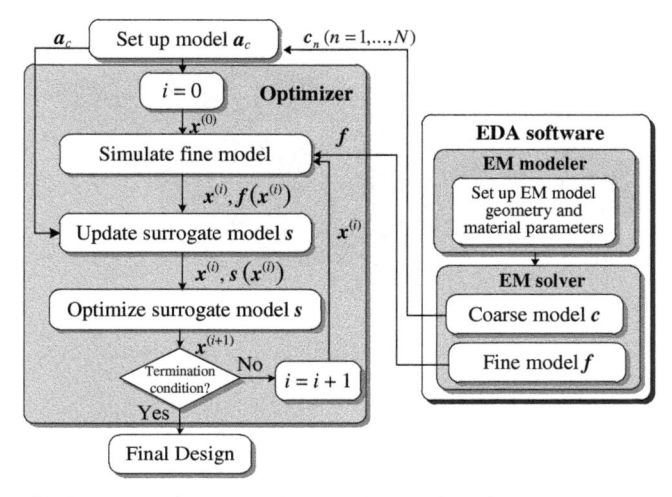

Fig. 12.3. SBO approach using the simulation-based superposition model for design of an antenna array excitation taper: a block-diagram.

reset at the beginning of each iteration. The flow of the entire design optimization approach is shown in Fig. 12.3.

The objective function H is defined as in (12.4). To ensure convergence, the algorithm (12.7) is embedded in the trust-region (TR) framework (Nocedal and Wright, 2006), i.e., the surrogate model optimization (12.7) is constrained to the vicinity of the current design $\mathbf{x}^{(i)}$ defined as $\|\mathbf{x} - \mathbf{x}^{(i)}\| \leq \delta^{(i)}$, where $\delta^{(i)}$, the TR radius at iteration i, is adjusted using the standard rules (Conn et al., 2000).

The design process shown in Fig. 12.3 is automated using an in-house interface between the EM solver (here CST MWS) and the optimization algorithm which is implemented in Matlab.

12.3. Results

In this section, we demonstrate the efficiency and scalability of the discussed technique. We consider several design optimization examples, including a 4×4- and 10×10-microstrip antenna arrays of the Cartesian lattice, as well as a 100-element microstrip antenna array of the hexagonal lattice. In all examples we want to keep the major lobe width as close as possible to that of the uniformly

280   *Simulation-Based Optimization of Antenna Arrays*

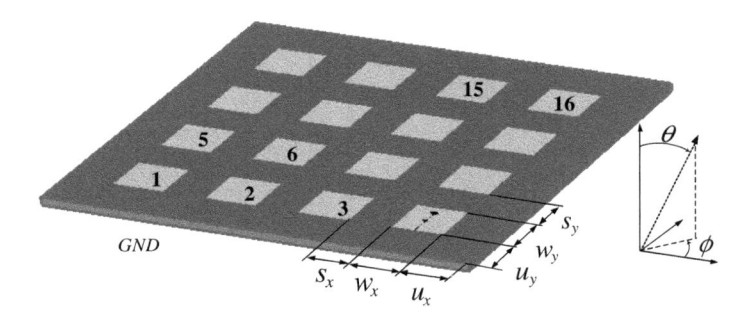

Fig. 12.4. Microstrip antenna array: geometry.

excited array; therefore, we adopt the uniform amplitude and phase excitation as the starting point of the optimization process.

12.3.1. *16-Element Cartesian lattice antenna array*

Consider a microstrip patch antenna array of Fig. 12.4. Each element is fed with a 50-ohm probe. Patch dimensions are $w_x = w_y = 9.15$ mm; spacing are $s_x = s_y = 8$ mm; all probes are 2.9 mm off the patch center in the y-direction. Substrate is a 1.575-mm thick layer of RT/duroid 5880 with finite lateral extends of $u_x = u_y = 9.15$ mm. The metal ground is modeled with infinite lateral extends.

Design requirements are the following: (a) suppress the SLL 10 GHz; (b) maintain the major lobe beamwidth and the peak realized gain at 10 GHz about those of the initial design; (c) keep active reflection coefficients $|S_m|$ of all ports under -10 dB at 10 GHz. Requirements (a) and (b) are formulated as minimization of the realized gain pattern at 10 GHz for zenith angles within the $[-90, -26]$- and $[26, 90]$-degree sectors. The design variables (32 in total) are amplitudes and phases of the signals exciting the array elements.

The high-fidelity model, $\mathbf{f}$, and the coarse model, $\mathbf{c}$, are simulated with the CST MWS transient solver on a 2.33-GHz 8 core CPU 8 GB RAM computer. Model $\mathbf{f}$ comprising 5,445,200 mesh cells takes 10 min 25 s for one simulation. One simulation of $\mathbf{c}$ (132,500 cells) needs only 33 s.

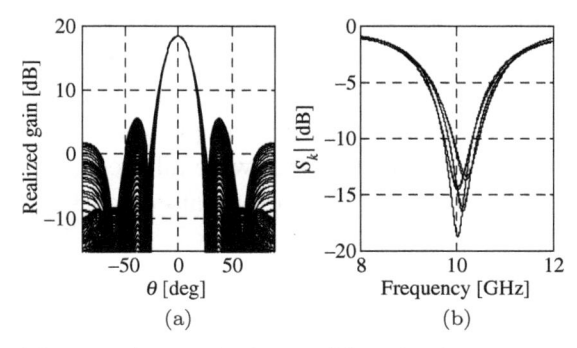

Fig. 12.5. Model $\mathbf{a}_f$ at the initial design: (a) realized gain pattern cuts (every 5 degrees in azimuth) at 10 GHz and (b) active reflection coefficients. Peak realized gain is 18.4 dB. SLL = −12.8 dB.

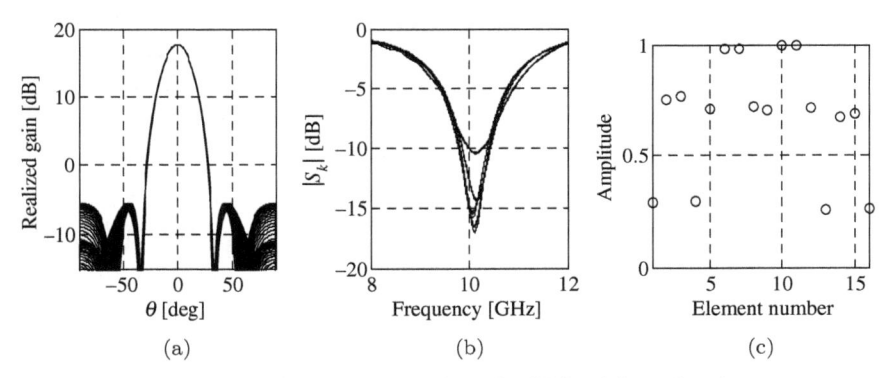

Fig. 12.6. $\mathbf{R}_f$ at the design obtained with SBO: (a) realized gain pattern cuts (every 5 degrees in azimuth) at 10 GHz; (b) active reflection coefficients; (c) excitation amplitudes. Peak realized gain is 17.7 dB. SLL = −23 dB. Design cost = model $\mathbf{a}_c$ setup cost $(16 \times \mathbf{c}) + 3 \times \mathbf{f} \approx 4 \times \mathbf{f}$.

The superposition low-fidelity model $\mathbf{a}_c$ is configured from 16 simulations of the coarse model $\mathbf{c}$ as described in Sec. 12.1. The superposition high-fidelity model $\mathbf{a}_f$ is configured from 16 simulations of the high-fidelity model $\mathbf{f}$ and used as a reference. The surrogate model $\mathbf{s}$ is set up from the superposition model $\mathbf{a}_c$ as described in Sec. 12.2.

Responses of the initial design, which is with the uniform amplitude and phase excitation, are shown in Fig. 12.5.

Optimization results are shown in Fig. 12.6. The SLL has been decreased from -12.8 dB of the initial design (Fig. 12.5(a)) to about -23 dB (Fig. 12.6(a)) trading only about 0.7 dB of the peak realized gain. Although all excitation amplitudes and phases of the signal incident at the element inputs were allowed to vary as design variables the optimizer found improvements essentially with a non-uniform amplitude excitation which is shown in Fig. 12.6(c) while the phases did not change noticeably. Optimization of the superposition model $\mathbf{a}_f$ gives the design of a similar quality, however, at a higher computational cost. Namely, the $\mathbf{a}_f$ model setup requires 16 high-fidelity simulations of the array high-fidelity model $\mathbf{f}$.

The design cost using the SBO approach that includes the surrogate model setup corresponds to only four high-fidelity simulations of the antenna array. The SBO approach cost comprises 16 simulations of the coarse-mesh model ($16 \times \mathbf{c} \approx 1 \times \mathbf{f}$) and three iterations of the algorithm (12.7) requiring one simulation of $\mathbf{f}$ per iteration.

12.3.2. 100-Element Cartesian lattice antenna array

Consider the antenna array of Fig. 12.1(a) with the following parameters: the substrate is a 1.575-mm thick layer of RT/duroid 5880; lateral extends of the substrate are $u_x = u_y = 9.15$ mm; the metal ground is assumed to be much larger than the substrate lateral dimensions so that the ground is modeled as an infinite metal plane; each microstrip patch is fed by a 50-ohm probe; patches' dimensions are $w_x = w_y = 9.15$ mm; probes are 2.9 mm off the patch center in the y-direction; separation between the patches is the same in both lateral directions, $s_x = s_y = 8$ mm.

For the array aperture described above the following design requirements are considered: (a) suppressing the SLL at 10 GHz, namely, minimizing the realized gain pattern at for zenith angles within the $[-90, -10]$- and $[10, 90]$-degree sectors; (b) keeping all active reflection coefficients $|S_m|$ under -10 dB at 10 GHz for all ports. The design variables are amplitudes and phases of the signals exciting the array elements (200 adjustable parameters in total).

Full-wave discrete models $\mathbf{f}$ and $\mathbf{c}$ are simulated using the CST MWS transient solver on a 2.33-GHz 8 core CPU 8 GB RAM computer. Each simulation of model **f** (11,320,960 mesh cells) takes 32 min 13 s. Model **c** (342,732 mesh cells) is simulated in only 2 min 35 s.

Given the aforementioned design requirements, the optimization algorithm started from the uniform design with the SLL of about -13 dB and arrived to the optimized design with the SLL of -18.5 dB in four iterations of the algorithm (12.7). Array responses corresponding to the initial and optimized designs are shown in Figs. 12.7 and 12.8(a)–(b), respectively. The SBO optimizer found improvements with a non-uniform amplitude excitation, shown in Fig. 12.8(c), while the maximal change of excitation phases is within a quarter percent. The total cost of the optimal design corresponds to only 12 simulations of **f** that includes data acquisition to set up model $\mathbf{a}_c$.

For comparison we also optimized the superposition model $\mathbf{a}_f$ which was built with 100 high-fidelity simulations of model **f**. The model $\mathbf{a}_f$ was optimized for the reduced SLL with and without constrains imposed on the reflection response. Radiation and reflection responses of these optimal designs, shown in Fig. 12.9, are similar in quality to the SBO design, shown in Fig. 12.8. However, the SBO design cost is ten times lower.

Figures 12.10 and 12.11 show the effect of the reflection control on the SLL. One notices that the absence of the constraints improves

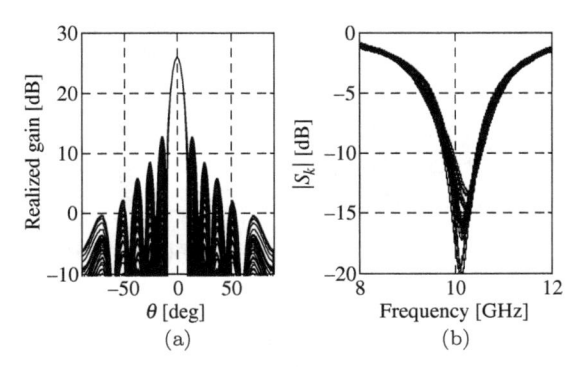

Fig. 12.7. Model $\mathbf{a}_f$ at the initial design: (a) realized gain pattern cuts (every 5 degrees in azimuth) at 10 GHz. Peak realized gain is 26 dB. SLL = -13 dB; (b) active reflection coefficients.

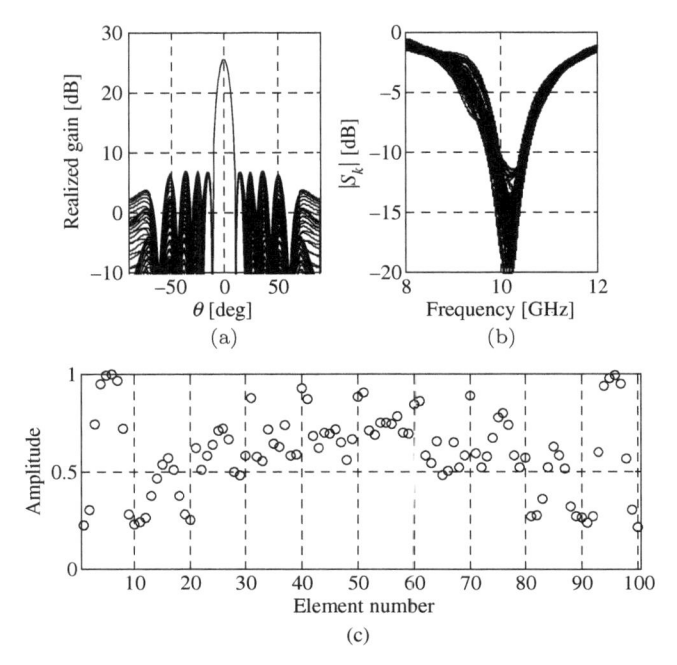

284 Simulation-Based Optimization of Antenna Arrays

Fig. 12.8. Model **f** at the SBO optimum obtained using constrains on the active reflection response: (a) realized gain pattern cuts (every 5 degrees in azimuth) at 10 GHz. Peak realized gain is 25.4 dB. SLL = −18.5 dB; (b) active reflection coefficients; (c) excitation amplitudes. Design cost is approximately 12 × **f**.

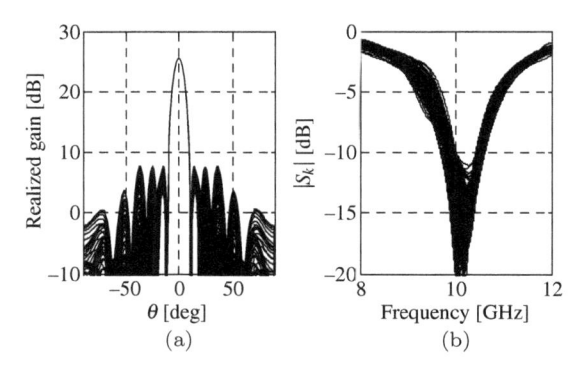

Fig. 12.9. Optimum of model $\mathbf{a}_f$ obtained with constrains on reflection: (a) realized gain pattern cuts (every 5 degrees in azimuth) at 10 GHz. Peak realized gain is 25.5 dB. SLL = −17.9 dB; (b) active reflection coefficients. Design cost is approximately 100 × **f**.

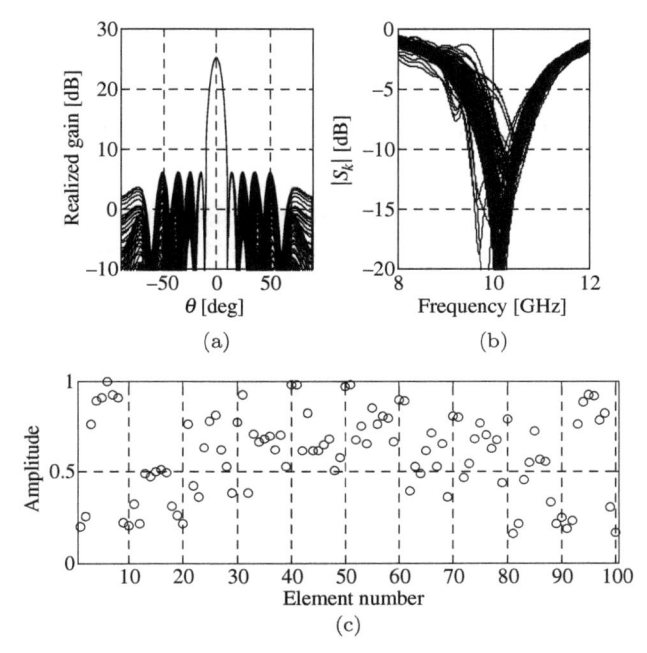

Design of Planar Microstrip Array Antennas   285

Fig. 12.10. Model **f** at the SBO optimum obtained without constrains on reflection: (a) realized gain pattern cuts (every 5 degrees in azimuth) at 10 GHz. Peak realized gain is 25.2 dB. SLL = −19 dB; (b) active reflection coefficients; (c) excitation amplitudes. Design cost is approximately $12 \times$ **f**.

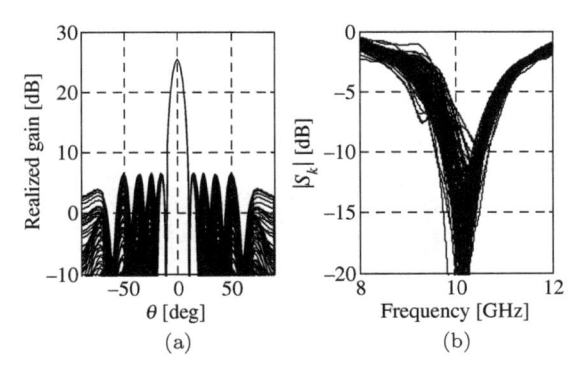

Fig. 12.11. Optimum of model $\mathbf{a}_f$ obtained without constrains on reflection: (a) realized gain pattern cuts (every 5 degrees in azimuth) at 10 GHz. Peak realized gain is 25.3 dB. SLL = −18.7 dB; (b) active reflection coefficients. Design cost is approximately $100 \times$ **f**.

the final SLL only marginally, in about 0.5 dB. Excitation amplitudes obtained without constrains on reflection are shown in Fig. 12.10(c).

12.3.3. 100-Element hexagonal antenna array

Consider an antenna array of Fig. 12.1(b) of the hexagonal lattice with the distance between patch centers of 17.15 mm ($s_x = 8$ mm and $s_y = 6.774$ mm) and offset $v_x = 8.575$ mm. Other relevant geometry and material parameters as well as the design requirements are the same as in the previous case of Sec. 12.3.2. Models **f** and **c** are simulated using the same software, solver, and computer as in the case of Sec. 12.3.2. Each simulation of **f** (12,876,173 mesh cells) takes 43 min 3 s. One run of **c** (511,056 cells) takes only 3 min 27 s. The array responses at the optimized design are shown in Fig. 12.12. The total cost of the optimal design corresponds to only 12 simulations of **f** including data acquisition to set up model $\mathbf{a}_c$.

A lack of reflection control in this case results in a design (shown in Fig. 12.13) with the marginally improved SLL and worse active reflection coefficients relative to the design obtained with constrains on reflection (shown in Fig. 12.12).

Controlling the active reflection coefficients in the process of sidelobe minimization turns to be also important from implementation standpoint as illustrated with Fig. 12.14 where the optimal excitation

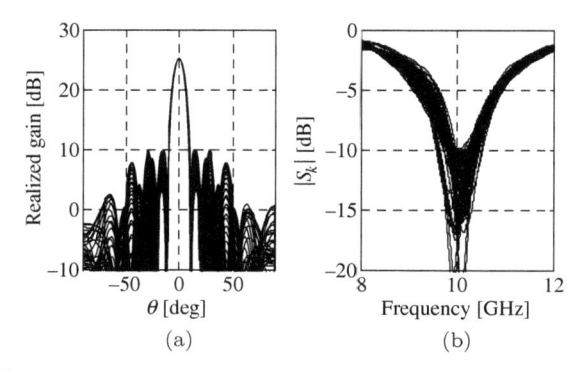

Fig. 12.12. Hexagonal lattice array at the SBO optimum obtained using constrains on the active reflection coefficients: (a) realized gain pattern cuts (every 5 degrees in azimuth) at 10 GHz. Peak realized gain is 25.1 dB. SLL = −15.3 dB; (b) active reflection coefficients. Design cost is approximately 12 × **f**.

amplitudes obtained with the constrains (plotted on panel (a)) are more convenient for implementation (due to their smaller span) than those obtained without the constrains (plotted on panel (b)).

12.4. Summary

The results of Sec. 12.3 indicate that the design optimization technique considered in this chapter provides reliable results in very low computational costs. The benchmark technique, exploiting the superposition model directly configured from the high-fidelity simulations, already exhibits significant computational complexity for medium-size arrays. The design speedup (with respect to the benchmark) offered by our approach is significant as indicated in Fig. 12.15. More importantly, the design speedup increases with the array size: the speedup is about one order of magnitude for array of 100 elements or higher.

The absolute design costs for the 100-element arrays considered in Secs. 12.3.2 and 12.3.3 are equivalent to about 12 simulations of the corresponding array high-fidelity EM models. The predicted relative speedup for large arrays approaches one as shown in Fig. 12.16. It means that the speedup is close to the ratio of the simulation times of the high- and low-fidelity EM models of the array under design.

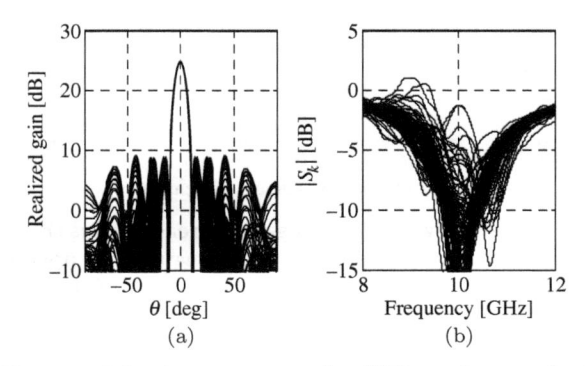

Fig. 12.13. Hexagonal lattice array at the SBO optimum obtained without constrains on the active reflection coefficients: (a) realized gain pattern cuts (every 5 degrees in azimuth) at 10 GHz. Peak realized gain is 24.8 dB. SLL = −15.7 dB; (b) active reflection coefficients. Design cost is approximately 12 × **f**.

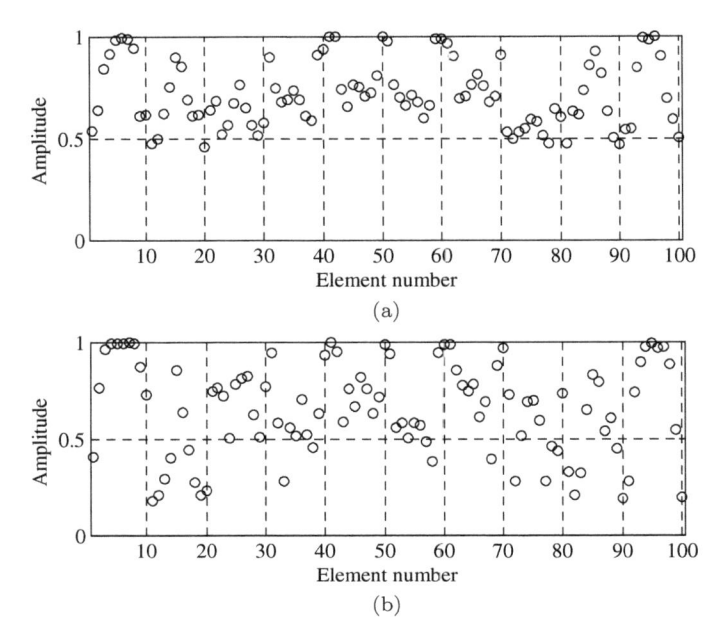

288   Simulation-Based Optimization of Antenna Arrays

Fig. 12.14. Excitation amplitudes of the 100-element hexagonal lattice array at the SBO optima obtained (a) with and (b) without constrains on the active reflection coefficients.

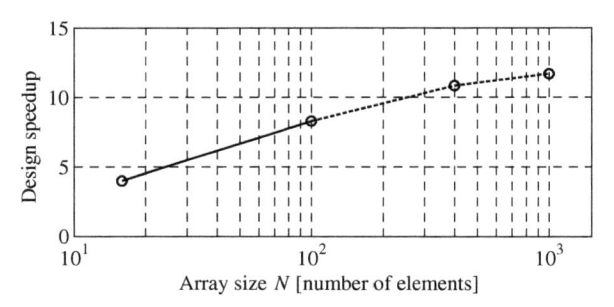

Fig. 12.15. Design speedup with the discussed approach versus the array size. The speedup is defined as the ratio of the cost of setting up the superposition model $\mathbf{a}_f$ (N high-fidelity simulations of the array) and the cost of the design procedure (N low-fidelity simulations of the array + a few high-fidelity simulations of the array due to the execution of the algorithm (12.7)). Solid line corresponds to the actual results (as presented in this chapter). Dashed line is an extrapolation assuming a small increase of the number of algorithm (12.7) iterations with growing value of N.

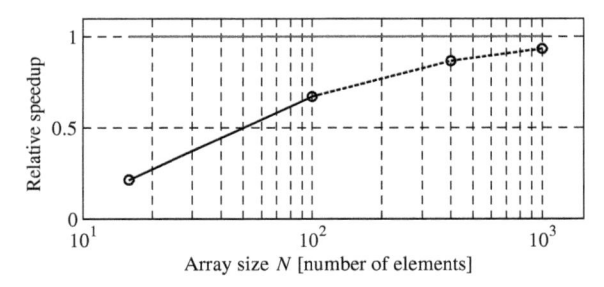

Fig. 12.16. Relative speedup of the discussed design approach versus the array size. The relative speedup is the design speedup divided by the time evaluation ratio between the high- and low-fidelity array EM model (the maximal theoretical relative speedup is one). Solid line corresponds to the actual results (as presented in this chapter). Dashed line is an extrapolation assuming a small increase of the number of algorithm (12.7) iterations with growing value of N. The horizontal line marks the maximum theoretical relative speedup.

Another important aspect of the discussed method is that the entire design flow is easy to automate. The optimization algorithm is implemented in Matlab whereas the discrete solver (here, CST MWS) is controlled through a dedicated Matlab-implemented socket. The presented approach might be of interest for researchers and engineers interested in low-cost simulation-driven design of real-world antenna arrays. Extensions may include optimization with phase-only adjustment as well as simultaneous adjustment of the array excitation taper and geometry.

Chapter 13

Design of Planar Arrays Using Radiation Response Surrogates

In this chapter, a technique for rapid simulation-based design of planar antenna arrays using radiation response correction is discussed. The method has been introduced in Chapter 10 for linear arrays. Here, it is generalized for more complex cases. In short, the methodology is based on establishing a suitable correction of the analytical array factor model of the array of interest. The corrected array factor is utilized as a fast surrogate allowing us to find the optimum element spacing and phase excitations at low computational cost. The correction terms are iteratively refined to account for design-dependent discrepancies between the array factor model and the EM-simulated model. The methodology is demonstrated with design of a 10-GHz 100-element planar array of microstrip patch antennas with a plastic cover, as well as a 28-element planar microstrip antenna array.

13.1. Design Optimization Methodology

In this section, a rigorous description of the design optimization methodology considered in this chapter is provided.

13.1.1. Problem formulation

Here we denote by $\mathbf{f}$ a radiation pattern of the full-wave EM-simulated model of the antenna array under design; $\mathbf{x}$ represents a vector of designable parameters such as element spacing as well as excitation amplitudes and phases. We consider the pattern as a function of the design variables $\mathbf{x}$, and over the zenith and azimuth angles, θ and φ such that $\mathbf{f}(\mathbf{x}) = \mathbf{f}(\mathbf{x}, \theta, \varphi)$. The design problem is defined as follows:

$$\mathbf{x}^* = \arg\min_{\mathbf{x}} U(\mathbf{f}(\mathbf{x})), \qquad (13.1)$$

where U is the objective function, e.g., corresponding to the sidelobe level (SLL).

13.1.2. Response correction of array factor model

Solving (13.1) directly is not a numerically efficient option because of a large number of objective function evaluations, discrete EM simulations in this work, each being already computationally expensive (typically, around 1 hour of CPU time for a 100-element array at fine discretization). Instead, we use an auxiliary model, which is an analytical array factor $\mathbf{a}(\mathbf{x}) = \mathbf{a}(\mathbf{x}, \theta, \varphi)$ (Balanis, 2006; Mailloux, 2005). The array factor $\mathbf{a}(\mathbf{x}, \theta, \varphi)$ is very fast; however, it is not accurate for angles off the main beam. Therefore, it cannot directly replace the EM-simulated pattern $\mathbf{f}$ in the design process. On the other hand, the array factor can be turned into a reliable surrogate model upon a suitable correction. Such correction should align the array factor pattern with the EM-simulated pattern, that is, account for anisotropy of the element pattern and difference of the element patterns resulted from coupling and proximity effects.

In order to reduce the discrepancies between the analytical and EM-simulated models, we introduce a response correction function $D(\mathbf{y}, \theta, \varphi)$ defined at a given design $\mathbf{y}$ as follows. Let $\varphi^s, s = 1, \ldots, N_\varphi$, be a set of angles uniformly distributed between 0 and 360 degrees (e.g., with 1-degree step). Let $p_{f.s}^k = [\theta_{f.s}^k \ r_{f.s}^k]$, $k = 1, \ldots, K_{f.s}$, be $K_{f.s}$ points extracted from the far-field pattern cut of

Design of Planar Arrays 293

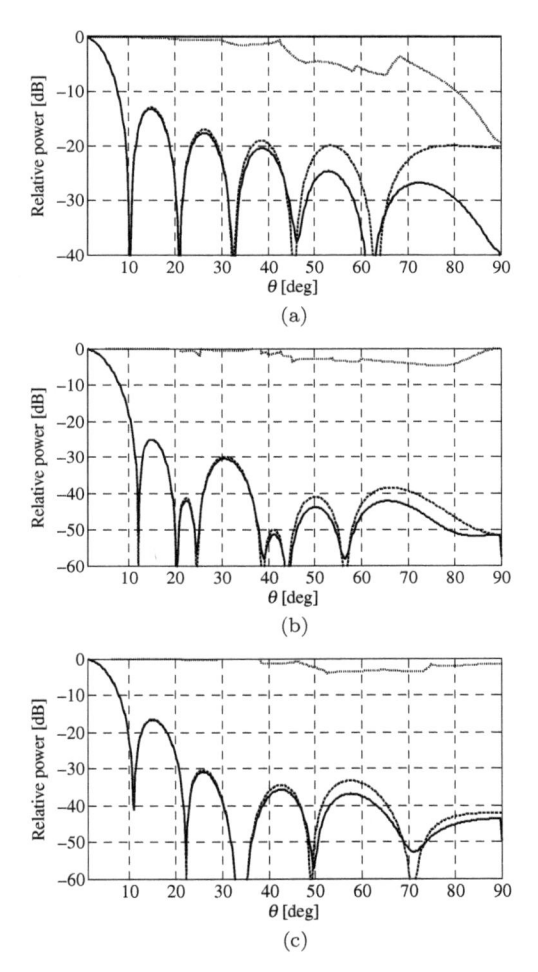

Fig. 13.1. Power pattern of the 100-element planar array antenna at 10 GHz at a certain design **x**, for three selected cuts of azimuth angles (a) $\varphi = 0°$, (b) $\varphi = 60°$, and (c) $\varphi = 110°$: (—) EM model **f**, (- - -) analytical array factor model **a**; (· · ·) correction function $D(\mathbf{x}, \theta, \varphi)$ extracted from the responses **f** and **a** (cf. (13.2)).

$\mathbf{f}(\mathbf{y}; \theta, \varphi_s)$, $\theta \in [0, 90]$ degrees (see Fig. 13.1). Here, $\theta_{f.s}^k$ is the angle and $r_{f.s}^k$ is the corresponding relative power value, for $\theta \in [0, 90]$ degrees as well as all local relative power maxima in the range $[0, 90]$ degrees. We denote by $p_{a.s}^k = [\theta_{a.s}^k \ r_{a.s}^k]$, $k = 1, \ldots, K_{a.s}$, similar points extracted from the array factor pattern cut. Then the correction function $D(\mathbf{y}, \theta, \varphi)$ is defined as

$$D(\mathbf{y}, \theta, \varphi) = E_f(\theta, \varphi) - E_a(\theta, \varphi) \tag{13.2}$$

with

$$E_f(\theta, \varphi) = F(\{\varphi_s, \theta_{f.s}^k, r_{f.s}^k\}_{s=1,\ldots,N\varphi; k=1,\ldots,K_{f.s}}, \theta, \varphi), \tag{13.3}$$

where $F(X, \theta, \varphi)$ represents the interpolation of the data set (defined on a discrete angle set $\{\varphi_s, \theta_{f.s}^k\}$ with corresponding values $r_{f.s}^k$) onto the entire angle plane $(\theta, \varphi) \in [0, 90] \times [0, 360]$. A definition similar to (13.3) holds for $E_a(\theta)$. Figure 13.1 shows the plot of $D(\mathbf{y}, \theta, \varphi)$ for selected values of φ, extracted from typical responses of **f** and **a**.

Having the correction function, one defines a surrogate model of the radiation pattern, **s**, as follows:

$$\mathbf{s}(\mathbf{x}, \theta, \phi) = \mathbf{a}(\mathbf{x}, \theta, \phi) + D(\mathbf{y}, \theta, \phi). \tag{13.4}$$

Clearly, the discrepancy between the array factor and EM-simulated model is dependent on **x**; therefore, the model $\mathbf{s}(\mathbf{y}) = \mathbf{s}(\mathbf{y}, \theta, \varphi)$ is only reliable in a certain vicinity of **y** and agrees perfectly with **f** at **y**, the design at which it was constructed.

13.1.3. *Optimization flow*

The optimization algorithm used here exploits the surrogate model (13.4) in the following iterative process:

$$\mathbf{x}^{(i+1)} = \arg\min_{\mathbf{x}} U(\mathbf{s}(\mathbf{x}^{(i)})), \tag{13.5}$$

where $\mathbf{x}^{(i)}$ is a sequence of approximate solutions to (13.1) with $\mathbf{x}^{(0)}$ being the initial design found by directly optimizing the array factor model **a**. Note that each iteration of (13.5) requires only one EM-simulation of the high-fidelity model **f** in order to update the correction function (13.2). The initial optimization of the array factor model normally performed as a global search which is realized here by means of a smart random search followed by gradient-based optimization with analytical derivatives (Koziel and Ogurtsov, 2012d).

13.2. Case Study I: 100-Element Microstrip Patch Antenna Array

Our first case study is an array of 100 microstrip patch antennas shown in Fig. 13.2. The substrate is a 1.58-mm thick layer of Rogers RT5880 with the finite lateral extensions (measured from the outer edges of the peripheral patches) equal to the patch size in the corresponding direction. The ground plane is modeled as infinite so that the array radiates in the upper half-space only. The 1.6-mm thick plastic cover is shown with a wireframe in Fig. 13.2. The top wall of the cover is $h_c = 9$ mm above the patches. The dielectric constant and loss tangent of the cover are 2.7 and 0.009 at 10 GHz.

Each microstrip patch is fed by a 50-ohm probe exciting the TM^z_{010} mode of the patch so that the YOZ-plane is the far-field E-plane.

The vector of design variables $\mathbf{x} = [s_x \; s_y \; a_1 \; \ldots \; a_{100} \; b_1 \; \ldots \; b_{100}]^T$ comprises element spacings (normalized to the free space wavelength) s_x and s_y as shown in Fig. 13.2, excitation amplitudes $a_1 \ldots a_{100}$, and excitation phases $b_1 \ldots b_{100}$. Other dimensions are kept fixed in the optimization process, they are: dimensions of the microstrip patches, $w_x = w_y = 9.2$ mm; feeding probe positions that are offset by $d_y = 2.8$ mm from the patch center in y-direction; and dimensions of the plastic cover.

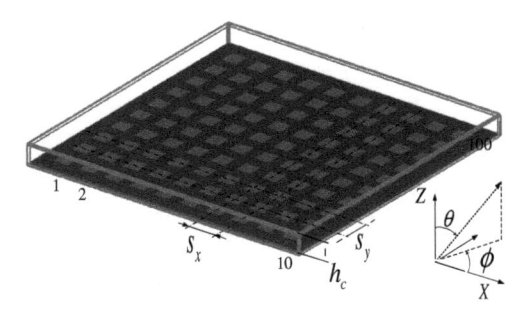

Fig. 13.2. A 100-element microstrip antenna array. The cover is shown with a wireframe.

The antenna array is modeled with CST MWS (CST, 2016). The discrete EM model comprises 12,265,344 hexahedral mesh-cells at the initial design and is simulated with the MWS transient solver (CST, 2016) in 28 min 30 s using an 8-core Intel Xeon 2.00 GHz CPU with 64 GB RAM computer.

The design objective is minimizing the SLL of the 3D pattern at 10 GHz while preserving the major lobe beamwidth of the initial design shown in Fig. 13.3.

Due to the array geometry and location of feeds the excitation taper (amplitudes ad phases) are set to be symmetrical with respect to the YOZ plane so that there are 50 independent amplitudes and 50 independent phases as design variables. The initial design x_{init} = $[0.56\ 0.56\ 1..1\ 0\dots 0]^T$ has the following pattern characteristics: SLL = -13.1 dB; the half power beamwidths in the XOZ plane (HPBW$_x$) of 9.1 degrees and that in the YOZ plane (HPBW$_y$) of 9.1 degrees; and full null beamwidths FNBW$_x$ = FNBW$_y$ = 20.5 degrees. The pattern of the design x_{init} is shown in Fig. 13.3.

The optimization process, described in Sec. 13.1, yields the final design at the total cost of only three simulations of the array EM model. The radiation response of the final design with SLL = -20 dB

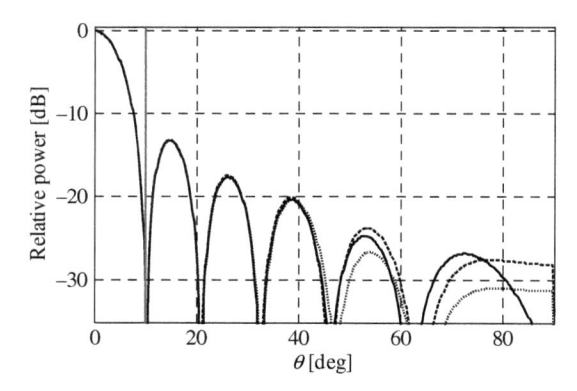

Fig. 13.3. Pattern cuts of the 100-element microstrip antenna array at the initial design at 10 GHz: the EM model for azimuth angles $\varphi = 0°(-)$, $90°(- -)$, and $270°(\cdots)$. SLL = -13.1 dB. Peak directivity is 26.0 dBi. The vertical line defines the major lobe.

Design of Planar Arrays 297

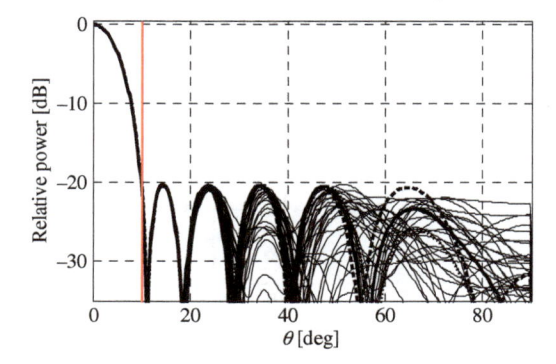

Fig. 13.4. Pattern cuts of the 100-element microstrip antenna array at the final design at 10 GHz: the EM model for azimuth angles $\varphi = 0°(—)$, $90°(- -)$, and $270°(\cdots)$, and for other azimuth angles with one degree increment (—). SLL $= -20.0$ dB. Peak directivity is 24.7 dBi. The vertical line defines the major lobe.

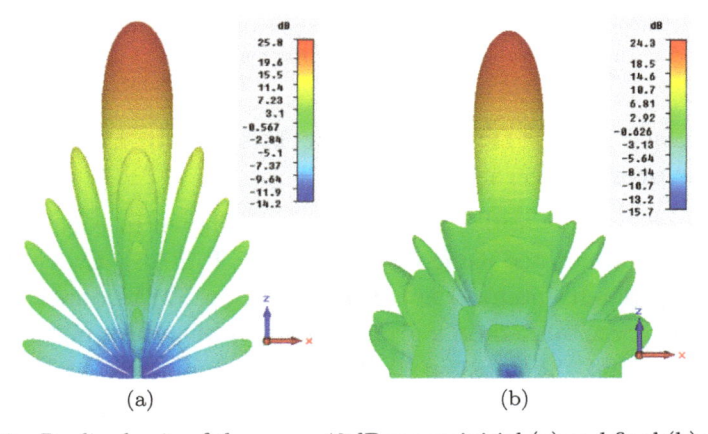

Fig. 13.5. Realized gain of the array, 40 dB range: initial (a) and final (b) designs.

is shown in Figs. 13.4 and 13.5. The final design has the HPBW_x and HPBW_y as those of the initial design while its $\text{FNBW}_x = \text{FNBW}_y = 22.5$ degrees are slightly widened. In the final design, 14 elements are switched off, i.e., with zero excitation amplitudes (cf. Fig. 13.6). Excitation phase angles of the active elements span from -16 to 0 degrees.

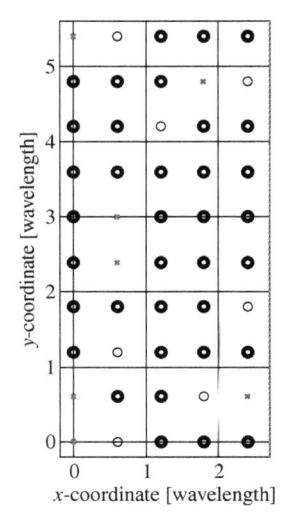

Fig. 13.6. A half of the lattice of the final design: elements with excitation amplitudes higher than 0.1 marked as thick circles; switched off elements (**x**); elements with excitation amplitudes lower than 0.1 of the maximum (∘). The vertical red dash-line shows the plane of symmetry.

13.3. Case Study II: 28-Element Microstrip Patch Antenna Array

Another example is a planar antenna array shown in Fig. 13.7. It is modeled and simulated using CST MWS (CST, 2016). Every element (microstrip patch antenna) is fed by a 50-ohm SMA connector. The connectors are mounted on a ground plane side and connected to the patches through the substrate with extension probes. At the initial design the patches are 9 mm × 9 mm. The microstrip substrate is a finite layer of TLP-5. The metal ground is modeled with infinite lateral extensions. The row and column elements' spacings are preset to 0.65 wavelengths. The lattice of the array is non-rectangular. The simulated pattern has the −15 dB SLL with the fully uniform excitation taper.

The design goal is to suppress SLL at 10 GHz for the major lobe pointed towards the zenith. The methodology of Sec. 13.1 is used as the optimization engine. Two design cases, referred to as Design I and II, are considered.

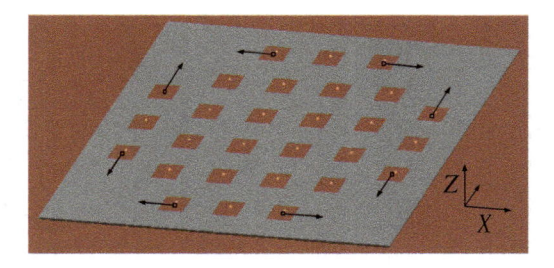

Fig. 13.7. 28-Element array of microstrip antennas.

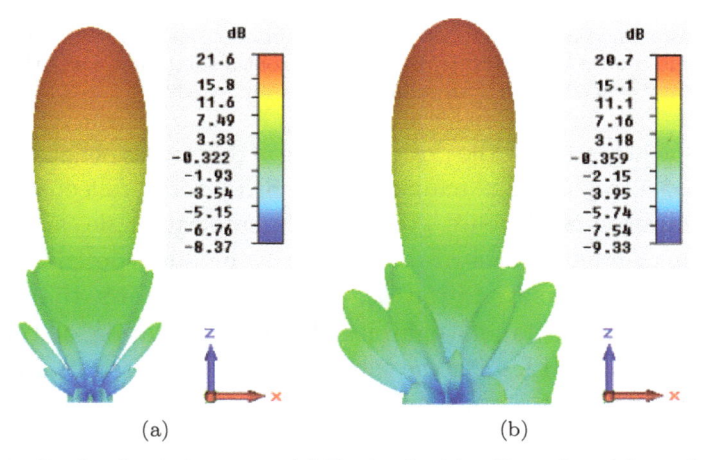

Fig. 13.8. Realized gain patterns: (a) Design I with adjusted positions of selected elements (SLL = −17.6 dB); (b) Design II, phase-only adjusted array (SLL = −17.5 dB).

Design I is with the fully uniform excitation taper and adjusted positions of the selected elements (marked in Fig. 13.7). The optimum is found with the elements equally shifted (in shown directions) by 3.5 mm from the initial position. The realized gain pattern of Design I with SLL = −17.6 dB is shown in Fig. 13.8(a). Design II is with the excitation phases as design variables. The realized gain pattern of Design II is shown in Fig. 13.8(b). For the major lobe pointed to the zenith Designs I and II are quite similar in terms of SLL, 3 dB HPBW, peak realized gain, and total reflected power at 10 GHz.

300 Simulation-Based Optimization of Antenna Arrays

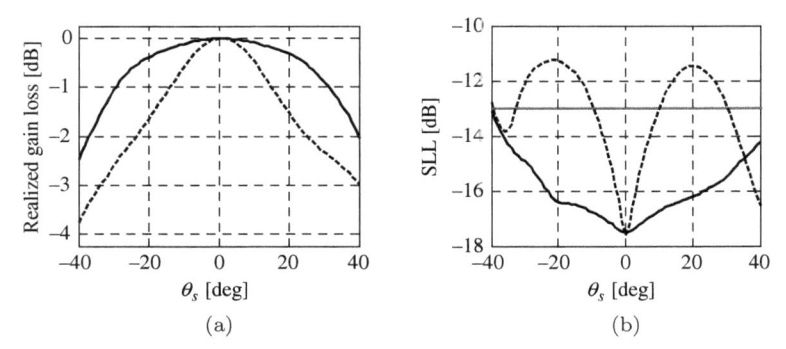

Fig. 13.9. Peak realized gain (a) and SLL (b) versus the scan angle for the mainbeam scanned in the plane of 45-degree azimuth: Design I (- - -) and Design II (—). θ_s is counted from the z-axis. The horizontal solid line marks the SLL of the uniformly spaced array with uniform excitation taper.

On the other hand, the designs perform differently when operated as phased arrays. Peak realized gain, SLL, total reflected power, active reflection coefficients, etc., for different scan angles are done at an EM high-fidelity level of description using superposition models (Koziel and Ogurtsov, 2015a–d). Comparison shows that Design II (with the phase taper) exhibits slower degradation of the peak realized gain and SLL with the scan angle as illustrated by Fig. 13.9. In particular the maximum zenith scan angle, θ_s, is estimated as 32.5 degrees for Design II and 17.7 degrees for Design I.

13.4. Summary

In this chapter, fast simulation-based design of planar antenna arrays using surrogate models obtained by means of a radiation response correction technique has been demonstrated. The discussed technique is an extension of the correction method previously considered in Chapter 10 in the context of linear arrays; here, it has been demonstrated for to minimizing sidelobes of planar arrays. The technique accounts for element coupling and proximity effects (e.g., presence of a cover) at the high-fidelity level of description by utilizing EM simulated data. The final example designs are obtained at the total costs equivalent to a few EM simulations of the array.

Chapter 14

Simulation-Based Design of Corporate Feeds for Low-Sidelobe Microstrip Linear Arrays

So far we discussed surrogate-assisted techniques for optimization of array apertures' dimensions and excitation tapers. Circuits realizing the required excitation taper, the feeds, stayed out of the scope of this book (i.e., the feeds were assumed to be separate design problems). At the same time, in modern antenna systems, in particular in printed-circuit arrays, the feeds are typically integrated with the aperture so that the post-manufacturing adjustment of the structure is hardly feasible or even impossible. In Chapter 8 we already demonstrated, for a case of microstrip subarrays, the importance and benefits of considering the aperture-feed structure of interest within a single simulation-based projects so that the contribution of the feed to the radiation response and mutual interactions of the elements as well as elements-to-feed interactions, all affecting the radiation and reflection responses of the structure, could be reliably evaluated and, thus, subsequently adjusted.

In principle, decomposition could be used to certain extend for printed-circuit integrated aperture-feed structures, e.g., as in Sec. IV of Koziel and Ogurtsov (2013b), by defining two simulation projects, one for the radiating part of the structure and another for the feed, both within the same CAD project, and then connecting them at the circuit level. Following this approach the feed can be tuned or optimized as a stand-alone EM simulated model. The excitation taper (complex waves incident at the aperture inputs) within the integrated structure can be numerically inferred as described in Sec. 2.5, with (2.42)–(2.50). Subsequently, the taper can be applied to the superposition-based model of the aperture (discussed in Sec. 9.3) to obtain the radiation response. Such an approach possesses a clear flow, involves contemporary array design practices, and is well justified for phased or switched beam applications. However, even when realized as a surrogate-assisted process and conducted with already dimensioned aperture, this approach is associated with the following challenges: necessity to configure the superposition-based model at the high-fidelity level of description with N simulations (required for an aperture of N-radiators without utilization of symmetry); necessity to combine the parasitic radiation of the feed with radiation of the aperture as complex-valued vector far-fields. Still, the results of such combination would not be, in general, identical to the far-field of the entire aperture-feed structure. Disregarding the parasitic feed radiation is a serious modeling approximation (Koziel et al., 2014d,e) in particular for low-sidelobe arrays (Pozar and Kaufman, 1990; Pozar, 1992). Furthermore, optimality of the feed topology needs to be addressed prior implementation. On the other hand, straightforward simulation-based optimization of the entire integrated array structure is numerically feasible only in cases of certain types of feeds; this problem will be addressed in this chapter.

In this chapter, we discuss and demonstrate alternative approaches to design of low-sidelobe linear arrays. The presented approaches aim at systematic CAD of low-sidelobe microstrip linear arrays with the aperture and feed integrated. The approaches are different from the contemporary decomposition approach: their design processes are conducted with just a few high-fidelity simulations of the entire structure. Contribution of the feed to the far-field is accounted for as well. All presented design processes start with identification of the optimal feed architectures. Then, simulation-based optimization of the feed and the radiating aperture is performed.

We will consider microstrip corporate feeds realizing non-uniform amplitude excitations for low-sidelobe linear array apertures. Two types of feeds, which are specific to each approach, will be considered: one type with equal power split junctions, and another with unequal power split junctions. For each type, we identify candidate feed architectures using numerical optimization of the corresponding fast models. Subsequently, the feeds are implemented as microstrip subcircuits within electromagnetic (EM) models of arrays. For the sake of comparison, the arrays are defined with the same linear aperture of microstrip patch antennas and implemented on the same microstrip substrates. Finally, the EM models are tuned — using simulation-based optimization techniques — to ensure an appropriate input reflection coefficient and minimum sidelobe levels. Selected optimal designs of array circuits with 12 microstrip patches and different feeds are manufactured and measured.

14.1. Sidelobe Reduction in Arrays Driven with Corporate Feeds Comprising Equal Power Split Junctions

It turns out that the usability of microstrip corporate feeds comprising the simplest equal-split T-junctions (with one low-impedance transforming section) has been somehow underestimated for low-sidelobe arrays. We demonstrate that with numerical optimization applied at different stages of the design process, utilization of such feeds can be a reliable option for microstrip linear arrays with reduced sidelobes (with sidelobe levels of about $-20\,\text{dB}$) unless an ultralow sidelobes ($-30\,\text{dB}$ and lower) are required.

14.1.1. *Approach justification*

Printed-circuit corporate feeds (CFs) are widely used with linear and planar antenna arrays, in particular, with microstrip antenna arrays (Munson, 1974; Hall and Hall, 1988; Levine *et al.*, 1989; Horng and Alexopoulos, 1993; Garg *et al.*, 2000; Balanis, 2005;

Mailloux, 2005; Hansen, 2009; Haupt, 2010). Such feeds provide structural simplicity of basic configurations, clarity of their circuit-based design and analysis, ease of manufacturing, and low costs. Feeds of uniform power distribution at the outputs, i.e., producing uniform excitation tapers are the commonly used architectures (Munson, 1974; Levine et al., 1989; Hall and Hall, 1988; Balanis, 2005). Such architectures, if comprise T-junctions or other three-port signal splitters, only allow excitation of array apertures with the number of elements being an integer power of two (Balanis, 2005). This limitation can be removed with unequal-split T-junctions or multi-port splitters (Hall and James, 1981; Levine and Shtrikman, 1989). On the other hand, utilization of such components for printed-circuit arrays is potentially associated with degraded reflection responses, stronger coupling, increased sensitivity to manufacturing tolerances, and degraded radiation characteristics (Hall and James, 1981; Hall and Hall, 1988).

Corporate feeds realizing predefined non-uniform excitations have been demonstrated for microstrip, stripline (Hall and James, 1981; Horng and Alexopoulos, 1993), and waveguide (Anthony and Zaghloul, 2009) implementations where optimality of the implemented feed architectures was not addressed. At the same time, necessity of EM simulations has been demonstrated not only for validation of the final design but also for design tuning (Hall and James, 1981; Horng and Alexopoulos, 1993; Anthony and Zaghloul, 2009; Jayasinghe et al., 2013). Wilkinson dividers (Wilkinson, 1960), branch line couplers, filters, and tunable attenuators improve feed characteristics at the expense of feed complexity, footprint, e.g., as well as (implicitly) strengthen requirements to quality of substrates and manufacturing tolerances (Pozar and Kaufman, 1990; Rigoland et al., 1996; Hu et al., 2011; Chen et al., 2017). Therefore, a question arises in the context of design of fixed-beam linear arrays, "Can we improve radiation and reflection responses' characteristics of such structures by staying with the simplest and thus compart types of power dividers in the feed by solving this task through detecting the optimal feed architectures and with subsequent simulation-driven optimization rather than introducing extra circuitry?" A positive answer is supported within the sections to follow, where we address systematic design of corporate feeds for broadside fixed-beam sum-pattern microstrip linear arrays. In particular, the formulated and conducted design process features the following:

(i) design specifications are imposed on the sidelobe level (SLL), major lobe beamwidth, and reflection coefficient at the input of the feed;
(ii) a linear array of interest can comprise any even number of elements;
(iii) the corporate feed is configured from the simplest elements which are three port T-junctions (matched at the input) of equal power split, uniform straight microstrip traces, and chamfered 90-degree microstrip bends;
(iv) CF architectures with the lowest SLL and the simplest routing are detected;
(v) cases of uniform and non-uniform array element spacings are considered as options before implementation;
(vi) the integrated array-feed circuit is optimized and validated using high-fidelity EM-simulations.

A study for an array with the aperture comprising 12 elements, microstrip patch antennas (MPAs), is used to describe and illustrate our point. Implementation, optimization, and measurement of selected cases, which are different in the feed architecture and realization, are provided.

14.1.2. *Design task, feed elements, and feed architectures*

Consider a linear array aperture, as depicted in Fig. 14.1, comprising an even number of elements. An estimation of the array radiation pattern is performed using the array factor (AF),

$$\text{AF}(\theta) = \sum_n a_n e^{j2\pi z_n \cos(\theta)} \tag{14.1}$$

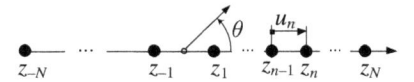

306 Simulation-Based Optimization of Antenna Arrays

Fig. 14.1. Linear array aperture with an even number of elements. Excitation amplitudes (elements' weights) and spacings are symmetrical with respect to the center.

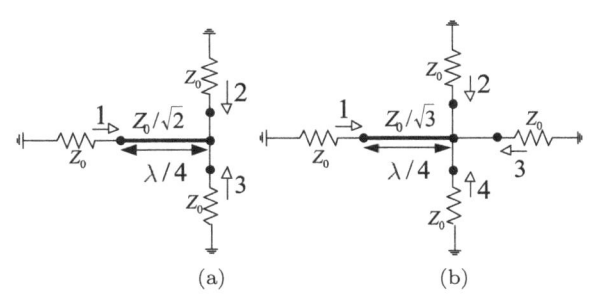

Fig. 14.2. Simplest junctions of equal power distribution for corporate feeds, schematics: (a) utilized 3-port junction; (b) 4-port junction.

where a_n stands for the nth element amplitude, z_n is the nth element location normalized to the operating wavelength, and θ is the elevation angle.

An array, with respect to which we search for improvements, is the broadside uniform linear array (ULA) (Balanis, 2005). For a given number of elements and fixed spacing, the uniform array aperture allows the simplest feed architecture, the highest directivity (besides the super-directive array apertures (Hansen, 2009)), the narrowest major lobe beamwidth, and the highest taper efficiency (Balanis, 2005; Mailloux, 2005; Haupt 2010) but the ULA SLLs are only about -13 dB. Our primary tasks are to minimize the SLL of the H-plane pattern and reflection coefficient at the input of the feed. The SLL should be minimized outside the null-to-null beamwidth of the corresponding ULA.

Feeds are to be configured with the simplest power distribution elements, T-junctions of the equal power split as shown in Fig. 14.2(a). Such a T-junction, being matched at the input (marked with 1), exhibits a quite high reflection loss at the output ports (marked with 2 and 3) and coupling of the output ports, namely

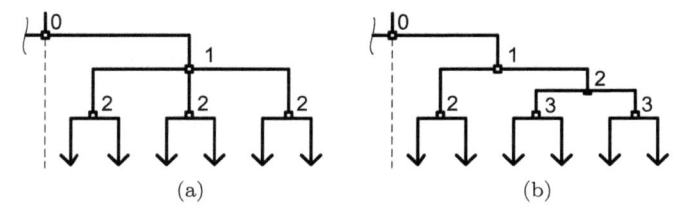

Fig. 14.3. Example feed architectures: (a) one realizing the uniform amplitude excitation where the outputs (marked with the arrowheads) belong to the same level 2. Notice a 4-port junction at level 1: it is not possible to stay only with 3-port junctions in this case; (b) one realizing a certain non-uniform amplitude excitation where the feed outputs belong to different levels, 2 and 3. Only halves of the respective feeds are shown due to the aperture symmetry depicted in Fig. 14.1 (Ogurtsov and Koziel, 2017a).

about 6 dB each. A 4-port junction of Fig. 14.2(b) also exhibits quite high coupling of the output ports, about 4.8 dB, and even higher reflection loss at the output ports, 3.5 dB.

To realize a non-uniform power distribution at the outputs of the feed, i.e., non-uniform aperture excitation and with a given number of array aperture elements, yet using equal-split T-junctions only, we consider the feed architectures with outputs belonging to different levels, as those outlined in Fig. 14.3. The paths from the common input (marked with 0) to the inputs of elements are assumed to be the same.

14.1.3. *Fast models of corporate feeds*

To configure the fast models of the corporate feeds, which we use to evaluate possible reduction in SLL for a given linear array aperture, we assume perfect matching at the junctions' inputs and no reflection from the array elements. Under such assumptions, the considered feed architectures realize non-uniform amplitude excitations of the array elements with amplitudes that can only take the following values:

$$a_n = 2^{-k_n/2}, \qquad (14.2)$$

where $n = 1, 2, \ldots, N$ indexes the elements for a half of the array (counting from the array center). Thus the total number of elements is $2N$. The integer k_n is the number of T-junctions within a half of the feed the signal passes reaching the nth element. Notice that the phase factor is suppressed in (14.2) assuming all paths being equal. The amplitudes of (14.2) are of the waves which are incident at the inputs of the array elements. The incident power at the input of the nth element is

$$p_n = \frac{|a_n|^2}{2} = 2^{-k_n-1}. \tag{14.3}$$

Conservation of power requires

$$\sum_{n=1}^{N} 2^{-k_n} = 1. \tag{14.4}$$

Equation (14.4) identifies all realizable architectures for a particular linear array with uniform as well as non-uniform spacings, i.e., it allows to detect possible sets of integers k_n, and, thus with (14.2), corresponding excitation tapers. Notice that a feed architecture is defined with the set of integers k_n up to a particular routing of traces. The fast model, which is used in the study presented below, is the AF (14.1) where the amplitudes take the form of (14.2) and obey (14.4).

First, consider the ULA design with the half-wavelength spacing, u_n. Its AF pattern with the -13.1 dB SLL is shown in Fig. 14.4. Note that a $2N$-element ULA aperture cannot be driven with a corporate

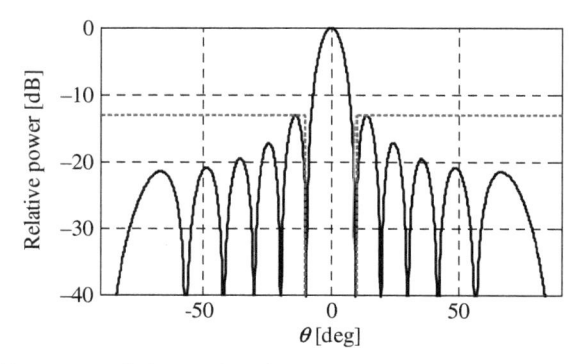

Fig. 14.4. AF pattern of the uniformly excited 12-element linear aperture with elements half-wavelength spaced. The SLL should be minimized over the sectors marked with the horizontal dash-lines.

feed consisting solely of 3-port T-junctions—at least one 4-port (Fig. 14.2(b)) is needed per half-array, e.g., as shown in Fig. 14.3(a). Differently from the ULA case, we study architectures of the type shown in Fig. 14.3(b) with T-junctions of Fig. 14.2(a). First, we search for improvements of the SLL with half-wavelength spacings. The SLL is defined beyond the null-to-null beamwidth of the ULA major lobe which occupies [−9.6°, 9.6°].

The number of candidate feed architectures grows very quickly not only with the aperture size $2N$ but also with the minimum incident power delivered to an array element. Notice that no closed-form expression can be derived for the number of architectures with constrained incident power. Assuming that the minimum incident power is 1/64 of the input power, the number of feeds is calculated as 3, 13, 75, 525, 1827, 5965, 18315, and 51885 for $N = 3$, 4, 5, 6, 7, 8, 9, and 10, respectively. Thus, there are 525 architectures for the 12-element aperture; however, only eight architectures are expected to have SLLs at least 2 dB better than the ULA. Table 14.1 lists the top five architectures with the lowest SLL. Figure 14.5 illustrates that from the top five architectures listed in Table 14.1 there are only two, number 1 and 4, which allow simple routing.

We also search for feed architectures providing reduced SLLs using non-uniform spacings as adjustable parameters. In this case,

Table 14.1. Top five corporate feed architectures for the half-wavelength spaced 12-element linear aperture. SLL is defined over [−90°, −9.6°] and [9.6°, 90°]. The corporate feeds utilize only equal-split T-junctions, 3-ports except ULA (Ogurtsov and Koziel, 2017a).

	SLL (dB)	Normalized amplitudes, elements' levels
ULA	−13.06	[1.000 1.000 1.000 1.000 1.000 1.000], NA
1*	−16.53	[1.000 1.000 0.707 0.707 0.707 0.707], [2 2 3 3 3 3]
2	−15.97	[1.000 0.707 1.000 0.707 0.707 0.707], [2 3 2 3 3 3]
3	−15.86	[1.000 1.000 0.707 1.000 0.500 0.500], [2 2 3 2 4 4]
4*	−15.73	[1.000 1.000 1.000 0.500 0.500 0.707], [2 2 2 4 4 3]
5	−15.72	[1.000 1.000 0.500 1.000 0.707 0.500], [2 2 4 2 3 4]

*Architectures without swapping of outputs.

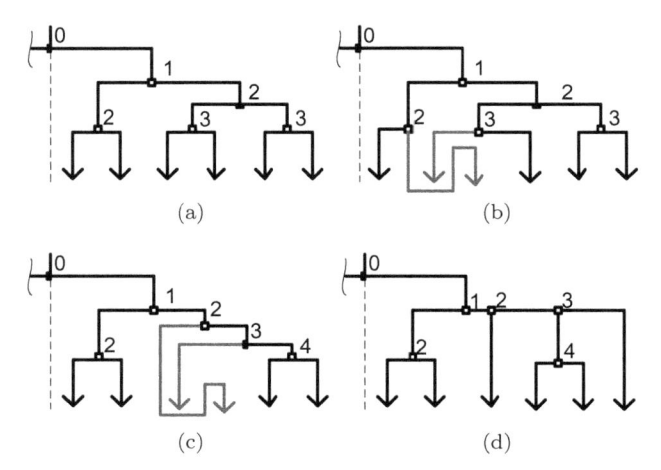

310 *Simulation-Based Optimization of Antenna Arrays*

Fig. 14.5. Corporate feed architectures for a 12-element broadside linear array aperture, top four listed in Table 14.1: (a) 1, (b) 2, (c) 3, and (d) 4. Only halves of the feeds are shown. Equal input-to-output paths are assumed (Ogurtsov and Koziel, 2017a).

the SLL is defined for the angles beyond $[-10°, 10°]$. This task is carried out using numerical optimization of the AF (14.1) over amplitudes satisfying (14.4). In addition, array aperture spacings are constrained within $[0.41, 0.60]$ wavelengths. The aperture length (the sum of the spacings) is also constrained to 5.5 wavelengths that is the length of the half-wavelength spaced ULA aperture. Optimization of the vectorized array factor and sorting has been done using Matlab (Matlab, 2010). Table 14.2 lists optimization results for the best five feed architectures in terms of the SLL and for the non-uniformly spaced ULA (with uniform excitation taper). Further reductions of the SLL can be obtained by subsequently removing upper bounds for the element spacings as well as restriction of the total aperture length. The results for these cases are summarized in Table 14.3 and 14.4.

Tables 14.2–14.4 also show that optimization of the non-uniformly spaced ULA (with the uniform excitation taper) delivers substantial improvement relative to the equally spaced ULA, namely 5.7 dB down for the cases of Tables 14.2 and 14.3, and about 7.0 dB down for the case of Table 14.4. These ULA designs, however, are not

Table 14.2. Top corporate feed architectures for non-uniformly spaced 12-element linear apertures with spacings constrained to [0.41, 0.60] and aperture lengths constrained by 5.5 wavelengths. SLL is over [−90°, −10°] and [10°, 90°]. Feeds 1–5 are only with equal-split T-junctions (Ogurtsov and Koziel, 2017a).

	SLL (dB)	Normalized element amplitudes, elements' levels, spacings
ULA	−18.77	[1.000 1.000 1.000 1.000 1.000 1.000], NA, [0.410 0.410 0.4315 0.4815 0.600 0.600]
1	−20.22	[1.000 0.707 1.000 0.707 0.707 0.707], [2 3 2 3 3 3], [0.479 0.435 0.441 0.534 0.520 0.571]
2*	−20.20	[1.000 1.000 0.707 0.707 0.707 0.707], [2 2 3 3 3 3], [0.488 0.505 0.454 0.438 0.526 0.573]
3	−19.49	[1.000 1.000 1.000 0.500 0.707 0.500], [2 2 2 4 3 4], [0.503 0.498 0.577 0.504 0.465 0.446]
4	−19.05	[1.000 1.000 1.000 0.500 0.500 0.707], [2 2 2 4 4 3], [0.517 0.488 0.562 0.491 0.415 0.523]
5*	−19.04	[1.000 1.000 1.000 0.707 0.500 0.500], [2 2 2 3 4 4], [0.478 0.516 0.574 0.564 0.432 0.419]

*Architectures without swapping of outputs.

Table 14.3. Top five corporate feed architectures for non-uniformly spaced 12-element linear apertures with the spacings constrained to [0.41, 0.60] and no constrains of the aperture lengths. SLL is defined over [−90°, −10°] and [10°, 90°]. Corporate feeds 1–5 are only with equal-split 3-port junctions (Ogurtsov and Koziel, 2017a).

	SLL (dB)	Normalized element amplitudes, elements' levels, spacings
ULA	−18.77	[1.000 1.000 1.000 1.000 1.000 1.000], NA, [0.410 0.410 0.4315 0.4815 0.600 0.600]
1*	−23.84	[1.000 1.000 1.000 0.707 0.500 0.500], [2 2 2 3 4 4], [0.544 0.527 0.595 0.595 0.561 0.591]
2	−22.26	[1.000 1.000 1.000 0.500 0.707 0.500], [2 2 2 3 4 3], [0.504 0.526 0.590 0.511 0.518 0.594]
3*	−20.77	[1.000 1.000 0.707 0.707 0.707 0.707], [2 2 3 3 3 3], [0.459 0.532 0.453 0.4246 0.550 0.595]
4	−20.64	[1.000 0.707 1.000 0.707 0.707 0.707], [2 3 2 3 3 3], [0.487 0.431 0.449 0.533 0.525 0.594]
5	−20.49	[1.000 0.707 1.000 1.000 0.500 0.500], [2 3 2 2 4 4], [0.533 0.430 0.494 0.596 0.590 0.588]

*Architectures without swapping of outputs.

Table 14.4. Top five corporate feed architectures for non-uniformly spaced 12-element linear array apertures with spacings constrained from below by 0.41. SLL is defined over $[-90°, -10°]$ and $[10°, 90°]$. Corporate feeds 1–5 are with only equal-split 3-port junctions (Ogurtsov and Koziel, 2017a,b).

	SLL (dB)	Normalized element amplitudes, elements' levels, spacings
ULA	−20.14	[1.000 1.000 1.000 1.000 1.000 1.000], NA, [0.420 0.417 0.462 0.478 0.626 0.693]
1*	−25.13	[1.000 1.000 1.000 0.707 0.500 0.500], [2 2 2 3 4 4], [0.534 0.532 0.611 0.629 0.552 0.621]
2	−23.80	[1.000 1.000 1.000 0.500 0.707 0.500], [2 2 2 4 3 4], [0.492 0.553 0.589 0.527 0.530 0.675]
3	−23.42	[1.000 1.000 0.500 1.000 0.707 0.500], [2 2 4 2 3 4], [0.453 0.573 0.418 0.512 0.693 0.692]
4	−22.88	[1.000 0.500 1.000 1.000 0.707 0.500], [2 4 2 2 3 4], [0.501 0.412 0.414 0.653 0.697 0.684]
5*	−21.63	[1.000 1.000 0.707 0.707 0.707 0.707], [2 2 3 3 3 3], [0.510 0.540 0.503 0.432 0.586 0.673]

*Architectures without swapping of outputs.

adopted for implementation and subsequent simulation-based tuning because their feeds would require 4-ports, as depicted in Fig. 14.3(a).

It is worth to note that only two feed architectures per every Tables 14.1–14.4 which can be realized without swapping of the outputs. Namely, an architecture with elements' levels [2 2 3 3 3 3], which is depicted in Fig. 14.5(a), appears in Tables 1–4, and it is the best one for the cases of Tables 14.1 and 14.2 in terms of SLL and with respect to realization, respectively. An architecture with elements' levels [2 2 3 4 4] can be also realized without swapping of the outputs and it appears in Tables 14.2–14.4. This architecture predicts the best designs for the cases of Tables 14.3 and 14.4 in terms of SLL. These two architectures are adopted for implementation, EM modeling, and tuning. AF patterns of the best designs are shown in Fig. 14.6.

One sees that the fast model described in this section can be quite useful for determining the best feed architectures comprising only equal-split T-junctions and aperture spacings with numerical

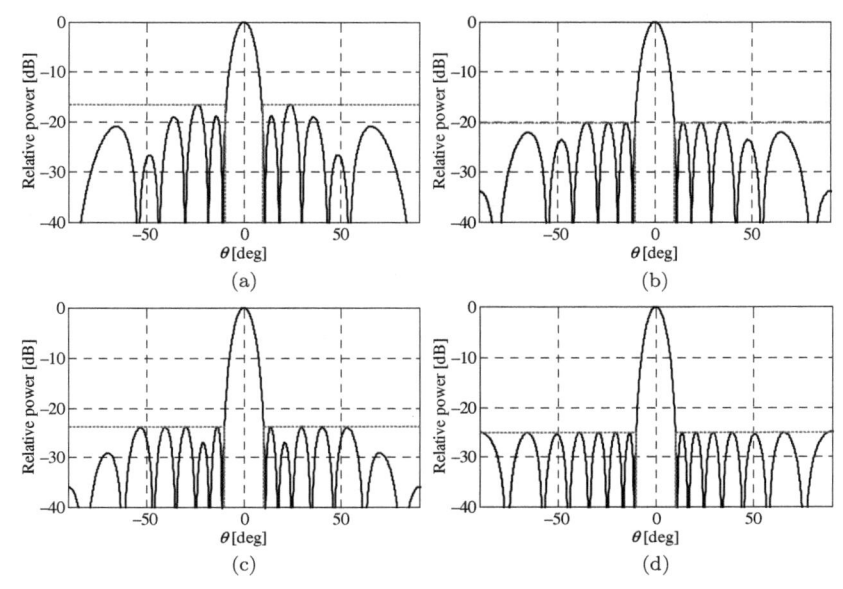

Fig. 14.6. AF patterns with the best corporate feed architectures utilizing only equal-split T-junctions for the cases of Tables 14.1–14.4: (a) Table 14.1, [2 2 3 3 3 3], D_{AF} = 10.7 dBi, HPBW = 9.0 degrees; (b) Table 14.2, [2 2 3 3 3 3], D_{AF} = 10.6 dBi, HPBW = 9.3 degrees; (c) Table 14.3, [2 2 2 3 4 4], D_{AF} = 11.0 dBi, HPBW = 8.7 degrees; (d) Table 14.4, [2 2 2 3 4 4], D_{AF} = 11.1 dBi, HPBW = 8.5 degrees. D_{AF} and HPBW stand for the array factor directivity and half-power beamwidth, respectively (Ogurtsov and Koziel, 2017a).

optimization providing the lowest SLL within the required (preassigned) range of angles. Such a study represents is an initial step of the process preceding implementation and EM-based tuning.

14.1.4. *Realization and simulation-based optimization of aperture-feed structures*

Selected designs, the ones with the feed element level [2 2 3 3 3 3] and [2 2 2 3 4 4], are implemented in CST Microwave Studio (CST, 2016) as C-band microstrip linear arrays with the integrated aperture and feed with the frequency bandwidth centered at 5.8 GHz. The configuration of the utilized equal-split junctions is shown in Fig. 14.7.

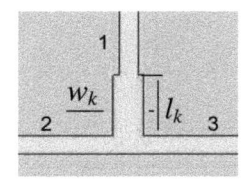

Fig. 14.7. An equal-split microstrip T-junction, the used configuration. The input is marked with 1. w_k and l_k are dimensions of the matching section.

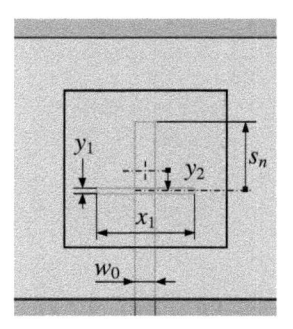

Fig. 14.8. An array aperture element, MPA. Dielectrics and metals are shown transparent.

The linear aperture comprises 12 microstrip patch antennas (MPAs) residing on a 1.574-mm TLP-5 substrate (Taconic TLP, 2013). Every MPA has a square patch, 14.85 mm × 14.85 mm and energized through a ground plane slot as shown in Fig. 14.8 where s_n stands for the length of the open-end microstrip of MPA element n. Size and locations of the slots are the same for all MPAs and parameterized as depicted in Fig. 14.8. The aperture is shown in Fig 14.9. The input of the feed is an edge-mount launcher (SMA connector, 2012). Figure 14.8 also shows fixing bolts which are included in the EM model.

The feed resides on a 0.762-mm RF-35 substrate (RF-35, 2018) on the back side of the circuit as shown in Figs. 14.10(a) and 14.10(b) for architectures [2 2 3 3 3 3] and [2 2 2 3 4 4], respectively. The feeds are parameterized in such a way that — regardless of the element spacing — the geometric paths to all elements are the same for the

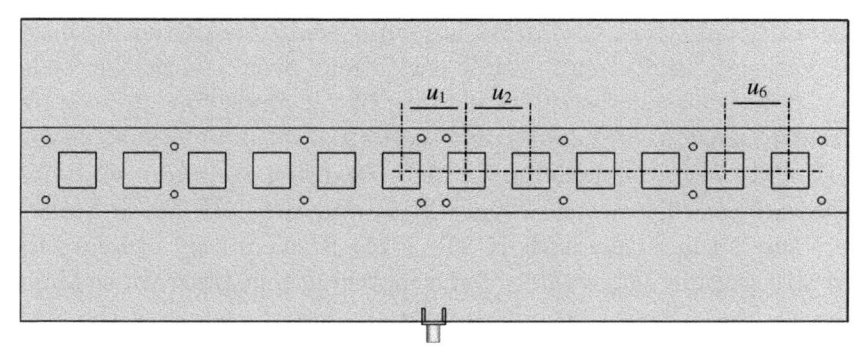

Fig. 14.9. Top view of the array circuit, the linear aperture of MPAs (Ogurtsov and Koziel, 2017a).

(a)

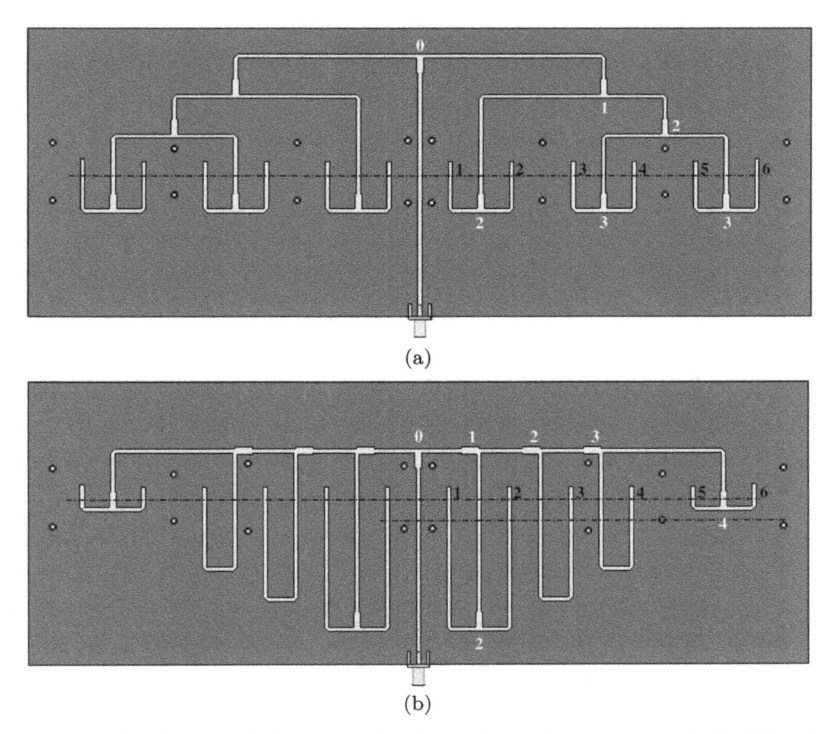

(b)

Fig. 14.10. Back view of the array circuits, microstrip corporate feeds (Ogurtsov and Koziel, 2017a,b): (a) realization of the feed architecture with the elements' levels [2 2 3 3 3 3]; (b) realization of the feed architecture with the elements' levels [2 2 2 3 4 4]. Feed levels and MPAs are marked with white and black integers, respectively (Ogurtsov and Koziel, 2017a).

initial designs; thus, the MPAs are supposed to be driven in phase, i.e., with non-uniform amplitude excitations produced by the feeds.

Connecting microstrip traces are of 50 ohms line impedance. Initial dimensions of the matching sections of the T-junctions, w_k and l_k (Fig. 14.7), are calculated for 5.8 GHz using standard analytical microwave circuit means (Pozar, 2011) and the CST MWS built-in transmission line calculator (CST, 2016) to accurately evaluate the effective permittivity values of the matching quarter wave sections; they are the same for all junctions at the initial designs of the feeds (Fig. 14.10).

Before being connected to a particular feed in the EM model, the aperture is pre-optimized for 5.8 GHz operation with the excitation of a particular feed applied to MPA microstrip inputs.

In the array model with the feed of Fig. 14.10 (a) there are 24 design variables in the setup of Table 14.1 (the case of uniform spacing) and 30 design variables in the setup of Tables 14.2 and 14.3 (the cases of non-uniform spacing). In the array model with the feed of Fig. 14.10(b) there are 26 design variables in the setup of Table 14.1 and 32 design variables in the setup of Tables 14.2 and 14.3. Out of those, nine variables control the MPAs, including 6 open-end extensions of the microstrips; 6 are center-to-center MPA spacings, and the rest, 14 for the feed of Fig. 14.10(a) and 16 for the feed of Fig. 14.10(b), controls geometry of the feeds including the transformer sections and chamfer widths of the connecting microstrip traces. Other dimensions are fixed. The array-feed circuits are simulated in the CST MWS environment (CST, 2016).

The initial designs are with the feeds dimensioned using the circuit rules; the simulated responses exhibit broadened major lobes, poor SLL, and high reflection coefficient, e.g., as shown in Figs. 14.11 and 14.12. At the same time, patterns (including SLLs) of the simulated aperture models, where the excitation amplitudes (listed in Tables 14.1–14.4) are applied to the MPA inputs, are very close (up to the MPA element pattern) to the fast models' patterns (shown in Fig. 14.6); also active reflection coefficients of the standalone aperture are centered at 5.8 GHz and better than -20 dB at 5.8 GHz. According to our numerical analysis using (2.42)–(2.50) and studies

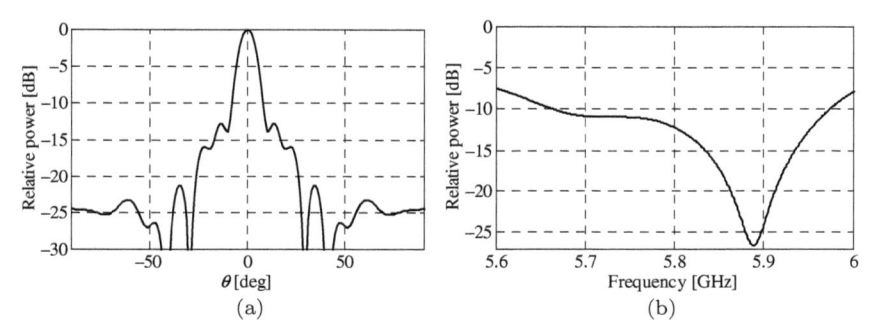

Fig. 14.11. Simulated array model with elements' levels [2 2 3 3 3 3] (feed of Fig. 14.10 (a)) at the initial design: (a) H-plane pattern at 5.8 GHz; (b) reflection coefficient. The responses are obtained using the CST MWS transient solver and finely discretized hexahedral mesh (Ogurtsov and Koziel, 2017a).

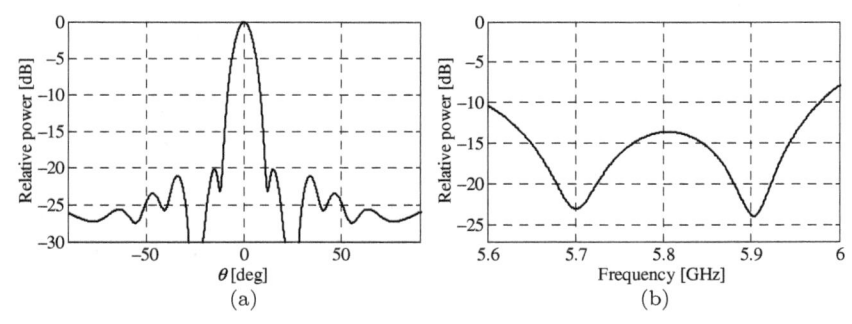

Fig. 14.12. Simulated array model with elements' levels [2 2 2 3 4 4] (feed of Fig. 14.10 (b)) at the initial design: (a) H-plane pattern at 5.8 GHz; (b) reflection coefficient. The responses are obtained using the CST MWS transient solver and finely discretized hexahedral mesh (Ogurtsov and Koziel, 2017a).

(Ogurtsov and Koziel, 2017b), poor responses of the initial designs are results of deviation of the incident wave amplitudes at the MPAs' inputs from nominal for a particular architecture (listed in Tables 14.1–14.4) as well as realized for a simulated standalone feed.

Simulation-based optimization is used to fix this problem without hardware complications, e.g., without Wilkinson dividers. Also note that simulation-based optimization is feasible for the entire array circuits (Figs. 14.9 and 14.10) in a part due to composition of the feeds (Fig. 14.10) which are configured with equal-split junctions

(Fig. 14.7) so that the EM models of the entire arrays are moderately graded, i.e., do not contain very fine elements.

The array models are optimized for the reduced H-plane pattern SLL and reflection coefficient response with the following requirements: (i) SLL is minimized over the [−90, −11]- and [11, 90]-degree sectors which is a slightly relaxed range compared to those of Tables 14.1–14.4; (ii) reflection coefficient at the common input (at the SMA connector) is to be kept under −15 dB within 5.75–5.85 GHz. Requirement (ii) is implemented by adding a penalty function to the primary objective (SLL). The penalty evaluates a relative violation of the $|S_{11}|$ level requirement within the aforementioned frequency range. To speed up simulation-based optimization, where the figures of interest are results of discrete EM simulations, we use the surrogate-based optimization tools (Koziel and Ogurtsov, 2014a,b). Cases different by constraints imposed on the spacings and the array aperture length are considered.

Selected optimization results for the array circuit with the elements' levels [2 2 3 3 3 3] and with constrains of Tables 14.1–14.3 are shown in Figs. 14.13–14.15. Substantial improvement of the SLLs predicted by the fast model (Tables 14.1–14.3) for this architecture is achieved. Reflection response is improved in comparison to the initial

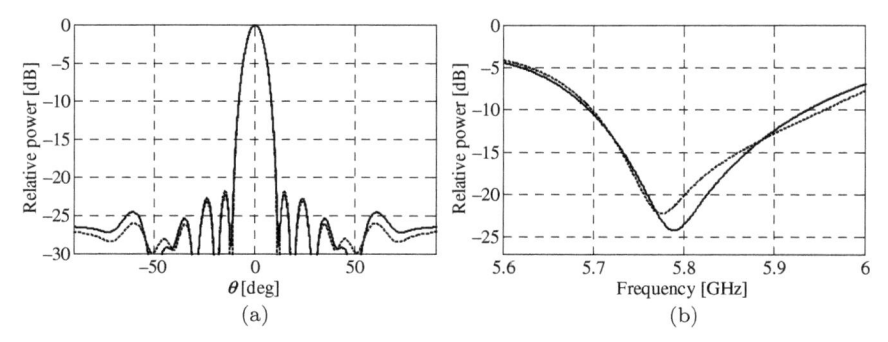

Fig. 14.13. Simulated array, the feed architecture with elements' levels [2 2 3 3 3 3] (feed of Fig. 14.10(a)): an optimized design with the half wavelength spaced aperture, the case of Table 14.1: (a) H-plane pattern at 5.8 GHz; (b) reflection coefficient. Responses are obtained using the transient solver with the hexahedral mesh (—) and the frequency domain solver with the tetrahedral mesh (- -) (Ogurtsov and Koziel, 2017a).

Simulation-Based Design of Corporate Feeds 319

design as well. It is worth mentioning that for the array model with the elements' levels [2 2 3 3 3 3] the lowest SLL, −22 dB, is achieved with uniform half-wavelength spacings. The cases of constrained non-uniform spacings show similar SLLs within −22 to −21 dB. The optimized non-uniform spacings are not noticeably deviated from

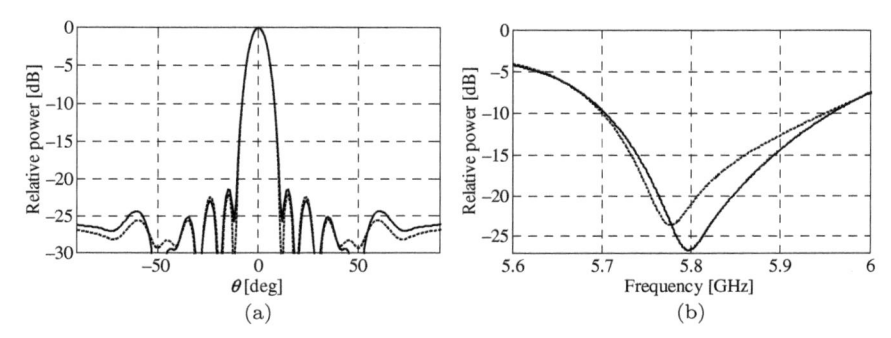

Fig. 14.14. Simulated array, the feed architecture with elements' levels [2 2 3 3 3 3] (feed of Fig. 14.10(a)): an optimized design with aperture dimensions constrained as in the case of Table 14.2: (a) H-plane pattern at 5.8 GHz; (b) reflection coefficient. Responses are obtained using the transient solver with the hexahedral mesh (—) and the frequency domain solver with the tetrahedral mesh (--) (CST, 2016; Ogurtsov and Koziel, 2017a).

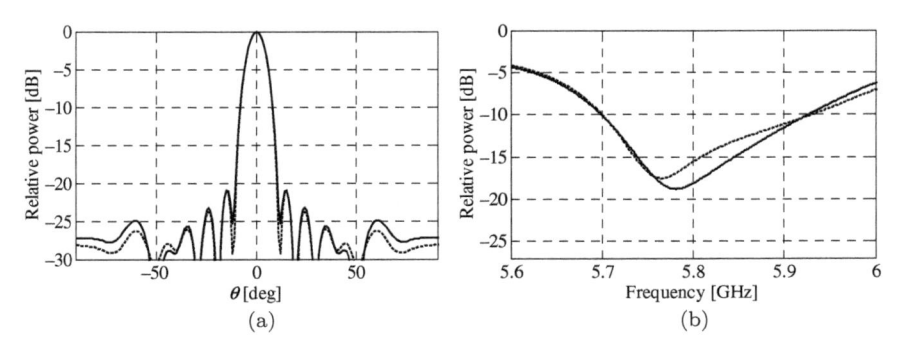

Fig. 14.15. Simulated array, feed architecture with elements' levels [2 2 3 3 3 3] (feed of Fig. 14.10(a)): an optimized design with aperture dimensions constrained as in the case of Table 14.3: (a) H-plane pattern at 5.8 GHz; (b) reflection coefficient. Responses are obtained using the transient solver with the hexahedral mesh (—) and the frequency domain solver with the tetrahedral mesh (--) (CST, 2016; Ogurtsov and Koziel, 2017a).

320 *Simulation-Based Optimization of Antenna Arrays*

the nominal half-wavelength values of 25.90 mm and all are within 25.80–25.93 mm. Thus, one can conclude that the major optimization effort has been made for tuning dimensions of the feed and the MPA feeding elements (the slot and the open-end microstrip stubs).

Optimization results under conditions of Tables 14.1–14.3 for the array circuit with the elements' levels [2 2 2 3 4 4] are shown in

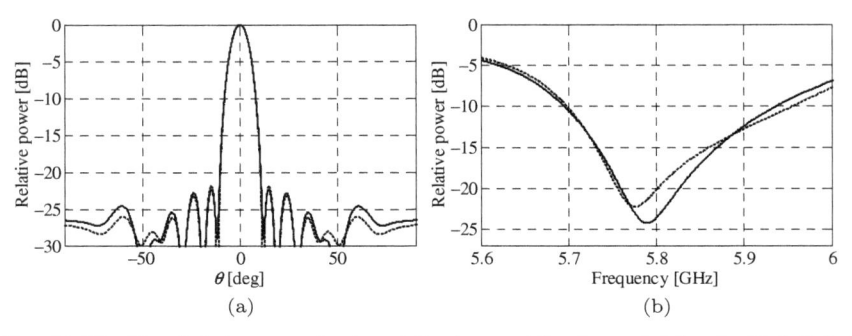

Fig. 14.16. Simulated array, feed architecture with elements' levels [2 2 2 3 4 4] (feed of Fig. 14.10(b)): an optimized design with the half wavelength spaced aperture, the case of Table 14.1: (a) H-plane pattern at 5.8 GHz; (b) reflection coefficient. Responses are obtained using the transient solver with the hexahedral mesh (—) and the frequency domain solver with the tetrahedral mesh (- -) (CST, 2016; Ogurtsov and Koziel, 2017a).

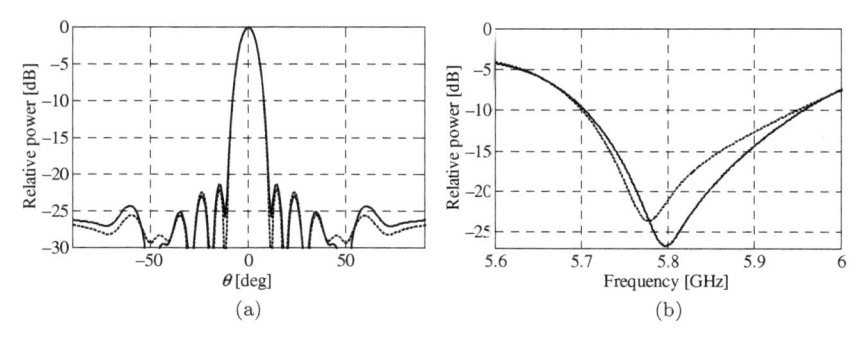

Fig. 14.17. Simulated array, feed architecture with elements' levels [2 2 2 3 4 4] (feed of Fig. 14.10(b)): an optimized design with aperture dimensions constrained as in the case of Table 14.2: (a) H-plane pattern at 5.8 GHz; (b) reflection coefficient. Responses are obtained using the transient solver with the hexahedral mesh (—) and the frequency domain solver with the tetrahedral mesh (- -) (Ogurtsov and Koziel, 2017a).

Figs. 14.16–14.18. Improvement of the initial design is achieved. The lowest SLL, -22 dB (the pattern shown in Fig. 14.18), is provided with non-uniform half-wavelength spacings and with no constrains of the aperture length. Other two cases show quite similar SLLs of about -21.5 dB. The optimized non-uniform spacings are not noticeably deviated from the nominal half-wavelength values of 25.90 mm and all are within 25.80–26.22 mm. Thus, similar to the architecture [2 2 2 3 4 4], the major optimization effort here was also in tuning the feed dimensions.

The presented study illustrates that, first, feed complexity is not increased with the discussed approach, instead, only simple equal-split T-junctions are utilized; thus, simulation-based optimization can be performed for the entire array circuit; second, optimization of the fast array factor-based model provides candidate feed architectures under various constrains imposed on an array aperture of interest; finally, simulation-driven design closure is essential as the final design stage. The flow of the discussed approach is summarized with the diagram of Fig. 14.19. One also needs to mention that absence of fine inclusions in the considered corporate feeds, such as thin high impedance sections of unequal power T-junctions or branch line couplers, chip resistors, etc. (Pozar, 2011;

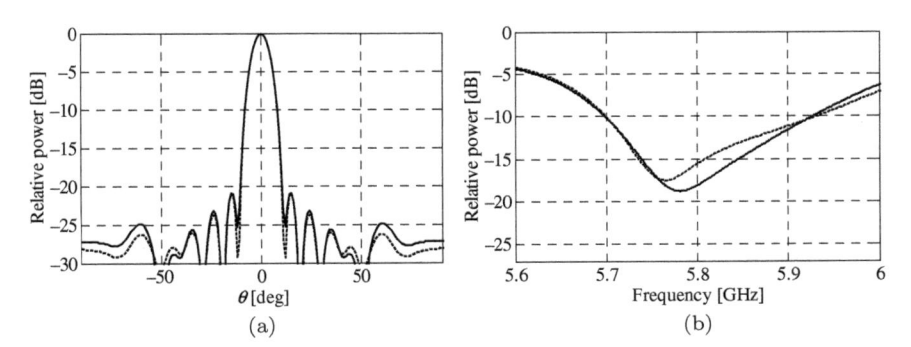

Fig. 14.18. Simulated array, feed architecture with elements' levels [2 2 2 3 4 4] (feed of Fig. 14.10(b)): an optimized design with aperture dimensions constrained as in the case of Table 14.3: (a) H-plane pattern at 5.8 GHz; (b) reflection coefficient. Responses are obtained using the transient solver with the hexahedral mesh (—) and the frequency domain solver with the tetrahedral mesh (--) (Ogurtsov and Koziel, 2017a).

322 *Simulation-Based Optimization of Antenna Arrays*

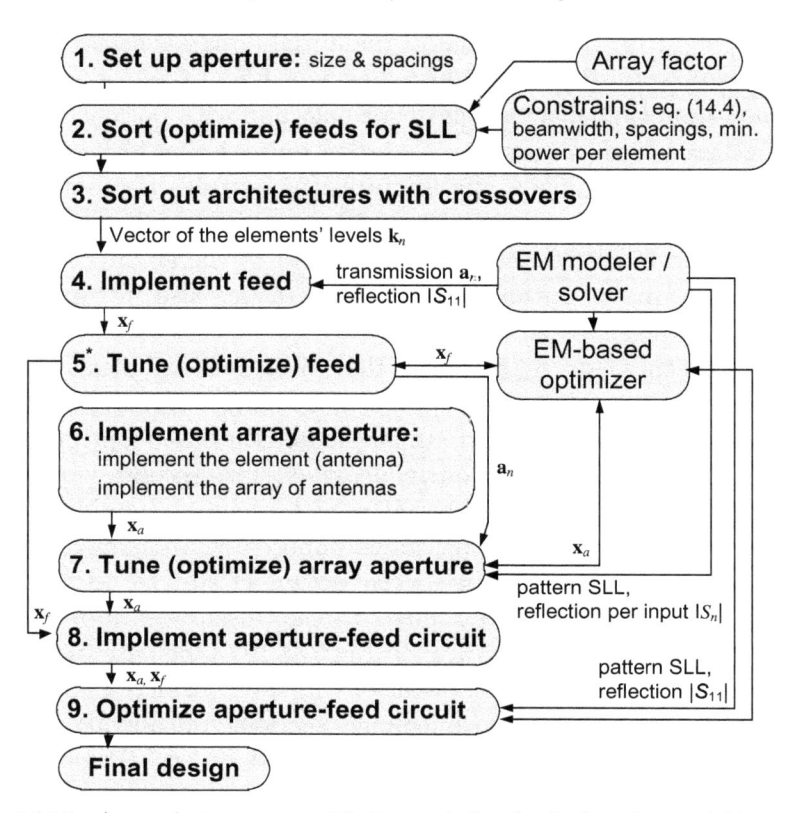

Fig. 14.19. Array design process. Vector $\mathbf{x}_f$ is for the feed design variables, $\mathbf{x}_a$ is for the aperture variables. Optimization of the feed, Step 5, can be implemented at Step 9, EM-based optimization of the integrated array circuit (Ogurtsov and Koziel, 2018).

Muraguchi *et al.*, 1983), makes the feeds potentially more tolerant to manufacturing non-idealities and deviations of substrates' parameters.

14.1.5. *Validation*

Layouts of selected optimized array circuits had been created from the EM-models (CST, 2016). Manufactured microstrip arrays are shown in Figs. 14.20–14.22. Nylon bolts fix the dielectric layers of two-side designs. All circuits are with 18 μm copper metallization.

The circuits shown in Figs. 14.20–14.22 had been measured in the anechoic chamber of Reykjavik University. The measured responses

Simulation-Based Design of Corporate Feeds 323

(a)

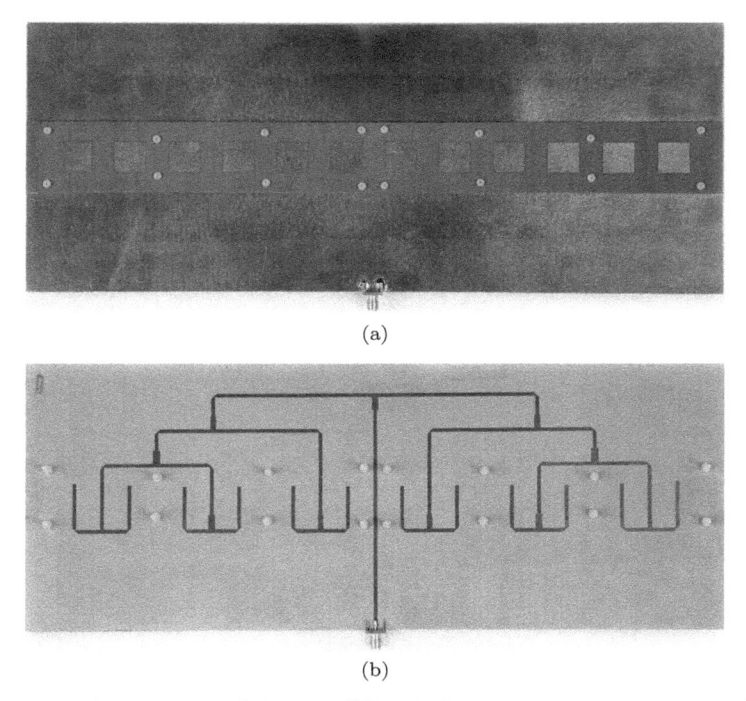

(b)

Fig. 14.20. Photographs of the 5.8-GHz 12-element microstrip array with the element levels [2 2 3 3 3 3]: (a) radiating aperture on two layers of TLP-5 (nominal thickness 1.574 mm in total). The TLP-5 layers are 329 mm × 35 mm. Microstrip patches are 14.85 mm × 14.85 mm; (b) feed on the 0.762 mm RF-35 layer with equal power split T-junctions. Radiating elements are energized through ground plane slots. The RF-35 layer and the ground plane is 329 mm × 133 mm (Ogurtsov and Koziel, 2018).

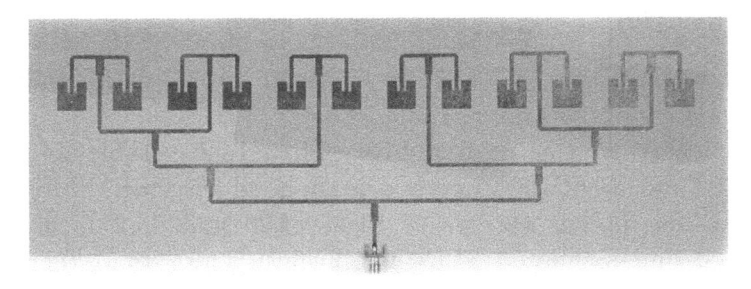

Fig. 14.21. Photograph of the 5.8-GHz 12-element microstrip array with the element levels [2 2 3 3 3 3]: one side realization on the 0.762 mm thick RF-35 layer. The microstrip RF-35 layer is 333 mm × 117 mm. The feed is with equal power split T-junctions. Microstrip patches are 13.74 mm × 13.80 mm (Ogurtsov and Koziel, 2018).

324 *Simulation-Based Optimization of Antenna Arrays*

(a)

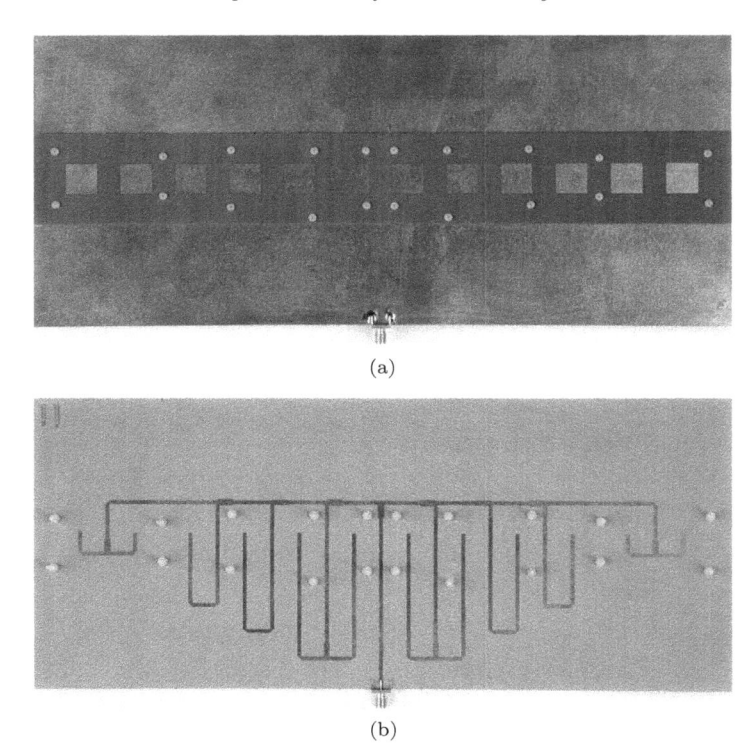

(b)

Fig. 14.22. Photographs of the 5.8-GHz 12-element microstrip array with the element levels [2 2 2 3 4 4]: (a) radiating aperture on two layers of TLP-5 (nominal thickness 1.574 mm in total). The TLP-5 layers are 329 mm × 45 mm. Microstrip patches are 14.85 mm × 14.85 mm; (b) feed on the 0.762 mm RF-35 layer with equal power split T-junctions. Radiating elements are energized through ground plane slots. The RF-35 layer and the ground plane is 329 mm × 145 mm (Ogurtsov and Koziel, 2018).

are shown in Figs. 14.23–14.25. Characteristics of the radiation patterns are summarized in Table 14.5.

According to the results, Figs. 14.23–14.25, Table 14.5, the measured and simulated responses are in good agreement. Moreover, the two-side designs (arrays of Figs. 14.20 and 14.22) show the measured and simulated SLLs significantly lower than their SLL estimated using the fast models. For instance, the design of Fig. 14.20 is with −22.9 dB of the measured SLL versus −16.5 dB estimated with the array factor and the element pattern. Similar improvement is obtained for the design of Fig. 14.22.

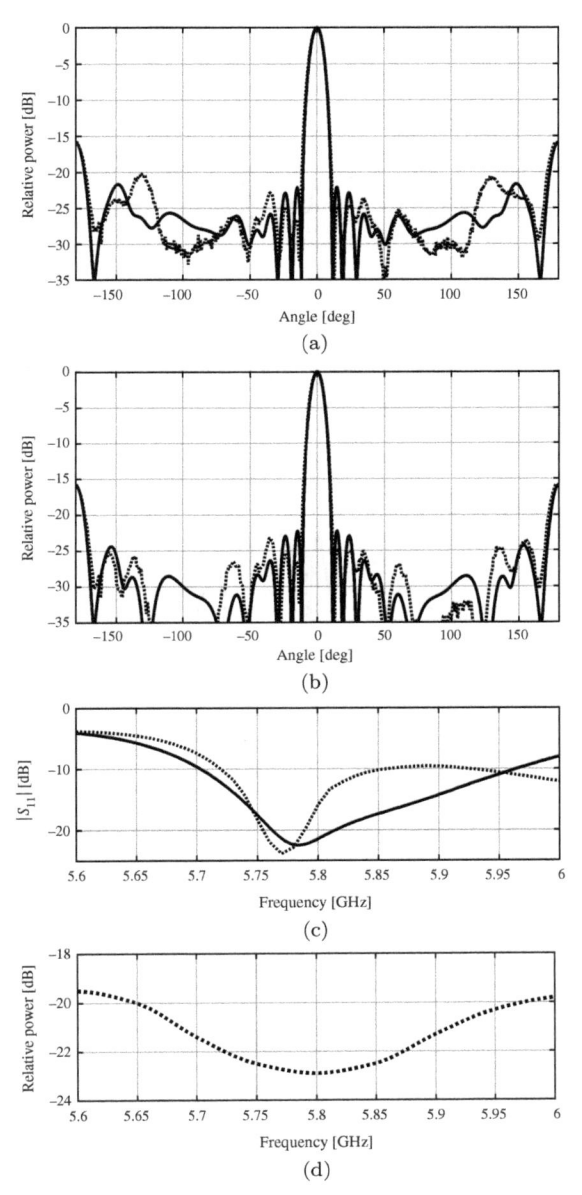

Fig. 14.23. Measured (···) and simulated (—) array of Fig. 14.20: H-plane (a) total (x-pol included), and (b) co-pol patterns at 5.8 GHz; (c) reflection coefficient; (d) measured SLL within the [−90, 90]-degree sector of the H-plane (Ogurtsov and Koziel, 2018).

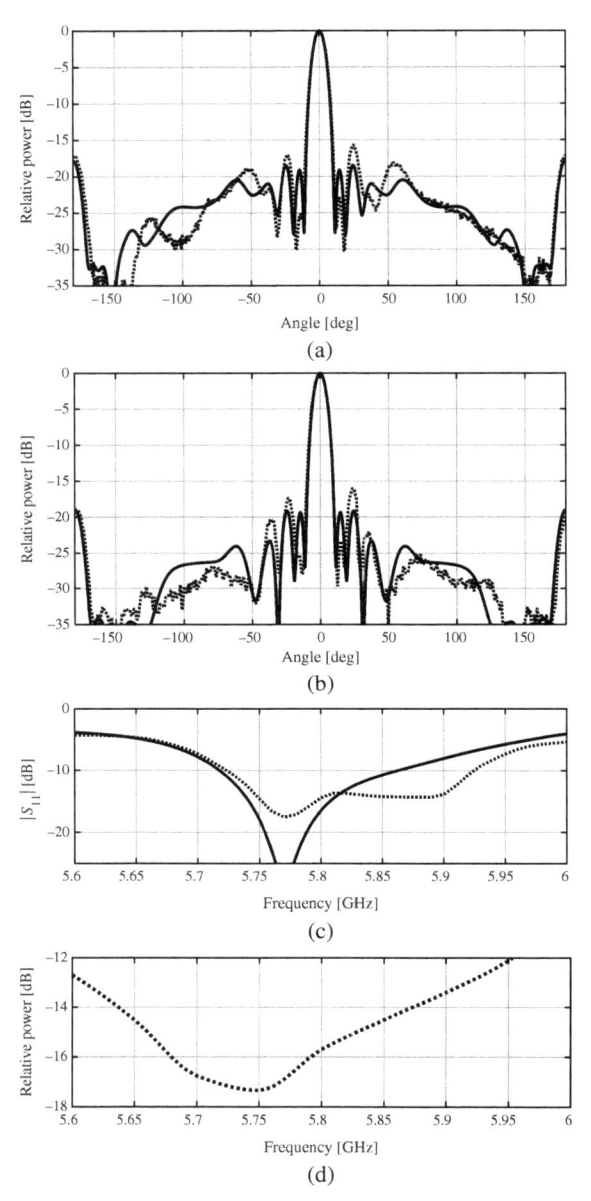

326 *Simulation-Based Optimization of Antenna Arrays*

Fig. 14.24. Measured ($\cdots$) and simulated (—) array of Fig. 14.21: H-plane (a) total (x-pol included), and (b) co-pol patterns at 5.8 GHz; (c) reflection coefficient; (d) measured SLL within the [−90, 90]-degree sector of the H-plane (Ogurtsov and Koziel, 2018).

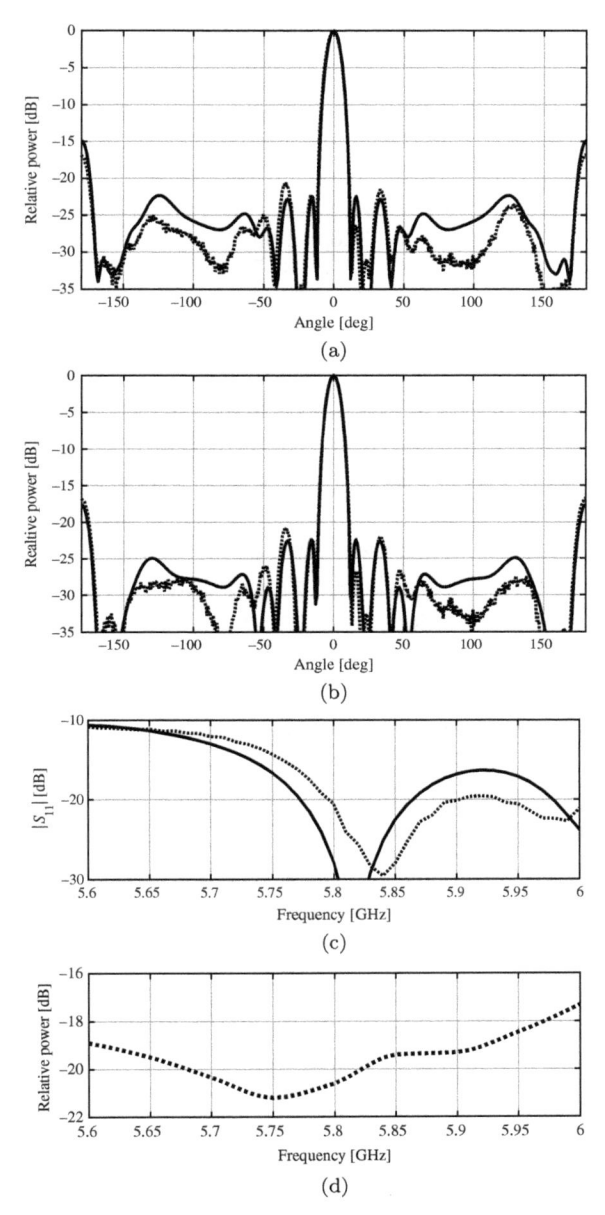

Fig. 14.25. Measured (···) and simulated (—) array of Fig. 14.22: H-plane (a) total (x-pol included), and (b) co-pol patterns at 5.8 GHz; (c) reflection coefficient; (d) measured SLL within the [−90, 90]-degree sector of the H-plane (Ogurtsov and Koziel, 2018).

Table 14.5. H-plane patterns at 5.8 GHz (Ogurtsov and Koziel, 2018).

Circuit of Figure	SLL [dB] Meas., Sim.	Co-pol SLL [dB] Meas., Sim.	Front-to-back ratio [dB] Meas., Sim.	HPBW [deg] Meas., Sim.
14.20	−22.9, −22.1	−23.2, −22.2	15.8, 15.8	−9.8, −9.4
14.21	−15.7, −18.5	−16.0, −18.6	17.0, 17.8	−9.6, −9.3
14.22	−20.6, −22.5	−22.7, −22.5	16.9, 14.9	−10.1, −9.8

A one-layer optimized design (array circuit of Fig. 14.21) shows about −16 dB of the measured SLL (Fig. 14.24, Table 14.5); that is close to the SLL value predicted with the fast model (i.e., before implementation). However, the radiation pattern and reflection coefficient of the array circuit have been significantly improved relative to those at the initial design, in −3 dB and −5 dB at 5.8 GHz, respectively. The higher SLL of this one-layer array, in comparison to the SLL of the two-side array sharing the same feed architecture, as well as the inaccurate prediction of the array factor model for this array are, in a significant part, due to the radiation from the feed in the forward direction (not accounted for by the array factor model). Thus, the simulated and measured responses justify the use of EM-based optimization in this case as well.

The discussed design approach including the presented type of microstrip corporate feeds can be a reliable option for low-sidelobe microstrip antenna arrays unless very low sidelobes are required.

14.2. Design of Low-Sidelobe Arrays Implementing Requited Excitation Tapers: The Case of Corporate Feeds Comprising Unequal-Split Junctions

Realization of low-sidelobe levels, i.e., lower than −20 dB, are challenging for printed-circuit arrays due to multiple affecting factors (Pozar and Kaufman, 1990; Hall and James, 1981). Such factors are difficult to identify upfront using simple means and then to account for in the process of modeling and design. The primary task in the design of low-sidelobe printed-circuit arrays is realization of an excitation of the array aperture elements resulting in the required sidelobe levels (SLLs). One should account not only for the signal distribution within the stand-alone feed but also for interactions of the feed and array elements, interactions of the array elements through the feed, emission from the feed, effects of the finite substrates, ground planes, connector, housing and radomes.

Microstrip corporate feeds with unequal power split junctions can realize non-uniform amplitude excitations required for low-sidelobe patterns. Series microstrip feeds of antenna arrays can be more economical in terms of the footprint and also realize low-sidelobe patterns to certain extent, e.g., as in Yin et al. (2017) and Afoakwa and Jung (2017). On the other hand, corporate feeds allow for a better control of amplitude and phase of aperture elements, thereby allowing designs and approaches developed for fixed-beam arrays to be extendable for phased, multi-beam, and shaped-beam arrays.

14.2.1. *Approach justification*

A reliable evaluation of low-SLL array patterns can only be obtained using discrete EM modelers'/solvers' projects. Such projects should be configured for the high-fidelity level of description of printed-circuits of different complexity, e.g., from a single microstrip power divider to the radiating aperture integrated with the feed. In the latter case, the circuit geometries are typically described by a large number of parameters. In conjunction with a high cost of individual simulations, adjustment of such array circuits using parametric optimization becomes a serious challenge. In particular, traditional simulation-based design using parameter sweeping is tedious, time-consuming, and hardly leads to optimal designs. Alternatively, utilization of standard optimization approaches and population-based metaheuristics (Wen *et al.*, 2016; Liu *et al.*, 2016; An *et al.*, 2017; Radivojević *et al.*, 2017; Salucci *et al.*, 2017; Liang *et al.*, 2017; You *et al.*, 2017; Yang *et al.*, 2017) turns to be of prohibitive numerical costs if it involves realistic array models accounting for EM interactions and factors affecting low-sidelobe patterns. On the other hand, we have demonstrated that the surrogate-based optimization (SBO) methodology can produce EM-based designs of array apertures, feeds, and integrated aperture-feed circuits at acceptable computational costs. However, here, the SBO methodology should be specifically elaborated for corporate feeds built of unequal-split junctions and integrated with apertures to address graded and integrated composition of circuits at hand and thus to ensure the required SLLs.

The surrogate-assisted process starts from the prototyping and element design, includes validation with EM simulation, and finally yields the apertures and feed layouts. Selected designs are validated experimentally.

14.2.2. Design process

Here, we consider linear arrays with an even number of radiators, as depicted in Fig. 14.1, producing the sum pattern. The arrays should be implemented with microstrip patch antennas. Design task includes realization of a required H-plane SLL and the matching at the input of the corporate feeds. The feeds will be implemented as with microstrip technology and configured with simple unequal-split T-junctions. For the sake of clarity, the results will be presented for 5.8 GHz 12-element linear arrays with half-wavelength spacing. The design process is outlined in the diagram of Fig. 14.26. The process can be described in general terms as follows.

Step 1 is a well-known routine of setting up the array aperture in major terms including the choice of the number of radiators, element spacings, and the excitation taper, a vector of excitation amplitudes. Radiation pattern properties, the SLL, first null beamwidth (FNBW), half-power beamwidth (HPBW), front-to back ratio (FBR), etc., of the aperture at this step are estimated using the array factor (AF).

At Step 2, the feed tree, providing the closest approximation of the targeted excitation taper, a_{target}, is searched using numerical optimization of the fast approximate model. The fast approximate model describes feeds in terms of the tree architecture and power splits (PSs) of the junctions. T-junctions' PSs are constrained to

Simulation-Based Design of Corporate Feeds 331

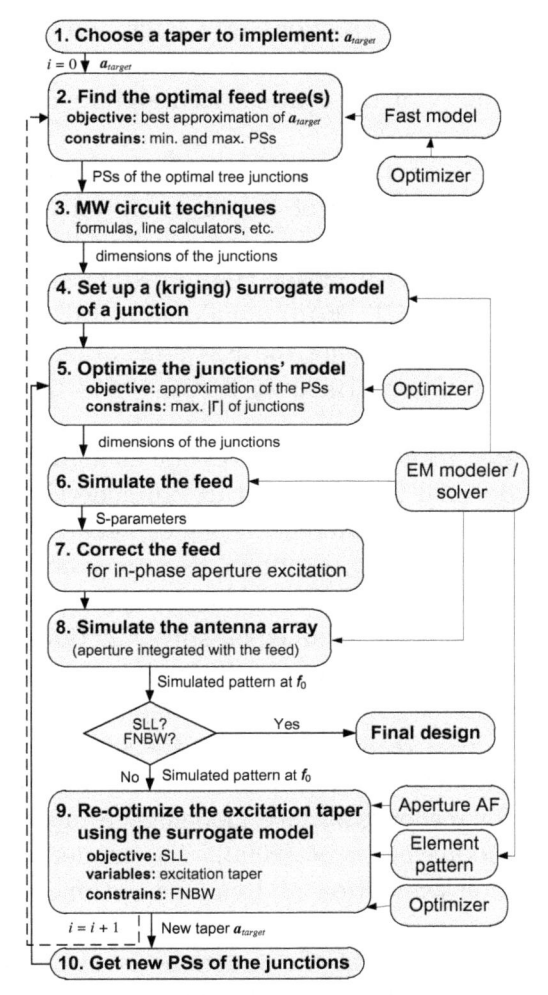

Fig. 14.26. Surrogate-assisted design of corporate feeds: a process diagram. PS stands for split ratio of a junction, $|\Gamma|$ denotes the reflection coefficients at the junctions' inputs, **a** stands for the excitation taper, f_0 is the pattern evaluation frequency. At Steps 8 and 9 the aperture (integrated with the feed) and the aperture element are assumed already designed. The fast model (used at Step 2) includes the aperture array factor (AF) which is explicitly depicted for Step 9. Iterations (Steps 5 through 10) were performed with the same feed architecture found at Step 2. A more general realization of the design process can search and implement a new optimal architecture after Step 9, i.e., at iterations of Steps 2–9. Two kinds of surrogate models are optimized, one at Steps 5 and another at Step 9.

avoid very thin microstrip sections in a later implementation. The outcome of this step is the optimal feed architecture and PSs of the junctions. The quality of the optimal feed architecture is evaluated using the array factor.

At Step 3, PSs of the optimal feed architecture, which are determined for the frequency of the array pattern evaluation, are cast into dimensions of junctions for a chosen circuit technology (e.g., microstrip) using analytical means (Pozar, 2011) and transmission-line calculators, e.g., (CST, 2016). The obtained dimensions should be subsequently adjusted using the EM simulations and numerical optimization. This is performed at Steps 4 and 5 with the SBO methodology.

Step 4 sets up a model for every junction configuration needed to implement the optimal feed architecture. In general, few junctions' configurations, which are different in angles between the junction's legs, might be needed, e.g., to obtain feeds of compact footprints. On the other hand, it is possible to configure the optimal feeds using the only one configuration of unequal-split junctions. In this case, only one model needs to be set up. Such a model is set up about the dimensions obtained at Step 3 using high-fidelity discrete simulations. The modeled responses include the simulated power splits and the reflection coefficient at the input, both over the frequency. We use kriging for response surface modeling (Simpson *et al.*, 2001; Kleijnen, 2009; Lophaven *et al.*, 2002). The model is subsequently used for T-junction evaluation and can be optimized at negligible computational cost.

At Step 5, the model of the junctions is optimized for every junction for the required PSs within the frequency band centered on the frequency at which the radiation pattern is evaluated. The maximum reflection coefficient at the junction's input is constrained, e.g., to -25 dB, over the frequency band. Because of the smoothness of the model gradient-based optimization is utilized (Matlab, 2010).

At Step 6, the model of the optimal feed is defined as a microstrip component. The configuration of the microstrip traces is set so that the feed can be integrated with the array aperture. The model is defined so that all the physical lengths from the common input to the output ports (input of aperture elements) are initially the same. The model of the feed is then simulated for complex transmission S-parameters with only the common input being active.

At Step 7, the phases of the simulated transmission S-parameters are used to correct the lengths of the microstrip traces to make excitation of the aperture elements all in phase. As a result of previously performed Step 5, the amplitudes of the simulated transmission S-parameters (the excitation amplitudes) are close to the required for the feed outputs terminated to the matched ports.

At Step 8, the EM model of the integrated antenna array consisting of the corrected feed and the aperture is simulated for the radiation pattern. The aperture should be already designed as a microstrip circuit so that (a) its radiation pattern, with the excitation taper applied directly to the microstrip inputs of the antennas, satisfies the design requirements in terms of the SLL and FNBW, and (b) active reflection coefficients at the antenna inputs are low, typically lower than -20 dB. Such low values of active reflection coefficient are needed to alleviate the aperture-feed interactions and, thus, minimize a deviation of the excitation taper from the required. Design of the aperture, including adjustment of the antenna dimensions, is performed using the SBO techniques similar to those discussed in Chapter 10.

If the simulated radiation pattern does not satisfy the requirements, in particular, exhibits increased sidelobes, shows inacceptable broadening of the major lobe, etc., the excitation taper is re-designed using optimization of the array aperture surrogate model. This process is depicted as Step 9 in Fig. 14.26; it is described in details in Sec. 14.2.2.2.

There are additional reasons (to fixing an unacceptable pattern) of the re-design Step 9. First, the sidelobe pattern is typically rolled down with angle due to the element pattern; thus, the SLL could be potentially made lower by making the sidelobe pattern more equally rippled. Second, even in case of an acceptable SLL, it is worth to make the SLL as low as possible (without increasing FNBW) from the implementation standpoint to secure a certain safety margin for the SLL.

The outcome of Step 9 is a new excitation taper. This new taper is cast to the new PSs of the feed junctions at Step 10. This conversion is performed using the fast model, already being utilized at Step 2. If one stays with the same architecture, the process is iterated starting from Step 5. At the same time, a more general realization of the process can be developed. Namely, the iterations would continue from Sep 2, i.e., search for a new feed architecture which is optimal for the new taper.

Steps 4, 5, and 9 of the design process are detailed below.

14.2.2.1. *Modeling and optimization of unequal-split junctions*

The primary issue concerning feed design for low-sidelobes is the necessity of repeated optimization of the T-junctions. In order to speed up the process, a kriging surrogate model of the T-junction is constructed and utilized for the feed optimization. Let $\mathbf{x}_T = [x_{T.1} \ldots x_{T.n_T}]^T$ denote a vector of feed dimensions. We also denote by $l_{T.j}$ and $u_{T.j}$, $j = 1, \ldots, n_T$, the lower and upper bounds for the dimension, respectively. The kriging model of S-parameters of the T-junction, $\mathbf{s}_T(x_T)$ is established using a set of training points $x_T^{(k)}$ ($k = 1, \ldots, N$) which are allocated within the interval determined by the aforementioned bounds. Here, a rectangular grid is utilized as a design of experiments technique.

Kriging interpolation is summarized below for the reader's convenience. Let $\mathbf{R}_T(\mathbf{x}_T^{(k)})$ be the simulated T-junction responses at the training designs. We use the notation $\mathbf{R}_T(\mathbf{x}) = [R_{T.1}(\mathbf{x}) \ldots R_{T.m}(\mathbf{x})]^T$ where the response vector components correspond to simulated S-parameters evaluated at m frequency points of interest. For simplicity, scalar notation $R_{T.j}(\mathbf{x})$ is utilized; however, in practice, real and imaginary parts of all relevant S-parameters are modeled independently. Ordinary kriging (Simpson et al., 2001) estimates deterministic function f as $f_p(\mathbf{x}) = \mu + \varepsilon(\mathbf{x})$, where μ is the mean of the response at base points, and ε is the error with zero expected value. The correlation structure is a function of a generalized distance between the base points. We use a Gaussian correlation function of the form

$$R(\mathbf{x}_T^{(i)}, \mathbf{x}_T^{(j)}) = \exp\left[\sum_{k=1}^{N} \theta_k |x_{T.k}^{(i)} - x_{T.k}^{(j)}|^2\right]. \quad (14.5)$$

Here, θ_k are correlation parameters (or hyperparameters) that are found during model identification; while $x_{T.k}^{(j)}$ are components of the base point $\mathbf{x}_T^{(j)}$.

The kriging-based coarse model $\mathbf{s}_T$ is defined as

$$\mathbf{s}_T(\mathbf{x}) = [s_{T.1}(\mathbf{x}) \ldots s_{T.m}(\mathbf{x})]^T, \quad (14.6)$$

where

$$s_{T.j}(\mathbf{x}) = \mu_j + \mathbf{r}^T(\mathbf{x})\mathbf{R}^{-1}(\mathbf{f}_j - \mathbf{1}\mu_j). \quad (14.7)$$

Here, $\mathbf{1}$ denotes an N-vector of ones,

$$\mathbf{f}_j = [R_{T.j}(\mathbf{x}_T^{(1)}) \ldots R_{T.j}(\mathbf{x}_T^{(N)})]^T, \quad (14.8)$$

$\mathbf{r}$ is the correlation vector between the point $\mathbf{x}$ and base points

$$\mathbf{r}^T(\mathbf{x}) = [R(\mathbf{x}, \mathbf{x}_T^{(1)}) \ldots R(\mathbf{x}, \mathbf{x}_T^{(N)})]^T \quad (14.9)$$

whereas $\mathbf{R} = [R(\mathbf{x}_T^{(i)}, \mathbf{x}_T^{(j)})]_{i,j=1,\ldots,N}$, is the correlation matrix between the base points. The mean $\mu_j = (\mathbf{1}^T\mathbf{R}^{-1}\mathbf{1})^{-1}\mathbf{1}^T\mathbf{R}^{-1}\mathbf{f}_j$. Correlation parameters θ_k are found by maximizing

$$-[N\ln(\sigma_j^2) + \ln|\mathbf{R}|]/2 \quad (14.10)$$

in which the variance $\sigma_j^2 = (\mathbf{f}_j - \mathbf{1}\mu_j)^T\mathbf{R}^{-1}(\mathbf{f}_j - \mathbf{1}\mu_j)/N$ and $|\mathbf{R}|$ are both functions of θ_k. The kriging model is implemented using the DACE Toolbox (Lophaven et al., 2002).

As in practice, the T-junction is described by a small number of parameters (here, four), a very accurate kriging surrogate can be established and directly utilized for optimization purposes with no further correction necessary. Given the required power split

p_k, $k = 1, \ldots, n$, where n is the number of junctions, the optimization process is formulated as

$$\mathbf{x}_T^{*(k)} = \arg\min_{\mathbf{x}_T} |P(s_T(\mathbf{x}_T)) - p(k)|, \tag{14.11}$$

where $P(s_T(\mathbf{x}_T))$ is the power split of the T-junction calculated from its kriging model for the vector of geometry parameters $\mathbf{x}_T$. The problem is constrained by imposing the conditions $l_{T.j} \leq x_{T.j} \leq u_{T.j}$, $j = 1, \ldots, n_T$. A standard gradient-based optimization is utilized for solving (14.11), *fmincon* (Matlab, 2010), because the kriging model is fast and smooth.

14.2.2.2. *Feed redesign for sidelobe minimization*

The initially designed feed needs to be re-designed to account for SLL degradation. Here, a customized surrogate-assisted procedure is developed and executed to realize the re-optimization process at a low computational cost. The procedure consists of several steps (here, **a** denotes the excitation taper, a vector of excitation amplitudes):

1. Simulate the pattern of the entire array (feed + aperture) denoted as $\mathbf{R}_f(\mathbf{a}^{(0)})$, where $\mathbf{a}^{(0)}$ stands for the initial excitation taper for which the feed has been designed;
2. Correct the analytical array factor (AF) model $\mathbf{R}_{AF}$ to align it with $\mathbf{R}_f(\mathbf{a}^{(0)})$;
3. Obtain updated excitation taper $\mathbf{a}^{\text{new}}$ by optimizing the corrected AF model;
4. Find power splits of T-junctions that realize the taper $\mathbf{a}^{\text{new}}$;
5. Re-optimize the T-junctions to realize power splits found at Step 4.

The critical step in the above procedure is correction of the AF model as well as its optimization. It should be noted that in the context of SLL minimization as well as controlling the major lobe width, namely the FNBW, one does not need to consider the entire pattern but only certain number of characteristic points, including the angle and level location of the first null as well as local sidelobe maxima as shown in Fig. 14.27(a). Figure 14.27(b) illustrates good

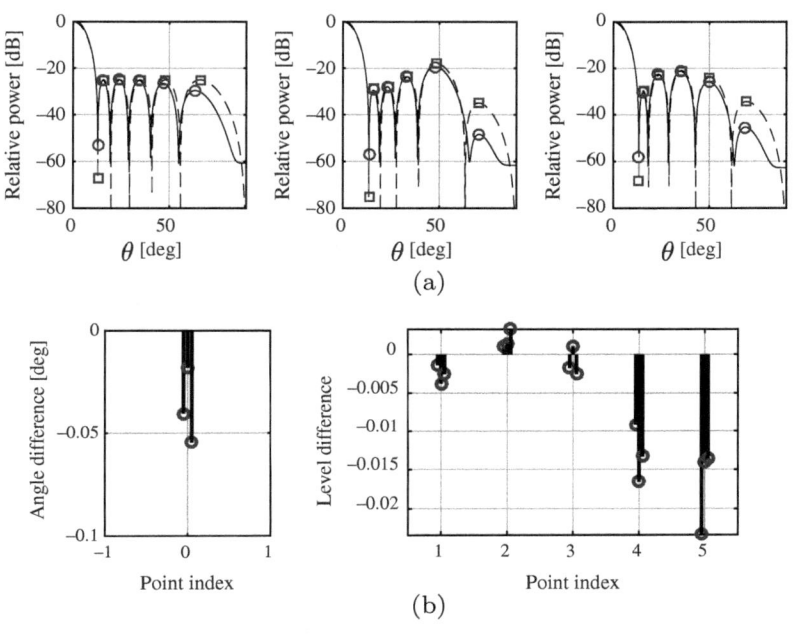

Fig. 14.27. Feature-based design concept illustrated for a 12-element microstrip linear array: (a) patterns and feature points for various excitation tapers; (b) differences of angle and level components of the feature points (first null angle on the left; local maxima levels in the linear scale on the right) showing good correlation of the models despite considerable absolute discrepancies.

correlation between the array factor and EM-simulated pattern of the integrated array circuit. That allows utilizing the AF model. The AF model is corrected based on the feature points.

Let $F(\mathbf{R}_f(\mathbf{a}))$ and $F(\mathbf{R}_{AF}(\mathbf{x}))$ denote the vectors of feature points of the EM-simulated and AF model patterns of the array, where F is a function extracting the feature points from the model response. We have $F = [p_0 \; p_1 \; \ldots \; p_K \; l_0 \; l_1 \; \ldots \; l_K]^T$, where p_j and l_j are angle and level components of the respective points ($j = 0$ corresponds to the first null, $j > 0$ are for pattern maxima). The design problem can be formulated as

$$\mathbf{a}^* = \arg\min_{\mathbf{a}} U(F(\mathbf{R}_f(\mathbf{a}))) \qquad (14.12)$$

where

$$U(F(\cdot)) = \max_{j=1,\ldots,K} l_j + \beta \left[\max(p - p_{\max}, 0)/p_{\max}\right]^2. \quad (14.13)$$

The first term in (14.13) is the SLL. The second term (a relative violation of the maximum major lobe beamwidth $p_{\max}$) is added to the primary objective upon violation of the condition $p \leq p_{\max}$.

Given good correlation between the AF and EM-simulated models, solving (14.12) is replaced by solving

$$\mathbf{a}^* = \arg\min_{\mathbf{a}} U(F(\mathbf{R}_{\mathrm{AF}}(\mathbf{a})) + \Delta(\mathbf{a}^{(0)})), \quad (14.14)$$

with

$$\Delta(\mathbf{a}^{(0)}) = F(\mathbf{R}_f(\mathbf{a}^{(0)})) - F(\mathbf{R}_{\mathrm{AF}}(\mathbf{a}^{(0)})). \quad (14.15)$$

The correction term (14.15) "translates" the feature points of the AF model into the points of the EM-simulated model of the entire array (aperture + feed). Note that the numerical cost of solving (14.14) is negligible. The procedure can be iterated if necessary. Each iteration requires only one simulation of the entire array.

14.2.3. *Realization of the design process: Numerical results and measurements*

Numerical results and experimental verification are presented here for 5.8 GHz twelve-element microstrip broadside linear array apertures with half-wavelength spacing. Design examples obtained using two versions of the process outlined in the diagram of Fig. 14.26 are described. One example is referred as a taper-oriented design. Another one is referred as an SLL-oriented design. Similarities and differences of the examples are detailed in the process of description. In all examples we adopt and implement an unequal-split T-junction, depicted in Fig. 14.28, within the feed half-tree. The central T-junction, connecting the feed half-trees, has the output legs inline.

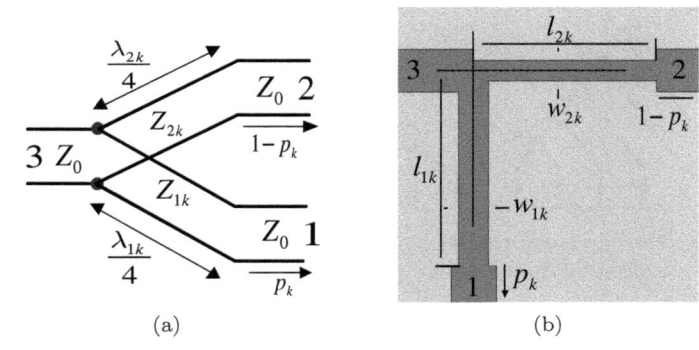

Fig. 14.28. T-junction: (a) schematic; (b) microstrip realization used in the half of the feed tree. Outputs are numbered with 1 and 2. Input is depicted with 3. Fraction of the total output power directed to output 1 of junction k is denoted with p_k.

14.2.3.1. *Example of a taper-oriented design*

First, we searched for feed architectures realizing Chebyshev excitations for 12-element arrays with the -30 dB and -35 SLLs (Stegen, 1953). The lower and upper bounds for the power split (PS) are set to 0.33 and 0.67, respectively, to avoid too narrow microstrip sections in the subsequent implementation of T-junctions. Optimization of the fast feed models has been implemented in Matlab (2010).

Two architectures, most accurately approximating the Chebyshev excitation tapers under the above listed PS constrains, are shown in Fig. 14.29. At the next step, the junctions of the best feeds are implemented in CST MWS (CST, 2016) as microstrip 3-ports (Fig. 14.28(b)) residing on the 0.762-mm RF-35 laminate (RF-35, 2018). The line impedance of inputs and outputs is 50 ohms. The junctions were tuned for the required power splits and matching at the input (denoted with 3 in Fig. 14.28(b)). Numerical optimization of EM models of the junctions was applied for tuning. Because of a low-dimensionality of the junction problem, gradient-based optimization embedded in the trust-region framework was utilized (Conn *et al.*, 2000). The primary objective was minimization of the reflection coefficient at the junction's input at 5.8 GHz. Power splits

340 *Simulation-Based Optimization of Antenna Arrays*

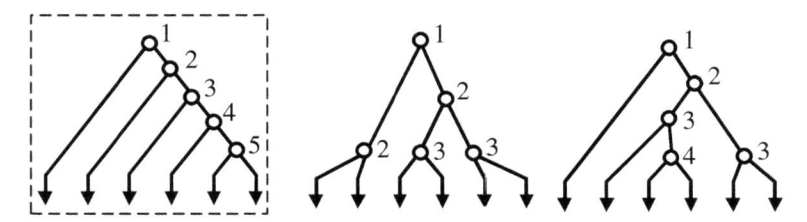

Fig. 14.29. Feed architectures without crossovers: the best three for the −35 dB Chebyshev excitation taper under power splits constrained within 0.33–0.67. The tree marked with the dashed-line box is the best also for the −30 dB SLL Chebyshev excitation under the same PS constrains. Halves of trees (down from the central equal-split junction) are shown due to the symmetry (Koziel and Ogurtsov, 2017; Ogurtsov and Koziel, 2018).

were controlled using the penalty function approach (Nocedal and Wright, 2006).

The corporate feeds were implemented as microstrip circuits comprising individually optimized T-junctions. Based on the simulated transmission parameters of the feeds the geometrical lengths from the common input to the ground plane slots of all antennas were corrected to have the transmission parameters all in phase. The uncorrected feed designed for −30 dB SLL had the 12-degree spread of phase shifts at the outputs. The uncorrected feed designed for −35 dB SLL had a similar spread of the phase shifts at the outputs, namely, 13 degrees. Subsequently, the feeds are included in the array circuits comprising the feed, the array aperture, and an SMA connector (SMA connector, 2012). The same aperture is used for the −30 dB SLL and −35 dB SLL cases. All planar components are with 18 μm copper metallization.

The aperture is on a 1.574-mm TLP-5 substrate (Taconic TLP, 2013). It consists of linearly polarized 5.8 GHz microstrip patch antennas operating at the dominant TM_{010} mode. Antenna inputs continue the feed. Antenna dimensions are parameterized as shown in Fig. 14.8. Lengths of the open-end microstrip sections, s_n, are element-specific. The line impedance of the inputs is 50 ohms. Before being connected to a particular feed in the EM model, the apertures had been tuned for 5.8 GHz using simulation-based optimization

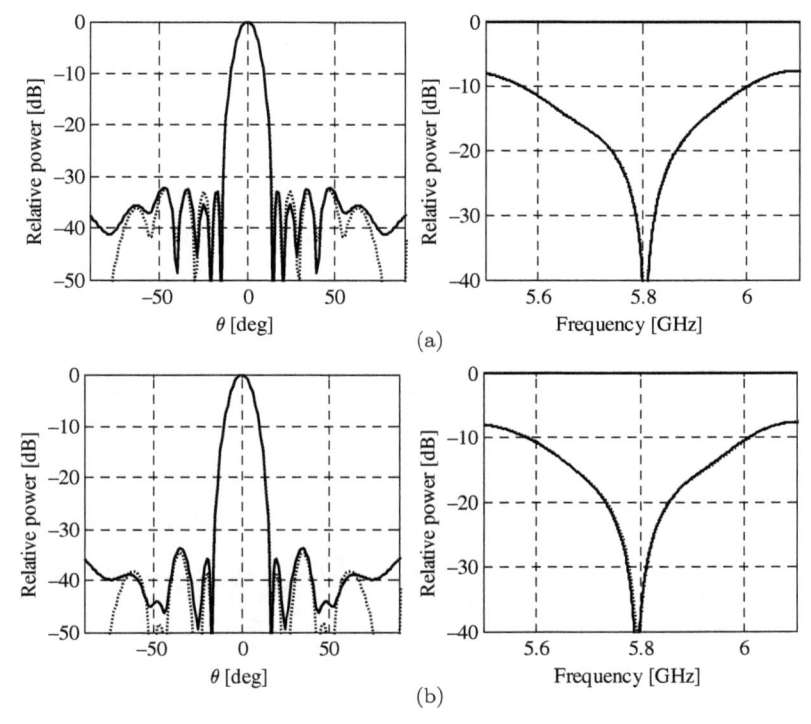

Fig. 14.30. Simulated array circuits with (—) finite and (···) infinite ground plane. H-plane power patterns (total fields) at 5.8 GHz (left) and reflection coefficient responses (right): (a) −30 dB SLL design; (b) −35 dB SLL design (Koziel and Ogurtsov, 2017).

(cf. Chapter 10). The entire circuits are validated with high-fidelity EM simulations.

Simulated responses of the designs having the linear aperture at the center of the ground plane and rather large foot-prints of the corporate feeds (due to extended wrapped microstrip sections feeding the outermost aperture elements) are shown in Fig. 14.30. The patterns of Fig. 14.30 illustrate the effects of the finite-ground plane and emission from the feed on low sidelobes, in particular for the −35 dB SLL case.

It is worth to note that the realized design process is in fact a version preceding to development of the process outlined in the

342 Simulation-Based Optimization of Antenna Arrays

Fig. 14.26 diagram. Namely, optimization of the T-junctions has been carried out using coarsely discretized models and the corresponding response-corrected surrogates of the T-junction, i.e., without setting up and using the kriging surrogates. In addition, feed redesign, Steps 9 and 10 of the Fig. 14.26 diagram, is not performed as described in Sec. 14.2.2.2, rather the low SLLs and matching were achieved with increasing the length of the outermost wrapped microstrip sections and enlarging the circuit foot-print. A more compact layout of the −30 dB SLL design was created from the EM-tuned model

(a)

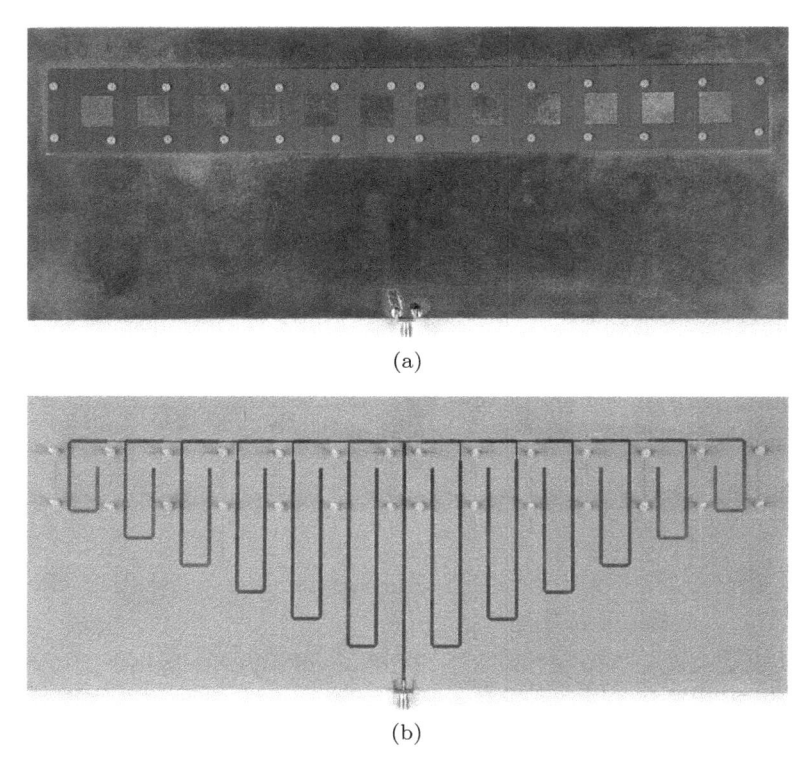

(b)

Fig. 14.31. Photographs of the 5.8 GHz 12-element array implementing the −30 dB SLL Chebyshev taper: (a) radiating aperture on two layers of TLP-5 (nominal thickness 1.574 mm in total). The TLP-5 layers are 330 mm × 40 mm. Microstrip patches are 14.85 mm × 14.85 mm; (b) realization of the feed architecture on the 0.762 mm RF-35 layer with unequal power split T-junctions. The RF-35 layer and the ground plane is 350 mm × 140 mm (Ogurtsov and Koziel, 2018).

(CST, 2016). The prototype (photographs shown Fig. 14.31) was measured in the Reykjavik University anechoic chamber.

Measured and simulated responses of the prototype are presented in Fig. 14.32. Radiation characteristics of the prototype are summarized in Table 14.6. Figure 14.23 shows that the increased SLL and back-lobes of the measured H-plane total power pattern are due to the cross-pol radiation which is mostly emitted by the feed according to our studies. According to the experiments, the circuit assembly inaccuracy and mechanical deformation, such as slight bending of the aperture, ground plane, and the feed, may significantly contribute to broadening of the measured major lobe for relative power below -20 dB. The above issues of the radiation pattern characteristics will be addressed using EM optimization of the array circuit as described in the next section.

14.2.3.2. *Example of an SLL-oriented design*

Results of the design process are described below by steps according to the description of Sec. 14.2.2 and the diagram flow of Fig. 14.26.

At Step 1, a 12-element -25 dB SLL Chebyshev excitation taper $\mathbf{a}_{\text{Target}} = [1.0000\ 0.9307\ 0.8031\ 0.6372\ 0.4572\ 0.4225]^T$ (Balanis, 2005) was adopted for implementation so that the array should produce the -25 dB SLL and have a 26-degree FNBW at 5.8 GHz. The array should also be matched at the input at 5.8 GHz.

At Step 2, the optimal feed architecture, shown in Fig. 14.33, providing the closest approximation to the array factor pattern with the excitation taper $\mathbf{a}_{\text{Target}}$, was found with the corresponding excitation taper of $\mathbf{a}^{(0)} = [1.0000\ 0.9308\ 0.8032\ 0.6372\ 0.4572\ 0.4225]^T$. In this case the search was performed with power splits (PSs) constrained to be within 0.4–0.6. The PSs of the optimal tree are found to be $\mathbf{p}^{(0)} = [0.5647\ 0.4484\ 0.5116\ 0.5358\ 0.5394]^T$. The meaning of the PS vector $\mathbf{p}$ is depicted in Fig. 14.28, i.e., p_k stands for the power going to the left output leg (numbered as 1, the closest to the center of the tree) of the kth junction.

At Step 3, the PSs are cast into the line impedances of the quarter wave transformer sections of the T-junctions, which are schematically

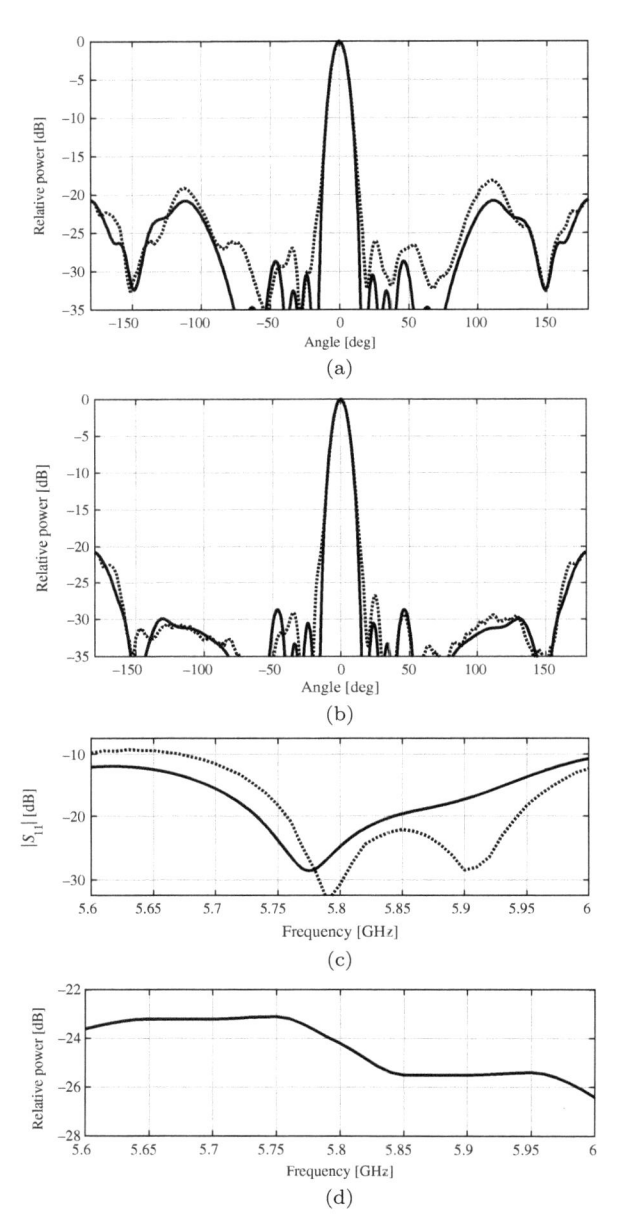

Fig. 14.32. Array of Fig. 14.31: measured (···) and simulated (—) H-plane (a) total (x-pol included) pattern and (b) co-pol pattern, both at 5.8 GHz; (c) reflection coefficient; (d) measured H-plane SLL within the [−90, 90]-degree sector (Ogurtsov and Koziel, 2018).

Table 14.6. H-plane pattern at 5.8 GHz (Ogurtsov and Koziel, 2018).

Circuit of Figure	SLL [dB] Meas., Sim.	Co-pol SLL [dB] Meas., Sim.	Front-to-back ratio [dB] Meas., Sim.	HPBW [deg] Meas., Sim.
14.31	−24.2, −26.8; −26.0[a], −28.7[a]	−26.8, −28.7	20.6, 20.7 18.2[b], 20.7[b]	−10.7, −10.8

[a]Measured and simulated SLL in $[-75°, 85°]$.
[b]Measured and highest back-lobes.

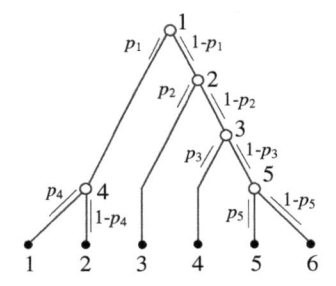

Fig. 14.33. A half of the optimal feed architecture, a tree with constrained PSs for the array factor pattern satisfying the design requirements. Junctions (○) and outputs going to the array aperture elements (•). The array symmetry plane is on the left.

shown in Fig. 14.28(a), using

$$Z_{1k} = Z_0/\sqrt{p_k}, \quad (14.16)$$

$$Z_{2k} = Z_0/\sqrt{1-p_k}, \quad (14.17)$$

where Z_{1k} and Z_{2k} stand for the line impedances of the transformer sections of a particular matched T-junction k required to provide a PS of p_k, a fraction of the total output power going to the output leg 1; Z_0 is the system impedance, a line impedance of the connecting microstrip traces, 50 ohms in this work; λ_{1k} and λ_{2k} are the wavelengths in the transformer sections. The line impedances of the transformer sections are within the 65–79 ohms range for the optimal feed architecture due to the constrained PSs.

The tree of Fig. 14.33 can be realized using solely one microstrip configuration which is shown in Fig. 14.28(b). The T-junctions were defined on a 0.762-mm Arlon AD 250 substrate (AD250C, 2018) with 17.5 µm metallization.

The EM model of the T-junctions was configured in the CST MWS environment (CST, 2016). The model is described using four design variables, which are dimensions of the transformer sections. At the frequency of interest, 5.8 GHz, these dimensions need to be adjusted for a particular PS per junction, p_k, and a low reflection coefficient at the input (marked with 3 in Fig. 14.28(b)).

For 5.8 GHz and with $0.4 \leq p_k \leq 0.6$, the MWS built-in transmission line calculator (CST, 2016) estimates the widths of the transformer section, w_{1k} and w_{2k}, be within 1.00–1.43 mm and the quarter wavelengths, l_{1k} and l_{2k}, be within 9.0–9.2 mm for the used substrate.

At Step 4, a kriging model was set up for frequencies from 5.5 GHz to 6.1 GHz, for 0.9 mm $\leq w_{1k}$, $w_{2k} \leq 1.5$ mm, and 8.0 mm $\leq l_{1k}$, $l_{2k} \leq 10.0$ mm using 1024 samples of high-fidelity simulations of the EM model driven at the input (port 3 in Fig. 14.28(b)). For the modeled ranges of frequencies and dimensions the error of the kriging model is below ±0.1 dB compared to the simulated S-parameters.

Simulated PSs at the dimensions calculated using Eqs. (14.16)–(14.17) and transmission line calculations noticeably deviate from the required PSs. For instance, for junction 1 with the required PS of $p_1 = 0.5647$ at $l_{11} = 9.098$ mm, $l_{21} = 9.167$ mm, $w_{11} = 1.374$ mm, and $w_{21} = 1.036$ mm, the simulated PS at the same dimensions was $p_1 = 0.5331$ at 5.8 GHz.

Therefore, at Step 5, the surrogate model was optimized for every junction for a particular required PS, listed at Step 2, over the frequencies from 5.75 GHz to 5.85 GHz. The maximum reflection coefficient at the junction's input was constrained to −25 dB over the frequencies from 5.70 GHz to 5.90 GHz. PSs have been adjusted at 5.8 GHz to the required values with optimization of the kriging surrogate. The optimized variables have been noticeably tuned from the initial values. For example, for junction 1 the new dimensions of the transformer sections were obtained as $l_{11} = 10.000$ mm (versus initial 9.098 mm), $l_{21} = 8.892$ mm (versus initial 9.167 mm), $w_{11} = 1.426$ mm (versus initial 1.374 mm), and $w_{21} = 1.068$ mm (versus initial 1.036 mm).

At Step 6, a microstrip feed of the optimal architecture (Fig. 14.33) has been defined with the optimized junctions' dimensions. The feed is defined as a part of the integrated array circuit shown in Fig. 14.34.

A straightforward adjustment of the excitation taper for the integrated structure using simulation-driven optimization of the feed dimensions as the simultaneous design variables, e.g., over the dimensions of the T-junctions (even with already designed and well matched aperture), represents a serious numerical challenge.

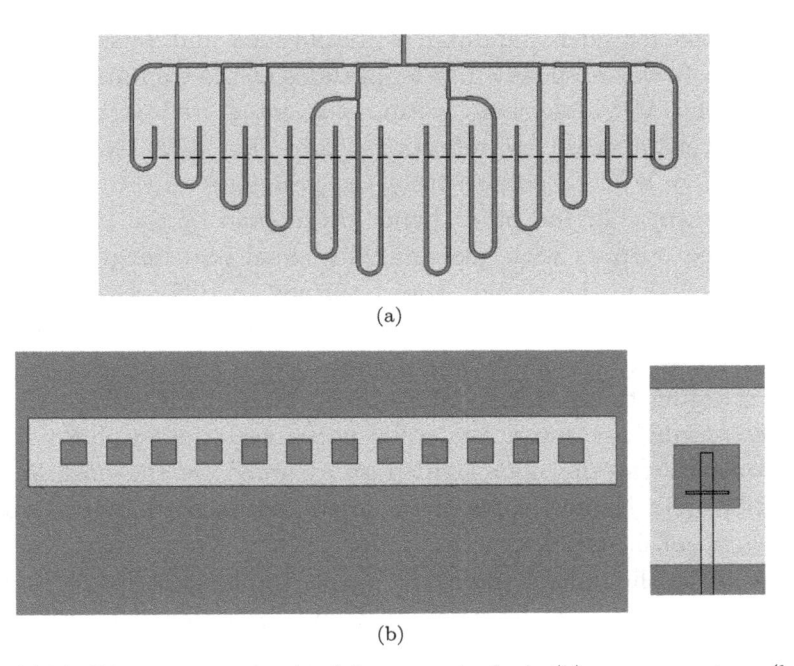

Fig. 14.34. Linear array circuit: (a) corporate feed; (b) array aperture (left) of microstrip pach antennas (MPAs) where every MPA (right) is fed with a microstrip through a ground plane slot, both denoted using the dot lines. The feed and MPAs share the same ground plane. The horizontal dash-line in (a) points to positions of the feed's output ports in the EM model at Step 6 as well as to positions of the MPA input ports in the EM model at the aperture optimization step.

This challenge is due to the fine power distribution the feed should realize since ensuring an accurate power distribution requires utilization of an accurate, and therefore, expensive discrete EM model of the integrated structure.

It worth to notice that the high-fidelity EM model should be simulated many times with a straightforward optimization approach, e.g., at least the number of design variables which is at least 20 for the feed of interest (five junctions with four dimensions each) not to mention the central equal-split junction and lengths of the wrapped microstrip sections. Consequently, to minimize the computational costs of the design yet to carry it out at the high-fidelity level of description, amplitudes and phases produced by the feed are adjusted in separate steps in this work.

Reasons for such handling of amplitudes and phases are the following. It was observed that amplitudes of the excitation signals, driving the MPA elements within the integrated structure, are destructively affected by reflections from the MPA elements and multiple low level reflections from the junctions on the way to the MPA element. On the other hand, the spread of the phase shifts at the feed outputs with geometrically equal path lengths are due to the fact that the signals travel through a different number of transformer sections and microstrip bends with different effective permittivities.

According to simulation of the array Step 6 the output signals arrived through geometrically equal paths were not in phase at the feed outputs (at the microstrip ports denoted by the dash-line in Fig. 14.34(a)). Namely, a spread of simulated phase shifts was about 14 degrees at 5.8 GHz.

To make the output signals in phase, the lengths of the 50-ohm microstrip traces were adjusted at Step 7 based on one performed simulation of the uncorrected feed. In this simulation the EM model included the MPA aperture which was inactive by absorbing the signals at the feed output ports. Correction of the microstrip traces were done using

$$\Delta l_{nm} = \frac{\theta_n - \theta_m}{\beta(f_0)} \qquad (14.18)$$

where n is for any one of five (out of six) traces to be corrected, m is for one trace, the reference, the length of which is not changed; θ_n and θ_m are the simulated phase shifts at the feed outputs (microstrip ports); $\beta(f_0)$ is the simulated propagation constant of the 50 ohms microstrip lines at the frequency of interest f_0, here 5.8 GHz; and Δl_{nm} is the increment (if calculated as a positive value) or decrement (if calculated as a negative value) of the 50-ohm microstrip section in the branch n (going to the nth MPA) (Fig. 14.35). Formula (14.18) is approximate; nevertheless, it reduces the spread of the simulated phase shifts, in this particular case, from 14 degrees to 3 degrees. The effect of the correction using (14.18) on the radiation pattern of the integrated structure with the outermost branch (here $m = 6$) chosen to be unadjusted is shown in Figs. 14.36 and 14.37. Other five branches were adjusted in the bended sections, that is, below the horizontal dash-line in Fig. 14.34(a).

Excitation amplitudes within the integrated structure are handled, first, through optimization of the radiation aperture, and,

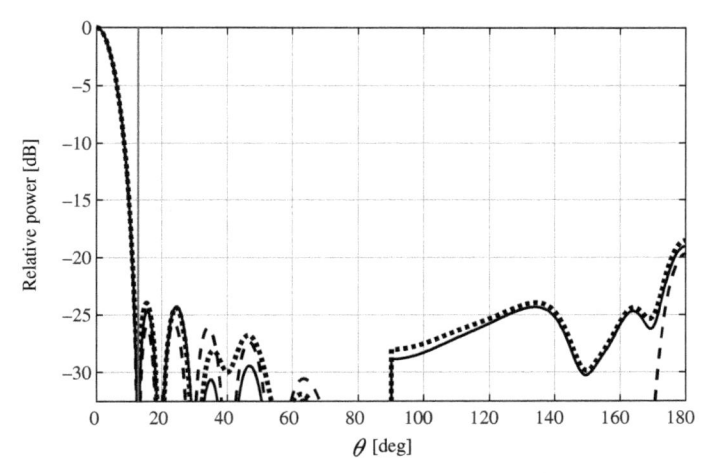

Fig. 14.35. Simulated H-plane power pattern at 5.8 GHz: the linear array (aperture and feed integrated) with corrected branch lengths (—); the linear array (aperture and feed integrated) with equal (uncorrected) path lengths (···); the linear array with the MPAs's inputs simultaneously excited with the 25 SLL Chebyshev taper (- -). The vertical line shows the first null angle of the array factor with the 25 SLL Chebyshev taper.

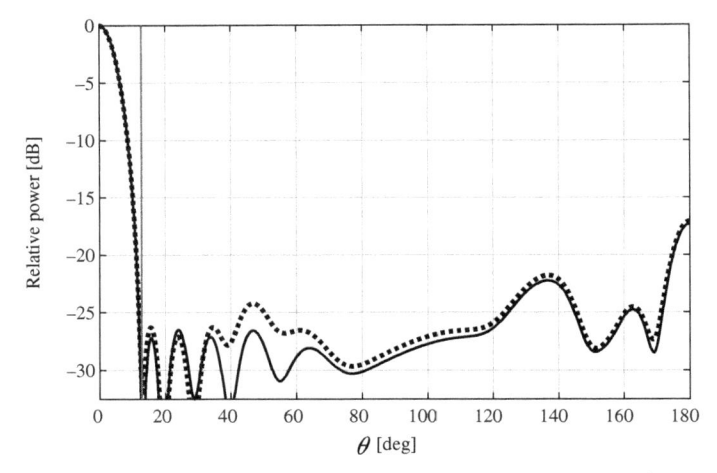

350 *Simulation-Based Optimization of Antenna Arrays*

Fig. 14.36. Simulated H-plane power pattern of the linear array (aperture and feed integrated) at 5.8 GHz after redesigning the feed for more equal sidelobes: −26.56 dB SLL for the [−90, 90]-degree sector with corrected path lengths (—); −24.23 dB SLL for the [−90, 90]-degree sector with uncorrected path lengths (· · ·). The vertical line shows the first null angle of the array factor with the 25 SLL Chebyshev taper.

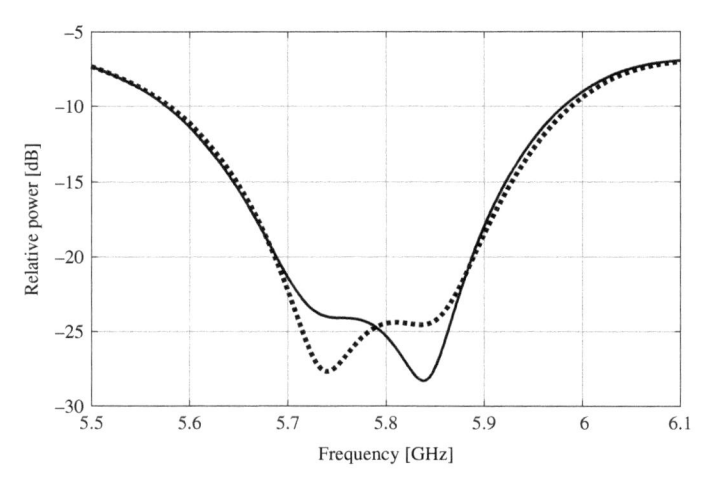

Fig. 14.37. Simulated reflection coefficient of the linear array (aperture and feed integrated) at the feed input after redesigning the feed for more equal sidelobes: with corrected path lengths (—); with uncorrected path lengths (· · ·).

second, for the integrated circuit as described further, at Steps 8 and 9.

The radiating aperture is defined as follows. The MPAs having the 17.5 μm metallization of the patches reside on a 1.524 mm Arlon AD 250 substrate (AD250C, 2018). MPAs center-to-center spacing is a half wavelength, 25.86 mm. Lateral extensions of the aperture substrate are finite, 40 mm × 336 mm.

The MPAs are energized through ground plane slots. All MPAs have identical dimensions. A stand-alone MPA had been dimensioned for 5.8 GHz operation; its dimensions are the initial values of the design variables for the aperture-embedded MPAs.

The MPA aperture, described by six-dimensional parameters having the most significant effect on the aperture reflection and radiation responses, has been adjusted for 5.8 GHz of operation using simulation-driven surrogate-based optimization (Koziel and Ogurtsov, 2018). In the process of aperture optimization, its EM model was simulated in the presence of the already dimensioned though inactive feed. The inputs of the MPAs, microstrip ports of the EM model, were placed on the feed substrate 20 mm from the patch centers. The ports were simultaneously excited with the taper a_{Target}, listed at Step 1.

The optimization objectives were active reflection coefficients at six MPA inputs and broadside realized gain. The active reflection coefficients should be minimized at 5.8 GHz, and the realized gain toward the zenith versus frequency should have its maximum at 5.8 GHz. These seven objectives were combined in a scalar cost function using a penalty function approach. The primary objective was minimization of the maximum of active reflection coefficients at the operating frequency; the penalty term was proportional to a (frequency-wise) deviation of the realized gain maximum from 5.8 GHz.

As a result of optimization, the dimensions of the aperture-embedded MPAs were set as follows: the patch size to 14.35 mm × 14.35 mm; the ground plane slot to 0.75 mm × 10.00 mm; the offset of the ground plane slots to 5 mm down off the patch center; and the length of the open-end microstrip stub (from the slot center) to 6.5 mm. Active reflection coefficients of the outermost MPAs are under -16.5 dB at 5.8 GHz, and for other MPAs are under -24.5 dB at 5.8 GHz. The H-plane pattern of the aperture is shown in Fig. 14.34 with the dash-line.

The entire structure including the optimized aperture, and driven through a common input of the feed, had been simulated at Sep 8. Simulation had been performed for the structure with the uncorrected and corrected feeds. The array structure had been modeled at this stage with infinite lateral extends of the ground plane. The H-plane patterns are shown in Fig. 14.34. The abrupt change of the relative power at the 90-degree angle in Fig. 14.34 is due to the decoupling plane of the simulated model. Contribution of the feed radiation to the H-plane patterns can be seen in Fig. 14.34 for angles from 90 to about 170 degrees. Figure 14.34 also shows that correction of the feed brings minor improvement of the SLL; however, it decreases the third and fourth sidelobes. This effect of decaying sidelobes is utilized at Step 9 in redesigning the feed for more equally rippled sidelobes, and thereby lowering the SLL.

At Step 9 the simulated H-plane pattern of the array is used to set up a fast surrogate model of the structure as described in Sec. 14.2.2.2. This fast surrogate is used to redesign the excitation taper for more equally rippled sidelobes. The result of optimization, the new excitation taper $\mathbf{a}^{(1)}$, had been cast to new power splits $\mathbf{p}^{(1)} = [0.5773\ 0.4593\ 0.5284\ 0.5055\ 0.4417]^T$ staying with the same feed tree. Thus, a new iteration started with Step 5, optimization of the kriging model of microstrip junctions for the new power splits.

A more generic realization of the process may include a search for a new optimal tree, i.e., with a new iteration cycle starting from Step 2, Fig. 14.26. It could be formally better in terms of realizing the new taper; however, the same feed architecture turned out to be optimal for the case considered here.

At Step 6 of the new iteration cycle, an SMA connector had been added to the EM model of the array at the common input of the feed. Additionally, the EM model had been simulated with finite lateral extends of the ground plane and substrate. The simulated H-plane

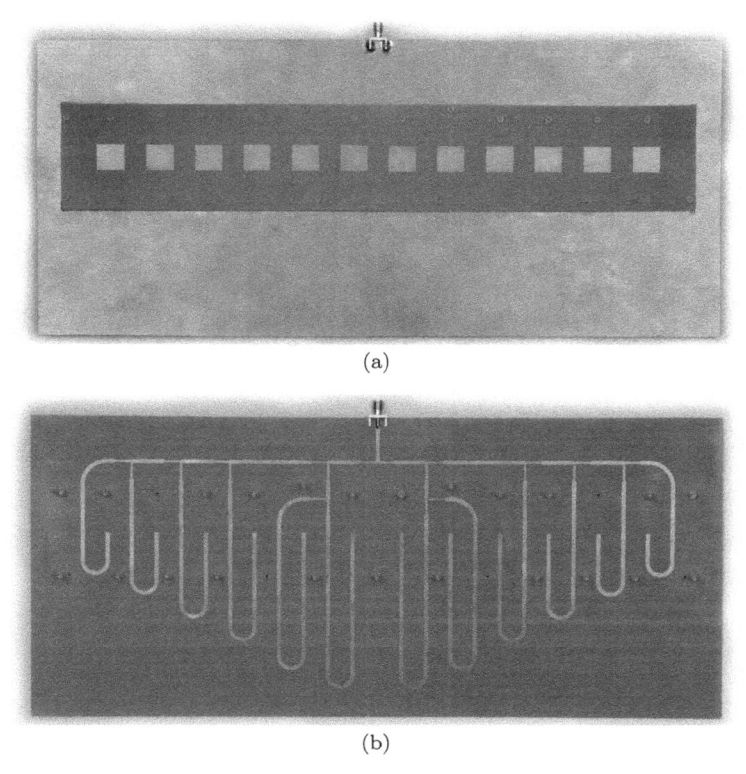

Fig. 14.38. Photographs of the fabricated array: (a) radiating aperture, lateral extensions are 60 mm × 336 mm; (b) feed. The feed substrate and ground plane (common to the feed and radiating elements of the aperture) lateral dimensions are 166 mm × 364 mm. The central T-junction is 25 mm from the upper substrate/ground edge (the flange of the edge-mount SMA connector).

pattern and the reflection coefficient are shown in Figs. 14.36 and 14.37, respectively. For this design, the effect of the path length correction turned to be significant (Fig. 14.36) due to the almost equally rippled sidelobes. The simulated SLL of the obtained design is −26.6 dB SLL for the [−90, 90]-degree sector. Thus, the redesign process has been concluded after the first iteration because the second one did not bring any further SLL improvement.

The final design has been fabricated and measured. Figure 14.38 shows the fabricated prototype. The measured H-plane total pattern and reflection coefficient are shown in Figs. 14.39 and 14.40,

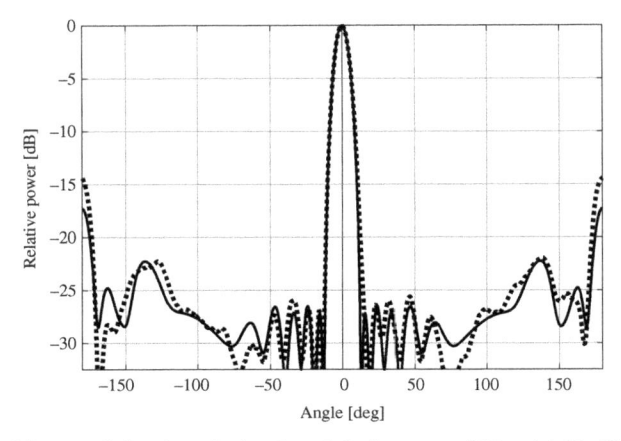

354 *Simulation-Based Optimization of Antenna Arrays*

Fig. 14.39. Measured (· · ·) and simulated (—) array of Fig. 14.38: H-plane total pattern at 5.8 GHz.

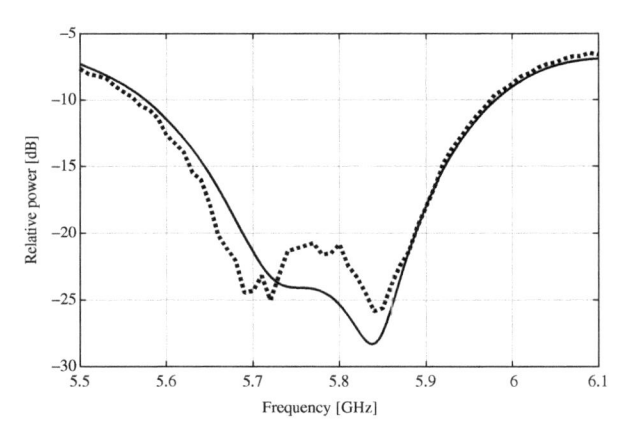

Fig. 14.40. Measured (· · ·) and simulated (—) array of Fig. 14.38: reflection coefficient.

respectively. The measured pattern exhibits certain broadening of the major lobe. In the same time, the measured and simulated patterns are consistent in the SLL, which is under −25 dB for the [−90, 90]-degree sector; thus, the described design process is justified.

We have described and demonstrated a comprehensive approach for design of microstrip corporate feeds of linear arrays that takes into account all relevant effects that—when not accounted for—lead to degradation of the array performance (in terms of sidelobe level, major lobe beamwidth, and reflection coefficient in the presented examples). Furthermore, due to very low computational cost (a few simulations of the array in general and only two simulations of the entire array in the presented example), the methodology is practical for handling high-fidelity EM simulation models of feeds and apertures of printed-circuit antenna arrays.

14.3. Summary

The material presented in this chapter demonstrated that surrogate-assisted approaches allow conducting design of low-sidelobe printed-circuit linear arrays in a systematic way, from design of a single radiating element up to design of the entire array circuit including the feed. It has been shown that the total computational costs of the designs obtained using surrogate-assisted techniques can be equivalent to only few high-fidelity simulations of an array at hand while various interactions within the array contributing to the radiation and reflection characteristics are accounted. All discussed design processes started from an identification of the optimal feed architectures which are strongly influenced by constrains imposed on the components to be subsequently implemented (e.g., aperture spacing, aperture size and length, minimal and maximal power splits). The presented results also illustrate the fact that the use of EM-simulations at the implementation steps is essential for obtaining reliable designs meeting performance requirements.

It has been demonstrated that designs with acceptable radiation and reflection characteristics can be obtained by utilizing the simplest configurations of power distribution elements of the corporate feeds, such as equal-split T-junctions or unequal-split T-junctions, rather than with complications of the feed circuitry; in this case surrogate-based optimization of realistic EM models of the array components and the entire array serves as a mean to adjust the components' as well as the entire array's characteristics in the presence of many affecting EM factors.

Chapter 15

Design of Linear Phased Array Apertures Using Response Correction and Surrogate-Assisted Optimization

In this chapter, we discuss and illustrate a simulation-driven surrogate-assisted approach to design of phase excitation tapers for linear phased array apertures. The design process is realized by a mean of surrogate-based optimization (SBO). SBO utilizes two surrogate models: a coarse-mesh surrogate of the array element for adjusting the array's active reflection responses, and a fast surrogate of the radiation pattern which is a corrected array factor pattern. Iterative referencing to high-fidelity simulated responses allows us to conduct the design process with respect to the near-field interactions, e.g., coupling, and other undesirable wave phenomena within the aperture. At the same time, the design process is made fast due to the utilized SBO technique.

The primary optimization objective of the considered example is the minimal SLLs at different scan angles. The optimization outcome is a set of phase tapers versus the scan angle, i.e., a table of excitation phases required to point the major lobe to the directions of interest and have the lowest SLLs possible. Another design objective is minimization of the maximum level of active reflection coefficients at the design frequency. Amplitudes of the excitation signals are set to be the same for all elements and for all scan angles of interest.

Performance and numerical cost of the presented approach are demonstrated by optimizing a 16-element linear array aperture of microstrip patch antennas (MPAs). Experimental verification had been carried out for the fabricated prototype of the optimized array aperture. A good agreement between the simulated radiation patterns and patterns constructed through superposition of the measured element patterns had been obtained.

Finally, a numerical study of the array model driven with optimal tapers justifies (in terms of SLL, grating lobe level, and realized gain scan loss) an application of the techniques for different scan angles.

15.1. Optimization Methodology

Optimization of the simulated antenna array EM model **f** can be accelerated by exploiting the SBO paradigm (Koziel and Ogurtsov, 2014a) and utilizing the analytical array factor model **a** as an underlying low-fidelity model that assumes ideal isotropic radiators of the aperture. We aim for the following two objectives: (i) simultaneous minimization of the pattern SLL at the operating frequency f_0 and centering the reflection responses on the operating frequency f_0, and (ii) obtaining phase excitation tapers which ensure the minimal SLLs in the process of scanning.

15.1.1. Element optimization

We optimize the EM model of the radiating element, $\mathbf{f}_e(\mathbf{x}_g)$, where $\mathbf{x}_g$ represents geometry parameters. The optimization process utilizes a corrected coarse-mesh model of the element, $\mathbf{c}_e$. Due to the narrow-band response of the antenna array, the best way to align $\mathbf{c}_e$ with $\mathbf{f}_e$ is frequency scaling $F(\omega) = \alpha_0 + \alpha_1 \omega$ (Bandler et al., 2004a), where α_0 and α_1 are parameters to be determined. The frequency scaled model is defined as

$$\mathbf{c}_{e.F}(\mathbf{x}) = [c_{e.F}(\mathbf{x_g}, F(\omega_1)), \ldots, c_{e.F}(\mathbf{x_g}, F(\omega_m))]^T, \tag{15.1}$$

where $\mathbf{c}_e(\mathbf{x}_g) = [c_e(\mathbf{x}_g, \omega_1), \ldots, c_e(\mathbf{x}_g, \omega_m)]^T$ is the reflection coefficient at frequencies ω_j, $j = 1, \ldots, m$; α_0, α_1 are obtained as

$$[\alpha_0^{(i)}, \alpha_1^{(i)}] = \arg\min_{[\alpha_0, \alpha_1]} \sum_{k=1}^{m} [f_e(\mathbf{x}_g, \omega_k) - c_e(\mathbf{x}_g, \alpha_0 + \alpha_1 \omega_k)]^2. \tag{15.2}$$

Then the optimal design is searched as

$$\mathbf{x}_g^{\text{opt}} = \arg\min_{\mathbf{x}_g} U_{S11}(\mathbf{c}_{e.F}(\mathbf{x}_g)), \tag{15.3}$$

where U_{S11} is the objective function defined as the reflection coefficient at the operating frequency f_0. In practice, up to three iterations (15.3) are sufficient to find the optimum of $\mathbf{f}_e$ with the scaling (15.1), (15.2) repeated at the beginning of each iteration.

15.1.2. *Correction of the array factor model*

Discrepancies between the models $\mathbf{f}$ and $\mathbf{a}$, evaluating the radiation pattern, can be reduced with a suitable response correction function $D(\theta)$. This function is defined at the initial design and with all excitation phases $\mathbf{x}_p$ being equal to zeros as follows.

Let $p_f^k = [\theta_f^k \ r_f^k]$, $k = 1, \ldots, K_f$, be K_f points extracted from the nominal high-fidelity model $\mathbf{f}$ response (cf. Fig. 15.1) where θ_f^k is the angle and r_f^k is the corresponding realized gain value, for $\theta = 0, +180$ degrees, and all local realized gain maxima in the range $[0, +180]$ degrees; $p_a^k = [\theta_a^k \ r_a^k]$, $k = 1, \ldots, K_a$, denote similar points extracted from the array factor model response (here $K_a = K_f$). Here $D(\theta)$ in decibels is defined as

$$D(\theta) = E_f(\theta) - E_a(\theta) \tag{15.4}$$

with $E_f(\theta) = F([\theta_f^1 \ldots \theta_f^{K_f}], [r_f^1 \ldots r_f^{K_f}], \theta)$, where $F(X, Y, x)$ is the interpolation of the data value vector Y (defined on a discrete argument set X) onto x. A similar definition holds for $E_a(\theta)$.

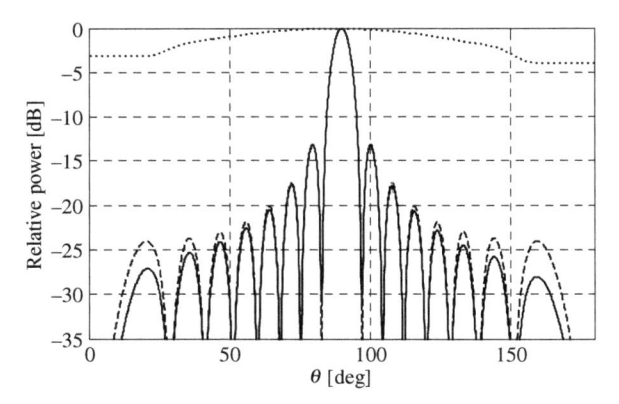

360 *Simulation-Based Optimization of Antenna Arrays*

Fig. 15.1. Power pattern of the 16-element array in the E-plane at 10 GHz at the initial design, $\theta_s = 0$ degree: (—) EM model **f**, and (- - -) analytical array factor model **a**; ($\cdots$) correction function D extracted from **f** and **a** (cf. (15.4)). $\theta = 90$ degrees is for the boresight direction.

Figure 15.1 shows the plot of $D(\theta)$ for the linear array of extracted from the responses of models **f** and **a**.

The surrogate model **s** is defined here using $D(\theta)$ as follows:

$$\mathbf{s}_s(\mathbf{x_p}, \theta) = \mathbf{a}(\mathbf{x_p}, \theta) + D(\theta), \tag{15.5}$$

where $\mathbf{x}_p = [p_1 \ p_2 \ldots p_N]$ is a vector of excitation phases. One should understand (10.19) as pattern summation in decibels. Note that correction D generally depends on the array geometry, the element's geometry $\mathbf{x}_g$, and the current phase taper $\mathbf{x}_p^{\text{curr}}$. Thus we have that $D(\theta) = D(\mathbf{x}_g, \mathbf{x}_p^{\text{curr}}, \theta)$. Consequently, we write $\mathbf{s}_s(\mathbf{x}_p; \mathbf{x}_g, \mathbf{x}_p^{\text{curr}}, \cdot) = \mathbf{a}(\mathbf{x}_p, \cdot) + D(\mathbf{x}_g, \mathbf{x}_p^{\text{curr}}, \cdot)$ to indicate its dependence on both $\mathbf{x}_g$ and $\mathbf{x}_p^{\text{curr}}$.

It is worth to notice that according to (15.4) and (15.5) the correction D is a numerical version of the effective array-embedded element pattern that relates the array aperture models of different fidelities.

15.1.3. *Design for scanning*

At this stage, we search for a phase taper ensuring the minimum SLL for a given scan angle θ_s. We start by optimizing the array for

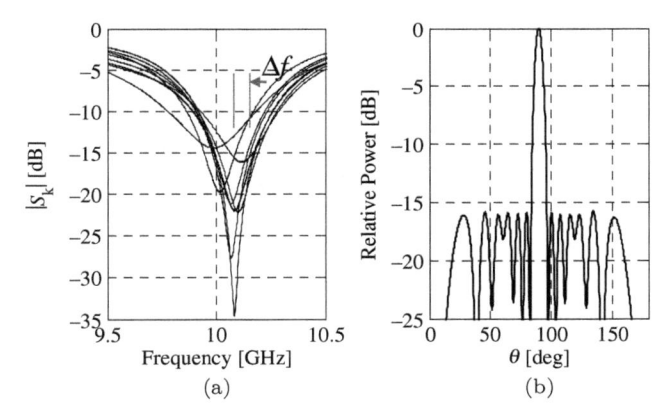

Fig. 15.2. Initially optimized EM array model **f**: (a) active reflection coefficients at all ports with the frequency shift Δf denoted; (b) E-plane radiation pattern.

$\theta_s = 0$, using the surrogate model (15.5) as follows:

$$\mathbf{x}_g^{\text{opt}} = \arg \min_{\mathbf{x}_p} U_{\text{SLL}}(\mathbf{s}(\mathbf{x}_p, \mathbf{x}_g)), \qquad (15.6)$$

where U_{SLL} evaluates the SLL for any given excitation $\mathbf{x}_p$. The random-search-initialized gradient-based algorithm is utilized at this point.

Because of mutual coupling within the aperture, simulated active reflection coefficients at $\mathbf{x}_p^{\text{opt}}$ differ from what we expect as illustrated in Fig. 15.2(a): reflection coefficients are jointly shifted in frequency by Δf. This can be eliminated by redesigning the array element so that its center frequency is moved from f_0 to $f_0 - \Delta f$. Optimization results in a new element geometry $\mathbf{x}_g^1$.

In the next step, the array model **f** is simulated at the element geometry $\mathbf{x}_g^1$ and the optimized phase taper $\mathbf{x}_p^{\text{opt}}$. At this point, the active reflection coefficients are centered around f_0 at the -10 dB level (cf. Fig. 15.3). The entire design process requires only four high-fidelity simulations of the array model. The computational cost of other operations is low and can be neglected compared to the cost of EM simulations.

Subsequently, the phase tapers for the scanned array are found by iterative identification of the response correction term (15.4) and surrogate-assisted optimization (15.5), (15.6). This is executed,

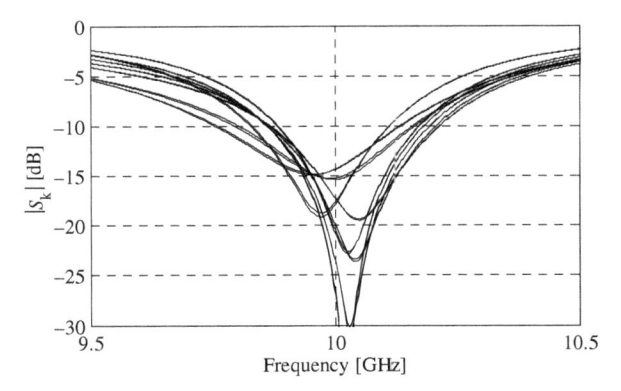

362 *Simulation-Based Optimization of Antenna Arrays*

Fig. 15.3. Active reflection coefficients (centered at the −10 dB level) of the array EM model **f** with re-optimized element $\mathbf{x}_g^1$.

separately, for each scan angle $\theta_s^{(k)}$ as follows (here, $\mathbf{x}_p^{\mathrm{opt},k}$ is the optimized phase taper corresponding to $\theta_s^{(k)}$):

1. Simulate the array model **f** at $\mathbf{x}_p^{\mathrm{opt},k-1}$ (progressive phase shift for $\theta_s^{(k)}$ is applied separately);
2. Calculate the correction term D (cf. (15.4));
3. Find the phase taper $\mathbf{x}_p^{\mathrm{opt},k}$ by optimizing the surrogate **s** (15.5) for the minimum SLL (corresponding to $\theta_s^{(k)}$) using (15.6).

The above procedure requires only one evaluation of the high-fidelity model **f** per scan angle.

15.2. Case Study: 16-Element Linear Phased Array

Consider a linear array of microstrip antennas shown in Fig. 15.4. We search for excitation phases (phase taper) of signals, which are incident at the elements' inputs to minimize the sidelobe level (SLL) in the E-plane. The active reflection coefficients of the array should be centered about 10 GHz. Amplitudes of the excitation signals are the same; center-to-center elements' patch spacing is a half wavelength.

15.2.1. *Element optimization*

The element is a microstrip patch antenna (MPA) Fig. 15.4(b). It is defined on a 0.76-mm thick layer of Taconic (RF-35, 2018).

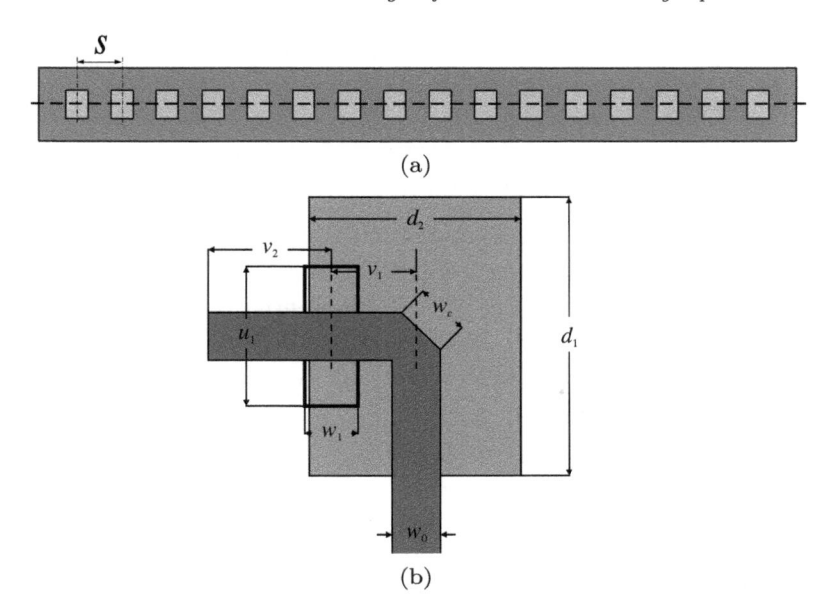

Fig. 15.4. A linear array of MPAs: (a) the aperture. The E-plane is denoted with the horizontal dash-line; (b) the element. Dielectric layers and the ground plane are shown transparent; the patch (light gray), slot (bold rectangle), open-end microstrip and input microstrip (dark gray) are contoured with solid lines.

Dimensional parameters are: patch size is $d_1 \times d_2$; the metal ground has a slot aperture of $w_1 \times u_1$; the slot center is v_1 relative to the patch center; a feed substrate is another 0.76-mm layer of RF-35; the input 50-ohm microstrip has the width of $w_0 = 1.7$ mm, an open-end stub of length v_2 terminates the feed and w_c is the chamfer length of the input microstrip. The MPA was simulated in CST Microwave Studio (CST, 2016).

The vector of design variables is $\mathbf{x}_g = [d_1\ d_2\ w_1\ u_1\ v_1\ v_2\ w_c]^T$. We use two EM models: the coarse-mesh model $\mathbf{c}_e$ (70,000 mesh cells, simulation time 30 s) and the fine model $\mathbf{f}_e$ (~1,000,000 cells, 10 min). The initial design is $\mathbf{x}^{\text{init}} = [9.0\ 7.4\ 2.0\ 6.0\ 2.5\ 6.0\ 1.5]^T$ mm. The optimized design $\mathbf{x}_g^{\text{opt}} = [9.53\ 7.29\ 2.90\ 5.50\ 1.91\ 5.50\ 1.48]^T$ mm was found in three iterations at the total cost corresponding to about 12 simulations of $\mathbf{f}_e$ (including 160 evaluations of $\mathbf{c}_e$, which corresponds to the CPU time of about 80 min, and four evaluations of the high-fidelity model, or 40 min), 2 hours in total. Reflection response of

15.2.2. Optimal excitation taper for the major lobe pointing broadside

At this stage we search for a set of excitation phases $\{\alpha_n\}_{n=1,...,N}$ of the linear array (Fig. 15.4(a)) so that the SLL of the initial power pattern is minimized (cf. Fig. 15.5). We assume symmetrical phase excitations with respect to the array center, i.e., $\alpha_{N+1-n} = \alpha_n$, $n = 1, \ldots, N/2$. In this work $N = 16$. Spacing s and excitation amplitudes are uniform.

The design variable vector is $\mathbf{x}_p = [\alpha_1 \; \alpha_2 \ldots \alpha_8]^T$. The uniform array with SLL ≈ -13.2 dB is the initial design, $\mathbf{x}_p^{\text{init}} = [0\; 0\; 0\; 0\; 0\; 0\; 0\; 0]^T$ degrees. The high-fidelity model of the array $\mathbf{f}$ was defined using CST MWS with about 10,000,000 mesh cells and simulated with the CST MWS transient solver in 76 min. The excitation phases for the major lobe pointing broadside and with the E-plane pattern minimum SLL $= -16.1$ dB was found at the optimum $\mathbf{x}_p^{\text{opt}} = [-58.1\; 0.0\; -95.9\; -70.7\; -71.4\; -64.4\; -65.6\; -59.9]^T$ degrees.

Active reflection coefficients at this design $\mathbf{x}_p^{\text{opt}}$ are shifted towards higher frequencies as shown in Fig. 15.2(a). A re-optimized element design was found at $\mathbf{x}_g^1 = [9.38\; 7.34\; 2.9\; 5.5\; 1.91\; 5.5$

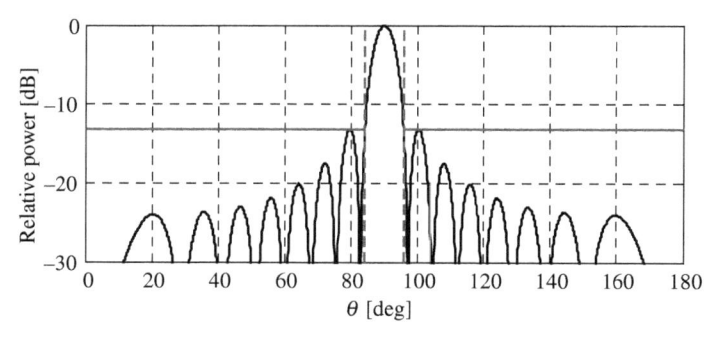

Fig. 15.5. Radiation pattern design specifications: horizontal (–) denotes the initial design's SLL (to be minimized); vertical (- - -) defines the major lobe sector (to be kept intact).

2.05 1.28]^T mm. The fine-tuned optimal design having SLL = −16.5 dB was found at $\mathbf{x}_p^{\text{opt},2}$ = [−58.5 0.0 −96.3 −71.1 −71.6 −64.7 −65.9 −60.3]^T degrees by repeating optimization (15.10) with the updated surrogate model.

The radiation pattern of $\mathbf{x}_p^{\text{opt},2}$ is shown in Fig. 15.8(a) with the solid line. Active reflection coefficients are shown in Fig. 15.3.

15.2.3. *Optimization for major lobe scanning*

The procedure of Sec. 10.5.1 in Chapter 10 has been applied to find the optimum tapers for the scan angles of 5, 10, ..., and 50 degrees. The results are listed in Table 10.7 of Chapter 10. It should be noted that the total phase excitation applied to a particular element for a given scan angle is the sum of the listed optimal value and the scan angle-dependent progressive phase shift (Mailloux, 2005) which is not listed in Table 10.7 of Chapter 10.

15.2.4. *Experimental validation of the optimal design*

The final design with geometry of $\mathbf{x}_g^{\text{opt}}$ has been manufactured (photograph shown in Fig. 15.6). Aperture-embedded element patterns, 16 in total with only one input fed at a time and the other terminated on matched loads, have been measured.

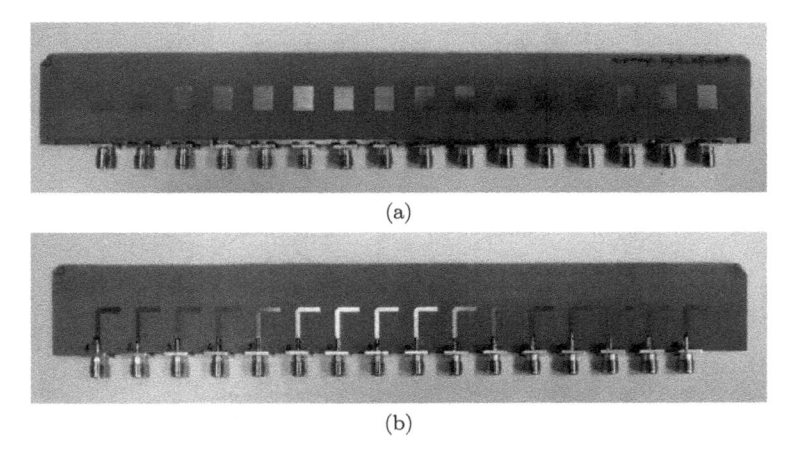

Fig. 15.6. Photograph of the fabricated array of MPAs: (a) front and (b) back.

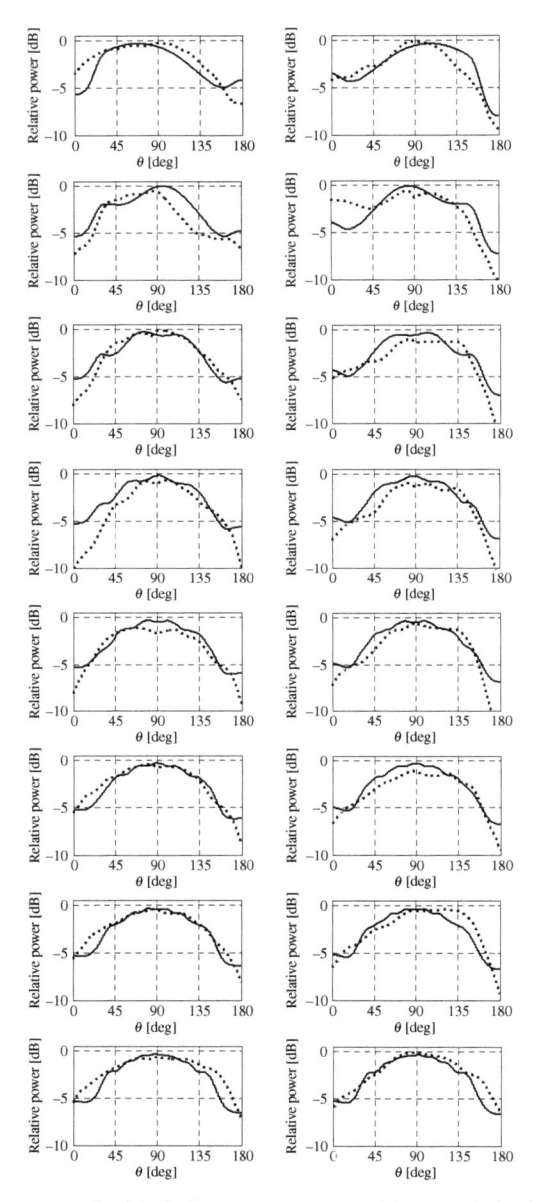

366 *Simulation-Based Optimization of Antenna Arrays*

Fig. 15.7. Aperture-embedded element patterns. Measured ($\cdots$) and simulated (—) patterns are normalized, respectively, to the maximal measured and simulated peak values. Element numbers are, from the top: 1 and 16 (outermost elements), 2 and 15, ..., 7 and 8 (central elements).

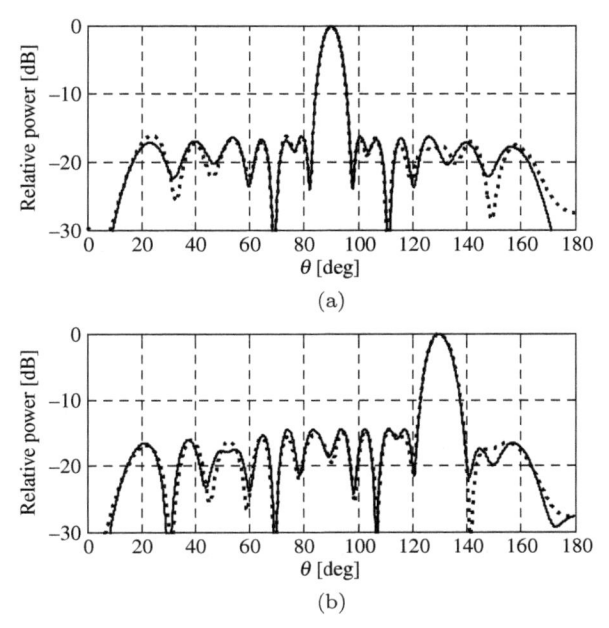

Fig. 15.8. Selected E-plane radiation patterns at 10 GHz of the final design of geometry $\mathbf{x}_g^{opt}$ with optimal phase tapers simulated with the high-fidelity model $\mathbf{f}$ (—) and evaluated using 16 measured radiation patterns (···): (a) for $\theta_s = 0$ degree (with phases corresponding to the scan angle of 0 degree. in Table 15.1); (b) for $\theta_s = 40$ degrees (with phases corresponding to the scan angle of 40 degrees. in Table 15.1).

The measured element patterns are shown in Fig. 15.7. The measured patterns have been combined using the optimal tapers $\mathbf{x}_p^{opt,2}$ and progressive phase shifts into the array radiation patterns for different scan angles. The radiation patterns for the scan angles of 0 and 40 degrees are shown in Fig. 15.8 with dot-lines. Excellent agreement of the simulated and evaluated patterns (configured from measured element patterns using the optimal phase tapers listed in Table 15.1) was observed for the ±50-degree sector of scan.

15.3. Optimal Design as the Phased Array

Appropriate interpolation allows us to utilize the results of Table 15.1 for any angle within the scan sector. The superposition model $\mathbf{f}_s$

Table 15.1. Optimal phase tapers*.

Scan Angle [deg]	Excitation phase* [deg] Elements							
	1(16)	2(15)	3(14)	4(13)	5(12)	6(11)	7(10)	8(9)
0	−58.5	0.0	−96.3	−71.1	−71.6	−64.7	−65.9	−60.3
5	−51.8	0.0	−97.0	−67.7	−70.2	−67.5	−61.4	−61.5
10	−48.6	0.0	−95.9	−64.1	−70.6	−63.7	−60.9	−59.1
15	−44.6	0.0	−92.4	−62.6	−67.5	−60.6	−57.6	−55.4
20	−40.4	0.0	−88.7	−60.7	−64.7	−58.9	−55.6	−53.6
25	−37.7	0.0	−87.1	−60.8	−63.3	−58.4	−54.7	−52.9
30	−36.1	0.0	−86.4	−61.3	−62.2	−58.1	−53.8	−52.0
35	−34.7	0.0	−85.8	−62.0	−61.7	−58.4	−53.5	−51.9
40	−33.7	0.0	−85.7	−62.8	−61.1	−58.4	−52.7	−51.2
45	−32.8	0.0	−85.1	−63.6	−61.0	−59.2	−53.1	−51.9
50	−32.6	−48.3	−78.2	0.0	−38.5	−38.9	−46.1	−38.9

*Listed values do not include the progressive phase shift.

of the aperture has been configured from 16 high-fidelity EM simulations (Koziel and Ogurtsov, 2014b) to evaluate performance of the linear array. The radiation and reflection responses have been evaluated with the interpolated phase tapers using the superposition model $\mathbf{f}_s$.

The following excitations have been compared: the interpolated set of the optimal phase tapers, the phase taper obtained as an optimum for the zero scan angle, and the all-uniform taper. The excitations were compared over the scan angles at 10 GHz in terms of the following figures: SLL (optimized figure), shown in Fig. 15.9; peak realized gain scan loss, shown in Fig. 15.10; beam broadening factor, shown in Fig. 15.11; and the total reflected power (adjusted with optimization for shifting minima of active reflection coefficients to 10 GHz), shown in Fig. 15.12.

Figure 15.9 indicates that excitation with the interpolated optimal phase tapers is superior over the other excitations as it allows keeping SLLs at much lower levels within the wide sector of scans. At the same time, Fig. 15.9 shows that with the excitation being optimal at broadside, the SLLs quickly degrade with scan angle.

Design of Linear Phased Array Apertures 369

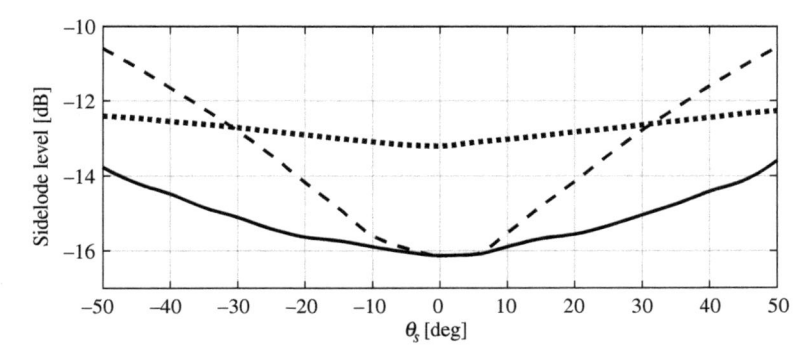

Fig. 15.9. SLL versus scan angle: with the interpolated set of the optimal phase tapers (——), with the phase taper obtained as an optimum for the zero scan angle (– –), with the uniform taper (· · ·).

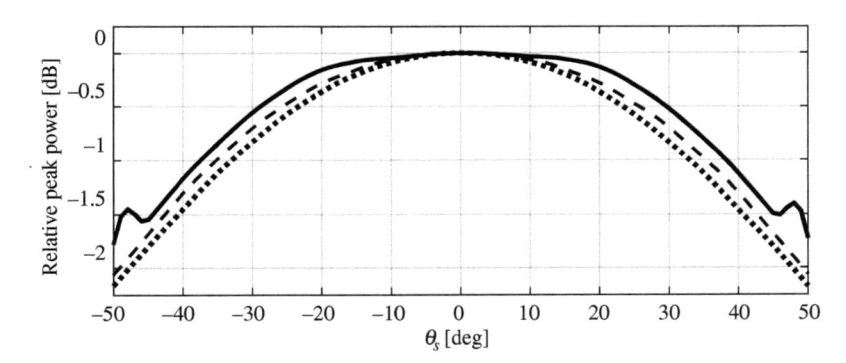

Fig. 15.10. Peak realized gain scan loss (relative to that at the zero scan angle that is 18 dB with the uniform excitation and 17 dB with the optimized excitation): the interpolated set of the optimal phase tapers (——); the phase taper obtained as an optimum for the zero scan angle (– –); uniform taper (· · ·).

This clearly justifies the optimization step described in Sec. 15.2.3 as well as the reliability of interpolation of the optimal phase tapers. Realized gain scan losses (cf. Fig. 15.10) are similar for all excitations up to the scan angles of ±50 degrees. In the same time, the interpolated set of the optimal phase tapers provides lower scan losses. The beam broadening with all excitations closely follows the large array limit (Mailloux, 2005). For all excitations, the total

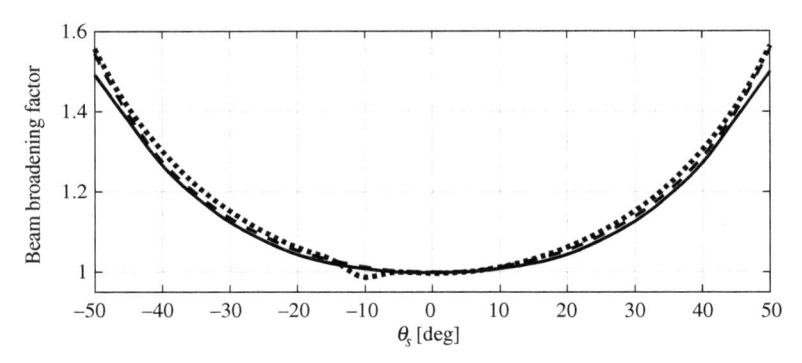

370 *Simulation-Based Optimization of Antenna Arrays*

Fig. 15.11. Beam broadening factor at 10 GHz versus scan angle: with the interpolated set of the optimal phase tapers (—), the phase taper obtained as an optimum for the zero scan angle (– –), uniform taper (· · ·).

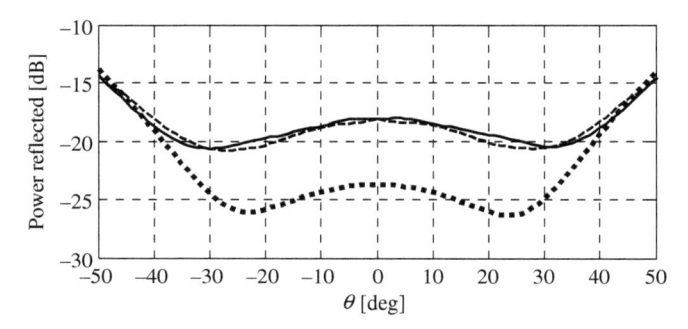

Fig. 15.12. Total reflected power at 10 GHz versus scan angle: with the interpolated set of the optimal phase tapers (—), the phase taper obtained as an optimum for the zero scan angle (– –), uniform taper (· · ·).

reflected power at the frequency of design, plotted in Fig. 15.12, stays under –15 dB almost up to the scan angles of ±50 degrees.

15.4. Summary

In this chapter, design of a linear array aperture for phased array applications using surrogate-assisted optimization techniques, including two response correction techniques, had been presented.

We demonstrated that utilization of variable-fidelity EM models and various kinds of surrogate modeling techniques, namely, coarse-mesh simulated models and response correction techniques, permits dramatic reduction of the design cost as well as reliable handling of both radiation pattern and active reflection coefficients.

Chapter 16

Fault Detection in Linear Arrays Using Response Correction Techniques

In this chapter, we discuss a surrogate-assisted technique for detecting faulty elements in a small linear microstrip array of patch antennas from samples of the array's far-field magnitude radiation pattern (here represented by realistic EM simulations). It is important that regardless of the array size, the method requires only one expensive full-wave entire-array simulation. This one simulation gives the accurate far-field magnitude pattern of the original defect-free array, and is used in conjunction with the defect-free array's analytical array factor to formulate a response correction function. This response correction function can then be used to construct an accurate approximation of the EM-simulated pattern of any arbitrary faulty array at very low cost. The low cost and high accuracy of approximations make possible an enumeration strategy for identifying the faulty elements, which would have been computationally prohibitive were EM-simulated patterns to be used. Furthermore, partial faults and measurement noise are addressed. Accuracies in detecting up to three faults (including partial ones) in arrays of

16 and 32 elements exceeded 97% under noise-free conditions, and were above 93% in the presence of 2 dB measurement noise.

16.1. Fault Detection in Array Antennas

The detection of faults in antenna arrays has received much attention in recent decades, especially as arrays are ubiquitous to radars, radiometry, communication systems, radio astronomy applications, and more. For arrays that cannot be brought to a laboratory for inspection and that do not have fault detection sensors integrated with the beamforming network, it is preferable to perform antenna diagnosis by means of the radiated far-field measurements (Bucci *et al.*, 2000). Furthermore, if the special requirements pertaining to measuring the complex far-field pattern (e.g., a reference phase signal) cannot be accommodated, the need arises for a fault detection methodology that relies on magnitude-only samples of the far-field (Bucci *et al.*, 2000).

This chapter is concerned with diagnosing faulty array elements of microstrip linear antenna arrays from magnitude-only samples of their far-field radiation patterns. In spite of the ubiquitous use of this kind of array, this topic has received scant attention in the literature. A likely reason is that, in general, studies concerned with fault detection based on magnitude-only pattern samples have predominantly relied on population-based metaheuristics, such as genetic algorithms (GAs) (Bucci *et al.*, 2000; Rodriguez-Gonzalez *et al.*, 2009; Rodriguez and Ares, 2000; Iglesias *et al.*, 2008). These approaches entail that a GA searches for the best match to a given (measured) faulty radiation pattern in the space of patterns due to all possible faulty excitation vectors. Even for relatively small arrays, there could be thousands of possible faulty patterns. Therefore, such a search is feasible only if the faulty patterns are inexpensive to compute, which generally is not the case for printed arrays on layered media. In the above GA-based studies, array elements have mostly been modeled as isotropic radiators that can be accounted for by an analytical array factor, sometimes in conjunction with the element pattern — however, mutual coupling is then neglected

(Bucci et al., 2000; Rodriguez and Ares, 2000; Iglesias et al., 2008). In Rodriguez-Gonzalez et al. (2009), array elements were specified to be wire dipoles which enabled mutual coupling to be accounted for by means of a readily calculated impedance matrix.

In Patnaik et al. (2007), the most cited study on fault detection in linear microstrip arrays, an alternative approach was followed: a neural network was used to construct a lookup table for finding up to three faults in a 16-element array based on magnitude-only far-field samples. The samples were obtained via full-wave EM simulations. The network was trained on all possible fault patterns and it is not clear whether it would have been able to predict the correct faulty elements for a previously unseen pattern had it been trained on a part of the patterns only. In any case, the cost of training data generation could become prohibitive as larger arrays are considered (Patnaik et al., 2007, required 696 EM simulations of the entire array).

Along the lines of this book, this chapter discusses a surrogate-assisted technique for detection of faults in realistic, EM-simulated antenna arrays using magnitude-only far-field samples of the far-field. The technique is specifically applied to small microstrip patch arrays (it is assumed that the number of faults in practice will typically be significantly less than the number of array elements, Rodriguez-Gonzalez et al., 2009). Both on–off faults and partial faults are considered. In the discussed technique, the only EM simulations (or alternatively field measurements) that are required involve samples of the far-field magnitude radiation pattern of the original, defect-free array; in other words, just one simulation of the array is required. Based on this information, a response correction strategy is formulated by which the analytical array factor of any faulty aperture can be obtained to serve as a fast surrogate model of the EM-simulated pattern. In particular, we exploit a customized formulation of single-point, additive response correction (Queipo et al., 2005), which seems to be appropriate given the type of misalignment between the analytical array factor model and the EM-simulated model. The surrogate is then used along with an enumeration technique to detect defective elements within a faulty array.

16.2. Fault Detection Methodology

The surrogate-based procedure for computationally efficiency fault detection in linear array antennas is outlined first. Demonstration examples are provided in Sec. 16.3.

16.2.1. *Surrogate model construction*

Consider an N-element linear array as shown in Fig. 16.1 where microstrip patches are excited individually with excitation $\mathbf{a}_{\text{nom}} = [a_{\text{nom},1}\ a_{\text{nom},2} \ldots a_{\text{nom},N}]^T$ representing the nominal defect-free low-sidelobe design. As in Bucci et al. (2000), Rodriguez-Gonzalez et al. (2009), Rodriguez and Ares (2000) and Iglesias et al. (2008), we assume in-phase excitation of the array elements.

We consider two models for the nominal array: the computationally inexpensive array factor assuming ideal isotropic radiators, i.e., model $\mathbf{R}_{c,\text{nom}}$; and an expensive EM-simulated radiation pattern of the array of the microstrip patches, denoted as $\mathbf{R}_{f,\text{nom}}$ (here, the notation $\mathbf{R}$ — with subscript that defines the particular model — denotes a vector of values of the magnitude-only coplanar far-field response (in dB) determined at elevation angles within the range $-90° \leq \theta \leq 90°$).

We use these models to define a response correction function $D(\theta)$ (to be used during fault detection) as follows. Let $p_f^k = [\theta_f^k\ r_f^k], k = 1, \ldots, K_f$ be K_f points corresponding to pattern maxima

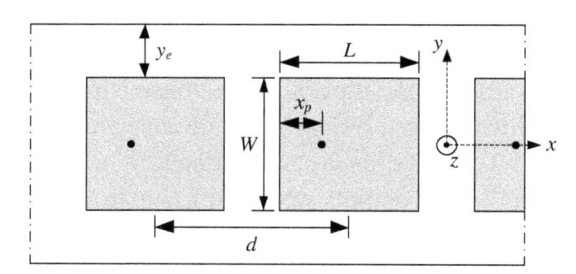

Fig. 16.1. Partial top view (in vicinity of array center) of an N-element microstrip patch array on a finite dielectric substrate with an infinite metal ground plane (not shown). Patches have identical lengths L and widths W, probe feed positions x_p (indicated by black dots), and spacings d.

extracted from the $\mathbf{R}_{f,\text{nom}}$ response, with θ_f^k the elevation angle and r_f^k the corresponding pattern value. We denote by $p_c^k = [\theta_c^k \ r_c^k]$, $k = 1, \ldots, K_c$, similar points (i.e., also corresponding to pattern maxima) extracted from the $\mathbf{R}_{c,\text{nom}}$ response obtained by using $\mathbf{a}_{\text{nom}}$ in the array factor. The response correction function $D(\theta)$ is defined in dB as

$$D(\theta) = E_f(\theta) - E_c(\theta) \qquad (16.1)$$

with

$$E_f(\theta) = F([\theta_f^1 \ldots \theta_f^{K_f}], [r_f^1 \ldots r_f^{K_f}], \theta), \qquad (16.2)$$

where $F(X, Y, x)$ represents the interpolation of the data value vector Y (defined on a discrete argument set X) onto x. A similar definition holds for $E_c(\theta)$. In other words, $E_f(\theta)$ is a function that interpolates the pattern maxima of the EM-simulated radiation pattern; likewise, $E_c(\theta)$ interpolates the pattern maxima of the array factor.

Consider now an arbitrary defective array described by its excitation vector $\mathbf{a}_F$, which is obtained by adjusting some of the elements of $\mathbf{a}_{\text{nom}}$ to reflect a faulty state. In accordance with the previous notation, we denote the defective array factor by $\mathbf{R}_c$, and its EM-simulated pattern by $\mathbf{R}_f$. We define a fast, inexpensive surrogate model $\mathbf{R}_s$ that can replace the computationally expensive model $\mathbf{R}_f$ in the detection process without compromising accuracy — in particular, we define $\mathbf{R}_s$ as the first-order correction of $\mathbf{R}_c$ by $D(\theta)$ in dB as

$$\mathbf{R}_s(\mathbf{a}_F, \theta) = \mathbf{R}_c(\mathbf{a}_F, \theta) + D(\theta). \qquad (16.3)$$

Thus, the fast surrogate for the pattern of a defective array is obtained by superimposing the correction function D on the response of its analytical array factor model. Note that the correction term is a function of angle θ but it does not depend on the defective array's excitation amplitudes $\mathbf{a}_F$. The correction term also can be recognized as the effective pattern of the array embedded element.

For the sake of illustration, consider the array shown in Fig. 16.2(a) along with the EM-simulated and array-factor-based

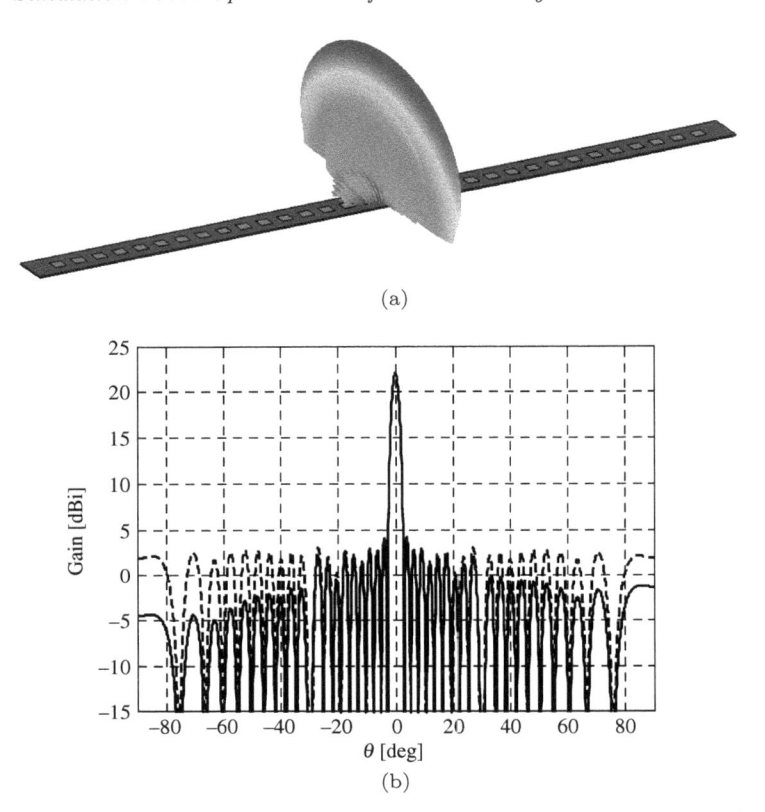

378 Simulation-Based Optimization of Antenna Arrays

Fig. 16.2. Linear array of 32 microstrip patch antennas on a dielectric substrate with finite lateral dimensions (the metal ground plane is not shown): (a) 3D-view and the simulated pattern (50 dB range); (b) radiation response in the E-plane at 10 GHz given excitation amplitudes optimized for minimum SLL: (—) EM model $\mathbf{R}_{f,\text{nom}}$, and (- - -) analytical array factor model $\mathbf{R}_{c,\text{nom}}$.

responses shown in Fig. 16.2(b). Note that despite the symmetry of the excitation, neither the EM-simulated response nor the correction function are fully symmetric with respect to θ. This is because the physical structure of the array is not symmetric due to the location of the feeding points (not shown in the picture).

Figure 16.3 shows the plot of $D(\theta)$ for the array of Fig. 16.2(a) extracted from the responses in Fig. 16.2(b). It should be noted that the response correction (16.3) ensures perfect matching between the corrected analytical model and the EM-simulation model $\mathbf{R}_f$ at the

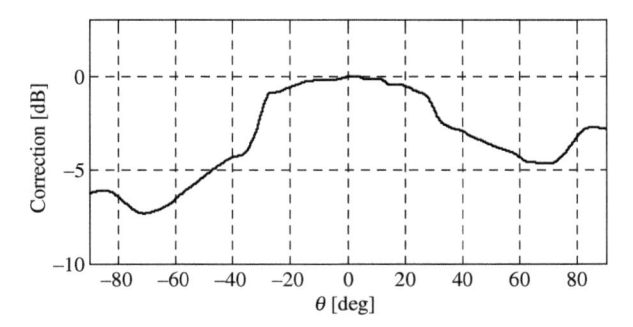

Fig. 16.3. Correction function D extracted from the SLL-optimized array responses and the corresponding array factor model (cf. (16.1)).

point at which the correction has been established. This is referred to as zeroth-order consistency (Alexandrov and Lewis, 2001).

Obviously, the specific relationships between the analytical array factor model and the full-wave EM one are dependent on a particular set of excitation amplitudes. However, despite relatively large discrepancies (in absolute terms) between the models, they are well correlated so that the surrogate model accuracy is expected to be preserved throughout the design space (here, for the entire range of excitation amplitude setups). Indeed, Fig. 16.4 shows that although the correction term is constant with respect to $\mathbf{a}_F$, generalization capability of the surrogate is excellent. One of the key aspects of this good generalization is the fact that the response correction is not based on the entire response (i.e., for all possible values of the θ angle) but only on the local maxima of the low- and high-fidelity model responses. In particular, the correction term (cf. Fig. 16.3) is a relatively smooth function despite the fact that the original responses are very rugged with extremely sharp minima.

As an example, consider the 32-element array of Fig. 16.2(b). Fault detection is based on 180 amplitude-only data from the EM simulation model of the faulty array. We consider a specific fault case (Case I) with elements 3, 5, and 8 disabled (on one side of the symmetry line of the array, and in symmetrical positions on the other side). In this example, fault identification is realized using

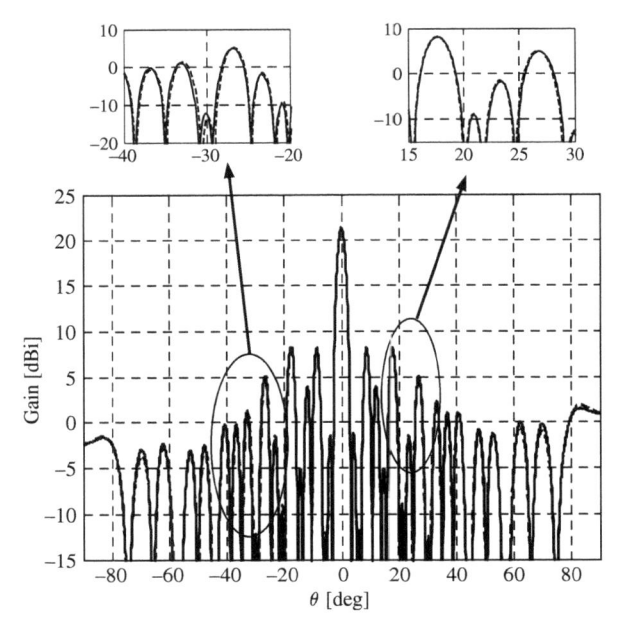

Fig. 16.4. Correction quality verification: radiation response of the EM model (—) of the array at a selected random excitation pattern, and the corresponding array factor model response (- - -) corrected using the D-term of Fig. 16.3. This plot indicates good generalization capability of the correction technique and confirms suitability of the technique for fast fault detection.

local search (specifically, a gradient-based algorithm, Nocedal and Wright, 2006) initialized through a smart random search (in order to avoid getting stuck in local optima) (Koziel and Ogurtsov, 2012d). Figure 16.5 shows the obtained radiation pattern: the match with the corresponding EM-simulated response is very good. Table 16.1 shows the fault detection results (due to space limitation and symmetry only half of the excitation vector is shown). The fault detection factors Q_k are defined as

$$Q_k = |a_k - b_k|/a_k, \qquad (16.4)$$

where b_k is the extracted kth excitation amplitude. The element is considered faulty if the corresponding Q value is close to 1, otherwise (when it is small, preferably close to zero) it is considered as working normally.

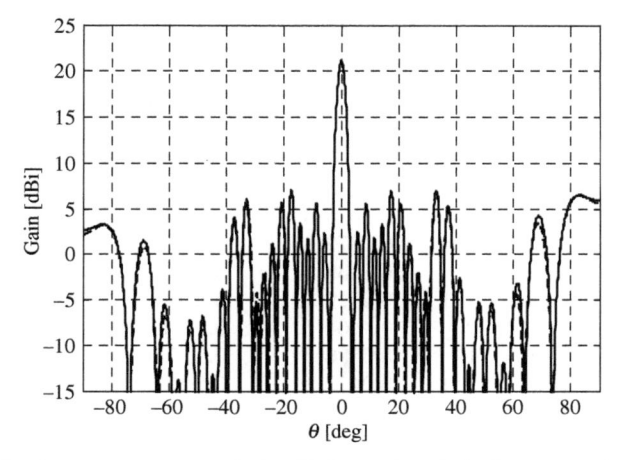

Fig. 16.5. Radiation response of the EM-simulated faulty array (—) and the corrected array factor model (- - -) upon completing the optimization-based fault detection (using amplitude-only data at 180 angles). Results for Case I (elements 3, 5, and 8 disabled).

It can be observed that all the faults have been identified correctly. Note that the entire process required only two EM simulations of the array (at the nominal and the faulty design). The method has also been verified for a number of other fault configurations. Table 16.2 shows the results for another case with elements 2, 3, and 11 disabled. Again, all the faults have been identified correctly.

16.2.2. Fault detection using fast enumeration

As noted earlier, fault detection using magnitude-only far-field samples is most often carried out by means of optimization with population-based metaheuristics such as genetic algorithms (GAs). These methods may be time-consuming, even when patterns of defective arrays are accounted for in the least expensive way, i.e., by their analytical array factors. Hence GA-based studies often explicitly test their methods on only a fraction of the possible configurations of faulty elements. Furthermore, in practice convergence to a global optimum is not necessarily guaranteed for these metaheuristics, including GAs.

Table 16.1. Array fault detection results (case I).

Excitations	Original excitation values[a]	Extracted excitation values[b]	Fault detection factor[c]	Fault detection result
a_1	0.9876	0.9875	0.0001	OK
a_2	0.3561	0.3575	0.0040	OK
a_3	$0.4728 \geq 0$	0.0372	**0.9212**	Faulty
a_4	0.2947	0.2439	0.1723	OK
a_5	$0.4608 \geq 0$	0.0419	**0.9091**	Faulty
a_6	0.5300	0.4464	0.1577	OK
a_7	0.4171	0.3596	0.1378	OK
a_8	$0.5336 \geq 0$	0.0412	**0.9228**	Faulty
a_9	0.6277	0.5140	0.1812	OK
a_{10}	0.4420	0.3813	0.1373	OK
a_{11}	0.5948	0.5291	0.1105	OK
a_{12}	0.7597	0.6482	0.1469	OK
a_{13}	0.5492	0.4888	0.1101	OK
a_{14}	0.5891	0.4683	0.2050	OK
a_{15}	0.7230	0.5967	0.1746	OK
a_{16}	0.5929	0.4983	0.1594	OK

Notes: [a]Faulty elements in this example are those with indices 3, 5, and 8 (set to zero in the faulty array case).
[b]Results of optimization-based fault detection.
[c]Fault detection factor as defined by (4).

In the present study, careful implementation allows us to realize fault detection through enumeration (exhaustive search) of the radiation patterns corresponding to all possible combinations of defective elements; here, involving one, two, and three faulty elements with on/off/partial faults as described in Sec. 16.3. (This approach is similar to the fault dictionary approach of Bandler and Salama, 1985). We prefer exhaustive search to population-based meta-heuristics, as the former is guaranteed to find the global optimum to the optimization problem. Exhaustive search — even for the relatively small arrays considered here — would not have been feasible if all faulty patterns had to be obtained via high-fidelity full-wave simulations of the corresponding degraded arrays. However, the low computational cost of our fast and accurate surrogate radiation

Table 16.2. Array fault detection results (case II).

Excitations	Original excitation values[a]	Extracted excitation values[b]	Fault detection factor[c]	Fault detection result
a_1	0.9876	0.9334	0.0549	OK
a_2	$0.3561 \geq 0$	0.0421	**0.8818**	Faulty
a_3	$0.4728 \geq 0$	0.0252	**0.9467**	Faulty
a_4	0.2947	0.2439	0.1724	OK
a_5	0.4608	0.4538	0.0152	OK
a_6	0.5300	0.5164	0.0257	OK
a_7	0.4171	0.3596	0.1379	OK
a_8	0.5336	0.5138	0.0371	OK
a_9	0.6277	0.5393	0.1408	OK
a_{10}	0.4420	0.3733	0.1554	OK
a_{11}	$0.5948 \geq 0$	0.0512	**0.9139**	Faulty
a_{12}	0.7597	0.6703	0.1177	OK
a_{13}	0.5492	0.4905	0.1069	OK
a_{14}	0.5891	0.4722	0.1984	OK
a_{15}	0.7230	0.6003	0.1697	OK
a_{16}	0.5929	0.5238	0.1165	OK

Notes: [a]Faulty elements in this example are those with indices 2, 3, and 11 (set to zero in the faulty array case).
[b]Results of optimization-based fault detection.
[c]Fault detection factor as defined by (4).

patterns $\mathbf{R}_s$, a substitute for these high-fidelity simulations, now makes exhaustive search possible.

Our enumeration approach can be explained as follows. Consider a defective array of which the EM-simulated magnitude-only far-field pattern $\mathbf{R}_{f,\text{defect}}$ is available at sampling angles $\theta_s = [\theta_{s.1} \ \theta_{s.2} \ldots \theta_{s.M}]$, and of which the faulty elements are unknown. Suppose that the total number of possible configurations of faulty elements is N_F, and that $\mathbf{a}_{F.k} = [a_{F.k,1} \ a_{F.k,2} \ldots a_{F.k,N}]$ is the element excitation vector corresponding to the kth fault configuration.

In the first step, a lookup matrix $\mathbf{U}$ consisting of vectors $\mathbf{R}_c(\mathbf{a}_{F.k}, \boldsymbol{\theta}_s)$, i.e., the array factors corresponding to all possible faulty apertures, is prepared. This is done only once for any given nominal array.

In the second step, fault detection takes place through identifying the index $k_{\min}$ of the correct fault configuration by means of

$$\begin{aligned} k_{\min} &= \underset{k \in \{1,\ldots,N_F\}}{\arg\min} \ ||\mathbf{R}_{f,\text{defect}}(\boldsymbol{\theta}_s) - \mathbf{R}_s(\mathbf{a}_{F.k},\boldsymbol{\theta}_s)|| \\ &= \underset{k \in \{1,\ldots,N_F\}}{\arg\min} \ ||[\mathbf{R}_{f,\text{defect}}(\boldsymbol{\theta}_s) - D(\boldsymbol{\theta}_s)] - \mathbf{R}_c(\mathbf{a}_{F.k},\boldsymbol{\theta}_s)||, \end{aligned} \tag{16.5}$$

where $D(\boldsymbol{\theta}_s)$ is obtained from (16.1)–(16.2) using cubic spline interpolation via the *interp1* function of Matlab. Array elements with "off" faults correspond to zero entries of the excitation vector $\mathbf{a}_{F.k\min}$. Other values (between zero and the nominal amplitude value) can be considered as well (partial faults); in this case, the lookup matrix has to include all the relevant array factors.

With careful implementation in Matlab (e.g., vectorization of operations to avoid "for" loops), setting up the lookup matrix $\mathbf{U}$ for a 32-element array with up to three partial faults takes less than 60 s on a 2-GHz 12 core dual CPU with 64 GB RAM. Equation (16.5) is executed as one command in a vectorized fashion, where the entire matrix $\mathbf{U}$ is compared to a matrix consisting of N_F identical copies of the vector $\mathbf{R}_{f,\text{defect}}(\boldsymbol{\theta}_s) - D(\boldsymbol{\theta}_s)$. This makes finding $k_{\min}$ extremely fast (here only a few milliseconds for a 32-element array).

Below, we evaluated our method against each and every possible configuration of one, two, and three faulty patches. This comprehensive numerical verification required the evaluation of tens of thousands of realistic faulty patterns. Such a study would be unfeasible if the expensive EM model were referred to each and every time.

In order to address this, we used superposition to obtain the far-field $\mathbf{E}(\theta, \varphi)$ associated with the kth configuration of faulty elements as $\mathbf{E}(\theta, \phi) = \sum_{n=1}^{N} a_{F.k,n} \mathbf{E}_n(\theta, \phi)$, where $a_{F.k,n}$ is the amplitude of the impressed voltage source exciting the nth port; and $\mathbf{E}_n(\theta, \phi)$ is the EM-simulated complex far-field of the array with only the nth element driven with the unity excitation. For evaluation purposes, we only used the magnitude of the total field $\mathbf{E}(\theta, \phi)$.

16.3. Numerical Results

We consider two nominal microstrip patch arrays with 16 and 32 elements (cf. Fig. 16.1). All arrays are implemented on a finite 1.575-mm-thick Rogers RT5880 dielectric substrate ($\varepsilon_r = 2.2$) which extends laterally beyond the patch edges by $x_e = 18.4$ mm in the x-direction and $y_e = 9.2$ mm in the y-direction. The patches have dimensions $L = W = 9.2$ mm (metallization is with 70 μm copper) and the spacing between their centers is $d = 15$ mm $\equiv 0.5$ free-space wavelengths at the operating frequency of 10 GHz. Each patch is independently fed by a wire probe containing an impressed voltage source, situated at a distance $x_p = 6.3$ mm from the leftmost patch edge. Hence the array geometry is not symmetrical in the x−direction relative to the array center.

High-fidelity models $\mathbf{R}_f$ for the nominal arrays were implemented in CST Microwave Studio and simulated with the transient solver. The number of mesh cells were as follows: ∼900,000 for the 16-element array (simulation time 8 min on a 2-GHz 12 core dual CPU with 64 GB RAM); and ∼1,700,000 for the 32-element array (simulation time 15 min).

16.3.1. *On–off faults*

For each of the considered arrays, the fault detection procedure described in Sec. 16.2 was followed. Test data consisted of the EM-simulated radiation patterns associated with all combinations of one, two, and three faulty elements that were possible for that array size (e.g., for the case $N = 32$ the number of test patterns was $n^* = \sum_{k=1,2,3} 32!/k!(32-k)! = 5488$). Faulty elements were represented by setting their excitations in the nominal excitation vector $\mathbf{a}_{\text{nom}}$ to zero.

The algorithm was furthermore tested for different values of M, the number of pattern samples at equally spaced sampling positions θ in a predefined range (here, $-64° \leq \theta \leq 64°$). Performance of the algorithm in the presence of measurement errors was also modeled by adding noise to the faulty test patterns. The noise was independent

Table 16.3. Percentage of all possible combinations of 1, 2, and 3 faults predicted correctly assuming on–off faults.

	M			
	$N = 16$ $(n^* = 696)$[a]		$N = 32$ $(n^* = 5488)$	
Δ_{noise} (dB)	$N + 1$[b]	$2N + 1$[b]	$N + 1$[b]	$2N + 1$[b]
0 (noise-free)	99.43	99.71	99.42	99.93
0.5	99.43	99.71	99.23	99.93
1.0	99.14	99.71	98.94	99.91
2.0	95.55	99.28	97.50	99.76

Notes: [a]n^*: number of test patterns.
[b]Number of equally spaced sample angles in range $-64° \leq \theta \leq 64°$.

and identically distributed over the interval $(-\Delta_{\text{noise}}, \Delta_{\text{noise}})$, with $\Delta_{\text{noise}} = 0.5, 1,$ and $2\,\text{dB}$.

Table 16.3 gives the percentage of fault configurations predicted correctly for the numbers of sampling angles $M = N + 1$ and $M = 2N + 1$, both without and with measurement noise. In the noise-free case, predictive accuracies exceeded 99.7% for both arrays when $M = 2N + 1$, and 99.4% when approximately a half of the samples were used ($M = N + 1$). Performance of the algorithm was relatively insensitive to noise for $M = 2N + 1$ with a negligible drop in the predictive accuracy. It was more sensitive for $M = N + 1$ with the accuracy decreases of about 4% and 2% for the 16 and 32 element arrays, respectively, at the maximum noise level of $2\,\text{dB}$.

Figures 16.6 and 16.7 provide graphical illustration of the method for each of the arrays when no noise was present. The subfigures labeled (a) show the EM-simulated model of the nominal (zero-defect) array, $\mathbf{R}_f(\mathbf{a}_{\text{nom}})$; the corresponding analytical array factor model $\mathbf{R}_c(\mathbf{a}_{\text{nom}})$; and the correction function. The subfigures labeled (b) and (c) compare EM-simulated faulty patterns $\mathbf{R}_f$, the corresponding corrected array factor models $\mathbf{R}_s$, and the analytical array factor models $\mathbf{R}_c$ for specific combinations of faulty elements. Agreement between $\mathbf{R}_f$ and $\mathbf{R}_s$ is generally good, especially over the sectors of elevation angles from which samples have been taken.

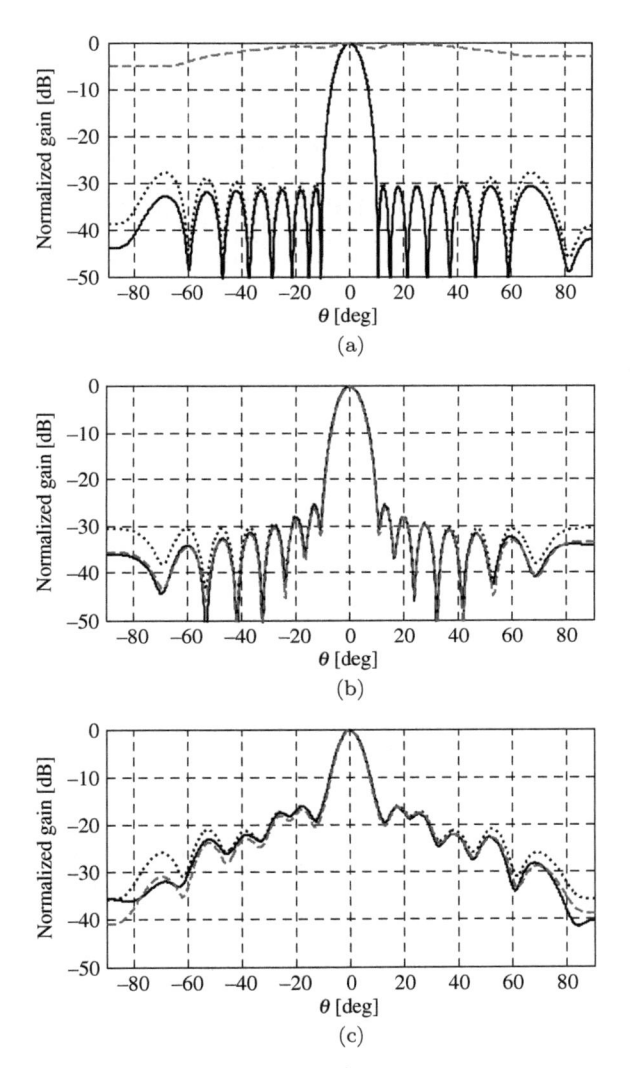

Fig. 16.6. 16-Element array example. (a) E-plane radiation responses given excitation amplitudes optimized for minimum SLL: EM model $\mathbf{R}_f$ (—), analytical array factor model $\mathbf{R}_c(\cdots)$, and correction function D (- - -). (b) E-plane radiation responses when element 1 is disabled: EM model $\mathbf{R}_f$ (—), analytical array factor model $\mathbf{R}_c(\cdots)$, and corrected analytical array factor model $\mathbf{R}_s$ (- - -). (c) E-plane radiation responses when elements 4 and 5 are disabled: EM model $\mathbf{R}_f$ (—), analytical array factor model $\mathbf{R}_c(\cdots)$, and corrected analytical array factor model $\mathbf{R}_s$ (- - -).

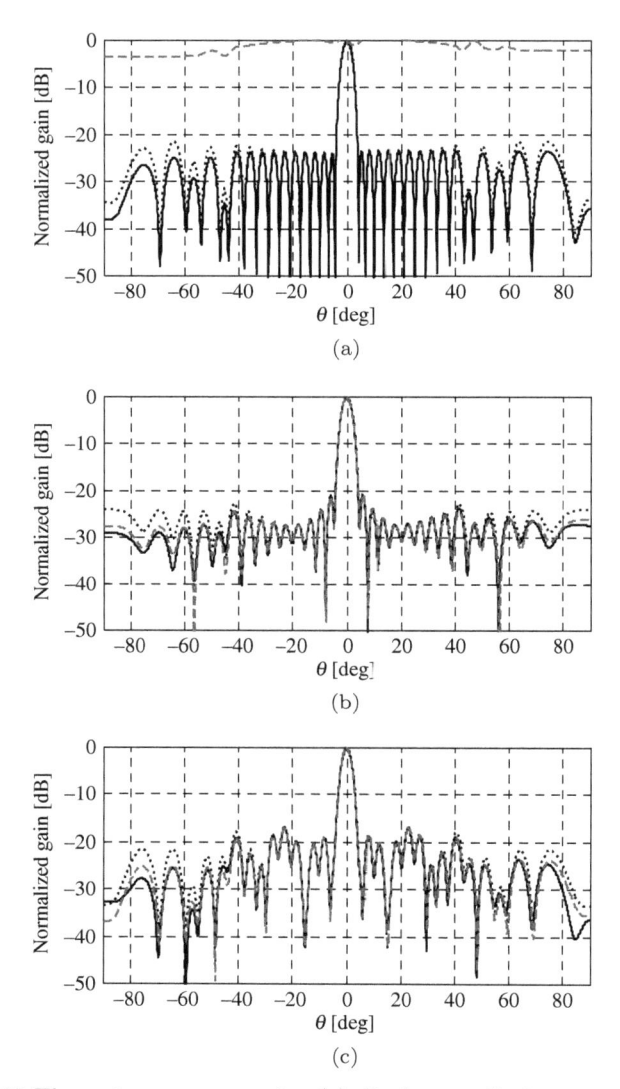

388 Simulation-Based Optimization of Antenna Arrays

Fig. 16.7. 32-Element array example. (a) E-plane radiation responses given excitation amplitudes optimized for minimum SLL: EM model $\mathbf{R}_f$ (—), analytical array factor model $\mathbf{R}_c(\cdots)$, and correction function D (- - -). (b) E-plane radiation responses when elements 1 and 3 are disabled: EM model $\mathbf{R}_f$ (—), analytical array factor model $\mathbf{R}_c(\cdots)$, and corrected analytical array factor model $\mathbf{R}_s$ (- - -). (c) E-plane radiation responses when elements 3, 8, and 9 are disabled: EM model $\mathbf{R}_f$ (—), analytical array factor model $\mathbf{R}_c(\cdots)$, and corrected analytical array factor model $\mathbf{R}_s$ (- - -).

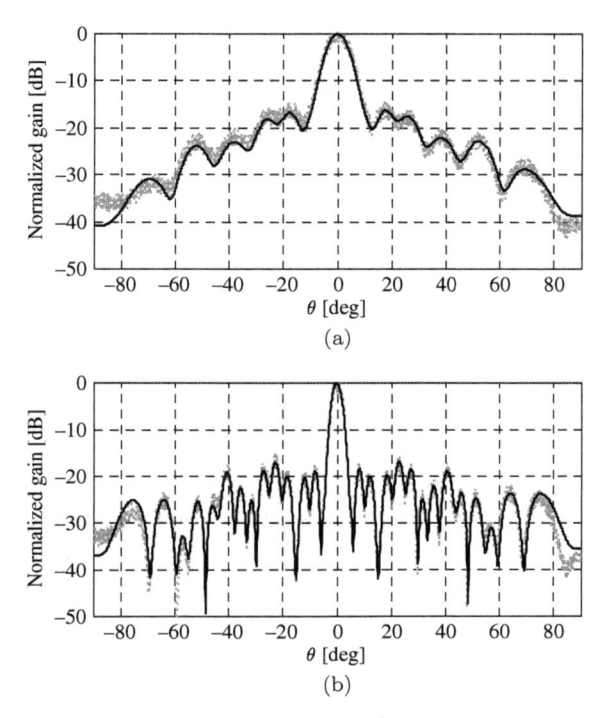

Fig. 16.8. Examples of radiation patterns of faulty arrays that were correctly identified in the presence of measurement noise ($\Delta_{\text{noise}} = 2\,\text{dB}$; $M = 2N + 1$). (a) E-plane radiation responses of 16-element array when elements 4 and 5 are disabled: EM model $\mathbf{R}_f$ ($\cdots$) with added noise, and corrected analytical array factor model $\mathbf{R}_s$ (—). (b) E-plane radiation responses of 32-element array when elements 3, 8, and 9 are disabled: EM model $\mathbf{R}_f$($\cdots$) with added noise, and corrected analytical array factor model $\mathbf{R}_s$ (—).

Figure 16.8 gives examples of specific instances of 16-element and 32-element faulty arrays that were correctly identified in the presence of noise with $\Delta_{\text{noise}} = 2\,\text{dB}$; EM-simulated faulty patterns with added noise and corresponding corrected array factor models are shown.

16.3.2. Partial faults

For each of the considered arrays, we assumed three possible discrete states for each excitation, namely, 100% working ("on"), faulty but still operational at 50% of its nominal amplitude, and "off". This caused the number of test cases to increase to $n^* = 4992$ ($N = 16$)

Table 16.4. Percentage of all possible combinations of 1, 2, and 3 faults predicted correctly when partial faults were assumed.

	M			
	$N = 16$ $(n^* = 4992)$[a]		$N = 32$ $(n^* = 41728)$[a]	
Δ_{noise} (dB)	$N+1$[b]	$2N+1$[b]	$N+1$[b]	$2N+1$[b]
0	96.374	99.099	94.095	97.417
1	93.770	98.297	89.884	95.744
2	83.053	94.151	82.046	93.000

Notes: [a]n^*: number of test patterns.
[b] Number of equally spaced sample angles in range $-64° \leq \theta \leq 64°$.

and $n^* = 41728$ ($N = 32$). Table 16.4 gives the percentage of all possible fault configurations predicted correctly for $M = N + 1$ and $M = 2N+1$, both without and with measurement noise. In the noise-free case, predictive accuracies exceeded 99% ($N = 16$) and 97% ($N = 32$) for $M = 2N + 1$, and 96% ($N = 16$) and 94% ($N = 32$) for $M = N + 1$. As expected, performance was more sensitive to noise than in the on–off case, with the best performance observed for $M = 2N + 1$, where accuracies under worst-case noise conditions ($\Delta_{\text{noise}} = 2\,\text{dB}$) were about 94% ($N = 16$), and 93% ($N = 32$).

16.4. Summary

In this chapter, a technique for accurate fault detection in small linear arrays of microstrip patches has been presented. Specifically, an accurate, computationally inexpensive surrogate model for calculating the far-field radiation patterns of all possible faulty arrays was developed. The surrogate model accounted for mutual coupling effects. The inexpensiveness of the surrogate model allowed us to accomplish fault detection through exhaustive search. It took less than a minute for calculation of all faulty patterns (i.e., setting up the lookup matrix) of the 32-element array with up to three partial faults and 41,782 possible faulty aperture configurations. Furthermore, having the lookup matrix set up, it took only few milliseconds to identify the fault configuration producing a particular test faulty pattern. Notice that setting up the lookup matrix needs to be done only once. In the same time computational costs of setting up the lookup matrix would be enormous if the faulty patterns were evaluated in a straightforward way using high-fidelity full-wave discrete simulations.

It is notable that constructing the surrogate model requires only a single high-fidelity full-wave simulation, namely, of the entire defect-free array. In contrast, the state-of-the-art in linear microstrip patch array fault detection (Patnaik et al., 2007) required as many full-wave simulations of the entire array as there were faulty array configurations (696 for a 16-element array with up to three faults). It was shown that the discussed method is robust in handling arrays which are twice larger than arrays considered in Patnaik et al. (2007), as well as in dealing with partial faults and measurement noise.

Chapter 17

Surrogate-Assisted Tolerance Analysis of Microstrip Linear Arrays with Corporate Feeds

In this chapter, we address the problem of fast statistical analysis of microstrip linear antenna arrays with corporate feeds using surrogate-assisted methods. A realistic design has to take into account manufacturing tolerances that normally lead to a degradation of array performance in terms of its circuit and radiation responses. Tolerances may be pertinent to array radiating elements, element spacing, geometry parameters of the corporate feed (resulting, among others in deviation of the excitation amplitudes and phases from the required nominal values), as well as deviations of parameters of the substrate, e.g., substrate local thickness and permittivity. The main performance figure affected by these parameter deviations is the sidelobe level (SLL) in case of low-sidelobe patterns. Estimation of the SLL spread requires statistical analysis that takes into account all of the aforementioned factors. Accurate evaluation of SLLs requires full-wave electromagnetic (EM) analysis of the array (and the feed if the entire system is of interest) which is very expensive in computational terms. Consequently, performing statistical analysis directly at the level of the EM array model is impractical. Here, we discuss simple techniques that permit speeding up the process of antenna array statistical analysis. These techniques include various surrogate modeling methods as well as correction techniques that are used to align responses of fast surrogates with responses of the respective EM models.

17.1. Manufacturing Tolerances in Linear Antenna Arrays with Corporate Feeds

Microstrip antenna arrays are described by a large number of geometry parameters pertinent to antenna elements, their spacing, corporate feeds, as well as material parameters of the dielectric substrates. All of these parameters are subject to uncertainties, typically due to fabrication inaccuracies. For the sake of further description, four types of parameters have been distinguished as described below:

- $\mathbf{x}_e = [x_{e.1} \ldots x_{e.Ne}]^T$ — geometry parameters of array elements, antennas;
- $\mathbf{x}_s = [x_{s.1} \ldots x_{s.Ns}]^T$ — element spacings;
- $\mathbf{x}_f = [x_{f.1} \ldots x_{f.Nf}]^T$ — geometry parameters of the corporate feed;
- $\mathbf{x}_m = [x_{m.1} \ldots x_{m.Nm}]^T$ — material parameters of the substrate, which, in practice, could be dielectric permittivity ε_r or permittivity and thickness h.

Parameter variation is described by appropriate probability distributions. For the sake of simplicity, these could be, e.g., independent normal distributions with zero mean and certain variance σ (e.g., 0.017 mm for a standard chemical etching process), or uniform with a specified maximum deviation. In reality, majority of parameter deviations are correlated (e.g., a variation of the gap between coupled transmission lines is negatively correlated to the line width variations, etc.). However, in the illustration examples of Sec. 17.4, independent distributions will be assumed.

17.2. Local Surrogate Modeling of Antenna Array Apertures

In this section, local surrogate modeling of antenna array apertures is described. In particular, we are interested in modeling SLL as a function of geometry and/or material parameter deviation of the array elements and their spacing from the nominal values. The surrogate is constructed using the array factor model involving the EM-simulated element pattern, further corrected using the response feature approach (Koziel, 2015).

17.2.1. Radiation pattern surrogate of array elements

We denote by $\mathbf{P}_{EM}(\mathbf{x}_e, \mathbf{x}_m, \theta)$ the EM-simulated radiation pattern of the array element, where $\mathbf{x}_e$ and $\mathbf{x}_m$ are the vectors of geometry and material parameters as defined in Sec. 17.1, and θ is the elevation angle. The local surrogate model $\mathbf{P}_s(d\mathbf{x}_e, d\mathbf{x}_m, \theta)$ of the pattern is a simple second-order polynomial without mixed terms defined as

$$\mathbf{P}_s(d\mathbf{x}_e, d\mathbf{x}_m, \theta) = [P_s(d\mathbf{x}_e, d\mathbf{x}_m, \theta_1) \ldots P_s(d\mathbf{x}_e, d\mathbf{x}_m, \theta_K)]^T, \quad (17.1)$$

where

$$P_s(d\mathbf{x}_e, d\mathbf{x}_m, \theta_j) = \lambda_{0.j} + \sum_{k=1}^{N_e} \lambda_{e.j.k} dx_{e.k} + \sum_{k=1}^{N_m} \lambda_{m.j.k} dx_{m.k}$$
$$+ \sum_{k=1}^{N_e} \lambda_{e.j.N_e+k} dx_{e.k}^2 + \sum_{k=1}^{N_m} \lambda_{m.j.N_m+k} dx_{m.k}^2$$

$$(17.2)$$

for $j = 1, \ldots, K$. Here, $d\mathbf{x}_e$ and $d\mathbf{x}_m$ are deviation vectors of $\mathbf{x}_e$ and $\mathbf{x}_m$, respectively. The coefficients $\lambda_{0.j}$, $\lambda_{e.j}$, and $\lambda_{m.j}$ are obtained by solving a linear regression problem of the form

$$\mathbf{P}_{EM}(\mathbf{x}_e^{(0)} + d\mathbf{x}_e^{(j)}, \mathbf{x}_m^{(0)} + d\mathbf{x}_m^{(j)}, \theta)$$
$$= \mathbf{P}_s(d\mathbf{x}_e^{(j)}, d\mathbf{x}_m^{(j)}, \theta), \quad j = 1, \ldots, N_B \quad (17.3)$$

in which $\mathbf{x}_e^{(0)}$ and $\mathbf{x}_m^{(0)}$ are nominal parameter values with respect to which the model is established, whereas $d\mathbf{x}_e^{(j)}$ and $d\mathbf{x}_m^{(j)}$,

$j = 1, \ldots, N_B$, are N_B training designs. Here, we use $N_B = 2(N_e + N_m) + 1$ with the training designs allocated according to star distribution (Cheng et al., 2006). Using this particular number of training point means that the surrogate (17.1) is interpolative. Furthermore, the model coefficients can be obtained analytically.

17.2.2. Local modeling of array aperture: Array factor model

The primary choice for the low-fidelity model of the array aperture radiation pattern is the analytical array factor model. The model will be denoted as $\mathbf{P}_{\text{AF}}(\mathbf{a}, \mathbf{x}_s, \mathbf{P}(\theta), \theta)$, where $\mathbf{a}$ is the excitation amplitude taper, $\mathbf{x}_s$ is a vector of element spacings, θ is the elevation angle, and $\mathbf{P}(\theta)$ is the element pattern at the angle θ.

The array factor model is composed with the radiation pattern surrogate of the array element $\mathbf{P}_s(d\mathbf{x}_e, d\mathbf{x}_m, \theta)$ in order to create the compound model

$$\mathbf{P}_{\text{AFS}}(\mathbf{a}, d\mathbf{x}_e, d\mathbf{x}_m, d\mathbf{x}_s, \theta) = \mathbf{P}_{\text{AF}}(\mathbf{a}, \mathbf{x}_s^{(0)} + d\mathbf{x}_s, \mathbf{P}_s(d\mathbf{x}_e, d\mathbf{x}_m, \theta), \theta) \quad (17.4)$$

in which $\mathbf{x}_s^{(0)}$ is the nominal element spacing vector, whereas $d\mathbf{x}_s$ is the spacing deviation vector.

The model (17.4) is a local model of the array aperture radiation pattern as a function of the excitation taper (here, excitation amplitudes), elevation angle, and deviations of the geometry and material parameters of the element as well as element spacing values (all with respect to their nominal values).

17.2.3. Local modeling of array aperture: Model correction using response features

In order to make the model (17.4) suitable for statistical analysis of EM models of array apertures, a correction has to be made which aligns the two models. Here, we utilize the response feature approach as explained in Fig. 17.1. Because SLL is determined by the local maxima of the sidelobes, it is sufficient to consider the levels of these maxima for the EM model, denoted as $F_f = [l_{f.1} \ldots l_{f.p}]^T$, and the

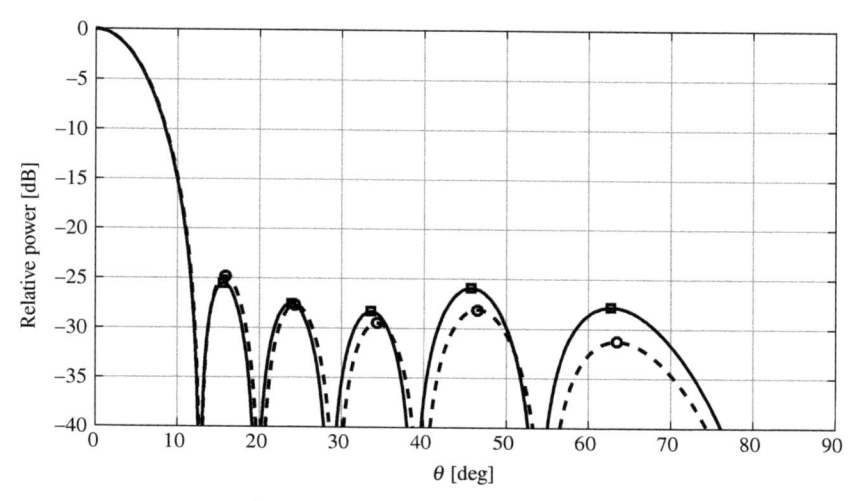

[Fig. 17.1 plot showing relative power [dB] vs θ [deg] from 0 to 90, with solid curve for EM-simulated array and dashed curve for array factor model, with square and circle feature markers.]

Fig. 17.1. Power patterns and response features of the EM-simulated array aperture (—) and the array factor model (17.4) (- - -) along with the corresponding feature points (squares for the EM model and circles for the array factor model).

array factor model, denoted as $F_{AF} = [l_{AF.1} \ldots l_{AF.p}]^T$, where p is the number of sidelobes. The models are (implicitly) functions of the excitation amplitudes (assuming in-phase excitation), elevation angle, and deviations of the geometry and material parameters of the element as well as element spacing values. The SLL value (assuming normalized patterns) can be then calculated as $SLL_f = \max\{j = 1, \ldots, p : l_{f.j}\}$ and $SLL_{AF} = \max\{j = 1, \ldots, p : l_{AF.j}\}$.

As indicated in Fig. 17.1, considerable local discrepancies between the array factor and EM model response can be observed. However, as demonstrated in Fig. 17.2, the models are very well correlated. Consequently, it is sufficient to employ a correction of the feature points so as to obtain the perfect alignment at the nominal design and utilize the corrected model to estimate the SLL value in its vicinity.

We adopt the following notation:

- $F_f(\mathbf{a}, d\mathbf{x}_e, d\mathbf{x}_m, d\mathbf{x}_s) = [l_{f.1} \ldots l_{f.p}]^T$ — sidelobe levels of the EM model (here, as a function of all relevant array variables);

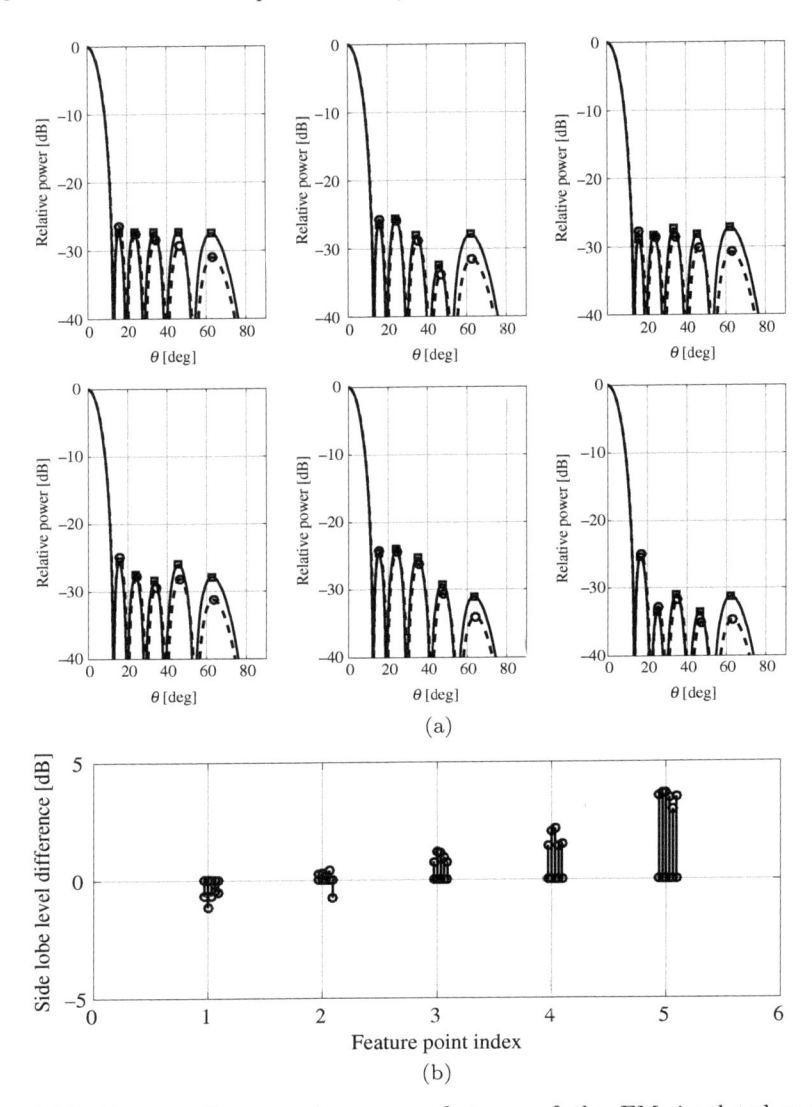

Fig. 17.2. Power patterns and response features of the EM-simulated array aperture (—) and the array factor model (17.4) (---) for various excitation amplitude tapers. Corresponding feature points (squares for the EM model and circles for the array factor model) exhibit good correlation throughout the plots: (a) power patterns for six amplitude tapers, (b) aggregated feature points.

- $F_{\mathrm{AF}}(\mathbf{a}, d\mathbf{x}_e, d\mathbf{x}_m, d\mathbf{x}_s) = [l_{\mathrm{AF}.1} \ldots l_{\mathrm{AF}.p}]^T$ — sidelobe levels of the array factor model (17.4) (here, as a function of all relevant array variables).

Using the above notations, the corrected array factor model is defined as

$$\mathrm{SLL}_s(\mathbf{a}, d\mathbf{x}_e, d\mathbf{x}_m, d\mathbf{x}_s),$$
$$= \max\{F_{\mathrm{AF}}(\mathbf{a}, d\mathbf{x}_e, d\mathbf{x}_m, d\mathbf{x}_s) + [F_f(\mathbf{a}, 0, 0, 0) - F_{\mathrm{AF}}(\mathbf{a}, 0, 0, 0)]\}, \tag{17.5}$$

where the correction term is the difference between the EM-simulated SLLs and the array factor SLLs at the nominal design.

17.3. Local Surrogate Modeling of Corporate Feeds

The local model of a corporate feed allows us to estimate the effect of feed manufacturing tolerances on the excitation amplitudes applied to the array aperture. Here, we use a very simple model, which is a first-order Taylor expansion. Let $A(\mathbf{x}_f)$ be the vector of excitation amplitudes obtained through EM simulation of the corporate feed with $\mathbf{x}_f$ being a vector of geometry parameters of the feed as explained in Sec. 17.1. The model is defined as follows:

$$s_a(d\mathbf{x}_f) = A(\mathbf{x}_f^{(0)}) + J_A(\mathbf{x}_f^{(0)}) \cdot d\mathbf{x}_f, \tag{17.6}$$

where $d\mathbf{x}_f$ is the deviation vector of feed dimensions and J_A is the feed Jacobian estimated using finite differentiation,

$$J_A(\mathbf{x}_f) = \left[\frac{A(\mathbf{x}_f + \mathbf{h}_f^{(1)})}{h_{f.1}} \ldots \frac{A(\mathbf{x}_f + \mathbf{h}_f^{(N_f)})}{h_{f.N_f}} \right], \tag{17.7}$$

with $h_{f.j}$ being the perturbation steps and $\mathbf{h}_f^{(j)} = [0 \ldots 0 \; h_{f.j} \; 0 \ldots 0]^T$ ($h_{f.j}$ on the jth position) being the perturbation vectors.

Using (17.6), the surrogate model SLL$_{sc}$ of the antenna array, SLL can be represented as

$$\text{SLL}_{sc}(d\mathbf{x}_f, d\mathbf{x}_e, d\mathbf{x}_m, d\mathbf{x}_s) = \text{SLL}_s(s_a(d\mathbf{x}_f), d\mathbf{x}_e, d\mathbf{x}_m, d\mathbf{x}_s) \tag{17.8}$$

The surrogate is established at the nominal design ($\mathbf{x}_f^{(0)}$, $\mathbf{x}_e^{(0)}$, $\mathbf{x}_m^{(0)}$, $\mathbf{x}_s^{(0)}$), and it is a function of parameter deviations $d\mathbf{x}_f, d\mathbf{x}_e, d\mathbf{x}_m, d\mathbf{x}_s$.

17.4. Case Study: 12-Element Microstrip Array

In this section, an illustration case of a 12-element microstrip linear array is discussed. The details of the array geometry are provided in Sec. 17.4.1. Section 17.4.2 contains numerical results.

17.4.1. *Array structure*

Our illustration example is a broadside 12-element microstrip linear array designed to operate at 5.8 GHz. The array design realizes the −27.3 dB SLL of the H-plane radiation pattern at 5.8 GHz and keeps the same major lobe null-to-null beamwidth as the initial design, which was configured with the −25 dB SLL Tschebyscheff excitation taper, [1.0000 0.9307 0.8031 0.6372 0.4572 0.4225]^T. The array comprises an aperture of microstrip patch radiators integrated with a microstrip feed.

Figure 17.3 shows the array element, a microstrip patch antenna (MPA). MPA patches reside on a 1.524-mm Arlon 250 D substrate. The MPA patches are with 17.5 μm copper metallization. The MPA is energized at the dominant TM$_{010}$ mode through a ground plane slot. The ground plane serves as a ground for the array elements and the microstrip corporate feed. The feed is configured on the other size of the ground on a 0.762-mm thick Arlon AD 250 substrate. The MPA input is a 50-ohm microstrip of the corporate feed. The microstrip is terminated with an open-end stub of the same trace width. The MPA produces a broadside radiation pattern and it had been designed for the minimum of the reflection coefficient at 5.8 GHz and maximum of the realized gain toward zenith at 5.8 GHz.

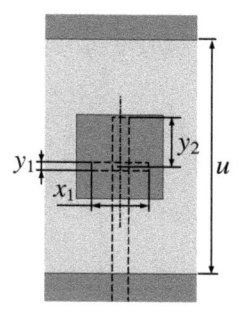

Fig. 17.3. Array element, a microstrip patch antenna (MPA): geometry. The MPA patch is 14.35 mm × 14.35 mm. Ground plane slot dimensions are $x_1 = 10.0$ mm and $y_1 = 0.75$ mm; the slot is 5.0 mm down off the patch center. Open-end microstrip stub length $y_2 = 6.5$ mm. Element spacing $s = 25.9$ mm, MPA substrate dimension $u = 40.0$ mm.

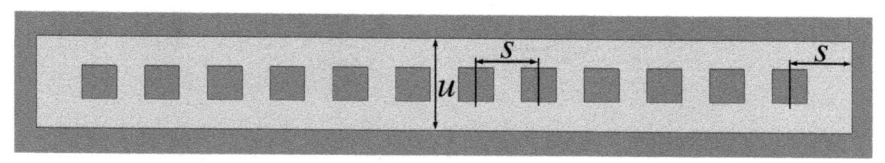

Fig. 17.4. Linear array aperture comprising 12 MPAs: geometry. Element spacing $s = 25.9$ mm, MPA substrate dimensions are 40 mm × 336 mm.

The linear array aperture comprising 12 MPAs is shown in Fig. 17.4. Patches of the MPAs are spaced a half-wavelength center-to-center. To energize the aperture with the required excitation taper a microstrip corporate feed had been designed as a part of the integrated structure. The corporate feed topology, schematic of the junctions, implementation of the junctions, and the feed layout are shown in Fig. 17.5. The junctions are connected within the feed with 50 ohms microstrips. The path lengths to all MPAs were set so that the MPAs would be driven in phase.

The tree of the corporate feed, shown in Fig. 17.5(a), with the power distribution of the junctions **p**, had been found with numerical optimization of a fast approximate model of the feed (specifically, a kriging interpolation model). T-junctions of unequal power split, shown in Figs. 17.5 (b) and 17.5(c) had been used to realize the signal

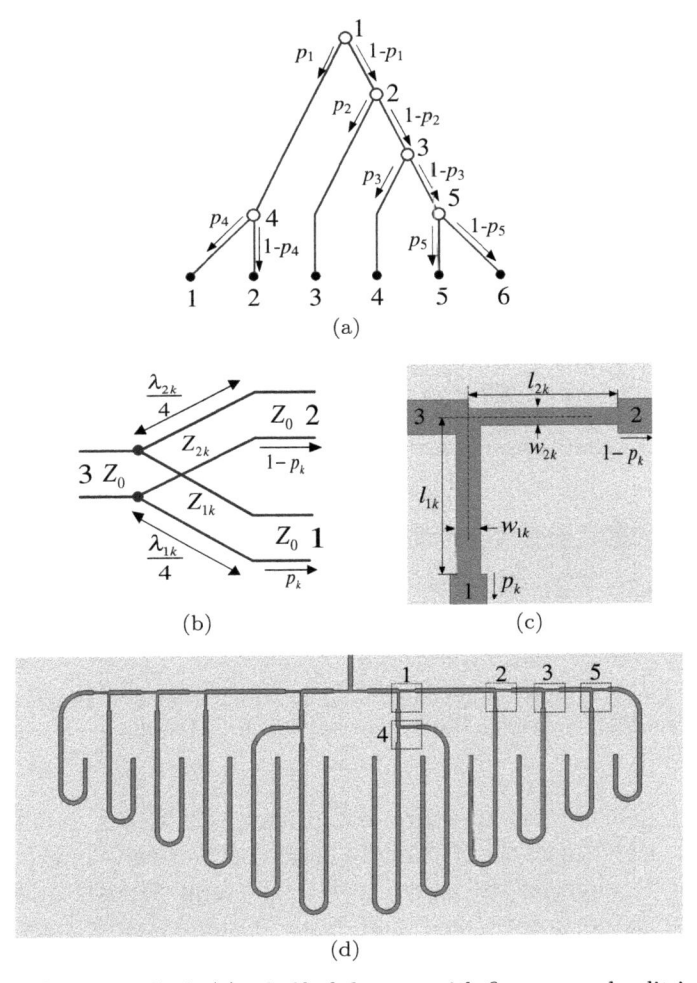

402 Simulation-Based Optimization of Antenna Arrays

Fig. 17.5. Corporate feed: (a) a half of the tree with five unequal-split junctions; (b) schematic of the T-junction k; (c) microstrip realization of T-junction k; (d) feed layout view. A fraction of the total output power propagating to the output 1 of a particular junction k is denoted with p_k.

distribution **p** within the feed. In the schematic

$$Z_{1k} = Z_0/\sqrt{p_k}, \tag{17.9}$$

$$Z_{2k}(k) = Z_0/\sqrt{1-p_k}, \tag{17.10}$$

where Z_0 is the line impedance of the connecting microstrips, here 50 ohms. Notice that the unequal power distribution within the feed had been realized with the same microstrip configuration of Fig. 17.5(c).

Dimensions of the microstrip junctions, shown in Fig. 17.5(c), resulting in the required power distribution **p**, had been found with optimization of the surrogate model. The surrogate has been set up using high-fidelity simulations of the discrete EM model of the T-junction allocated on a rectangular grid and, subsequently, approximated using kriging. Initial dimensions, calculated from the parameters of the schematic, quarter wavelengths and line impedances of the transformer sections, had been adjusted in the process of optimization.

17.4.2. *Results and discussion*

The results of statistical analysis have been gathered in Tables 17.1–17.4. There are four cases considered, assuming different probability distributions of the manufacturing tolerances, specifically: Gaussian with zero mean and variance of 0.017 mm (Case 1), Gaussian with zero mean and variance of 0.033 mm (Case 2), uniform with maximum deviation of 0.05 mm (Case 3), and uniform with maximum deviation of 0.1 mm (Case 4). In all cases, distribution of dielectric permittivity is the same as mentioned above in terms of absolute numbers (see Figs. 17.6–17.9).

The nominal SLL, realized by the design, is −27.3 dB. The analysis has been executed using the surrogate model described earlier in this chapter and 1000 random samples. It can be observed

Table 17.1. Results of statistical analysis of the 12-element array assuming Gaussian probability distribution with 0.017 mm variance.

Tolerances considered	SLL [dB] Average	Worst	Standard deviation
Element geometry/material	−27.3	−27.3	0.001
Element spacing	−27.3	−27.2	0.01
Feed geometry	−26.7	−25.3	0.34
All components	−26.7	−25.3	0.33

Table 17.2. Results of statistical analysis of the 12-element array assuming Gaussian probability distribution with 0.033 mm variance.

Tolerances considered	SLL [dB]		
	Average	Worst	Standard deviation
Element geometry/material	−27.3	−27.2	0.002
Element spacing	−27.2	−27.1	0.02
Feed geometry	−26.2	−23.7	0.60
All components	−26.2	−23.6	0.61

Table 17.3. Results of statistical analysis of the 12-element array assuming uniform probability distribution with maximum deviation of 0.05 mm.

Tolerances considered	SLL [dB]		
	Average	Worst	Standard deviation
Element geometry/material	−27.3	−27.3	0.002
Element spacing	−27.2	−27.2	0.014
Feed geometry	−26.3	−24.7	0.52
All components	−26.3	−24.7	0.52

Table 17.4. Results of statistical analysis of the 12-element array assuming uniform probability distribution with maximum deviation of 0.1 mm.

Tolerances considered	SLL [dB]		
	Average	Worst	Standard deviation
Element geometry/material	−27.3	−27.2	0.004
Element spacing	−27.2	−27.1	0.027
Feed geometry	−25.5	−22.3	0.91
All components	−25.5	−22.5	0.89

that SLL is insensitive to variations of the element geometry as well as element spacing (assuming practical levels of parameter deviations), whereas it is highly sensitive to the feed geometry. Even for small deviations (Case 1), degradation of SLL can be significant (around 0.6 dB in terms of the mean value and almost 1.5 dB for the

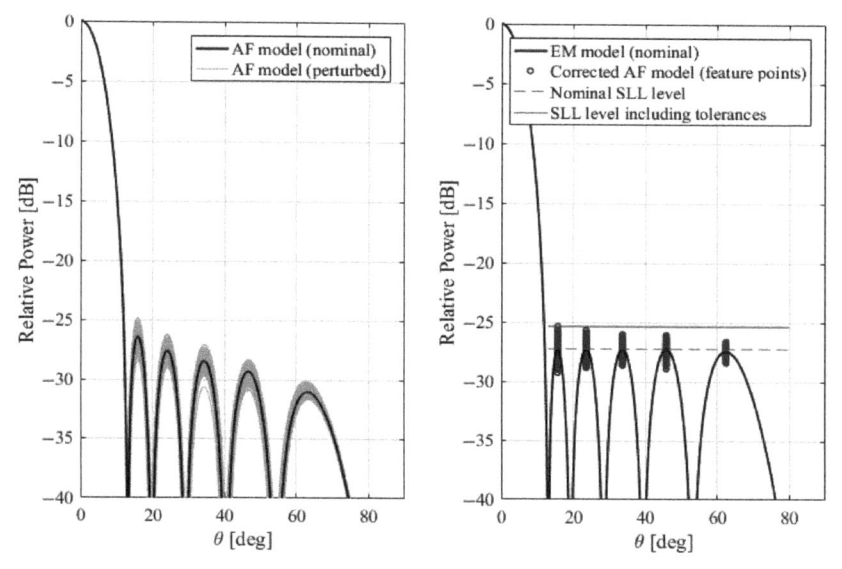

Fig. 17.6. Visualization of statistical analysis of the 12-element microstrip array assuming Gaussian probability distribution with 0.017 mm variance (Case 1) for all components (element geometry, dielectric permittivity, element spacing, and feed geometry).

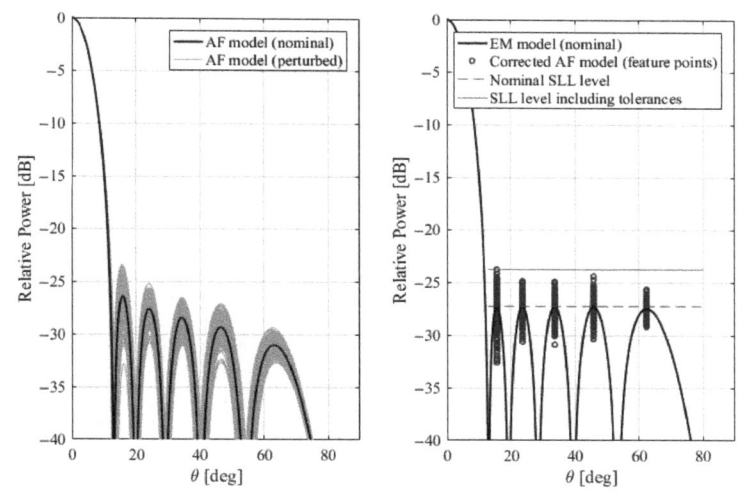

Fig. 17.7. Visualization of statistical analysis of the 12-element microstrip array assuming Gaussian probability distribution with 0.033 mm variance (Case 2) for all components (element geometry, dielectric permittivity, element spacing, and feed geometry).

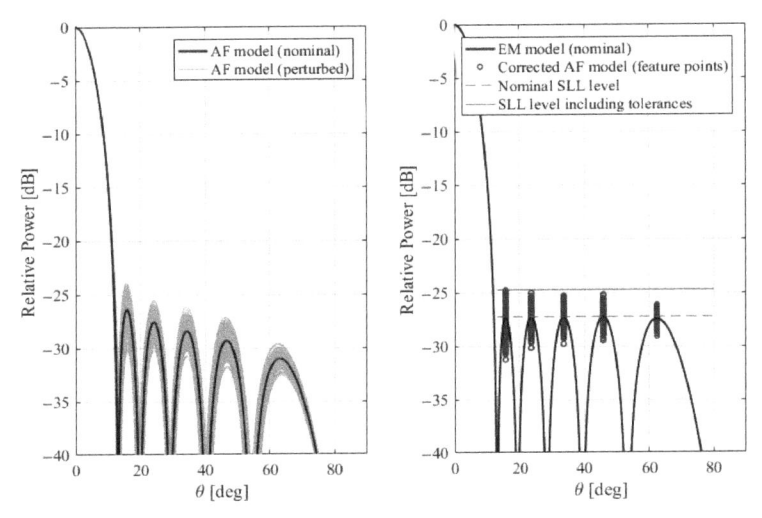

406 Simulation-Based Optimization of Antenna Arrays

Fig. 17.8. Visualization of statistical analysis of the 12-element microstrip array assuming uniform probability distribution with 0.05 mm maximum deviation (Case 3) for all components (element geometry, dielectric permittivity, element spacing, and feed geometry).

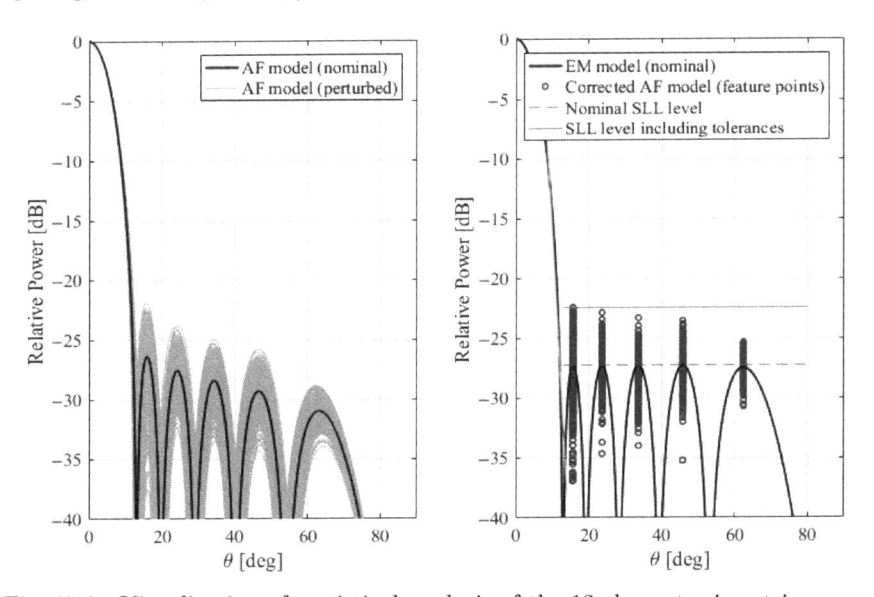

Fig. 17.9. Visualization of statistical analysis of the 12-element microstrip array assuming uniform probability distribution with 0.1 mm maximum deviation (Case 4) for all components (element geometry, dielectric permittivity, element spacing, and feed geometry).

worst case). For larger deviations, SLL degradation can be as high as a few dB.

It should be emphasized that the presented technique is computationally cheap. Construction of the surrogate model requires $2(N_e + N_m) + 1$ full-wave simulations of the array element, N_f simulations of the feed, and only one simulation of the aperture.

Chapter 18

Discussion and Recommendations: Prospective Look

Full-wave electromagnetic (EM) analysis has become a fundamental tool in the design of antennas and antenna arrays. Contemporary EM solvers feature great flexibility and versatility in terms of structure meshing, analysis setup, result post-processing, and are capable of accounting for various EM effects such as coupling of elements of apertures and feeds, undesirable radiation from feeds, excitation of higher-order modes, leakage of fields into substrates, environmental interactions (e.g., due to the presence of radomes, housing, installation fixtures, connectors), as well as dispersion, anisotropy, and nonlinear properties of materials. All of this is necessary to permit reliable evaluation of the array performance. However, accuracy comes at the expense of considerable computational costs. A single simulation may be as long as a few dozen of minutes or even a few hours depending on the array complexity and the modeling and simulation arrangements. Clearly, it is manageable in case of design verification but becomes a bottleneck for simulation-driven design, especially when a large number of analyses are necessary. Typical examples of such tasks include parametric optimization, statistical analysis, yield-driven design, fault detection, etc. For practical reasons (primarily, reduction of the computational overhead), various simplifications are exercised, such as utilization of analytical models (e.g., array factors assuming isotropic radiators), decomposition (e.g., separate treatment of the feed and the aperture), or leaving out some of design objectives at certain design stages (e.g., controlling the reflection level while designing the elements, without further correction upon assembling the aperture). All of these simplifications help maintaining the design time within acceptable frames but also lead to suboptimal performance of the final array designs. On the other hand, to yield the best possible design, accounting for all relevant effects (element coupling, non-perfect matching, feed radiation, etc.) is mandatory. Yet, the computational cost of such comprehensive treatment within the EM-based design frameworks is prohibitive when using conventional methods.

This book has had several purposes. On the one hand, our objective was to present and handle the antenna array design task as an optimization problem. We discussed how to control various design specifications through appropriate formulation of the cost functions and constraints, as well as outlined a variety of computational models available in the design of antenna arrays and feeding structures, with the ultimate model being a full-wave EM simulation one. At the same time, the challenges related to simulation-driven array optimization have been elaborated on.

The second objective was providing a background material concerning conventional numerical optimization methods, both local and global, which — in some situations — might be useful as stand-alone techniques for solving antenna design subproblems, or, in the context of this book, as building blocks of more involved design frameworks, primarily surrogate-based ones. The third and the core goal of the book was comprehensive introduction to surrogate-assisted optimization methods and their applications to design of antenna arrays. For the sake of making the manuscript self-contained, the coverage of basic material on surrogate modeling and optimization has been provided. On the other hand, a number of surrogate-based optimization (SBO) techniques tailored for antenna array problems have been described and illustrated using real-world structures.

As indicated in Chapter 5, SBO algorithms may involve data-driven surrogates, physics-based surrogates, or a combination thereof. In case of antenna and antenna array problems, a selection of a particular approach largely depends on availability of the underlying low-fidelity models. For certain problems (e.g., aperture or feed design), analytical array factor models, equivalent networks, or EM-based superposition models may constitute good candidates for fast surrogates. For other tasks, where a low-dimensional formulation is possible (e.g., with decomposition-based corporate feed design), approximation models may be a better choice. Occasionally, it might be useful to resort to coarse-mesh EM-simulation models (e.g., when simultaneous handling of radiation patterns and reflection responses is required as function of the excitation tapers as well as array aperture dimensions). In some cases, it is also possible to exploit a particular structure of the array response at hand (e.g., feature-based optimization and related methods). Selection of an appropriate range of modeling and optimization "building blocks" should be preceded by a careful analysis of the system responses, design space dimensionality, computational costs of the models, design constraints, as well as the types and the number of performance figures to be handled in the design process.

The modeling and optimization methods presented in this book have been comprehensively demonstrated using a large number of antenna elements, subarrays, linear and planar patch arrays, as well as corporate feeds. In many cases, numerical results have been supported by experimental validation of the fabricated prototypes. The most typical problem considered was sidelobe level reduction with the design variables being the excitation amplitudes and phases, element dimensions, and/or element spacing. Other design tasks illustrated in the book included improvement of reflection responses, null controlled pattern design, enhancement of array directivity, multi-objective design, fault detection, selection of optimal feed architecture, feed implementation, as well as statistical analysis of arrays. It has been demonstrated for all considered cases that the surrogate-assisted methods can considerably reduce the computational cost of the design process or make it practically acceptable in the first place.

There are several conclusions that can be drawn from the results presented in the book:

- Utilization of surrogate modeling techniques allows for reliable and accelerated optimization of antenna arrays. The computational cost very much depends on the problem complexity (the number of adjustable parameters and objectives, availability of faster representations of the structure at hand, etc.). Nevertheless, the cost can be usually kept at the level equivalent to a few dozen of simulations of the high-fidelity EM model of the structure at hand.
- Several major SBO techniques are available for constructing algorithms for antenna array optimization; they are discussed at a generic level in Chapter 5 and in the context of particular design problems in Chapters 7–17. The choice of the most suitable approach is problem dependent and requires certain experience as well as working knowledge on numerical optimization.
- Surrogate-assisted methods involving physics-based surrogates are normally more efficient but require careful selection of the underlying low-fidelity models (e.g., coarse-mesh simulations, or analytical array factor models). Some discussions and examples on low-fidelity model selection have been provided in Chapters 6 and 7.
- Data-driven models are typically useful as auxiliary surrogates (e.g., for repetitive solving of smaller and low-dimensional sub-problems such as EM-driven T-junction design, cf. Chapter 14) or if the model domain can be restricted to reduce the training data acquisition costs to acceptable levels.
- Surrogate-assisted methods exhibit good scaling properties, particularly those working with physics-based models. In particular, the relationship between the computational cost and the problem dimensionality is practically attractive, close to linear or at most quadratic. This means that the methods can potentially be applied for even more complex antenna array design problems.
- Development of universal optimization techniques that could be suitable for solving a wide range of antenna array design problems does not seem to be possible. Depending on the situation, e.g.,

the types of responses and design specifications to be handled (radiation figures, circuit reflection coefficients), the types of input parameters (excitation tapers, dimensions of radiators and feeds), individualized design process flows should be developed as demonstrated throughout the book, often combining different types of surrogate models.

There are several practical aspects of the modeling and optimization methodologies considered in this book. These include implementation, automation, generality, as well as a potential for further developments. While in most cases, the book provides sufficient information to replicate the discussed algorithms and frameworks, certain level of working knowledge concerning numerical optimization as well as high-level programming is necessary. This applies to optimization routines, procedures for surrogate model construction, but also interfacing commercial solvers (the last topic is outside the scope of this work and has not been addressed).

It should also be noted that low-level procedures for solving optimization subproblems (such as optimization of local surrogate models, etc.) are widely available through various programming environments. In particular, all the algorithms presented in this book have been implemented in Matlab (MATLAB, 2010) with extensive use of the Matlab Optimization Toolbox. In terms of design automation, certain decisions are up to the user (e.g., selection of particular low-fidelity models, control parameters of the algorithm, approximation technique for constructing a data-driven surrogate) and may influence the algorithm performance. In general, integration of the optimization procedures into fully automated design frameworks is still an open research problem.

It should be reiterated that rigorous EM-driven design optimization of antenna arrays is a very challenging task. As mentioned on several occasions throughout various chapters of the book, conventional methods (especially those involving population-based metaheuristics) tend to be prohibitively expensive in most practical situations. Traditional alternatives, such as replacing EM simulations by simpler representations (e.g., array factor models, schematics) with no further correction, or utilization of hands-on methods (parameter sweeping), lead to compromising accuracy or suboptimal results.

This book demonstrates that design processes aided by surrogate modeling techniques, variable-fidelity EM simulations, analytical and approximation models, as well as various levels of system decomposition permits conducting design closure of antenna arrays (as well as other tasks normally requiring massive EM analyses) in a computationally tractable manner. The authors believe that the presented material will be helpful for the readers interested in simulation-driven optimization of antenna arrays as well as for the readers dealing with computationally expensive models in other engineering disciplines. The authors hope that this book will not only provide various ready-to use algorithms and design procedures but also inspire the readers to develop their own surrogate-assisted methods.

We conclude with a few remarks on the perspectives of surrogate-assisted techniques for design of antenna arrays. Surrogate-based algorithms are capable of dramatic acceleration of array optimization processes and yielding reliable designs. Algorithms involving physics-based surrogates of antenna arrays are preferred because such models (especially those based on variable-fidelity EM simulations) exhibit excellent generalization capabilities resulting in good scalability of the corresponding algorithms with respect to dimensionality of the design space, which is directly related to the complexity of the antenna array structure under design. On the other hand, the best results can be obtained by customizing the surrogate-assisted design procedures for particular types of structures (apertures and feeds), application areas, performance specifications to be handled, as well as for kinds of design variables involved (dimensions, material parameters, excitation tapers). All of this gives rise to a trade-off between efficiency and generality of the design procedures. That is, involving problem-specific knowledge, selecting the best suitable surrogate (data-driven or physics-based), appropriate combination of numerical routines, as well as suitable approaches to handling responses and constraints, all necessarily leads to lower numerical costs of design procedures and improved reliability of the obtained designs, yet it all narrows down the scope of problems that can be solved using a particular design framework. To our understanding such the trade-off can only be efficiently managed with design process automation.

As of now, successful application of surrogate-assisted methods still strongly depends on the user experience. Clearly, this is a serious bottleneck for widespread acceptance of surrogate-based procedures amongst the antenna professionals. In order to alleviate these challenges, further research on surrogate-assisted design process automation is necessary. Such research should aim at development of dedicated software tools where the most of the technicalities related to optimal choice of modeling options, algorithm selection, and control parameters could be handled automatically in the design process and require minimal user intervention. Finally, it is expected that surrogate-assisted techniques will become more and more practical for antenna array design with further development of computing hardware, specialized acceleration units, as well as computational EM software.

References

Abbass, H.A., Sarker, R., Newton, C. (2001). PDE: A Pareto-frontier differential evolution approach for multi-objective optimization problems, *Proc. Congress Evol. Comp.*, **2**, 971–978.

AD250C (2018). Data Sheets, Rogers Corporation, Chandler, AZ.

Afoakwa, S., Jung Y.-B. (2017). Wideband microstrip comb-line linear array antenna using stubbed-element technique for high sidelobe suppression, *IEEE Trans. Ant. Prop.*, **65**(10), 5190–5199.

Afshinmanesh, F., Marandi, A., Shahabadi, M. (2008). Design of a single-feed dual-band dual-polarized printed microstrip antenna using a Boolean particle swarm optimization, *IEEE Trans. Ant. Prop.*, **56**, 1845–1852.

Agilent (2013). Agilent E5071C ENA Network Analyzer. Data Sheet, Agilent Technologies, Santa Clara, CA, USA.

Alander, J.T., Zinchenko, L.A., Sorokin, S.N. (2004). Comparison of fitness landscapes for evolutionary design of dipole antennas, *IEEE Trans. Ant. Prop.*, **52**, 2932–2940.

Alba, E., Marti, R. (eds.) (2006). *Metaheuristic Procedures for Training Neural Networks*, Springer.

Alexandrov, N.M., Lewis, R.M. (2001). An overview of first-order model management for engineering optimization, *Opt. Eng.*, **2**, 413–430.

Alexandrov, N.M., Dennis, J.E., Lewis, R.M., Torczon, V. (1998). A trust region framework for managing use of approximation models in optimization, *Struct. Multidisciplinary Optim.*, **15**, 16–23.

Aljibouri, B., Lim, E., Evans, H., Sambell, A. (2000). Multiobjective genetic algorithm approach for a dual-feed circular polarised patch antenna design, *Electronics Lett.*, **36**, 1005–1006.

Aljibouri, B., Sambell, A., Sharif, B. (2008). Application of genetic algorithm to design of sequentially rotated circularly polarised dual-feed microstrip patch antenna array, *Electronics Lett.*, **44**, 708–709.

Allen, L.J., Diamond, B.L. (1966). Mutual coupling in antenna arrays, Technical Report, EDS-66-443, Lincoln Lab, MIT.

An, S., Yang, S., Ren, Z. (2017). Incorporating light beam search in a vector normal boundary intersection method for linear antenna array optimization, *IEEE Trans. Magnetics*, **53**(6), ASN: 7001304.

Andrés, E., Salcedo-Sanz, S., Monge, F., Pérez-Bellido, A.M. (2012). Efficient aerodynamic design through evolutionary programming and support vector regression algorithms, *Int. J. Expert Syst. Appl.*, **39**, 10700–10708.

Angiulli, G., Cacciola, M., Versaci, M. (2007). Microwave devices and antennas modelling by support vector regression machines, *IEEE Trans. Magnetics*, **43**, 1589–1592.

Anthony, T.K., Zaghloul, A.I. (2009). Designing a 32 element array at 76GHz with a 33dB Taylor distribution in waveguide for a radar system, *Proc. IEEE Antennas and Propagation Society International Symposium*, North Charleston, SC, USA, pp. 1–4.

Antoniou, A., Lu, W.-S. (2007). *Practical Optimization: Algorithms and Engineering Applications*, Springer US.

Ares-Pena, F.J., Rodriguez-Gonzalez, J.A., Villanueva-Lopez, E., Rengarajan, S.R. (1999). Genetic algorithms in the design and optimization of antenna array patterns, *IEEE Trans. Ant. Prop.*, **47**, 506–510.

Babu, B.V., Jehan, M.M.L. (2003). Differential evolution for multi-objective optimization, *Proc. Congress Evol. Comp.*, **4**, 2696–2703.

Bäck, T. (1996). *Evolutionary Algorithms in Theory and Practice*, Oxford University Press, New York.

Bäck, T., Fogel, D.B., Michalewicz, Z. (eds.) (2000). *Evolutionary Computation 1: Basic Algorithms and Operators*, Institute of Physics Publishing, Bristol.

Bakr, M.H., Bandler, J.W., Biernacki, R.M., Chen, S.H., Madsen, K. (1998). A trust region aggressive space mapping algorithm for EM optimization, *IEEE Trans. Microwave Theory Tech.*, **46**, 2412–2425.

Bakr, M.H., Bandler, J.W., Georgieva, N.K., Madsen, K. (1999). A hybrid aggressive space-mapping algorithm for EM optimization, *IEEE Trans. Microwave Theory Tech.*, **47**, 2440–2449.

Balanis, C.A. (2005). *Antenna Theory*, 3rd edn., Wiley-Interscience.

Bandler, J.W., Biernacki, R.M., Chen, S.H., Grobelny, P.A., Hemmers, R.H. (1994). Space mapping technique for electromagnetic optimization, *IEEE Trans. Microwave Theory Tech.*, **42**, 2536–2544.

Bandler, J.W., Biernacki, R.M., Chen, S.H., Hemmers, R.H., Madsen, K. (1995). Electromagnetic optimization exploiting aggressive space mapping, *IEEE Trans. Microwave Theory Tech.*, **43**, 2874–2882.

Bandler, J.W., Cheng, Q.S., Dakroury, S.A., Mohamed, A.S., Bakr, M.H., Madsen, K., Søndergaard, J. (2004a). Space mapping: the state of the art, *IEEE Trans. Microwave Theory Tech.*, **52**, 337–361.

Bandler, J.W., Cheng, Q.S., Gebre-Mariam, D.H., Madsen, K., Pedersen, F., Søndergaard, J. (2003). EM-based surrogate modeling and design exploiting implicit, frequency and output space mappings, *IEEE Int. Microwave Symp. Digest*, Philadelphia, PA, pp. 1003–1006.

Bandler, J.W., Cheng, Q.S., Nikolova, N.K., Ismail, M.A. (2004b). Implicit space mapping optimization exploiting preassigned parameters, *IEEE Trans. Microwave Theory Tech.*, **52**, 378–385.
Bandler, J.W., Koziel, S., Madsen, K. (2006). Space mapping for engineering optimization, *SIAG/Optim. Views-and-News Special Issue on Surrogate/Derivative-free Optim.*, **17**(1), 19–26.
Bandler, J.W., Salama, A.E. (1985). Fault diagnosis of analog circuits, *Proc. IEEE*, **73**(8), 1279–1325.
Bandler, J.W., Seviora, R.E. (1972). Wave sensitivities of networks, *IEEE Trans. Microwave Theory Tech.*, **20**, 138–147.
Bansal, J.C., Singh, P.K., Saraswat, M., Verma, A., Jadon, S.S., Abraham, A. (2011). Inertia weight strategies in particle swarm optimization, *World Congress Nature Biologically Inspired Comp.*, pp. 633–640.
Bates, D.M., Watts, D.G. (1988). *Nonlinear Regression and Its Applications*, John Wiley & Sons, New York.
Beachkofski, B., Grandhi, R. (2002). Improved distributed hypercube sampling, *Structures, Structural Dynamics, Materials Conf.*, Paper AIAA 2002-1274.
Bekasiewicz, A., Koziel, S. (2015). Structure and computationally-efficient simulation-driven design of compact UWB monopole antenna, *IEEE Ant. Wireless Prop. Lett.*, **14**, 1282–1285.
Bekasiewicz, A., Koziel, S. (2016). Cost-efficient design optimization of compact patch antennas with improved bandwidth, *IEEE Ant. Wireless Prop. Lett.*, 270–273.
Bekasiewicz, A., Koziel, S. (2017). A structure and EM-driven design of novel compact UWB slot antenna, *IET Microwaves, Ant. Prop.* **11**(2), 219–223.
Bekasiewicz, A., Koziel, S., Leifsson, L. (2014a). Computationally efficient multi-objective optimization of an experimental validation of Yagi–Uda antenna, *Int. Conf. Simulation Modeling Methodologies, Technologies Appl.*, Vienna, pp. 798–805.
Bekasiewicz, A., Koziel, S., Zieniutycz, W. (2014b). Design space reduction for expedited multi-objective design optimization of antennas in highly-dimensional spaces, in S. Koziel, L. Leifsson, X.-S. Yang (eds.) *Solving Computationally Expensive Engineering Problems: Methods and Applications*, Springer, New York, pp. 113–147.
Bertsekas, D.P. (1982). *Constrained Optimization and Lagrange Multiplier Methods*, Academic Press, New York.
Bevelacqua, P.J., Balanis, C.A. (2007). Optimizing antenna array geometry for interference suppression, *IEEE Trans. Ant. Prop.*, **55**, 637–641.
Beyer, H.-G. (2001). *The Theory of Evolution Strategies*, Springer-Verlag, Berlin.
Beyer, H.-G., Schwefel, H.-P. (2002). Evolution strategies: A comprehensive introduction, *Natural Computing*, **1**, 3–52.
Binh, T.T., Korn, U. (1997). Multiobjective evolution strategy for constrained optimization problems, *Proc. IMACS World Congress Scientific Comp., Modelling Applied Math.*, pp. 357–362.
Bischof, C., Corliss, G., Griewank, A. (1993). Structured second- and higher-order derivatives through univariate Taylor series, *Opt. Methods Soft.*, **2**, 211–232.

Booker, A.J., Dennis, J.E., Frank, P.D., Serafini, D.B., Torczon, V., Trosset, M.W. (1999). A rigorous framework for optimization of expensive functions by surrogates, *Structural Opt.*, **17**, 1–13.

Broyden, C.G. (1965). A class of methods for solving nonlinear simultaneous equations, *Math. Comp.*, **19**, 577–593.

Bucci, O.M., Capozzoli, A., D'Elia, G. (2000). Diagnosis of array faults from far-field amplitude-only data, *IEEE Trans. Ant. Prop.*, **48**, 647–652.

Burke, G.J., Pogio, A.J. (1981). Numerical electromagnetic code (NEC)—Method of moments, Technical document 116. Naval Ocean Systems Center, San Diego.

Caorsi, S., Massa, A., Pastorino, M. (2000). A computational technique based on a real-coded genetic algorithm for microwave imaging purposes, *IEEE Trans. Geoscience Remote Sens.*, **38**, 1697–1708.

Chakraborty, U. (2008). *Advances in Differential Evolution*, Studies in Computational Intelligence, Springer.

Chamaani, S., Abrishamian, M.S., Mirtaheri, S.A. (2010). Time-domain design of UWB Vivaldi antenna array using multiobjective particle swarm optimization, *IEEE Ant. Wireless Prop. Lett.*, **9**, 666–669.

Chen, J.-S. (2007). CPW-fed compact slot-ring coupled square patch antennas, *IEEE Ant. Prop. Society Int. Symp.*, pp. 2297–2300.

Chen, F.-C., Hu, H.-T., Li, R.-S., Chu, Q.-X., Lancaster, M.J. (2017). Design of filtering microstrip antenna array with reduced sidelobe level, *IEEE Trans. Ant. Prop.*, **65**(2), 903–908.

Cheng, M.Y., Prayogo, D. (2014). Symbiotic organism search: A new metaheuristic optimization algorithm, *Comput. Struct.*, **139**, 98–112.

Cheng, Q.S., Bandler, J.W., Koziel, S. (2010). Space mapping design framework exploiting tuning elements, *IEEE Trans. Microwave Theory Tech.*, **58**, 136–144.

Cheng, Q.S., Koziel, S., Bandler, J.W. (2006). Simplified space mapping approach to enhancement of microwave device models, *Int. J. RF Microw. Comput.-aided Eng.*, **16**, 518–535.

Chew, W.C., Jin, J.-M., Michielssen, E., Song, J. (eds.) (2000). *Fast and Efficient Algorithms in Computational Electromagnetics*, Artech House, Norwood, MA.

Chu, Q.X., Mao, C.X., Zhu, H. (2013). A compact notched band UWB slot antenna with sharp selectivity and controllable bandwidth, *IEEE Trans. Ant. Prop.*, **61**(8), 3961–3966.

Chung, Y.S., Cheon, C., Park, I.-H., Hahn, S.Y. (2001). Optimal design method for microwave device using time domain method and design sensitivity analysis. II. FDTD case, *IEEE Trans. Magnetics*, **37**, 3255–3259.

Clerc, M. (2006). *Particle Swarm Optimization*, ISTE, London.

Clerc, M., Kennedy, J. (2002). The particle swarm—explosion, stability, and convergence in a multidimensional complex space, *IEEE Trans. Evol. Comp.*, **6**, 58–73.

Coello Coello, C.A., Lamont, G.B., van Veldhuizen, D.A. (2007). *Evolutionary Algorithms for Solving Multi-objective Problems*, 2nd edn., Springer.

Conn, A.R., Gould, N.I.M., Toint, P.L. (2000). *Trust Region Methods*, MPS-SIAM Series on Optimization, MPS-SIAM, Philadelphia.

Conn, A.R., Scheinberg, K., Vincente, L.N. (2009). *Introduction to Derivative-Free Optimization*, MPS-SIAM Series on Optimization, MPS-SIAM, Philadelphia.

Couckuyt, I., (2013). *Forward and Inverse Surrogate Modeling of Computationally Expensive Problems*, PhD thesis, Ghent University.

CST (2016). CST Microwave Studio, CST AG, Bad Nauheimer Str. 19, D-64289 Darmstadt, Germany.

Da, Y., Xiurun, G. (2005). An improved PSO-based ANN with simulated annealing technique, *Neurocomputing*, **63**, 527–533.

Datta, T., Misra, I.S. (2009) A comparative study of optimization techniques in adaptive antenna array processing: The bacteria-foraging algorithm and particle-swarm optimization, *IEEE Ant. Prop. Mag.*, **51**(6), 69–81.

David, H. (1994). A Markov Chain analysis of genetic algorithms with a state dependent fitness function, *Complex Syst.*, **8**, 407–417.

Davis, C. (1954). Theory of positive linear dependence, *Amer. J. Math.*, **76**, 733–746.

De Jong, A.K. (1975). *An Analysis of the Behavior of a Class of Genetic Adaptive Systems*, PhD thesis, University of Michigan.

Deb, K. (2001). *Multi-objective Optimization Using Evolutionary Algorithms*, John Wiley & Sons, Chichester.

Deb, K., Pratap, A., Agarwal, S., Meyarivan, T. (2002). A fast and elitist multi-objective genetic algorithm: NSGA-II, *IEEE Trans. Evol. Comp.*, **6**, 182–197.

Deb, A., Roy, J.S., Gupta, B. (2014) Performance comparison of differential evolution, particle swarm optimization and genetic algorithm in the design of circularly polarized microstrip antennas, *IEEE Trans. Ant. Prop.*, **62**(8), 3920–3928.

Devabhaktuni, V.K., Yagoub, M.C.E., Zhang, Q.J. (2001) A robust algorithm for automatic development of neural-network models for microwave applications, *IEEE Trans. Microwave Theory Tech.*, **49**, 2282–2291.

Ding, D., Wang, G. (2013). Modified multiobjective evolutionary algorithm based on decomposition for antenna design, *IEEE Trans. Ant. Prop.*, **61**, 5301–5307.

Ding, M., Jin, R., Geng, J., Wu, Q., Yang, G. (2008). Auto-design of band-notched UWB antennas using mixed model of 2D GA and FDTD, *Electronics Lett.*, **44**, 257–258.

Director, S.W., Rohrer, R.A. (1969). The generalized adjoint network and network sensitivities, *IEEE Trans. Circuit Theory*, **16**, 318–323.

Doğan, B., Ölmez, T. (2015). A new metaheuristic for numerical function optimization: Vortex search algorithm, *Inf. Sci.*, **293**, 125–145.

Dorigo, M., Stutzle, T. (2004). *Ant Colony Optimization*, MIT Press, Cambridge.
Draper, N.R., Smith, H. (1998). *Applied Regression Analysis*, 3rd edn., Wiley-Interscience, New York.
Duman, E., Uysal, M., Alkaya, A.F. (2012). Migrating birds optimization: A new metaheuristic approach and its performance on quadratic assignment problem, *Inf. Sci.*, **217**, 65–77.
Echeverría, D., Hemker, P.W. (2005). Space mapping and defect correction, *Comput. Methods Appl. Math.*, **5**, 107–136.
Echeverría, D., Hemker, P.W. (2008). Manifold mapping: A two-level optimization technique, *Computing Vis. Sci.*, **11**, 193–206.
Eiben, A.E., Smith, J.E. (2003). *Introduction to Evolutionary Computing*, Springer, Berlin.
Elliott, R.C. (2003). *Antenna Theory and Design*, revised edn., John Wiley & Sons, Hoboken.
Elragal, H.M., Mangoud, M.A., Alsharaa, M.T. (2011). Hybrid differential evolution and enhanced particle swarm optimization technique for design of reconfigurable phased antenna arrays, *IET Microwaves, Ant. Prop.* **5**(11), 1280–1287.
FEKO (2015). Altair Development S.A. (Pty) Ltd, 32 Techno Avenue, Technopark Stellenbosch 7600, South Africa.
Fletcher, R. (1987). *Practical Methods of Optimization*, 2nd edn., John Wiley & Sons, New York.
Fletcher R., Reeves, C.M. (1964). Function minimization by conjugate gradients, *Computer J.*, **7**, 149–154.
Fong, S., Wong, R., Vasilakos, A.V. (2016). Accelerated PSO swarm search feature selection for data stream mining big data, *IEEE Trans. Services Comput.*, **9**, 33–45.
Forrester, A.I.J., Keane, A.J. (2009). Recent advances in surrogate-based optimization, *Prog. Aerospace Sci.*, **45**, 50–79.
Forrester, A.I.J., Sobester, A., Keane, A.J. (2007). Multi-fidelity optimization via surrogate modeling, *Proc. Roy. Soc. A*, **463**, 3251–3269.
Forrester, A.I.J., Sobester, A., Keane, A.J. (2008). *Engineering Design via Surrogate Modelling: A Practical Guide*, John Wiley & Sons, Hoboken.
Gao, S., Luo, Q., Zhu, F. (2014). *Circularly Polarized Antennas*, Wiley–IEEE Press, Chichester.
Garg, R., Bhartia, P., Bahl, I., Ittipiboon, A. (2000). *Microstrip Antenna Design Handbook*, Artech House Publishers, Norwood, MA.
Gazi, V., Passino, K.M. (2004). Stability analysis of social foraging swarms, *IEEE Trans. Syst., Man, Cybernetics, Part B: Cybernetics*, **34**, 539–557.
Geem, Z.W., Kim, J.H., Loganathan, G.V. (2001). A new heuristic optimization algorithm: harmony search, *Simulation*, **76**, 60–68.
Geisser, S. (1993). *Predictive Inference*, Chapman & Hall.
Georgieva, N.K., Glavic, S., Bakr, M.H., Bandler, J.W. (2002). Feasible adjoint sensitivity technique for EM design optimization, *IEEE Trans. Microwave Theory Tech.*, **50**, 2751–2758.

Ghassemi, M., Bakr, M., Sangary, N. (2013). Antenna design exploiting adjoint sensitivity-based geometry evolution, *IET Microwaves Ant. Prop.*, **7**, 268–276.
Giunta, A.A. (1997). *Aircraft Multidisciplinary Design Optimization Using Design of Experiments Theory and Response Surface Modeling Methods*, PhD thesis, Virginia Polytechnic Institute and State University.
Giunta, A.A., Eldred, M.S. (2000). Implementation of a trust region model management strategy in the DAKOTA optimization toolkit, *Proc. Symp. Multidisciplinary Analysis Opt.*, AIAA-2000-4935, Long Beach, CA.
Giunta, A.A., Wojtkiewicz, S.F., Eldred, M.S. (2003). Overview of modern design of experiments methods for computational simulations, *Aerospace Sci. Meeting Exhibit*, Paper AIAA 2003-0649.
Glubokov, O., Koziel, S. (2014a). Substrate integrated waveguide microwave filter tuning through variable-fidelity feature space optimization, *Int. Rev. Progress Appl. Comput. Electromagnetics*, pp. 1–5.
Glubokov, O., Koziel, S. (2014b). EM-driven tuning of substrate integrated waveguide filters exploiting feature-space surrogates, *IEEE Int. Microwave Symp.*, **7**, 1–3.
Goldberg, D.E. (1989). *Genetic Algorithms in Search, Optimization and Machine Learning*, Addison-Wesley Longman Publishing.
Golub, G.H., van Loan, Ch.F. (1996). *Matrix Computations*, 3rd edn., Johns Hopkins University Press.
Gorissen, D., Couckuyt, I., Laermans, E., Dhaene, T. (2010b). Multiobjective global surrogate modeling, dealing with the 5-percent problem, *Eng. Comput.*, **26**, 81–98.
Gorissen, D., Crombecq, K., Couckuyt, I., Dhaene, T., Demeester, P. (2010a). A surrogate modeling and adaptive sampling toolbox for computer based design, *J. Machine Learning Res.*, **11**, 2051–2055.
Goudos, S., Zaharis, Z., Kampitaki, D., Rekanos, I., Hilas, C. (2009). Pareto optimal design of dual-band base station antenna arrays using multi-objective particle swarm optimization with fitness sharing, *IEEE Trans. Mag.*, **45**, 1522–1525.
GPU Computing Guide. (2016). CST Studio Suite. CST AG, Bad Nauheimer Str. 19, D-64289 Darmstadt, Germany.
Gunn, S.R. (1998). Support vector machines for classification and regression, Technical Report, School of Electronics and Computer Science, University of Southampton.
Gupta, K.C., Garg, R., Chadha, R. (1981). *Computer-Aided Design of Microwave Circuits*, Artech House, Norwood, MA.
Hall, P.S., Hall, C.M. (1988). Coplanar corporate feed effects in microstrip patch array design, *IEE Proc. H Microvaves, Antennas Propag.*, **135**, 180–186.
Hall, P.S., James, J.R. (1981). Design of microstrip antenna feeds. Part 2: Design and performance limitations of triplate corporate feeds, *IEE Proc. H Microwaves, Optics Antennas*, **128**(1), 26–34.

Hammersley, J.M. (1960). Monte-Carlo methods for solving multivariable problems, *Ann. New York Acad. Sci.*, **86**, 844–874.
Han, S.P., (1977). A globally convergent method for nonlinear programming, *J. Optim. Theory Appl.*, **22**, 297–309.
Hanninen, I. (2012). Optimization of a reflector antenna system, Whitepaper, Computer Simulation Technology AG.
Hansen, E., Walster, G.W. (2003). *Global Optimization Using Interval Analysis*, 2nd edn., Marcel Dekker, New York.
Hansen, N., Arnold, D.V., Auger, A. (2015). Evolution Strategies, in J. Kacprzyk, W. Pedrycz (eds.) *Springer Handbook of Computational Intelligence*, Springer, Berlin, pp. 871–898.
Hansen, N., Kern, S. (2004). Evaluating the CMA evolution strategy on multimodal test functions, in X. Yao *et al.* (eds.) *Parallel Problem Solving from Nature*, Springer, Berlin, pp. 282–291.
Hansen, R.C. (2007). Phased arrays, in J.L. Volakis (ed.) *Antenna Engineering Handbook*, 4th edn., McGraw-Hill.
Hansen, R.C. (2009). *Phased Arrays Antennas*, 2nd edn., Wiley, Hoboken, NJ.
Harrington, R.F. (1993). *Field Computation by Moment Method*, reprint edn., Wiley-IEEE Press, Piscataway, NJ.
Haupt, R.A. (2010). *Antenna Arrays: A Computational Approach*, Wiley, Hoboken, NJ.
Haupt, R.L. (1995). Comparison between genetic and gradient-based optimization algorithms for solving electromagnetic problems, *IEEE Trans. Magnetics*, **31**(3), 1932–1935.
Haykin, S. (1998). *Neural Networks: A Comprehensive Foundation*, 2nd edn., Prentice-Hall, Upper Saddle River.
Hesser, J., Männer, R. (1991). Towards an optimal mutation probability for genetic algorithms, in H.-P. Schwefel, R. Männer (eds.) *Parallel Problem Solving from Nature*, Springer, Berlin, pp. 23–32.
HFSS (2016). ANSYS, Inc., Southpointe, 2600 ANSYS Drive, Canonsburg, PA 15317.
Holland, J.H. (1975). *Adaptation in Natural and Artificial Systems: An Introductory Analysis with Applications to Biology, Control, and Artificial Intelligence*, MIT Press, Cambridge.
Hong, J.-S., Lancaster, M. (2001). *Microstrip Filters for RF/Microwave Applications*, John Wiley & Sons, Hoboken.
Hooke, R., Jeeves, T.A. (1961). Direct search solution of numerical and statistical problems, *J. ACM*, **8**(2), 212–229.
Horn, J. (1993). Finite Markov chain analysis of genetic algorithms with niching, *Proc. Int. Conf. Gentetic Alg.*, pp. 110–117.
Horng, T.-S., Alexopoulos, N.G. (1993). Corporate feed design for microstrip arrays, *IEEE Trans. Ant. Prop.*, **41**(12), 1615–1624.
Horst, R., Tuy, H. (1996). *Global Optimization: Deterministic Approaches*, 3rd edn., Springer, Berlin.

Horst, R., Pardalos, P.M., Thoai, N.V. (2000). *Introduction to Global Optimization*, 2nd edn., Kluwer Academic Publishers, Dordrecht.

Hosder, S., Watson, L.T., Grossman, B., Mason, W.H., Kim, H. (2001). Polynomial response surface approximations for the multidisciplinary design optimization of a high speed civil transport, *Opt. Eng.*, **2**, 431–452.

Hu, Y.-J., Ding, W.-P., Cao, W.-Q. (2011). Broadband circularly polarized microstrip antenna array using sequentially rotated technique, *IEEE Antennas Wireless Propag. Lett.*, **10**, 1358–1361.

Huang, L., Gao, Z. (2012). Wing-body optimization based on multi-fidelity surrogate model, *Int. Congress Aeronautical Sci.*, Brisbane, Australia.

Iglesias, R., Ares, F., Fernandez-Delgado, M., Rodriguez, J., Bregains, J., Barro, S. (2008). Element failure detection in linear antenna arrays using case-based reasoning, *IEEE Antennas Propaga. Mag.*, **50**, 198–204.

Jackson, D.R. (2007). Microstrip antennas, Chapter 7, in J.L. Volakis (ed.) *Antenna Engineering Handbook*, 4th edn. McGraw-Hill, New York, NY, USA.

Jacobs, J.H., Etman, L.F.P., van Keulen, F., Rooda, J.E. (2004). Framework for sequential approximate optimization, *Struct. Multidisc. Opt.*, **27**, 384–400.

Jacobs, J.P. (2012). Bayesian support vector regression with automatic relevance determination kernel for modeling of antenna input characteristics, *IEEE Trans. Ant. Prop.*, **60**, 2114–2118.

Jacobs, J.P. (2016). Characterization by Gaussian processes of finite substrate size effects on gain patterns of microstrip antennas, *IET Microwaves, Ant. Prop.*, **10**(11), 1189–1195.

Jacobs, J.P., Koziel, S. (2014). Cost-effective global surrogate modeling of planar microwave filters using multi-fidelity Bayesian support vector regression, *Int. J. RF Microw. Comput.-aided Eng.*, **24**, 11–17.

Jacobsson, P., Rylander, T. (2010). Gradient-based shape optimisation of conformal array antennas, *IET Microwaves, Ant. Prop.*, **4**(2), 200–209.

Jang, C.H., Hu, F., Li J., Zhu, D. (2016) Low-redundancy large linear arrays synthesis for aperture synthesis radiometers using particle swarm optimization, *IEEE Trans. Ant. Prop.* **64**(6), 2179–2188.

Jaulin, L., Kieffer, M., Didrit, O., Walter, E. (2001). *Applied Interval Analysis*, Springer, London.

Jayasinghe, J., Anguera, J., Uduwawala, D. (2013). Genetic algorithm optimization of a high-directivity microstrip patch antenna having a rectangular profile, *Radioengineering*, **22**(3), 700–707.

Jin, J.-M. (2008). *Finite Element Analysis of Antennas and Arrays*, Wiley-IEEE Press, Hoboken, NJ.

Jin, J.-M. (2014). *Finite Element Method in Electromagnetics*, 3rd edn., Wiley-IEEE Press, Hoboken, NJ.

Jin, N., Rahmat-Samii, Y. (2005). Parallel particle swarm optimization and finite-difference time-domain (PSO/FDTD) algorithm for multiband and wideband patch antenna designs, *IEEE Trans. Ant. Prop.*, **53**, 3459–3468.

Jin, N., Rahmat-Samii, Y. (2006). A novel design methodology for aperiodic arrays using particle swarm optimization, *Nat. Radio Sci. Meeting Dig.*, Boulder, CO.

Jin, N., Rahmat-Samii, Y. (2007). Advances in particle swarm optimization for antenna designs: real-number, binary, single-objective and multiobjective implementations, *IEEE Trans. Ant. Prop.*, **55**, 556–567.

Jones, D., Schonlau, M., Welch, W. (1998). Efficient global optimization of expensive black-box functions, *J. Global Opt.*, **13**, 455–492.

Journel, A.G., Huijbregts, C.J. (1981). *Mining Geostatistics*, Academic Press.

Kabir, H., Wang, Y., Yu, M., Zhang, Q.J. (2008). Neural network inverse modeling and applications to microwave filter design, *IEEE Trans. Microwave Theory Tech.*, **56**, 867–879.

Kao, Y.-T., Zahara, E. (2008). A hybrid genetic algorithm and particle swarm optimization for multimodal functions, *Appl. Soft Comput.*, **8**, 849–857.

Karamalis, P.D., Kanatas, A.G., Constantinou, P. (2009). A genetic algorithm applied for optimization of antenna arrays used in mobile ratio channel characterization devices, *IEEE Trans. Instrumentation Meas.* **58**(8), 2475–2487.

Kempel, L.C. (2007). Computational electromagnetics for antennas, in J.L. Volakis (ed.) *Antenna Engineering Handbook*, 4th edn., McGraw-Hill, New York.

Kennedy, J. (1997). The particle swarm: social adaptation of knowledge, *Proc. Int. Conf. Evolutionary Comp.*, pp. 303–308.

Kennedy, J., Eberhart, R.C., Shi, Y. (2001). *Swarm Intelligence*, Academic Press, London.

Kerkhoff, A.J., Ling, H. (2007). Design of a band-notched planar monopole antenna using genetic algorithm optimization, *IEEE Trans. Ant. Prop.*, **55**, 604–610.

Kibria, S., Islam, M.T., Yatim, B. (2014) New compact dual-band circularly polarized universal RFID reader antenna using ramped convergence particle swarm optimization, *IEEE Trans. Ant. Prop.* **62**(5), 2795–2801.

Kirkpatrick, S., Gelatt, C.D., Vecchi, M.P. (1983). Optimization by simulated annealing, *Science, New Series*, **220**, 671–680.

Kleijnen, J. (2008). *Design and Analysis of Simulation Experiments*, Springer, Berlin.

Kleijnen, J.P.C. (2009). Kriging metamodeling in simulation: A review, *European J. Operational Res.*, **192**, 707–716.

Koehler, J.R., Owen, A.B. (1996). Computer experiments, in S. Ghosh, C.R. Rao (eds.) *Handbook of Statistics*, Elsevier Science B.V. pp. 261–308.

Kogure, H., Kogure, Y., Rautio, J.C. (2011). *Introduction to Antenna Analysis Using EM Simulators*, Artech House.

Kolda, T.G., Lewis, R.M., Torczon, V. (2003). Optimization by direct search: New perspectives on some classical and modern methods, *SIAM Rev.*, **45**, 385–482.

Kowalski, M.E., Jin, J.-M. (2000). Determination of electromagnetic phased-array driving signals for hyperthermia based on a steady-state temperature criterion, *IEEE Trans. Microwave Theory Tech.*, **48**(11), 1864–1873.

Koziel, S. (2009). Multi-fidelity optimization of microwave structures using response surface approximation and space mapping, *Appl. Comp. EM Soc. J.*, **24**, 601–608.

Koziel, S. (2010a). Shape-preserving response prediction for microwave design optimization, *IEEE Trans. Microwave Theory Tech.*, **58**, 2829–2837.

Koziel, S. (2010b). Adaptively adjusted design specifications for efficient optimization of microwave structures, *Prog. EM Res. B*, **21**, 219–234.

Koziel, S. (2010c). Shape-preserving response prediction for microwave circuit modeling, *IEEE MTT-S Int. Microwave Symp. Dig.*, Anaheim, CA, pp. 1660–1663.

Koziel, S. (2010d). Efficient optimization of microwave structures through design specifications adaptation, *IEEE Int. Symp. Ant. Prop.*, Toronto, Canada.

Koziel, S. (2010e). Multi-fidelity multi-grid design optimization of planar microwave structures with Sonnet, *Int. Rev. Prog. Applied Comp. EM*, pp. 719–724.

Koziel, S. (2012). Accurate low-cost microwave component models using shape-preserving response prediction, *Int. J. Numerical Modelling*, **25**, 152–162.

Koziel, S. (2015). Fast simulation-driven antenna design using response-feature surrogates, *Int. J. RF Microw. Comput.-aided Eng.*, **25**(5), 394–402.

Koziel, S., Bandler, J.W. (2007a). Coarse and surrogate model assessment for engineering design optimization with space mapping, *IEEE MTT-S Int. Microwave Symp. Dig.*, Honolulu, HI, pp. 107–110.

Koziel, S., Bandler, J.W. (2007b). Space-mapping optimization with adaptive surrogate model, *IEEE Trans. Microwave Theory Tech.*, **55**, 541–547.

Koziel, S., Bandler, J.W. (2015). Reliable microwave modeling by means of variable-fidelity response features, *IEEE Trans. Microwave Theory Tech.*, **63**, 4247–4254.

Koziel, S., Bekasiewicz, A. (2015a). Fast EM-driven size reduction of antenna structures by means of adjoint sensitivities and trust regions, *IEEE Ant. Wireless Prop. Lett.*, **14**, 1681–1684.

Koziel, S., Bekasiewicz, A. (2015b). Fast simulation-driven feature-based design optimization of compact dual-band microstrip branch-line coupler, *Int. J. RF Microw. Comput.-aided Eng.*, DOI: 10.1002/mmce.20923.

Koziel, S., Bekasiewicz, A. (2016a). Variable-fidelity design optimization of antennas with automated model selection, *European Antenna and Propagation Conf.*, pp. 1–5.

Koziel, S., Bekasiewicz, A. (2016b). Rapid design optimization of antennas using variable-fidelity EM models and adjoint sensitivities, *Engineering Comput.* **33**(7), 2007–2018.

Koziel, S., Bekasiewicz, A. (2016c). Low-cost multi-objective optimization and experimental validation of UWB MIMO antenna, *Engineering Comput.* **33**(4), 1246–1268.

Koziel, S., Bekasiewicz, A. (2016d). *Multi-objective Design of Antennas Using Surrogate Models*, World Scientific, London.

Koziel, S., Leifsson, L. (2012a). Knowledge-based airfoil shape optimization using space mapping, *AIAA Applied Aerodynamics Conf.*, Paper AIAA 2012-3016.

Koziel, S., Leifsson, L. (2012b) Response correction techniques for surrogate-based design optimization of microwave structures, *Int. J. RF Microw. Comput.-aided Eng.*, **22**, 211–223.

Koziel, S., Leifsson, L. (2012c). Multi-fidelity airfoil shape optimization with adaptive response prediction, *AIAA/ISSMO Multidisciplinary Analysis and Optimization Conf.* 2012–5454.

Koziel, S., Leifsson L. (eds.) (2013a). *Surrogate-Based Modeling and Optimization. Applications in Engineering*, Springer, Berlin.

Koziel, S., Leifsson, L. (2013b). Multi-point response correction for reduced-cost EM-simulation-driven design of antenna structures, *Microwave Opt. Tech. Lett.*, **55**(9) 2070–2074.

Koziel, S., Michalewicz, Z. (1998). A decoder-based evolutionary algorithm for constrained parameter optimization problems, in T. Bäck, A.E. Eiben, M. Schoenauer, H.-P. Schwefel, (eds.) *Proc. Parallel Problem Solving from Nature*, Lecture Notes in Computer Science, Vol. 1498, Springer-Verlag, Berlin, pp. 231–240.

Koziel, S., Ogurtsov, S. (2011a). Simulation-driven design in microwave engineering: Application case studies, in X.S. Yang, S. Koziel (eds.) *Computational Optimization and Applications in Engineering and Industry*, Studies in Computational Intelligence, Vol. 359, Springer, Berlin, pp. 57–98.

Koziel, S., Ogurtsov, S. (2011b). Rapid design optimization of antennas using space mapping and response surface approximation models, *Int. J. RF Microw. Comput.-aided Eng.*, **21**, 611–621.

Koziel, S., Ogurtsov, S. (2011c). Simulation-driven design in microwave engineering: Methods, in S. Koziel, X.S. Yang (eds.) *Computational Optimization, Methods and Algorithms*, Studies in Computational Intelligence, Springer.

Koziel, S., Ogurtsov, S. (2012a). Model management for cost-efficient surrogate-based optimization of antennas using variable-fidelity electromagnetic simulations, *IET Microwaves, Ant. Prop.*, **6**, 1643–1650.

Koziel S., Ogurtsov, S. (2012b). Fast simulation-driven design of microwave structures using improved variable-fidelity optimization technique, *Eng. Opt.*, **44**, 1007–1019.

Koziel, S., Ogurtsov, S. (2012c). Reduced-cost design optimization of antenna structures using adjoint sensitivity, *Microwave Opt. Technol. Lett.*, **54**, 2594–2597.

Koziel, S., Ogurtsov, S. (2012d). Linear antenna array synthesis using gradient-based optimization with analytical derivatives, *IEEE Int. Symp. Antennas Prop.*, Chicago, IL, USA, July 8–12.

Koziel, S., Ogurtsov, S. (2013a). Rapid optimization of omnidirectional antennas using adaptively adjusted design specifications and kriging surrogates, *IET Microwaves, Ant. Prop.*, **7**, 1194–1200.

Koziel, S., Ogurtsov, S. (2013b). Multi-objective design of antennas using variable-fidelity simulations and surrogate models, *IEEE Trans. Ant. Prop.*, **61**(12), 5931–5939.
Koziel, S., Ogurtsov, S. (2014a). *Antenna Design by Simulation-Driven Optimization*, Springer, Berlin.
Koziel, S., Ogurtsov, S., (2014b). Phase-spacing optimization of linear microstrip antenna arrays by EM-based superposition models, *Loughboroh Antenna Propag. Conf.* (LAPC), 26–30.
Koziel, S., Ogurtsov, S. (2015a). Phase-spacing optimization of linear microstrip antenna arrays using simulation-based surrogate superposition models, *Int. J. RF Microw. Comput.-aided Eng.*, **25**(6), 536–547.
Koziel, S., Ogurtsov, S. (2015b). Simulation-based design of microstrip linear antenna arrays using fast radiation response surrogates, *IEEE Antennas Wireless Propag. Lett.*, **14**, 759–762.
Koziel, S., Ogurtsov, S. (2015c). Rapid design of microstrip antenna arrays by means of surrogate-based optimization, *IET Microwaves, Ant. Prop.*, **9**(5), 463–471.
Koziel, S., Ogurtsov, S. (2015d). Fast simulation-driven optimization of planar microstrip antenna arrays using surrogate superposition models, *Int. J. RF Microw. Comput.-aided Eng.*, **25**(5), 371–381.
Koziel, S., Ogurtsov, S. (2017). On systematic design of corporate feeds for Chebyshev microstrip linear arrays, in *Proc. 2017 IEEE Antennas Propagat. Society Int. Symposium.* San Diego, CA, USA, July 2017, pp. 1–2.
Koziel, S., Ogurtsov, S. (2018). Rapid design closure of linear microstrip antenna array apertures using response features, *IEEE Antennas Wireless Prop. Lett.*, **17**(4), 645–648.
Koziel, S., Bandler, J.W., Madsen, K. (2006). A space mapping framework for engineering optimization: Theory and implementation, *IEEE Trans. Microwave Theory Tech.*, **54**, 3721–3730.
Koziel, S., Cheng, Q.S., Bandler, J.W. (2008a). Space mapping, *IEEE Microwave Mag.*, **9**, 105–122.
Koziel, S., Bandler, J.W., Madsen, K. (2008b). Quality assessment of coarse models and surrogates for space mapping optimization, *Opt. Eng.*, **9**, 375–391.
Koziel, S., Bandler, J.W., Madsen, K. (2009). Space mapping with adaptive response correction for microwave design optimization, *IEEE Trans. Microwave Theory Tech.*, **57**, 478–486.
Koziel, S., Bandler, J.W., Cheng, Q.S. (2010a). Robust trust-region space-mapping algorithms for microwave design optimization, *IEEE Trans. Microwave Theory Tech.*, **58**, 2166–2174.
Koziel, S., Cheng, Q.S., Bandler, J.W. (2010b). Implicit space mapping with adaptive selection of preassigned parameters, *IET Microwaves, Ant. Prop.*, **4**, 361–373.
Koziel, S., Bandler, J.W., Cheng, Q.S. (2010c). Adaptively constrained parameter extraction for robust space mapping optimization of microwave circuits, *IEEE MTT-S Int. Microwave Symp. Dig.*, pp. 205–208.

Koziel, S., Echeverría Ciaurri, D., Leifsson, L. (2011a). Surrogate-based methods, in S. Koziel, X.-S. Yang (eds.) *Computational Optimization, Methods and Algorithms*, Springer, Berlin, pp. 33–59.

Koziel, S., Bandler, J.W., Cheng, Q.S. (2011b). Tuning space mapping design framework exploiting reduced EM models, *IET Microwaves, Ant. Prop.*, **5**, 1219–1226.

Koziel, S., Leifsson, L., Couckuyt, I., Dhaene, T. (2013a). Reliable reduced cost modeling and design optimization of microwave filters using co-kriging, *Int. J. Numerical Modelling: Electronic Devices and Fields*, **26**, 493–505.

Koziel, S., Ogurtsov, S., Bandler, J.W., Cheng, Q.S. (2013b). Reliable space-mapping optimization integrated with EM-based adjoint sensitivities, *IEEE Trans. Microwave Theory Tech.*, **61**, 3493–3502.

Koziel, S., Ogurtsov, S., Couckuyt, I., Dhaene, T. (2013c). Variable-fidelity electromagnetic simulations and co-kriging for accurate modeling of antennas, *IEEE Trans. Ant. Prop.*, **61**, 1301–1308.

Koziel, S., Yang, X.S., Zhang, Q.J. (eds.) (2013d). *Simulation-Driven Design Optimization and Modeling for Microwave Engineering*, Imperial College Press, London.

Koziel, S., Bekasiewicz, A., Zieniutycz, W. (2014a). Expedited EM-driven multi-objective antenna design in highly-dimensional parameter spaces, *IEEE Ant. Wireless Prop. Lett.*, **13**, 631–634.

Koziel, S., Ogurtsov, S., Cheng, Q.S., Bandler, J.W. (2014b). Rapid electromagnetic-based microwave design optimisation exploiting shape-preserving response prediction and adjoint sensitivities, *IET Microwaves, Ant. Prop.*, **8**, 775–781.

Koziel, S., Bekasiewicz, A., Kurgan, P. (2014c). Rapid EM-driven design of compact RF circuits by means of nested space mapping, *IEEE Microwave Wireless Comp. Lett.*, **24**, 364–366.

Koziel, S., Ogurtsov, S., Zieniutycz, W., Sorokosz, L. (2014d) Expedited design of microstrip antenna subarrays using surrogate-based optimization, *IEEE Ant. Wireless Prop. Lett.*, **13**, 635–638.

Koziel, S., Ogurtsov, S., Zieniutycz, W., Sorokosz, L. (2014e). Simulation-driven design of microstrip antenna subarrays, *IEEE Trans. Ant. Prop.*, **62**(7), 3584–3591.

Koziel, S., Ogurtsov, S., Zieniutycz, W., Bekasiewicz, A. (2015). Design of a planar UWB dipole antenna with an integrated balun using surrogate-based optimization, *IEEE Ant. Wireless Prop. Lett.*, **14**, 366–369.

Koziel, S., Bekasiewicz, A., Leifsson, L. (2016). Rapid EM-driven antenna dimension scaling through inverse modeling, *IEEE Ant. Wireless Prop. Lett.*, **15**, 714–717.

Kraus, J.D. (1988). *Antennas*, McGraw-Hill, New York.

Kroese, D.P., Taimre, T., Botev, Z.I. (2011). *Handbook of Monte Carlo Methods*, John Wiley & Sons, Hoboken.

Kuhn, H.W., Tucker, A.W. (1951). Nonlinear programming, in J. Neyman (ed.) *Proc. Berkeley Symp. Mathematical Statistics Probability*, University of California Press, Berkeley, pp. 481–492.

Kukkonen, S., Lampinen, J. (2004). An extension of generalized differential evolution for multi-objective optimization with constraints, *Parallel Problem Solving from Nature*, pp. 752–761.

Lai, M.-I., Jeng, S.-K. (2006). Compact microstrip dual-band bandpass filters design using genetic-algorithm techniques, *IEEE Trans. Microwave Theory Tech.*, **54**, 160–168.

Laurenceau, J., Sagaut, P. (2008). Building efficient response surfaces of aerodynamic functions with kriging and cokriging, *AIAA J.*, **46**, 498–507.

Leary, S., Bhaskar, A., Keane, A. (2003). Optimal orthogonal-array-based latin hypercubes, *J. Appl. Statist.*, **30**, 585–598.

Leifsson, L., Koziel, S. (2015a). *Simulation-Driven Aerodynamic Design Using Variable-Fidelity Models*, Imperial College Press, London.

Leifsson, L., Koziel, S. (2015b). Variable-resolution shape optimization: Low-fidelity model selection and scalability, *Int. J. Mathematical Modeling Numerical Opt.*, **6**, 1–21.

Levin, D. (1998). The approximation power of moving least-squares, *Math. Comput.*, **67**, 1517–1531.

Levine, E., Malamud, G., Shtrikman, S., Treves, D. (1989). A study of microstrip array antennas with the feed network, *IEEE Trans. Ant. Prop.*, **37**, 426–437.

Levine, E., Shtrikman, S. (1989). Optimal designs of corporate-feed printed arrays adapted to a given aperture, *Proc. Sixteenth Conf. Electrical and Electronics Engineers in Israel*, Tel-Aviv, Israel, pp. 1–4.

Li, W.T., Shi, X.W., Hei, Y.Q., Liu, S.F., Zhu, J. (2010). A hybrid optimization algorithm and its application for conformal array pattern synthesis, *IEEE Trans. Ant. Prop.*, **58**, 3401–3406.

Liang; S. Feng, T. Sun, G. (2017). Sidelobe-level suppression for linear and circular antenna arrays via the cuckoo search-chicken swarm optimisation algorithm, *IET Microwaves, Ant. Prop.*, **11**(2), 209–218.

Lim, W.H., Isa, N.A.M. (2014). An adaptive two-layer particle swarm optimization with elitist learning strategy, *Information Sci.*, **273**, 49–72.

Liu, B., Aliakbarian, H., Zhongkun, M., Vandenbosch, G.A.E., Gielen, G., Excell, P. (2014a). An efficient method for antenna design optimization based on evolutionary computation and machine learning techniques, *IEEE Trans. Ant. Prop.*, **62**, 7–18.

Liu, B., Zhang, Q., Gielen, G. (2014b) A Gaussian process surrogate model assisted evolutionary algorithm for medium scale expensive black box optimization problems, *IEEE Trans. Evol. Comp.*, **18**, 180–192.

Liu, H., Zhao, H. Li, W., Liu, B. (2016). Synthesis of sparse planar arrays using matrix mapping and differential evolution, *IEEE Ant. Wireless Prop. Lett.*, **15**, 1905–1908.

Liu, J., Han, Z., Song, W. (2012). Comparison of infill sampling criteria in kriging-based aerodynamic optimization, *Int. Congress Aeronautical Sci.*, pp. 1–10.

Lizzi, L., Viani, F., Azaro, R., Massa, A. (2008). A PSO-driven spline-based shaping approach for ultra-wideband (UWB) antenna synthesis, *IEEE Trans. Ant. Prop.*, **56**, 2613–2621.

Lophaven, S.N., Nielsen, H.B., Søndergaard, J. (2002). *DACE: A Matlab kriging toolbox*, Technical University of Denmark.

Luenberger, D.G. (2003). *Linear and Nonlinear Programming*, 2nd edn., Kluwer Academic Publishers, Boston.

Madavan, N.K. (2002). Multiobjective optimization using a pareto differential evolution approach, *Congress Evol. Comp.*, **2**, 1145–1150.

Mailloux, R.J. (2005). *Phased Array Antenna Handbook*, 2nd edn., Artech House, Boston.

Makarov, S. (2002). *Antenna and EM Modeling with Matlab*, Wiley.

Maloney, J., Smith, G., Thiele, E., Gandhi, O., Chavannes, N., Hagness, S. (2005). Antennas, Chapter 14, in A. Taflove, S.C. Hagness (eds.) *Computational Electrodynamics: The Finite-Difference Time-Domain Method*, 3rd edn., Artech House, Norwood, MA.

Manica, L., Rocca, P., Poli, L., Massa, A. (2009). Almost time-independent performance in time-modulated linear arrays, *IEEE Ant. Wireless Prop. Lett.*, **8**, 843–846.

Marheineke, N., Pinnau, R., Reséndiz, E. (2012). Space mapping-focused control techniques for particle dispersions in fluids, *Opt. Eng.*, **13**, 101–120.

Matheron, G. (1963). Principles of geostatistics, *Economic Geology*, **58**, 1246–1266.

Matlab, Version 7.8 (2010). The MathWorks, Inc., 3 Apple Hill Drive, Natick, MA 01760-2098.

McKay, M., Conover, W., Beckman, R. (1979). A comparison of three methods for selecting values of input variables in the analysis of output from a computer code, *Technometrics*, **21**, 239–245.

Meng, J., Xia, L. (2007). Support-vector regression model for millimeter wave transition, *Int. J. Infrared and Milimeter Waves*, **28**(5), 413–421.

Mezura-Montes, E., Velázquez-Reyes, J., Coello Coello, C.A. (2006a). A comparative study of differential evolution variants for global optimization, *Conf. Genetic Evolutionary Comp.*, pp. 485–492.

Mezura-Montes, E., Reyes-Sierra, M., Coello Coello, C.A. (2006b). Multi-objective optimization using differential evolution: A survey of the state-of-the-art, in U.K. Chakraborty (ed.) *Advances in Differential Evolution*, pp. 173–196.

Mezura-Montes, E., Miranda-Varela, M.E., Gómez-Ramón, R.C. (2010). Differential evolution in constrained numerical optimization: An empirical study, *Inform. Sci.*, **180**, 4223–4262.

Michalewicz, Z. (1996). *Genetic Algorithms + Data Structures = Evolution Programs*, Springer, Berlin.

Minsky, M.I., Papert, S.A. (1969). *Perceptrons: An Introduction to Computational Geometry*, MIT Press, Cambridge.

Munson, R.E. (1974). Conformal microstrip antennas and microstrip phased arrays, *IEEE Trans. Ant. Prop.*, **22**, 74–78.

Muraguchi, M., Yukitake, T., Naito, Y. (1983). Optimum design of 3-dB branch-line couplers using microstrip lines, *IEEE Trans. Microwave Theory Tech.*, **31**, 674–678.

Nelder, J.A., Mead, R. (1965). A simplex method for function minimization, *Computer J.*, **7**, 308–313.

Nickabadi, A., Ebadzadeh, M.M., Safabakhsh, R. (2011). A novel particle swarm optimization algorithm with adaptive inertia weight, *Appl. Soft Comput.*, **11**, 3658–3670.

Nielsen, H.B. (1999). Damping parameter in Marquardt's method, IMM DTU. Report IMM-REP-1999-05.

Nocedal, J., Wright, S. (2006). *Numerical Optimization*, 2nd edn., Springer, New York.

Noel, M.M. (2012). A new gradient based particle swarm optimization algorithm for accurate computation of global minimum, *Appl. Soft Comput.*, **12**, 353–359.

Noman, N., Iba, H. (2008). Accelerating differential evolution using an adaptive local search, *IEEE Trans. Evol. Comp.*, **12**(1), 107–125.

Ogurtsov, S., Koziel, S. (2011). Design of microstrip to substrate integrated waveguide transitions with enhanced bandwidth using protruding vias and EM-driven optimization, *Int. Rev. Prog. Appl. Comp. EM*, pp. 91–96.

Ogurtsov, S., Koziel, S. (2017a). Systematic approach to sidelobe reduction in linear antenna arrays though corporate-feed-controlled excitation, *IET Microwaves, Ant. Prop.*, **11**(6), 779–786.

Ogurtsov, S., Koziel, S. (2017b). Sidelobe reduction in linear microstrip arrays driven through microstrip corporate feeds, *Proc. Int. Applied Computational Electromagnetics Society (ACES) Symp.*, Italy, pp. 1–2.

Ogurtsov, S., Koziel, S. (2018). On alternative approaches to design of corporate feeds for low-sidelobe microstrip linear arrays, *IEEE Trans. Ant. Prop.*, DOI: 10.1109/TAP.2018.2823915.

Ogurtsov, S., Koziel, S., Cheng, Q.S. (2017). Design of compact microstrip branchline couplers for broadband circularly polarized DRAs, *Int. Applied Computational Electromagnetics Society (ACES) Symp.*, Suzhou, China.

O'Hagan, A. (1978). Curve fitting and optimal design for predictions, *J. Royal Statist. Soc. B*, **40**, 1–42.

Okoshi, T. (1985). *Planar Circuits for Microwaves and Lightwaves*, Springer.

Palmer, K., Tsui, K.-L. (2001). A minimum bias Latin hypercube design, *IIE Trans.*, **33**, 793–808.

Pan, G.W. (2003). *Wavelets in Electromagnetics and Device Modeling*, Wiley–IEEE Press.

Patnaik, A., Choudhury, B., Pradhan, P., Mishra, R.K., Christodoulou, C. (2007). An ANN application for fault finding in antenna arrays, *IEEE Trans. Ant. Prop.*, **55**, 775–777.

Pattipati, B., Sankavaram, C., Pattipati, K.R. (2011). System identification and estimation framework for pivotal automotive battery management system characteristics, *IEEE Trans. Syst., Man, Cybernetics, Part C*, **41**, 869–884.

Pavlidis, N.G., Parsopoulos, K.E., Vrahatis, M.N. (2005). Computing Nash equilibria through computational intelligence methods, *J. Comp. Appl. Math.*, **175**, 113–136.

Pérez, V.M., Renaud, J.E., Watson, L.T. (2002). Interior point sequential approximate optimization methodology, *Proc. Symp. Multidisciplinary Analysis Opt.*, Atlanta, GA, AIAA-2002-5505.

Petko, J.S., Werner, D.H. (2007). An autopolyploidy-based genetic algorithm for enhanced evolution of linear polyfractal arrays, *IEEE Trans. Ant. Prop.*, **55**, 583–593.

Petosa, A. (2007). *Dielectric Resonator Antenna Handbook*, Artech House, Norwood, MA.

Polak, E., Ribiére, G. (1969). Note sur la convergence de méthodes de directions eonjuguées, *Rev. Fr. Inform. Rech. Opér.*, **16**, 35–43.

Poli, L., Rocca, P., Oliveri, G., Massa, A. (2014). Failure correction in time-modulated linear arrays, *IET Radar, Sonar Nav.*, **8**, 195–201.

Pozar, D.M. (1992). Microstrip antennas. *Proc. IEEE*, **80**(1), 79–81.

Pozar, D.M. (2011). *Microwave Engineering*, 4th edn., Wiley, Hoboken, NJ.

Pozar, D.M., Kaufman, B. (1990). Design considerations for low sidelobe microstrip arrays, *IEEE Trans. Ant. Prop.*, **38**(8), 1176–1185.

Price, K., Storn, R.M., Lampinen, J.A. (2005). *Differential Evolution: A Practical Approach to Global Optimization*, Springer, Berlin.

Priess, M., Koziel, S., Slawig, T. (2011). Surrogate-based optimization of climate model parameters using response correction, *J. Comp. Sci.*, **2**, 335–344.

QPar (2012). WBHDP0.9-18S Dual Polarized Horn Antenna. Product catalog. QPar Angus Ltd., Leominster, Herefordshire, U.K.

Queipo, N.V., Haftka, R.T., Shyy, W., Goel, T., Vaidynathan, R., Tucker, P.K. (2005). Surrogate based analysis and optimization, *Prog. Aerospace Sci.*, **41**, 1–28.

Radivojević, V.M., Rupčić, S., Grgić, K. (2017). Radiation pattern optimisation of an antenna array on the spherical surface by using a varying number of optimisation parameters, *IET Microwaves, Antennas Prop.*, **11**(13), 1846–1853.

Rajo-Iglesias, E., Quevedo-Teruel, O. (2007). Linear array synthesis using an ant-colony-optimization-based algorithm, *IEEE Antenna Propag. Mag.*, **49**(2), 70–79.

Rao, S., Wilton, D., Glisson, A. (1982). Electromagnetic scattering by surfaces of arbitrary shape, *IEEE Trans. Ant. Prop.*, **30**(3), 409–418.

Rashedi, E., Nezamabadi-pour, H., Saryazdi, S. (2009). GSA: A gravitational search algorithm, *Inf. Sci.*, **179**, 2232–2248.

Rasmussen, C.E., Williams, C.K.I. (2006). *Gaussian Processes for Machine Learning*, MIT Press, Cambridge.

Rautio, J.C., Harrington, R.F. (1987). An electromagnetic time-harmonic analysis of shielded microstrip circuits, *IEEE Trans. Microwave Theory Tech.*, **35**, 726–730.

Rayas-Sanchez, J.E. (2004). EM-based optimization of microwave circuits using artificial neural networks: The state-of-the-art, *IEEE Trans. Microwave Theory Tech.*, **52**, 420–435.

Rayas-Sanchez, J.E. (2016). Power in simplicity with ASM: tracing the aggressive space mapping algorithm over two decades of development and engineering applications, *IEEE Microwave Mag.*, **17**(4), 64–76.

Rechenberg, I. (1965). Cybernetic solution path of an experimental problem, Royal Aircraft Establishment, Library Translation Number 1122, Farnborough, UK.

Redhe, M., Nilsson, L. (2004). Optimization of the new Saab 9-3 exposed to impact load using a space mapping technique, *Struct. Multidisc. Opt.*, **27**, 411–420.

RF-35 (2018). ORCER RF-35. Data Sheet, Taconic, 136 Coonbrook Rd., Petersburgh, N.Y. 12138, USA.

Rigoland, P., Drissi, M., Terret, C., Gadenne, P. (1996). Wide-band planar arrays for radar applications, *Proc. IEEE Int. Symp. Phased Array Systems and Technology*, Boston, MA, USA, pp. 163–167.

Robinson, T.D., Eldred, M.S., Willcox, K.E., Haimes, R. (2008). Surrogate-based optimization using multifidelity models with variable parameterization and corrected space mapping, *AIAA J.*, **46**, 2316–2326.

Rocca, P., Oliveri, G., Massa, A. (2011). Differential evolution as applied to electromagnetics, *IEEE Ant. Prop. Mag.*, **53**, 38–49.

Rocca, P., Poli, L., Oliveri, G., Massa, A. (2012). Adaptive nulling in time-varying scenarios through time-modulated linear arrays, *IEEE Ant. Wireless Prop. Lett.*, **11**, 101–104.

Rodriguez, J.A., Ares, F. (2000). Finding defective elements in planar arrays using genetic algorithms, *Prog. Electromagn. Res.*, **29**, 25–37.

Rodriguez-Gonzalez, J.A., Ares-Pena, F., Fernandez-Delgado, M., Iglesias, R., Barro, S. (2009). Rapid method for finding faulty elements in antenna arrays using far field pattern samples, *IEEE Trans. Ant. Prop.*, **57**, 1679–1683.

Rojo-Alvarez, J.L., Camps-Valls, G., Martinez-Ramon, M., Soria-Olivas, E., Navia-Vazquez, A., Figueiras-Vidal, A.R. (2005). Support vector machines framework for linear signal processing, *Signal Processing*, **85**, 2316–2326.

Rosenbrock, H.H. (1960). An automatic method for finding the greatest or least value of a function, *Computer J.*, **3**, 175–184.

Roux, W.J., Stander, N., Haftka, R.T. (1998). Response surface approximations for structural optimization, *Int. J. Numerical Methods Eng.*, **42**, 517–534.

Roy, G.G., Das, S., Chakraborty, P., Suganthan, P.N. (2011). Design of non-uniform circular antenna arrays using a modified invasive weed optimization algorithm, *IEEE Trans. Ant. Prop.*, **59**, 110–118.

RT/duroid 6006/6010 Laminate (2011). Data Sheet, Rogers Corporation. Chandler, AZ.

Salucci, M., Gottardi, G., Anselmi, N., Oliveri, G. (2017). Planar thinned array design by hybrid analytical-stochastic optimisation, *IET Microwaves, Antennas Propag.*, **11**(13), 1841–1845.

Sans, M., Selga, J., Rodriguez, A., Bonache, J., Boria, V.E., Martin, F. (2014). Design of planar wideband bandpass filters from specifications using a two-step aggressive space mapping (ASM) optimization algorithm, *IEEE Trans. Microwave Theory Tech.*, **62**, 3341–3350.

Santner, T.J., Williams, B., Notz, W. (2003). *The Design and Analysis of Computer Experiments*, Springer.

Schantz, H. (2005). *The Art of Science of Ultra-Wideband Antennas*, Artech House, Norwood, MA.

Schwefel, H.-P. (1968). Experimentelle optimierung einer zweiphasendüse tell 1, AEG Research Institute Project MHD-Staustrahlrohr 11034/68, Technical report 35.

Schwefel, H.-P. (1988). Collective intelligence in evolving systems, in W. Wolff, C.-J. Soeder, F.R. Drepper (eds.) *Ecodynamics*, Springer, Berlin, pp. 95–100.

Schwefel, H.-P., Rudolph, G. (1995). Contemporary evolution strategies, in F. Moran, A. Moreno, J.J. Merelo, P. Chacon (eds.) *Advances in Artificial Life*, Springer, Berlin, pp. 891–907.

Secmen, M., Demir, S., Alatan, L., Civi, O.A., Hizal, A. (2006). A compact corporate probe fed antenna atrray, *Proc. First Euro. Conf. Antennas Propag. (EuCAP)*, pp. 1–4.

Selleri, S., Mussetta, M., Pirinoli, P., Zich, R.E., Matekovits, L. (2008). Differentiated meta-PSO methods for array optimization, *IEEE Trans. Ant. Prop.*, **56**, 67–75.

Shaker, G.S.A., Bakr, M.H., Sangary, N., Safavi-Naeini, S. (2009). Accelerated antenna design methodology exploiting parameterized Cauchy models, *J. Progress EM Res. B*, **18**, 279–309.

Shelokar, P.S., Siarry, P., Jayaraman, V.K., Kulkarni, B.D. (2007). Particle swarm and ant colony algorithms hybridized for improved continuous optimization, *App. Math. Comput.*, **188**, 129–142.

Sim, C.Y.D., Chang, M.H., Chen, B.Y. (2014) Microstrip-fed ring slot antenna design with wideband harmonic suppression, *IEEE Trans. Ant. Prop.*, **62**(9) 4828–4832.

Simpson, T.W., Pelplinski, J.D., Koch, P.N., Allen, J.K. (2001). Metamodels for computer-based engineering design: survey and recommendations, *Eng. Comput.*, **17**, 129–150.

Sivasubramani, S., Swarup, K.S. (2011). Multi-objective harmony search algorithm for optimal power flow problem, *Int. J. Electrical Power Energy Syst.*, **33**, 745–752.

SMA connector (2012). Part no. 142-0701-881, Catalog, Johnson/Emerson Connectivity, p. 1179.

Smola, A.J., Schölkopf, B. (2004). A tutorial on support vector regression, *Statist. Comp.*, **14**, 199–222.

Sobieszczanski-Sobieski, J., Haftka, R.T. (1997). Multidisciplinary aerospace design optimization: Survey of recent developments, *Struct. Opt.*, **14**, 1–23.

Sonnet EM (2016). Sonnet Software, Elwood Davis Road 100, North Syracuse, NY 13212.
Special Issue (2007). Special issue on synthesis and optimization techniques in electromagnetics and antenna system design, *IEEE Trans. Ant. Prop.*, **55**(3), part 1, 518–781.
Steer, M.B., Bandler, J.W., Snowden, C.M. (2002). Computer-aided design of RF and microwave circuits and systems, *IEEE Trans. Microwave Theory Tech.*, **50**(3), 996–1005.
Stegen, R.J. (1953). Excitation coefficients and beamwidths of Tschebyscheff arrays, *Proc. IRE*, **41**(11), 671–1674.
Storn R., Price, K. (1997). Differential evolution—a simple and efficient heuristic for global optimization over continuous spaces, *J. Global Opt.*, **11**, 341–359.
Suzuki, J. (1995). Finite Markov chain analysis of genetic algorithms, *IEEE Trans. Syst. Man. Cybernetics.*, **25**, 1995.
Søndergaard, J. (2003) *Optimization Using Surrogate Models—By the Space Mapping Technique*, PhD thesis, Informatics and Mathematical Modelling, Technical University of Denmark, Lyngby.
Taconic TLP (2013). Technical Data Sheet, Taconic International Ltd, Ireland.
Taflove, A., Hagness, S.C. (2005). *Computational Electrodynamics: The Finite-difference Time-Domain Method*, 3rd edn., Artech House, Norwood, MA.
Talbi, E.-G. (2009). *Metaheuristics: From Design to Implementation*, John Wiley & Sons, Hoboken.
Tamura, K., Yasuda, K. (2011). Spiral dynamics inspired optimization, *J. Adv. Comp. Intelligence Intelligent Informatics*, **15**(8), 1116–1122.
Tan, K., Khor, E., Lee, T. (2005). *Multiobjective Evolutionary Algorithms and Applications*, Springer, London.
Tapia, R.A. (1978). Quasi-Newton methods for equality constrained optimization: Equivalence of existing methods and a new implementation, in O. Mangasarian, R. Meyer, S. Robinson (eds.) *Nonlinear Programming*, Academic Press, New York, pp. 125–164.
Toal, D.J.J., Keane, A.J. (2011). Efficient multipoint aerodynamic design optimization via cokriging, *J. Aircraft*, **48**, 1685–1695.
Toivanen, J.I., Makinen, R.A.E., Jarvenpaa, S., Yla-Oijala, P., Rahola, J. (2009). Electromagnetic sensitivity analysis and shape optimization using method of moments and automatic differentiation, *IEEE Trans. Ant. Prop.*, **57**(1), 168–175.
Toropov, V.V., Filatov, A.A., Polynkin, A.A. (1993). Multiparameter structural optimization using FEM and multipoint explicit approximations, *Struct. Optim.*, **6**, 7–14.
Tu, S., Cheng, Q.S., Zhang, Y., Bandler, J.W., Nikolova, N.K. (2013). Space mapping optimization of handset antennas exploiting thin-wire models, *IEEE Trans. Ant. Prop.*, **61**, 3797–3807.
van Laarhoven, P.J., Aarts, E.H. (1987). *Simulated Annealing: Theory and Applications*, Springer.

Volakis, J.L. (ed.) (2007). *Antenna Engineering Handbook*, 4th edn., McGraw-Hill, New York.

Volakis, J.L., Chen, C., Fujimoto, K. (2010). *Small Antennas: Miniaturization Techniques and Applications*, McGraw-Hill, New York.

Vouvakis, M.N., Schaubert, D.H. (2011). Vivaldi antenna arrays, Chapter 3, in F.B. Gross (ed.) *Frontiers in Antennas: Next Generation Design & Engineering*, McGraw-Hill, New York.

Waterhouse, R.B. (2003). *Microstrip Patch Antennas: A Designer's Guide*, Kluwer Academic Publishers, Norwell, MA.

Wen, Y.Q., Wang, B.Z., Ding, X. (2016). A wide-angle scanning and low sidelobe level microstrip phased array based on genetic algorithm optimization, *IEEE Trans. Ant. Prop.* **64**(2), 805–810.

Weyland, D. (2015). A critical analysis of the harmony search algorithm—How not to solve sudoku, *Oper. Res. Perspect.*, **2**, 97–105.

Wild, S.M., Regis, R.G., Shoemaker, C.A. (2008). ORBIT: Optimization by radial basis function interpolation in trust-regions, *SIAM J. Sci. Comput.*, **30**, 3197–3219.

Wilkinson, E.J. (1960). An N-way hybrid power divider, *IRE Trans. Microwave Theory Tech.*, **8**(1), 116–119.

Wolpert, D.H., Macready, W.G. (1997). No free lunch theorems for optimization, *IEEE Trans. Evol. Comp.*, **1**, 67.

Wolpert, D.H., Macready, W.G. (2005). Coevolutionary free lunches, *IEEE Trans. Evol. Comp.*, **9**(6), 721–735.

XFDTD (2014). Remcom, Inc., South Allen 315, Suite 416, State College, PA 16801.

Xia, L., Xu, R.M., Yan, B. (2007). LTCC interconnect modeling by support vector regression, *Prog. EM Res.*, **69**, 67–75.

Yang, F., Yang, S., Chen, Y., Qu, S. (2017). A joint optimization approach for the synthesis of large 4-D heterogeneous antenna arrays, *IEEE Trans. Ant. Prop.*, **65**(9), 4585–4594.

Yang, X.S. (2005). Engineering optimization via nature-inspired virtual bee algorithms, in *IWINAC 2005*, Lecture Notes in Computer Science, Vol. 3562, Springer, pp. 317–323.

Yang, X.S. (2008). *Nature-Inspired Metaheuristic Algorithms*, Luniver Press.

Yang, X.-S. (2010a). *Engineering Optimization: An Introduction with Metaheuristic Applications*, John Wiley & Sons, Hoboken.

Yang, X.-S. (2010b). A new metaheuristic bat-inspired algorithm, in J.R. González, D.A. Pelta, C. Cruz, G. Terrazas, N. Krasnogor (eds.) *Nature Inspired Cooperative Strategies for Optimization*, Springer, Berlin, pp. 65–74.

Yang, X.-S. (ed.) (2014). *Cuckoo Search and Firefly Algorithm: Theory and Applications*. Springer International Publishing.

Ye, K.Q. (1998). Orthogonal column Latin hypercubes and their application in computer experiments, *J. Amer. Statist. Assoc.*, **93**, 1430–1439.

Yelten, M.B., Zhu, T., Koziel, S., Franzon, P.D., Steer, M.B. (2012). Demystifying surrogate modeling for circuits and systems, *IEEE Circuits Syst. Mag.*, **12**, 45–63.

Yeung, S.H., Man, K.F., Luk, K.M., Chan, C.H. (2008). A trapeizform U-slot folded patch feed antenna design optimized with jumping genes evolutionary algorithm, *IEEE Trans. Ant. Prop.*, **56**, 571–577.

Yin, J., Wu, Q., Yu, C., Wang, H., Hong, W. (2017). Low-sidelobe-level series-fed microstrip antenna array of unequal interelement spacing, *IEEE Antennas Wireless Propag. Lett.*, **16**, 1695–1698.

You, P., Liu, Y., Chen, S.-L., Xu, K.D., Li, W., Liu, Q.H. (2017). Synthesis of unequally spaced linear antenna arrays with minimum element spacing constraint by alternating convex optimization, *IEEE Antennas Wireless Propag. Lett.*, **16**, 3126–3130.

Zhang, K., Han, Z. (2013). Support vector regression-based multidisciplinary design optimization in aircraft conceptual design, *AIAA Aerospace Sciences Meeting*, AIAA paper 2013–1160.

Zhang, Y., Nikolova, N.K., Meshram, M.K. (2012). Design optimization of planar structures using self-adjoint sensitivity analysis, *IEEE Trans. Ant. Prop.*, **60**, 3060–3066.

Zhigljavsky, A., Žilinskas, A. (2008). *Stochastic Global Optimization*, Springer US.

Zitzler, E., Thiele, L. (1999). Multiobjective evolutionary algorithms: A comparative case study and the strength Pareto approach, *IEEE Trans. Evol. Comp.*, **3**, 257–271.

Index

A

AA apertures, 204
AA CAD, 204
absolute design costs, 287
absolute gain, 16
absorbing boundaries, 139, 141, 209, 211
accelerated optimization, 412
accelerated PSO, 80
acceleration constants, 77
accurate evaluator, 234
active constraints, 48
active reflection coefficients, 25, 32, 212, 234–237, 239, 242–243, 245–247, 251, 253–254, 264, 266, 268–270, 274, 276–278, 281, 283–285, 316, 333, 357, 361–362, 364, 368, 371
active reflection response, 231, 242, 268, 270, 275, 277–278, 284
active reflections coefficients, 27
active set methods, 48
adaptive meshing, 52, 140
adaptive meshing techniques, 3, 52
adaptive response correction, 131, 143
adaptive response prediction, 131
adaptive sampling, 114
adaptive sampling techniques, 114

adaptively adjusted design specifications (AADS), 115, 121–123
additional suppression of SLL, 264
additive response, 145, 186
additive response correction, 129, 146–147, 186, 267, 277, 375
adjoint sensitivities, 5, 57, 85
adjustable parameter, 30, 412
aerodynamic design, 99
aerodynamic shape optimization, 65
affine frequency scaling, 237
affinely independent, 55
affinely independent points, 55
agents, 77
Aggressive SM (ASM), 116–117
air, 18
algorithm convergence, 75, 81
algorithm parameters, 76
amplitude tapers, 398
analytical array factor (AF) models, 82, 88, 204, 240, 396
analytical derivatives, 215, 230, 235
analytical formulas, 91
analytical gradient, 241
analytical model, 2, 23, 86, 102, 231, 240, 265–269, 275, 277, 410
analytical model of the array, 263
analytical model of the radiation response, 264

441

442 Index

analytical second-order derivatives, 43
analytically intractable, 58
anechoic chamber, 322, 343
angular sector, 185
anisotropy of substrates, 140
ANN architecture, 99
ant colony optimization, 4, 77, 81
antenna(s), 7, 11, 14, 17, 32, 35, 57, 65, 85, 87–88, 91, 102, 128–129, 133, 140, 170, 201, 208–209, 211, 409
antenna analysis, 13
antenna array analysis, 11, 14
antenna array apertures, 2, 7, 17, 207, 254, 395
antenna array CAD, 1, 7, 14
antenna array characteristics, 13
antenna array circuits, 7, 29, 31, 210
antenna array components, 208
antenna array design, 7, 26, 29–30, 58, 82, 133, 206, 410, 415
antenna array design problems, 12, 235, 412
antenna array design process, 2–3
antenna array elements, 1
antenna array excitation taper, 279
antenna array far-field, 205
antenna array model accuracy, 210
antenna array models, 1–2, 8, 13–14, 271
antenna array optimization, 6, 412
antenna array problems, 209, 410
antenna array radiation figures, 14, 17
antenna array simulation, 12
antenna array statistical analysis, 394
antenna array structures, 1
antenna array theory, 14
antenna arrays (AAs), 1, 4–5, 8, 15, 17, 20, 22, 24, 31–33, 35, 52, 57–60, 65, 85, 88, 91, 102, 145, 176, 203, 208, 236, 292, 296, 375, 409–414
antenna arrays with strong element coupling, 134
antenna characteristics, 134
antenna circuits, 11–12

antenna components, 12
antenna design, 57, 185
antenna design cases, 89
antenna design process, 135
antenna design subproblems, 410
antenna diagnosis, 374
antenna elements, 20, 411
antenna engineering, 8, 131
antenna footprint, 151–152
antenna input, 17
antenna models, 22
antenna modules, 210
antenna optimization, 5, 121, 134
antenna radiation efficiency, 17
antenna reflection, 137
antenna size, 34, 145, 152
antenna structures, 52, 134
antenna subarray, 8
antenna systems, 11
antenna technology, 30
antenna theory, 1
antenna-feed structure, 13
antennas and antenna arrays, 82
aperture, 7, 14, 27, 203, 407
aperture design, 212
aperture dimensions, 207
aperture element, 211
aperture excitation, 27
aperture footprint, 32
aperture optimization, 351
aperture periphery, 206
aperture scattering matrix, 27
aperture spacings, 207
aperture variables, 322
aperture-embedded element patterns, 365–366
aperture-feed circuit, 3, 7, 204
aperture-feed modules, 208
aperture-feed structure, 9, 301–302
apparent impedance, 25
apparent input impedance, 25
approximate model, 401
approximation model management optimization (AMMO), 115, 119

Index 443

approximation models, 6, 52, 91, 411, 414
approximation surrogates, 6
approximation-based models, 115
approximation-based surrogate, 115
arithmetic, 70
arithmetic crossover, 64
arithmetic recombination, 79
arithmetical averaging, 67
array aperture, 1, 4, 9, 18, 23–24, 26, 29, 31, 205–206, 212, 231, 254, 347, 358, 399, 411
array aperture element, 27, 314
array aperture EM model, 27
array aperture implementation, 29
array aperture model, 24
array aperture radiation pattern, 396
array aperture scattering matrix, 27
array aperture scattering parameters, 22
array aperture surrogate model, 333
array circuits, 203, 207, 210–211, 315, 317–318, 320–322, 328
array components, 1, 210
array design problem, 255
array design process, 322
array directivity, 215, 411
array element, 2, 7, 22, 204, 206, 395, 407
array element weights, 13
array EM model, 362
array embedded elements, 15
array environment, 2
array factor (AF), 3, 18, 20–21, 23, 26, 205–207, 213, 215, 231, 233, 235, 240, 242, 244, 246, 248–250, 253, 263, 266, 292, 305, 332, 337, 373–377, 381, 383, 397, 410
array factor (AF) model, 60, 206–207, 235, 237, 291, 293–294, 328, 336, 358–360, 375, 377–380, 386–388, 395–399, 411–413
array factor of a linear array, 214
array factor optimization, 4

array factor pattern, 292–293
array factor-based models, 206–207
array factor-based surrogate models, 207
array feeds, 2, 13, 32, 207
array lattices, 23
array models, 318
array of microstrip antennas, 299
array of MPAs, 365
array optimization, 234, 255, 410, 414
array pattern, 258
array radiation, 231, 270
array radiation apertures, 208
array radiation pattern, 305
array radiation pattern optimization, 121
array spacing, 206
array structures, 85
array synthesis, 215, 235
array-embedded single-patch responses, 276
array-feed circuits, 316
arrays, 57
arrays elements, 183
arrays of antennas, 11
artificial neural networks (ANNs), 6, 98
ASM algorithm, 117
assembly imperfections, 154
attractiveness parameter, 80
augmentation, 50
augmented Lagrangian, 50
augmented Lagrangian methods, 50
automated design frameworks, 413
automated mesh adaptation, 209
auxiliary low-fidelity model, 257
auxiliary model, 292
auxiliary surrogates, 412
average gain, 171
axial ratio, 139, 176, 207
azimuth angle, 19

B

back radiation level, 12
back-to-front ratio, 32

444 Index

bacteria foraging algorithm, 81
balun, 145–146
bandstop filter example, 122
barrier method, 49
basis functions, 95–96, 139
bat algorithm, 62, 81
beam, 12, 15
beam broadening, 369
beam broadening factor, 368, 370
beam scanning operation, 231
beam steering, 11
beam width, 2, 11, 206, 212
bee algorithm, 81
best feeds, 339
BFGS formula, 45
binary coding, 68
binary strings, 68
binary tournament selection, 63
bird schooling, 76
bit-flip mutation, 70
bit-string representation, 68–69
bit-string-based genetic algorithms, 74
boundary, 48
box constraints, 71
branch and bound methods, 60
broadside linear array apertures, 213
Broyden update, 117
Broyden–Fletcher–Goldfarb–Shanno (BFGS), 44
brute-force optimization, 4
building blocks, 74
built-in numerical optimization routines, 12
built-in optimizers, 208

C

CAD, 14
CAD tools, 7, 30
candidate set, 64
candidate design, 31, 110
Cartesian coordinates, 19
Cartesian lattice, 273, 275, 279
cavity models, 175
center frequency, 140

center of gravity, 55
CF architectures, 305
characteristic impedance, 25
characteristic points, 120, 122–123, 336
characteristic wavelength, 157
Chebyshev excitation, 339–340
Chebyshev excitation taper, 343
Chebyshev taper, 342, 349–350
chromosome, 70
chromosome length, 70
circuit equivalents, 7
circuit evaluators, 12
circuit models, 3
classification, 98
CMA-ES, 65
co-kriging, 101
co-pol patterns, 325–327
co-simulations, 12
coarse discretization, 139
coarse meshes, 211
coarse model, 31, 129, 133, 206–207, 211, 262, 280–281
coarse-discretization (low-fidelity) model, 194
coarse-discretization array model, 277
coarse-discretization electromagnetic simulations, 83
coarse-discretization model, 139, 262, 265–266, 268
coarse-discretization simulations, 88, 102, 201, 271
coarse-mesh EM analysis, 88
coarse-mesh EM model, 253
coarse-mesh EM simulations, 247
coarse-mesh EM-simulation models, 411
coarse-mesh model, 198, 236, 241, 275, 358, 363
coarse-mesh simulated models, 371
coarse-mesh simulations, 231, 233, 412
coarse-mesh simulation-based superposition model, 241
coarse-mesh surrogate, 357

coarsely discretized models, 211, 342
coarsely discretized simulated models, 211
combinatorial optimization, 69
combinatorial problems, 68
combined arithmetic–discrete recombination, 79
commercial EM packages, 209
commercial EM solvers, 5
commercial simulation software packages, 86
commercial simulation tools, 4
commercial solvers, 52
common ground plane, 184
common input, 12
communication technologies, 11
compact antennas, 168
compact UWB slot antenna, 149
complex amplitude, 14
complex antenna arrays, 87
complex excitation coefficient, 14
complex far-field pattern, 374
complex far-fields, 231
complex S-parameters, 231
complex taper, 233
component-wise division, 164
component-wise multiplication, 77, 127
compound model, 396
computational budget, 92
computational complexity, 46, 100
computational cost, 29, 57, 82, 133, 159, 210, 212, 240, 253, 409, 411–412
computational cost of adjusting the active reflection coefficients, 268
computational cost of the low-fidelity model, 257
computational domain, 139
computational electromagnetic methods, 204
computational electromagnetics (CEM), 1, 3, 11, 86, 116, 208–209
computational EM models, 208
computational methods, 12

computational models, 32, 58, 157, 205
computational simplifications, 139
computationally expensive model, 377
computer simulation, 52
computer simulation models, 86
computer-aided design (CAD), 1, 12
computers, 12
computing hardware, 210
conjugate gradients, 3, 46
conjugate-gradient methods, 40–41, 46
connectors, 1–2, 13, 58, 85, 154
conservation of power, 308
constrained metaheuristic optimization, 82
constrained minima, 48
constrained minimum, 48–49
constrained optimization, 33, 46–48, 50
constrained optimization problems, 51
constrained problem, 47, 49–50
constrained PSs, 345
constrained SBO, 114
constrained solution, 50
constraints, 34, 47–48
constraint handling, 49
continuous global optimization, 77
continuous optimization, 67–68
continuous optimization problems, 65, 68
continuous search spaces, 67
continuously differentiable, 48, 86
contour plots, 49
contraction, 57
control parameters, 79, 413
controlling the active reflection coefficients, 286
conventional methods, 57
conventional numerical optimization, 85
conventional numerical optimization methods, 410

446 *Index*

conventional optimization algorithms, 58
conventional optimization methods, 5
conventional optimization techniques, 3, 57
convergence, 38, 52, 57, 86, 89–90, 103, 110, 119, 127
convergence of the algorithm, 71
convergence plot, 129
convergence problems, 118
convergence rate, 38, 46, 80
convergence safeguards, 90
convex combinations, 55, 64
convex hull, 55
coordinate system, 18
coplanar waveguide (CPW), 156
corporate feed architectures, 309–313
corporate feed design, 411
corporate feed topology, 401
corporate feeds (CFs), 9, 183, 188–190, 194, 201, 207, 303–307, 309, 311–312, 321, 329–331, 340–341, 347, 355, 393–394, 399–402, 411
corrected analytical array factor model, 387–389
corrected analytical model, 378
corrected array factor, 291
corrected array factor (AF) model, 336, 381, 386, 389, 399
corrected array factor pattern, 357
corrected feed, 333
corrected model, 397
corrected path lengths, 350
corrected simulation-based superposition model, 237
corrected space mapping, 131
correction function, 119, 293–294, 360, 377–379, 386–388
correction matrix, 120
correction of the AF model, 336
correction of the microstrip traces, 348
correction techniques, 32, 276, 380, 394

correction term, 291, 338, 362, 377, 379
correlation analysis, 157
correlation function, 97
correlation matrix, 97, 335
correlation parameters, 97, 335
correlation vector, 335
cost function, 37, 42
cost function improvement, 54
cost function value, 55
cost of the optimal design, 283, 286
coupling, 22, 204
covariance matrix, 65–66, 97
covers, 1
CPU cost, 34
critical scan angle, 249
cross-validation, 100–101, 171
crossover, 64, 69–70
crossover operator, 68
crossover probability, 78
CST MWS, 238, 296, 298, 313, 316, 339, 346, 363–364, 385
CST MWS environment, 241
CST MWS transient solver, 241, 245, 280, 282, 317, 364
cubic spline interpolation, 384
cuckoo search, 62
current iteration point, 51
curse of dimensionality, 6, 91
cutting plane methods, 60

D

DACE Toolbox, 335
damped Newton method, 43–44
data-driven modeling, 88, 91, 94, 115, 160, 412
data-driven surrogate models, 109
data-driven surrogates, 91–92, 102, 109–110, 113, 411, 413
DE algorithm, 78
DE scheme, 78
decomposition, 301
defect-free array, 375
defected ground structure, 164
defective array, 377

Index 447

defective elements, 375
densely filled matrices, 209
derivative data, 52
derivative information, 52–53
derivative-free, 3, 33
derivative-free downhill simplex procedure, 55
derivative-free global optimization methods, 61
derivative-free methods, 8, 57–58
derivative-free optimization, 51
derivative-free optimization techniques, 52
derivative-free stochastic search algorithm, 77
derivative-free techniques, 59
derivatives, 51, 119
descent algorithm, 38
descent direction, 36, 38, 43
descent methods, 37, 39, 41, 46
design automation, 87, 413
design constraints, 411
design cost, 229, 244, 281, 284–287
design cost using the SBO approach, 282
design frameworks, 410, 415
design goal, 298
design objectives, 296, 410
design of antenna arrays, 414
design of experiments (DOE), 92, 164
design of microstrip corporate feeds, 354
design optimization, 60, 135, 160, 253, 279
design optimization problem, 276
design optimization process, 138
design parameters, 47
design problem, 234, 337
design process, 23, 330, 411
design refinement, 173
design requirements, 280, 282–283
design space, 6, 61, 64, 76, 90, 92–93, 102, 109, 115, 124, 134, 143, 170, 210
design space dimensionality, 411

design space exploitation, 114
design space reduction, 171, 173
design specifications, 12, 122–123, 215, 234, 262, 268, 305, 413
design speedup, 287–289
design time cost, 251
design variable vector, 244, 258
design variables, 30, 239, 244, 255, 280, 282, 363, 414
design verification, 136
design-of-experiment plan, 112
designable parameters, 234
determination coefficient, 157
deterministic computer simulations, 93
deterministic methods, 60
deterministic selection, 78
deterministic truncation selection, 66
deviation vector, 395, 399
deviations of the geometry and material parameters, 396–397
dielectric losses, 140
dielectric permittivity, 106
dielectric resonator, 104
dielectric resonator antennas (DRAs), 104, 106, 134
dielectric substrates, 141, 394
differential evolution (DE), 4, 60, 62, 64–65, 68, 77–79, 81
dimensionality of the problem, 6
dipole, 146
dipole element, 170
direct gradient optimization, 216
direct methods, 58
direct optimization, 34
direct optimization methods, 58
direct search, 130
direction of interest, 16
directivity, 16–17, 215, 218, 262, 266
directivity maximization, 215
directivity pattern, 136, 217–218, 220, 258, 261, 264–266, 269–270
directivity pattern optimization, 262, 266, 268
discontinuous, 82

448 Index

discontinuous functions, 86
discrete, 70
discrete array model, 275
discrete electromagnetic (EM) models, 3, 208, 254
discrete electromagnetic (EM) solvers, 12, 276
discrete EM modelers, 204
discrete EM simulations, 204
discrete EM simulators, 136–137
discrete full-wave electromagnetic (EM) simulations, 13
discrete full-wave electromagnetic models, 12
discrete full-wave EM simulation, 134
discrete full-wave EM solvers, 133
discrete full-wave modeler, 30
discrete full-wave simulations, 137
discrete full-wave solvers, 29
discrete simulations, 27
discrete sources, 140
discretization, 58
discretization density, 138
discretization fidelity, 32
dispersion, 210
distance, 164
diversity of the population, 74
DOE strategies, 93
Dolph–Chebyshev array, 263, 269
Dolph–Chebyshev design, 263, 269
dominant mode of the microstrip patches, 238

E

12-element array, 342
E-plane, 262–263, 269
E-plane half-power beamwidth (HPBW), 239, 269
E-plane patterns, 190
effect of the reflection control on the SLL, 283
effective array-embedded element pattern, 360
effective pattern of the array embedded element, 377

effective peak directivity per element, 260
efficiency, 2, 17, 32
efficient global optimization (EGO), 114–115
electric far-field, 15, 205
electric field, 14–15
electrical engineering, 99, 108, 116
electromagnetic (EM) analysis, 12, 102, 393, 414
electromagnetic (EM) models, 33, 183
electromagnetic simulations, 82
electromagnetic solvers, 1
electromagnetics, 13
element design, 29
element far-field in the E-plane, 233
element geometry, 404
element pattern, 205, 207, 263, 367, 374, 396
element power pattern, 27
element radiation intensity, 20
element reflection coefficient, 27
element spacings, 13, 292, 394, 396, 404
elements, 12
elements coupling, 15
elevation angle, 35
elitism, 72
EM array model, 361, 394
EM models, 3, 27, 32, 146, 151, 156, 169–170, 177, 188–189, 192, 194, 197, 210, 212, 231, 255, 296–297, 303, 333, 358, 363, 394, 396–398
EM model response, 397
EM models of array apertures, 255, 396
EM problems, 209
EM simulation cost, 58
EM simulation model, 57, 379
EM simulation-based, 58
EM simulations, 29, 34, 106, 108, 126, 148, 186, 212, 292, 375
EM simulators, 136
EM solver, 3, 22, 58, 185, 211
EM-based design frameworks, 410

EM-based designs, 330
EM-based optimization, 5, 322, 328
EM-based superposition models, 411
EM-driven design, 86, 145, 271
EM-driven design optimization, 413
EM-simulated antenna model, 161
EM-simulated array aperture, 397–398
EM-simulated complex far-field, 384
EM-simulated element pattern, 395
EM-simulated far-field, 274
EM-simulated high-fidelity model, 207
EM-simulated model, 291, 338, 375, 378, 386
EM-simulated pattern, 292, 337, 373, 375, 377
EM-simulated radiation pattern, 376–377, 385, 395
EM-simulated radiation response, 266
EM-simulated response, 3, 5, 380
EM-simulation results, 83
EM-simulation-driven design, 4
EM-simulations, 355
EM-tuned model, 342
embedded elements, 261–262
end-fire designs, 216
end-fire gain, 170
end-fire radiation, 221
engineering, 11
engineering problem, 75
engineering systems, 52
enumeration, 382–383
enumeration technique, 375
equal power split, 305–306
equal power split junctions, 303
equal power split T-junctions, 323–324
equal-split T-junctions, 307, 321, 355
equal-split junctions, 313, 317
equal-split microstrip T-junction, 314
equal-split T-junctions, 303, 309, 311–313

equivalent circuit, 106
equivalent network, 102, 411
equivalent surface currents, 209
error, 100
error back-propagation algorithm, 98
ES algorithm, 66
etching, 154
Euclidean distance, 80
evaluation time of the high-fidelity model, 257
evaluation times of the high- and low-fidelity models, 257
evanescent thickness, 140
evolution strategies (ES), 60–61, 65–67, 81
evolutionary algorithms (EAs), 61–62, 64–65, 74–75, 81
exact line search, 39–40
excitation, 383
excitation (power distribution) network, 29
excitation amplitude taper, 396
excitation amplitudes, 2, 13, 85, 210, 214, 229, 234, 260, 266, 268, 281, 284–285, 288, 292, 297–298, 316, 336, 377–380, 387, 393, 397, 399, 411
excitation amplitudes as design variables, 268
excitation amplitudes within the integrated structure, 349
excitation coefficient, 234
excitation phase shifts as design variables, 269
excitation phases, 233, 235, 241, 244, 297, 357, 359–360, 362, 364
excitation phasor, 15
excitation set, 24
excitation taper, 9, 13, 25, 207, 212–213, 273–275, 296, 301–302, 330, 333–334, 336–337, 343, 347, 352, 396, 401, 411, 413–414
exhaustive search, 382–383, 390
expansion, 57
expansion coefficients, 162–163

expected improvement, 91, 113
expected model error, 113
expensive EM-simulations, 131
experimental validation, 411
exploitation, 6, 74, 109, 115
exploitative, 64, 70
exploitative models, 6
exploitative operators, 64
exploration, 6, 74, 109, 114
explorative, 69
explorative models, 6
exploratory, 64
exploratory move, 54
exploratory operators, 64
exponential, 96
exponential crossover, 79
extraction process, 118
extreme Pareto-optimal designs, 171

F

fabrication inaccuracies, 394
factorial designs, 93
factorial DOEs, 93
far-field(s), 15–16, 18, 29, 179, 205, 212, 234
far-field in the E-plane, 233
far-field magnitude pattern, 373
far-field of the isolated MPA element, 248
far-field radiation patterns, 390
far-zone, 15, 18
far-zone sphere, 16
fast approximate model, 330
fast array factor-based model, 321
fast array model, 253
fast computing, 13
fast estimator, 234
fast model, 244, 307–308, 312, 318
fast models' patterns, 316
fast surrogate, 256, 377, 394, 411
fast surrogate model, 352, 375
fault configuration, 384, 386, 391
fault detection, 374–376, 379–385, 390–391, 409, 411
fault dictionary approach, 382

fault identification, 379
fault patterns, 375
faults in antenna arrays, 374
faulty apertures, 383
faulty array configurations, 391
faulty arrays, 373, 381, 389–390
faulty elements, 373, 375, 381–386
faulty patterns, 374, 382, 384, 386, 389–391
faulty state, 377
faulty test patterns, 385
FBO procedure, 156
feasibility of individuals, 69
feasibility of solutions, 76
feasible domain, 47
feasible point, 50
feasible region, 47–49
feasible region boundary, 49
feature location, 126
feature model gradients, 126
feature point set, 130
feature points, 127, 130, 157, 337–338, 397–398
feature-based design, 337
feature-based model, 126
feature-based optimization (FBO), 116, 124, 130, 159, 165, 177, 180, 411
feature-based optimization algorithm (FBO), 127–128, 155, 181
feature-based surrogates, 102
feed, 1–2, 4, 7, 13, 18, 27, 31, 85, 201, 204, 211, 301–305, 309, 314, 317, 407, 409, 413
feed architecture, 302–310, 312, 315, 319–321, 328, 334, 340, 342, 352, 411
feed design, 411
feed design for low-sidelobes, 334
feed design variables, 322
feed footprint, 32
feed geometry, 404
feed junctions, 334
feed networks, 2
feed optimization, 334

Index 451

feed ports, 27
feed radiation, 352
feed scattering matrix, 27
feed tree, 330
feeding circuit, 203
feeding elements, 134
feeding structures, 410
FEKO, 209
FEM algorithms, 209
FEM-based solvers, 209
fenotype, 68
filter, 107
final design, 143, 260, 263–264, 269
fine model, 135, 363
fine model response, 120
finite differentiation, 52, 127, 399
finite grounds, 1
finite-difference time-domain (FDTD) method, 208–209
finite-element, 139
finite-element analysis, 52
finite-element method (FEM), 139, 208–209
finite-volume transient solvers, 52
firefly, 79–80
firefly algorithm (FA), 60, 62, 65, 79–81
first-order consistency, 42, 89, 103, 113
first-order consistency conditions, 119
first-order continuously differentiable, 36
first-order correction, 377
first-order necessary conditions, 47
first-order response surface models, 258
first-order stationary point, 39
first-order Taylor expansion, 36, 42, 399
first-order Taylor model, 127
fitness, 62, 68, 71
fitness value, 62, 71, 76
fixed-beam arrays, 23, 329
fixed-beam linear arrays, 304

fixed-beam apertures, 205, 207, 212
fixed-beam planar array apertures, 9
Fletcher–Reeves conjugate-gradient method, 41
floating-point numbers, 65
floating-point representation, 67–68, 70, 74
four-patch subarray, 184
fractional bandwidth, 126, 128–129
free-space propagation, 205
frequency, 35
frequency domain, 209, 319
frequency domain methods, 140
frequency domain solver, 318–321
frequency of interest, 18
frequency scaling, 108, 129, 145, 147, 186, 277, 358
frequency scaling parameters, 147
frequency-scaled model, 186, 237, 358
frequency-scaled parameters, 186
frequency-scaled surrogate, 108
front-to-back ratios, 143
full-wave analysis, 11
full-wave discrete electromagnetic modelers/solvers, 12
full-wave discrete EM solvers, 206
full-wave discrete models, 276, 282
full-wave discrete solvers, 1, 14, 208
full-wave electromagnetic (EM) analysis, 33
full-wave electromagnetic (EM) simulation, 2, 82, 203, 208
full-wave EM models, 214
full-wave EM simulation models, 213, 292
full-wave EM simulations, 2, 24
full-wave methods, 13
full-wave simulations, 22, 407
full-wave superposition models, 24
function value, 38
function-approximation models, 6
functional, 99
functional landscape, 59

G

gain, 11, 16–17, 108, 141, 172
gain ratio, 42, 89
Gaussian, 80, 97, 403
Gaussian correlation function, 335
Gaussian kernels, 100
Gaussian probability distribution, 403–405
Gaussian process regression (GPR), 6, 95, 100
Gaussian processes, 100
Gaussian random process, 97
general-purpose desktops, 12
generalization, 100, 102
generalization capability, 101, 106, 113
generalization error, 92, 171
generation gap, 73
generation gap models, 73
generational models, 73
generic SBO scheme, 256, 260
generic SM surrogate, 118
genetic algorithms (GAs), 4, 61–62, 65, 68–71, 74, 77, 81, 169, 203, 374, 381
genetic and evolutionary algorithms, 60
genetic material, 74
genetic operations, 69, 76
genotype, 68, 70
geometry constraints, 34
geometry dimensions, 260
geometry parameters, 394, 399
global accuracy, 102, 110, 113–114
global best, 76
global methods, 90
global optimization, 6, 8, 52, 58–60, 85, 99, 102, 109, 114
global optimization methods, 97
global response correction, 103
global search, 59, 87, 90, 294
global search algorithm, 90
global search properties, 82
global solution, 67, 75
globally accurate surrogates, 114
globally convergent, 43
GPR models, 100
GPR-based surrogates, 100
gradient, 35–36, 38, 128
gradient estimation, 157
gradient optimization, 215
gradient search, 213, 215
gradient-based, 8, 57
gradient-based algorithms, 5, 33, 53, 59, 86, 380
gradient-based methods, 3
gradient-based optimization, 35, 57, 332, 336, 339
gradient-based optimization algorithms, 32, 52
gradient-based optimization with analytical derivatives, 294
gradient-based routines, 61
gradient-based schemes, 58
gradient-based search, 42, 52, 85
gradient-based techniques, 35, 52
gradient-search with analytical derivatives, 229
graphical user interface, 208
graphics processing unit (GPU), 1, 208–209
grating lobe level (GLL), 247, 249–250
grating lobes, 212, 248–249
gravitational search algorithm, 81
gray coding, 68
grid, 53
grid sampling, 94
ground plane, 140, 164, 206
ground-plane rectangular slot, 178
ground-plane slot, 156
grounded substrate, 189

H

H-plane, 155
H-plane HPBW, 155, 262–263, 269
H-plane beamwidth, 197
H-plane pattern, 319–320, 328, 345, 352
H-plane power patterns, 341

H-plane radiation pattern, 190
H-plane SLL, 330
H-plane total pattern, 354
half-power beamwidth (HPBW), 12, 26, 247, 250, 263
half-wavelength, 220
half-wavelength (HW) spacings, 216
Hammersley sampling, 94
hardware acceleration, 208
hardware-accelerated FDTD, 208
harmony search, 81
Hessian, 36, 42–44
heuristics, 68
hexagonal lattice, 273, 275, 279, 286
hexagonal lattice array, 286–288
hexahedral meshes, 138
hidden layers, 98
high-fidelity computational models, 210
high-fidelity data, 101–103, 110, 112
high-fidelity design, 123
high-fidelity discrete model, 245
high-fidelity discrete simulations, 332
high-fidelity EM level of description, 212
high-fidelity EM model, 58, 253, 255–256, 348, 412
high-fidelity EM simulation models, 355
high-fidelity EM simulations, 31, 85, 145–146, 176, 305, 341, 368
high-fidelity full-wave discrete simulations, 391
high-fidelity full-wave simulation, 391
high-fidelity level of description, 82, 348
high-fidelity model(s), 5, 7, 24, 87–89, 92, 101–104, 106, 108, 110, 112, 115–116, 118, 120, 128, 134, 138–139, 143, 146–148, 151, 156, 158, 164, 170, 173, 185–186, 194, 198, 202, 210–211, 231, 234, 236–237, 241, 243–247, 256–257, 260–261, 264–265, 267, 277–278, 280–281, 294, 359, 362–364, 367, 385
high-fidelity model data, 87, 110
high-fidelity model gradients, 89
high-fidelity model of the array aperture, 246
high-fidelity model optimum, 119, 134, 243, 247
high-fidelity model response, 87, 120, 122
high-fidelity model space, 117
high-fidelity Pareto-optimal set, 172
high-fidelity simulation data, 6
high-fidelity simulation model, 87
high-fidelity simulations, 263, 270, 283, 287, 302, 346, 361, 383, 403
high-fidelity simulations of the array, 288
high-fidelity superposition model, 241, 246
higher-order modes, 204
higher-order derivatives, 42
Hooke–Jeeves direct search, 53–54
Hooke–Jeeves procedure, 55
horn antenna, 199
housing, 2, 58
HPBW behavior in the large array limit, 250
HPBW versus the scan angle, 250
HPBWs, 263, 269
HPBWs of this Dolph–Chebyshev design, 263
HW arrays, 217, 219–220, 224–228
HW designs, 216, 218–223, 229
HW spacing, 221
hybrid methods, 40
hypercube, 164, 171
hyperparameters, 97, 335
hyperplane, 162

I

ε-insensitive loss function, 99
identity matrix, 25, 43, 66
ill-conditioned, 43
impedance bandwidth, 27, 32, 104

impedance matrix, 25
impedance transformer, 196
implicit space mapping, 106, 107, 117
inactive, 48
incident power, 309
incident waves, 14, 23, 25
independent distributions, 394
individuals, 65, 70–72, 76
inequality constraints, 47–48
infeasible individuals, 76
infill criteria, 6, 92, 101, 110, 114
infill points, 113
infill samples, 92
infill strategy, 87, 113
initial design, 265, 283, 317
initial Pareto set, 171
initial search vectors, 55
initial solution, 59
input parameters, 413
input SM, 116–117
input space mapping (ISM), 106
input transmission lines, 25
installation fixture, 2
integer representation, 69
integrated antenna array, 333
integrated antenna array structures, 2
integrated aperture-feed circuits, 330
integrated apertures, 9
integrated array circuit, 211, 322, 347
integrated array structure, 302
integrated array-feed circuit, 305
interior point methods, 49
intermediate, 70
interpolated phase tapers, 368
interpolation, 258
interpolation techniques, 6
interval methods, 60
intrinsic impedance, 15
inverse multiquadric, 96
inverse of Hessian, 44
inverse quadratic, 96
isolated element, 27, 261
isotropic radiator, 17
iterative CEM solvers, 209
iterative correction, 256

iterative exploration, 115
iterative process, 294
iterative SBO process, 146
iterative scheme, 238

J

Jacobian, 89, 117, 120, 127, 399
junctions, 306

K

Karush–Kuhn–Tucker or KKT
    conditions, 47
kernel approach, 99
knots, 12
kriging, 6, 92, 94, 96–97, 113, 332,
    334, 403
kriging interpolation, 164–165, 334
kriging interpolation model, 401
kriging model, 113, 165, 335–336,
    346, 352
kriging model of S-parameters of the
    T-junction, 334
kriging predictor, 97
kriging surrogate model, 334
kriging surrogates, 109, 335
kriging-based coarse model, 335
kriging-based SBO, 115

L

Lagrange multipliers, 48, 51
Lagrangian, 50
Lagrangian function, 48
laptops, 210
large array limit, 369
large voids, 209
large-step design perturbations, 258
Latin hypercube sampling (LHS), 94,
    171
layouts, 12
least-square solution, 95, 163
Levenberg–Marquardt (LM)
    algorithm, 44
line impedance, 25, 164
line search, 38–39, 41, 51
line segment, 55

Index 455

linear antenna array apertures, 8, 213
linear aperture, 308–309, 311, 314–315
linear array(s), 9, 19, 21, 213, 216, 223, 230, 305, 330, 338, 349, 354, 362–364, 376, 378
linear array aperture, 213, 232, 251, 305–307, 310, 312, 358, 370, 401
linear array circuit, 347
linear array of microstrip patch antennas, 231
linear array pattern synthesis, 214
linear arrays of microstrip patches, 390
linear combination, 96
linear convergence, 38
linear convergence rate, 41
linear correction, 120
linear end-fire array, 214
linear microstrip arrays, 373, 375
linear model, 127, 130
linear phased array apertures, 9
linear regression, 99, 104
linear regression models, 157
linear regression problem, 395
linear responses, 15
linear system, 95
linearly independent, 55
local algorithms, 59
local approximation model, 36, 110
local expansion, 41
local maxima, 122–123
local maximum/minimum, 31
local methods, 52
local minima, 55
local minimizer, 36–37
local minimum, 58
local model, 396, 399
local optimization, 89–90, 110
local optimum, 47, 59, 89
local search, 112, 216, 380
local search methods, 77
local sidelobe maxima, 336
local surrogate models, 395, 413
localized variations, 97

lookup matrix, 383–384, 390–391
lossless dielectrics, 186
lossless materials, 211
low computational costs, 287
low profile antenna array, 11
low sidelobe apertures, 2
low sidelobe levels, 328
low sidelobe patterns, 329, 393
low SLLs, 188
low-cost surrogate-assisted approach, 270
low-dimensional formulation, 411
low-dimensional spaces, 94
low-fidelity (coarse) EM models, 103, 133
low-fidelity (or coarse) models, 6
low-fidelity antenna models, 133, 140
low-fidelity discrete model, 245
low-fidelity EM model, 171
low-fidelity EM-simulated models, 133, 211
low-fidelity model(s), 6, 8, 87, 89, 102, 106–108, 115, 118, 122–123, 128–130, 133–135, 138–141, 143, 146–148, 151, 157, 159, 170, 173, 177, 181, 185, 236, 253, 257–258, 267, 278, 358, 396, 411–413
low-fidelity model alignment, 122
low-fidelity model correction, 106, 121
low-fidelity model domain, 116
low-fidelity model evaluations, 129
low-fidelity model LPW, 158–159
low-fidelity model responses, 120, 122, 138
low-fidelity model selection, 412
low-fidelity model setups, 133
low-fidelity model subjects, 143
low-fidelity responses, 143
low-fidelity simulation model, 156
low-fidelity simulations of the array, 288
low-fidelity superposition model, 245
low-order polynomials, 95, 110
low-profile arrays, 183
low-sidelobe arrays, 302–303

456 Index

low-sidelobe linear arrays, 302–303
low-sidelobe microstrip antenna arrays, 328
low-sidelobe microstrip linear antenna arrays, 9
low-sidelobe microstrip linear arrays, 302
low-sidelobe patterns, 27, 329
low-sidelobe printed-circuit linear arrays, 355
low-sidelobe printed-circuit arrays, 329
low-SLL array patterns, 329
lower and upper bounds, 34, 47, 170

M

magnetic symmetry wall, 179
magnitude-only far-field pattern, 383
magnitude-only far-field samples, 375, 381
magnitude-only pattern samples, 374
magnitude-only samples, 374
main-beam, 188
major beams, 218
major lobe beamwidth, 190, 192, 218, 305
major lobe null-to-null beamwidth, 266
manifold, 162–164
manifold mapping (MM), 90, 115, 119
manual parameter sweeps, 184
manufacturing tolerances, 393, 399, 403
Markov Chain analysis, 75
matched loads, 23
matching, 262
matching correction, 262
material dispersion, 140
material libraries, 208
material losses, 210
material parameter deviation, 395
material parameters, 13, 161
material science, 11
mathematical modeling, 13

Matlab, 238, 240, 275, 279, 289, 310, 339
Matlab *fminimax*, 215, 235
Matlab optimization toolbox, 413
Matlab-implemented socket, 289
matrix inversion, 208
matrix operations, 46
maximal active reflection coefficient, 232
maximum and minimum spacing, 241
maximum deviation, 403
maximum directivity, 16, 20–22, 262
maximum likelihood, 97
maximum radiation intensity, 16
Maxwell's equations, 208
mean, 97
mean square error, 114
measured element patterns, 358
measured H-plane SLL, 344
measured H-plane total pattern, 353
measured H-plane total power pattern, 343
measured pattern, 354
measured SLL, 325–328
measurement noise, 373–374, 386, 389–391
mechanical, 12
merit function, 31, 34–35
mesh, 140
mesh density, 157
mesh quality, 211
mesh step, 135
mesh topology, 3, 52
mesh-cells, 12
metaheuristic algorithms, 58, 61, 64–65, 86, 204
metaheuristic optimization, 203
metaheuristics, 4, 53, 61, 68, 72, 82, 169, 381
metal losses, 140
metallization, 140, 210
method of moments (MoM), 208–209
microstrip antenna array, 273, 275, 279–280, 295–297, 394

microstrip antenna array apertures, 253
microstrip antennas, 183, 362
microstrip array, 322–324, 405–406
microstrip bandstop filter, 120
microstrip circuit, 27
microstrip corporate feeds, 9, 207, 303, 315, 328–329, 400–401
microstrip feed, 184, 347, 400
microstrip input, 170
microstrip junctions, 403
microstrip line, 164, 176
microstrip linear antenna arrays, 374, 393
microstrip linear arrays, 9, 303, 305, 313, 400
microstrip patch, 254, 264, 295, 376
microstrip patch antenna (MPA), 135, 174–176, 179–180, 183–184, 188–190, 192, 231–232, 254, 265–266, 270, 291, 295, 298, 314–315, 330, 340, 347, 358, 362–363, 378, 400–401
microstrip patch antenna arrays, 273
microstrip patch arrays, 375, 385
microstrip patch radiators, 400
microstrip path, 264
microstrip stub, 146
microstrip subarrays, 183–184
microstrip substrate, 188, 199
microstrip substrate height, 106
microstrip technology, 207
microstrip-line components, 107
microstrip-to-coplanar strip (CPS) transition, 146, 170
microstrip-to-SIW transition, 123
microwave CAD, 12
microwave circuit analysis, 22
microwave circuit design optimization, 133
microwave circuit techniques, 27
microwave components, 210
microwave design, 136
microwave engineering, 6, 13, 88, 116–117

microwave filters, 7, 106
microwave frequencies, 206
milling, 154
minimax, 216
minimax optimization task, 215
minimax optimization task with upper and lower specifications, 235
minimax-type of specifications, 122
minimization of the realized gain pattern, 280
minimization problem, 92, 255
minimizer, 43
minimum, 36, 50
minimum incident power, 309
minimum sidelobe designs, 213
model, 30, 41
model coarseness, 211
model coefficients, 96
model correction, 129
model fidelity, 135, 143
model fidelity selection, 143
model parameters, 97
model selection process, 143
model validation, 91, 101
modeling, 11
modeling flow, 92
modern DOE, 93
MoM algorithms, 209
MoM-based solvers, 209
moment method solvers, 139
monomials, 95
Monte-Carlo methods, 60
moving least squares (MLS), 100
MPA inputs, 316
MPA microstrip inputs, 316
MPA patches, 400
MPA spacings, 188, 316
MPAS configurations, 188
multi-band structures, 121
multi-layer ANN, 98
multi-layer feed-forward network, 98
multi-objective, 12
multi-objective antenna optimization, 170
multi-objective design, 411

458   Index

multi-objective design optimization, 168
multi-objective evolutionary algorithm, 171
multi-objective optimization, 82, 169, 173–174
multi-objective problem, 185
multi-point response correction, 106
multi-dimensional rational approximation, 6, 95
multi-dimensional space, 76
multi-disciplinary activity, 13
multi-level fast multipole method (MLFMM), 209
multi-modal, 52, 82, 85
multi-modal optimization problems, 79
multi-modal problems, 77
multiple local optima, 4, 59, 61, 85, 204
multiple-use models, 110
multiplicative response correction, 237, 277
multiquadric, 96
mutation, 64, 67, 70, 78
mutation differential, 78
mutation operators, 64–65, 70, 72, 76
mutation rate, 74
mutual coupling, 22, 26–27, 32, 374–375, 390
MWS transient solver, 296

N

n-dimensional space, 55
n-simplex, 55
narrow fractional bandwidth, 183
narrow impedance bandwidth, 174
narrow-band antenna, 126
natural representation, 68
near-field(s), 15, 22
near-field interactions, 24, 233
Nelder–Mead, 3
Nelder–Mead algorithm, 53, 55, 57
nested space mapping, 131
network architecture, 98

network models, 204
network training, 98
neural network architecture, 98
neural networks, 92, 94, 375
neuron, 98
neuron weights, 98
Newton, 42
Newton algorithm, 43–44
Newton method, 43–44
Newton-type methods, 46
no free lunch (NFL) theorems, 75
noisy, 82
noisy cost functions, 55
noisy objective functions, 61
nominal design, 400
nominal parameter values, 395
nominal values, 23
non-convex, 52
non-differentiable, 82
non-singular, 43
non-uniform amplitude excitation(s), 264, 282–283, 303, 307, 316, 329
non-uniform distributions, 71
non-uniform excitations, 304
non-uniform phases, 238
non-uniform spacings, 238, 309, 319
non-uniformly spaced array, 245
non-uniformly spaced ULA, 310
non-uniqueness, 116
nonlinear equality, 47
nonlinear least-squares regression, 98
nonlinear materials, 208
nonlinear minimization problem, 146, 184
nonlinear regression problem, 118, 148, 237, 277
nonlinear system, 117
non-uniform aperture excitation, 307
non-uniform power distribution, 307
normal distributions, 394
normal probability distribution, 64
normalized distance, 164
normalized pattern, 223–228
null controlled pattern, 230
null locations, 11

Index 459

null controlled pattern design, 411
null-to-null beamwidth, 12
number of high-fidelity simulations, 236
number of objective function evaluations, 215, 235
numerical costs, 266, 414
numerical derivatives, 58
numerical efficiency, 273
numerical noise, 3, 32, 52–53, 258
numerical optimization, 3, 7–8, 13, 29, 33, 233, 310, 330, 413
numerical optimization techniques, 8
numerical validation, 12

O

objective, 48
objective function, 3, 34–36, 44, 51–53, 58–59, 62, 71, 126, 146, 151, 184–185, 234–235, 246, 255–256, 276, 279, 292
objective function defined, 359
objective function domain, 34
objective function evaluations, 204
objective function evaluator, 58
objective function values, 79, 55
observation point, 14
offspring, 64–65, 73
omnidirectional, 148
omnidirectional element radiation, 234
omnidirectional radiation, 155
one-iteration SBO approach, 260
one-point response correction, 106
one-shot approach, 110
operating bandwidth, 152
operating condition, 161
operating frequencies, 12, 156, 161, 164, 166, 202
opportunistic manner, 54
optimal architecture, 331, 347
optimal designs, 220, 222, 359
optimal excitation amplitudes, 287
optimal feed, 332

optimal feed architectures, 9, 207, 304, 332, 343, 345, 355
optimal performance of the SBO process, 257
optimal phase shifts, 222
optimal phase tapers, 367–370
optimal solution, 49, 74
optimality conditions, 47
optimization, 114, 211, 215, 217–219, 221, 225–228, 231, 234, 240, 244, 253, 256, 258, 262–263, 266, 358, 365
optimization algorithm, 30, 53, 58, 74, 125, 138, 152, 289
optimization cost, 86, 88–89, 143, 165, 215
optimization methods, 411
optimization of antennas, 60
optimization of array factor models, 214, 294
optimization of EM models, 339
optimization of planar arrays, 274
optimization of the analytical model, 233
optimization of the directivity pattern, 264
optimization of the kriging surrogate, 346
optimization of the radiation response, 267
optimization of the superposition model, 233, 282
optimization of the T-junctions, 334, 342
optimization path, 130
optimization problem, 3, 61, 87, 410
optimization process, 36, 40, 51, 124, 151, 173, 201, 204, 296, 336
optimization results, 282
optimization run, 76
optimization setup, 30
optimization subproblems, 413
optimization task, 276
optimization techniques, 7, 412
optimization-based design, 203

Index

optimized design, 262, 283, 286, 318–321, 328
optimized excitation amplitudes, 269
optimized excitation phases, 270
optimized junctions, 347
optimized patterns, 227
optimized T-junctions, 340
optimum, 38, 44
optimum design, 115, 188
optimum element spacing, 291
optimum of the fast model, 241
optimum of the high-fidelity model, 241
optimum of the simulation-based surrogate model, 241
optimum tapers, 365
optional random search, 235
orthogonal array sampling, 94
output space mapping, 119, 171, 267

P

parallel computing, 54
parallelization, 60
parameter deviations, 394, 400, 404
parameter extraction (PE), 116
parameter space, 148
parameter sweeping, 13, 177, 181, 329, 414
parameter variation, 394
parameter-sweep-based optimization, 4
parameters sweeps, 3
parametric optimization, 329, 409
parametric SM, 117
parasitic emission, 204
parasitic feed radiation, 302
parasitic radiation, 184, 302
parasitic surface waves, 206
parent individuals, 62, 64–65, 67
parents, 73
Pareto fronts, 171
Pareto optimal set, 171
Pareto set, 168
partial faults, 373, 375, 384, 390–391
partially-mapped crossover, 76

particle(s), 76–77
particle swarm optimizers (PSO), 4, 60, 62, 64–65, 68, 76–77, 80–81, 169, 203
patch antenna, 189, 258, 373
path length correction, 353
pattern, 229, 258, 358
pattern maxima, 376–377
pattern nulls, 2, 223, 229
pattern optimization, 253
pattern search, 33, 53, 130
pattern search algorithm, 52–53, 129–130, 148, 258
pattern search method, 53–54
pattern specifications, 230
pattern search, 59
pattern search techniques, 3
peak directives of the final designs, 263
peak directivity, 196, 206, 217–222, 227–228, 239, 254, 260–262, 266, 269–270
peak directivity maximization, 218
peak realized gain, 2, 12, 32, 35, 212, 232, 239, 242, 285, 300
peak realized gain scan loss, 248
peak realized gain versus the scan angle, 248
peak the realized gain, 249
penalized cost function, 49
penalty, 185
penalty coefficient, 185
penalty factor, 49, 152, 185, 235, 262
penalty function, 185, 202, 234–235, 239, 246, 276
penalty function approach, 49, 340, 351
penalty method, 50–51
penalty term, 50
perfect conductors, 186
performance figures, 12, 34
periodic boundary conditions, 24, 27, 206
permittivity, 162, 394
personal computers, 210

perturbation steps, 399
perturbation vectors, 399
perturbed designs, 258, 260
phase excitation tapers, 231, 357–358
phase excitations, 364–365
phase optimization, 246
phase shift, 18–19, 210, 215, 270
phase shifts at the optimal designs, 221
phase taper, 300, 357, 360–362, 368–370
phase–spacing optimization, 231, 240
phase–spacing synthesis, 233
phase-tapers, 247
phased apertures, 15, 27
phased array(s), 8, 232, 300, 357
phased array antennas, 25, 246
phased array apertures, 22, 207, 212
phasor, 14
physical measurements, 93
physics based-surrogate models, 8
physics-based, 83
physics-based models, 211, 412
physics-based SBO, 115, 118
physics-based SBO algorithms, 130
physics-based SBO methods, 131
physics-based surrogate modeling, 103, 133
physics-based surrogate-assisted algorithms, 121
physics-based surrogates, 6, 102–103, 108, 115, 411–412, 414
planar antenna array apertures, 15
planar antenna arrays, 273, 291, 298
planar apertures, 18–19
planar array antenna, 293
planar array aperture, 8, 264
planar arrays, 212, 270, 274, 291
planar inverted-F antenna (PIFA), 128
planar microstrip antenna array, 253, 291
planar microstrip array, 264
planar UWB dipole antenna, 145
planar Yagi antenna, 170

Polak–Ribiére method, 41
polynomial approximation, 92
polynomial regression, 94–95
population, 62, 65–66, 74, 78
population models, 72
population-based algorithms, 60
population-based approaches, 60
population-based metaheuristic algorithms, 62, 77, 81
population-based metaheuristics, 33, 52, 59–61, 68, 75, 82, 86, 102, 114, 169, 174, 213, 329, 374, 381–382, 413
population-based methods, 65, 82, 204, 215, 235
population-based optimization, 61, 81
population-based search, 74, 216
population-based search algorithm, 63
positions, 76
positive definite, 37, 43–44
positive definite matrix, 51
positive semidefinite, 37
positive spanning directions, 53
power density, 16
power distribution, 401, 403
power distribution networks, 29
power pattern, 244, 263, 293
power split (PS), 330, 332, 335–336, 339–340, 343, 352
power split of the T-junction, 336
preconditioning, 208
predicted high-fidelity model response, 121
prediction capability of the surrogate, 257
prediction capability of the surrogate model, 236
prediction error, 101
prediction–correction scheme, 109
predictive power, 92
premature convergence, 63, 71, 73–74
primary objective, 185, 202, 235
printed antennas, 140
printed apertures, 1

462 Index

printed Yagi antenna, 140
printed-circuit antenna arrays, 355
printed-circuit antenna structures, 145
printed-circuit arrays, 301, 328
printed-circuit subarrays, 201
printed-circuit technology, 174
printed-circuit ultra-wideband (UWB), 134
probability distributions, 394, 406
probability of improvement, 114
problem dimensionality, 412
problem-specific knowledge, 91, 130
progressive phase shift, 22, 217, 219, 223–228
propagation constant, 18
prototype, 146
prototyping, 87
proximity effects, 204
PS constrains, 340
pseudo-random sampling, 94
pseudoinverse, 95, 119
PSO algorithm, 76–77
pure random search, 216

Q

quadratic convergence, 38
quadratic programming, 99
quadratic programming subproblem, 51
quarter-wave transformer, 170, 190
quasi-analytical models, 270–271
quasi-Monte Carlo sampling, 94
quasi-Newton, 3
quasi-Newton algorithms, 45–46
quasi-Newton methods, 42, 44
quasi-Newton step, 117

R

radar, 11
radial basis function (RBF) interpolation, 96
radial basis functions, 6, 94, 96
radially symmetric functions, 96
radiated power, 32

radiating aperture, 2, 211–212, 303, 351
radiating elements, 85, 183
radiation, 134, 137, 211, 231
radiation aperture, 204
radiation directions, 15
radiation element, 14, 20
radiation figures, 15, 18, 26–27
radiation from the feed, 328
radiation intensity, 15–17, 20, 35, 205
radiation of the high-fidelity model, 277
radiation pattern, 2, 11–12, 15, 30, 32, 35, 88, 148, 155, 176, 178, 180–181, 190, 205–207, 213, 234, 236, 238, 251, 267, 273, 324, 328, 333, 364–365, 367, 373–375, 380
radiation pattern cut, 184
radiation pattern nulls, 12
radiation pattern of the array, 278
radiation pattern of the array factor, 246
radiation pattern optimization, 267
radiation pattern surrogate, 396
radiation pattern synthesis, 13, 59
radiation pattern synthesis of linear arrays, 231
radiation response(s), 140, 188, 207, 278
radiation response correction, 291, 300
radiation response optimization, 266
radiator, 13
radome covers, 13
radomes, 1, 85
random deviation, 71
random initialization, 77–78, 216
random number, 71
random process, 97
random sampling, 164, 407
random search, 74–75, 80, 216
random search as an initial synthesis step, 216
random selection, 66
random walk, 80

Index   463

random-search-based initialization, 213
randomization parameter, 80
re-optimization, 336
real-world antenna design, 82
realized (absolute) gain, 17
realized gain, 15–17, 22, 154, 177, 186, 188, 192, 196, 234, 280, 283–285, 299, 359
realized gain in the E-plane, 232, 242–243, 245, 247
realized gain pattern, 281, 283–285, 299
realized gain responses, 135
realized gain scan loss, 247, 249, 358, 368–369
realized gain scan loss versus the scan angle, 249
receiving element, 11
recessed microstrip line, 176
recombination, 72, 76, 78
recombination operators, 64–66, 82
rectangular grid, 53, 403
rectangular patch, 156, 176
reduced SLL, 190, 240, 244, 283, 309
reduction of SLLs, 269
reduction of the computational overhead, 410
reference designs, 106, 157, 164–165
refined grid, 53
refined Pareto sets, 171
reflection, 57, 211
reflection characteristics, 125, 134, 152
reflection coefficient, 12, 17, 22, 32, 35, 88, 106, 134, 139, 146, 148, 152, 156, 172, 176, 178, 180–181, 184–186, 188, 190, 192, 194, 202, 231, 239, 258–261, 264–265, 273, 305, 319–320, 325–328, 341, 365
reflection coefficient adjustment, 268
reflection coefficient responses, 141
reflection loss, 2
reflection response adjustment, 268

reflection responses, 13, 24, 134, 159, 199, 201, 203, 212, 258, 262, 268, 270–271, 411
reflection responses of the final design, 269
reflector, 209
region of attraction, 59
regression coefficients, 98
regression models, 98
relative dielectric, 161
relative permittivity, 164
relative speedup, 289
relative speedup for large arrays, 287
relaxed convergence criteria, 186
relaxed simulation termination criteria, 140
relocation strategy, 110, 112–113
residual energy, 52, 140
resonant frequency formula, 177, 181
response correction, 9, 103, 119, 147, 213, 236, 243, 251, 370–371, 378–379
response correction function, 292, 359, 373, 376–377
response correction methods, 7
response correction techniques, 9, 119, 145
response correction term, 361
response feature approach, 395–396
response features, 124–125, 127, 397–398
response scaling, 104
response surface approximation models, 91
response surface approximations (RSAs), 88, 109
response surface methodology, 109
response surface model, 211
response surface surrogate, 134
response vector, 255
response-corrected surrogates of the T-junction, 342
response surface modeling, 332
RMS errors, 165
Rosenbrock function, 40–41

464 *Index*

roulette wheel selection, 71
rugged functional landscapes, 52

S

S-parameter error, 140
S-parameters, 108, 206, 212, 274, 333–334, 346
saddle point, 43
SADEA framework, 115
SBO algorithm, 5, 88–90, 103, 143, 146, 184, 188, 193, 256–257, 267
SBO approach, 256
SBO design cost, 283
SBO methodology, 197, 330, 332
SBO methods, 211
SBO numerical costs, 211
SBO optimum, 284–287
SBO paradigm, 5, 358
SBO procedures, 88, 259
SBO process, 103, 134, 143, 152, 192, 210
SBO scheme, 266
SBO techniques, 412
SBO-assisted design process, 211
scalability, 273
scalar objective functions, 34
scalarization, 185
scaled response, 237, 277
scaling parameters, 96, 100, 108, 237, 277
scaling procedure, 236–237, 277–278
scan angle, 15, 21, 212, 233, 248–250, 300, 357–358, 360, 362, 365, 367–370
scan loss, 249–250
scan reflection coefficient, 25
scanned array beam, 22
scanning, 247
scanning-beam array apertures, 24
scatterers, 209
scattering matrix, 27
scattering parameters, 24, 30
scattering problems, 209
scattering waves, 25
schemata theorem, 74

schematics, 413
search direction, 40, 51, 54
search process, 55
search region, 112
search space, 34, 66–67, 69, 77–78
second-order derivatives, 36, 44
second-order polynomial, 95
second-order Taylor approximation, 43
second-order Taylor expansion, 42
sector beam pattern, 230
selection, 62, 71
selection pressure, 63, 71–74, 81
selection schemes, 71
self-adaptation, 66, 74
self-organization, 79
sensitivity analysis, 157
sensitivity data, 120
sensitivity information, 57
sequential approximate optimization (SAO), 110, 112–113
sequential exploration, 54
sequential quadratic programming, 3, 51
sequential sampling, 101, 114
sequential sampling techniques, 88
sequential-quadratic programming (SQP), 215
sequential-quadratic programming (SQP) algorithm, 52, 235
shape-preserving response prediction (SPRP), 115, 120–121
sidelobe levels (SLLs), 2, 11, 24, 26, 32, 34, 134, 183, 185, 188, 190, 192, 201–202, 206, 216–222, 227–228, 232, 234, 239, 243, 247, 249–250, 254, 263, 266, 269–270, 276, 292, 296, 300, 303, 305–307, 329, 357–358, 360–362, 364–365, 368–369, 378, 387–388, 393, 395–397, 399–400, 403–404, 411
sidelobe minimization, 286
sidelobe patterns, 329, 333
sidelobe reduction, 121, 219, 222, 274
sidelobe suppression, 263

sidelobes, 210, 212, 262, 303
sidelobes of planar arrays, 300
signal-to-noise level, 32
signal-to-noise ratio, 12
simplex, 55, 57
simplex vertices, 55
simplified physics, 211
simplified physics models, 88
simulated annealing, 60, 77
simulated array, 318–321
simulated array circuits, 341
simulated array model, 317
simulated H-plane pattern, 352
simulated H-plane power pattern, 349
simulated H-plane power pattern of the linear array, 350
simulated phase shifts, 349
simulated propagation constant, 349
simulated radiation pattern, 232, 333
simulated reflection coefficient of the linear array, 350
simulated scattering parameters, 25
simulation bandwidth, 140
simulation models, 87
simulation process, 52
simulation time, 140, 245
simulation-based antenna array design, 7, 29
simulation-based design, 5, 8, 13, 31, 214, 255, 291, 329
simulation-based design of planar antenna arrays, 300
simulation-based design processes, 206
simulation-based methods, 205
simulation-based model of the array aperture, 233–234
simulation-based models, 15, 233
simulation-based optimization, 18, 27, 30–31, 133, 189, 207, 303, 317–318, 321, 340
simulation-based optimization of antenna arrays, 273

simulation-based superposition models, 8–9, 88, 212–213, 240, 246, 250–251, 274, 279
simulation-based surrogate, 231
simulation-based surrogate model, 231, 242, 244
simulation-based surrogate-assisted antenna array design, 211
simulation-based techniques, 86
simulation-based tuning, 24, 27, 190, 192
simulation-based validation, 26
simulation-driven design, 33, 86–87, 201, 253, 256, 273, 409
simulation-driven design closure, 321
simulation-driven optimization, 35, 231, 305, 414
simulation-driven optimization of the feed, 347
simulation-driven optimization processes, 210
simultaneous excitation, 24, 277
single ring circular array, 19
single-layer MPA, 176
single-objective optima, 171
single-unit perceptron, 98
singular value decomposition, 119
sinusoidal, 96
slack variables, 99
SLL degradation, 407
SLL minimization, 223, 336
SLL reduction, 215, 217–218, 220–222
SLL suppression, 215, 229
SLL-oriented design, 338
SLL/GLL, 250
slope parameter, 98
slot-energized MPA, 176
slot-ring coupled patch antenna, 155
SM algorithm, 129
SM surrogate, 118, 129
SM surrogate model, 118
SM transformations, 118
SMA connector, 148, 151, 174, 176, 180, 199
SMA edge-mount connector, 146

Index

smart random initialization, 215
smart random search, 229, 294, 380
smart random search algorithm, 216
smooth, 36
smooth objective functions, 52
social learning, 76
soft line search, 39–41, 45
software packages, 11
software tools, 86
soldering, 154
space exploration, 114
space mapping (SM), 6, 90, 115–118, 129–130
space mapping correction, 143
space-filling DOEs, 92, 94
sparse arrays, 59
sparse system matrices, 209
specification levels, 123
specifications, 11
split-sample method, 101
SPRP surrogate, 121
stand-alone antenna, 205
standard chemical etching process, 394
standard deviation, 74
standard PSO, 80
star distribution, 93, 396
starting point, 57
starting point for the optimization process, 236
stationary point, 36
statistical analysis, 157, 393–394, 396, 404–406, 409, 411
statistical infill criteria, 91
statistical lower bound, 114
steady-state models, 73
steepest descent direction, 36, 40
steepest descent method, 40, 44
stencil, 53
stencil size, 53
stochastic convergence, 67
stochastic methods, 60
stochastic processes, 75
stochastic search operators, 74
stochastic selection, 71, 74

stopping criteria, 112
strongly active, 48
suboptimal results, 414
subarray(s), 29, 183–184, 188–189, 194, 201–202, 411
subarray configuration, 190
substrate, 128, 146, 161, 164, 170, 183, 188
substrate dielectric permittivity, 162
substrate permittivities, 164, 166
substrate thick, 107
substrate-integrated cavity antenna, 108
sufficient condition, 36
sum pattern, 330
superlinear convergence, 38
superposition, 206, 233, 240, 273–274, 358, 384
superposition high-fidelity model, 281
superposition low-fidelity model, 281
superposition model, 233, 235–236, 243, 281, 283, 287–288, 300, 367–368
superposition of simulated radiation and reflection responses, 231
superposition principle, 14, 29
superposition-based model, 302
support vector machines, 6
support vector regression (SVR), 95, 99–100
surrogate(s), 5, 34, 83, 86–87, 89, 101, 103–104, 106–107, 109–110, 112, 114, 118–119, 133, 147, 161, 166, 186, 256, 267, 379
surrogate accuracy, 102
surrogate management framework (SMF), 115
surrogate model, 5, 7, 31, 57, 86–88, 90, 92, 95, 101, 106, 110, 113, 116, 119–120, 147, 152, 171, 173, 185, 210, 231, 237–238, 243–244, 246, 249–250, 256–257, 264, 267, 278, 281, 294, 331, 346, 357, 360–361, 365, 377, 390–391, 403, 406–407, 413

surrogate model accuracy, 379
surrogate model construction, 92, 413
surrogate model correction, 248
surrogate model domain, 164
surrogate model evaluation, 89
surrogate model of the radiation pattern, 294
surrogate model optimization, 238, 279
surrogate model optimum, 113, 242–243, 246–247
surrogate modeling, 145, 251, 371, 395, 410, 412
surrogate modeling flow, 91
surrogate modeling methods, 394
surrogate modeling process, 92
surrogate modeling techniques, 86, 96, 414
surrogate model domain, 163
surrogate superposition models, 247
surrogate-assisted algorithms, 86, 115
surrogate-assisted approaches, 212, 355
surrogate-assisted CAD methods, 9
surrogate-assisted design, 331
surrogate-assisted design methods, 60
surrogate-assisted design procedures, 414
surrogate-assisted design process automation, 415
surrogate-assisted fault detection, 9
surrogate-assisted methods, 9, 34, 82, 87, 393, 411–412, 414–415
surrogate-assisted optimization, 9, 58, 86, 110, 113–114, 133, 148, 207, 211, 251, 361, 370
surrogate-assisted optimization methods, 410
surrogate-assisted optimization techniques, 145
surrogate-assisted procedures, 169, 336
surrogate-assisted process, 330

surrogate-assisted techniques, 9, 53, 373, 375, 414–415
surrogate-based algorithms, 8, 89, 414
surrogate-based design process, 276
surrogate-based methods, 8, 58
surrogate-based modeling, 8
surrogate-based optimization (SBO), 5–6, 8, 29, 52, 86–90, 176, 201, 213, 246, 253, 255, 267, 269, 318, 330, 333, 351, 355, 357
surrogate-based optimization (SBO) algorithm, 184
surrogate-based optimization (SBO) techniques, 271, 410
surrogate-based optimization frameworks, 57
surrogate-based optimization methodology, 201
surrogate-based optimization process, 87
surrogate-based procedures, 415
swarm, 76
swarm behavior, 76
symmetry wall, 255
synthesis problem, 216
system decomposition, 414
system responses, 34, 411
system-specific knowledge, 122
systematic CAD technique, 13

T

T-junctions, 304–307, 309, 316, 321, 330, 332, 334–336, 339, 343, 345–347, 401–403
tangent plane, 120
taper-oriented design, 338
target agent, 78
termination condition, 55, 57, 141
termination criteria, 31, 53
test set, 101
tetrahedron, 55
thermal, 12
thin plate spline, 96
three-stage SBO approach, 240
through-feed-coupling, 23–24

468 Index

time evaluation ratio, 236
time-domain finite-difference method, 140
time-domain solvers, 140
tolerance analysis, 9
total cost, 134, 270
total efficiency, 176
total far-field, 15
total near-field, 15
total power incident, 17
total radiated power, 16
total reflected power, 212
tournament selection, 62, 64, 72
tournament size, 72
TR radius, 238, 279
training data, 164, 171
training data acquisition, 92, 102
training data acquisition costs, 412
training designs, 396
training point, 95, 396
training samples, 112, 171
training set, 101
transient solver, 135, 140, 146, 194, 246, 318–321, 385
transient solver of CST MWS, 257
translation vectors, 120
transmission coefficient, 27, 120
transmission line, 25, 175
transmission line calculators, 208, 346
transmission parameters, 340
transmission-line-based components, 7
traveling salesman problem, 69
trend function, 97
triangle, 55
trust region, 127
trust-region size, 127
trust-region (TR) framework, 41–42, 89, 110, 113, 127, 238, 258, 279, 339
trust-region methodology, 119
trust-region methods, 52
trust-region radius, 42, 89
Tschebyscheff excitation taper, 400
TSP, 76
tuning, 29

tuning SM, 7
tuning space mapping, 131
two-objective problem, 276
two-parent crossover, 69

U

ULA aperture, 308
ULA design, 308
uncertainties, 93
unconstrained, 33
unconstrained minimum, 48
unconstrained optimization, 46, 55, 82
unconstrained problems, 50
uncorrected feed, 340, 348
uncorrected path lengths, 350
unequal power distribution, 403
unequal power split, 401
unequal power split T-junctions, 342
unequal power split junctions, 303, 329
unequal-split junctions, 330, 332, 402
unequal-split microstrip junctions, 211
unequal-split T-junctions, 304, 330, 338, 355
uniform amplitude excitation, 214
uniform amplitude-phase-spacing design, 232
uniform array, 266
uniform excitation amplitudes, 270
uniform excitation taper, 299, 304, 310
uniform grid sampling, 94
uniform linear array (ULA), 306
uniform phase shift, 218
uniform probability distribution, 78, 80, 164, 403–404, 406
uniform sampling, 77
uniform sampling techniques, 94
uniform spacing, 217–219, 229, 238, 244
uniform taper, 369–370
uniformly continuous, 39

uniformly distributed random numbers, 77
uniformly spaced antenna array, 245
unique minimizer, 43
unit excitation, 18
updating formula, 41, 44
UWB antenna, 148
UWB band, 154
UWB dipole antenna, 145
UWB slot antenna, 148

V

vacuum, 18
variable-fidelity electromagnetic (EM) simulations, 253
variable-fidelity EM models, 8, 251, 276, 371
variable-fidelity EM simulation models, 145
variable-fidelity EM simulations, 184, 270, 414
variance, 97
vector of design variables, 295
vector operations, 46
vector space, 55

vector-based evolutionary algorithm, 77
vectorized array factor, 310
velocity, 76
velocity vectors, 76
voltage standing wave ratio (VSWR), 170
VSVR, 171

W

waveguide ports, 140, 146
wavelength, 18, 21, 140–141
weakly active, 48
weighting factors, 118
weights of the array elements, 214
wire dipoles, 27
Wolfe conditions, 39

Y

Yagi antenna, 170, 173–174
yield-driven design, 409

Z

zeroth-order consistency, 89–90, 103, 186, 267, 278, 379